D1165017

Human
Tumor
Viruses

Human Tumor Viruses

EDITED BY

Dennis J. McCance

Department of Microbiology and Immunology
University of Rochester School of Medicine and Dentistry
Rochester, NY 14624

**ASM
PRESS** WASHINGTON, D.C.

Library of Congress Cataloging-in-Publication Data

Human tumor viruses / edited by Dennis J. McCance.
     p.   cm.
  Includes bibliographical references and index.
  ISBN 1-55581-130-2
  1. Oncogenic viruses. 2. Viral oncogenesis. I. McCance, Dennis
J.
  [DNLM: 1. Neoplasms—etiology. 2. Oncogenic Viruses—
pathogenicity. 3. Tumor Virus Infections—microbiology. QZ 202
H9185 1998]
  QR372.O6H87 1998
  616.99′4071—dc21
  DNLM/DLC
  for Library of Congress   98-12185
                                          CIP

CONTENTS

CONTRIBUTORS

George R. Beck, Jr.   Fels Institute for Cancer Research and Molecular Biology, Temple University School of Medicine, 3307 North Broad Street, Philadelphia, Pennsylvania 19140

Yuan Chang   Department of Pathology, Columbia University College of Physicians & Surgeons, 630 West 168th Street, New York, New York 10032

Laura Diamondstone   Ischemia Research and Education Foundation, 250 Executive Boulevard, #3400, San Francisco, California 94134-3306

Alison A. Evans   Fox Chase Cancer Center, 7701 Burholme Avenue, Philadelphia, Pennsylvania 19111

Graham R. Foster   Department of Medicine, Imperial College School of Medicine at St. Mary's, Norfolk Place, London W2 1NY, United Kingdom

Laimonis A. Laimins   Department of Microbiology-Immunology and Department of Biochemistry, Molecular Biology and Cell Biology, Northwestern University, 303 East Chicago Avenue, Chicago, Illinois 60611

Paul H. Levine   Division of Cancer Epidemiology and Genetics, Viral Epidemiology Branch, National Institutes of Health, Bethesda, Maryland 20892, and The George Washington University Medical Center, Washington, DC 20037

David Liebowitz   Marjorie B. Kovler Viral Oncology Laboratories, Department of Medicine, Section of Hematology/Oncology and Virology, University of Chicago, 910 East 58th Street, Chicago, Illinois 60637

W. Thomas London   Fox Chase Cancer Center, 7701 Burholme Avenue, Philadelphia, Pennsylvania 19111

Richard Longnecker   Microbiology-Immunology, Northwestern University Medical School, 303 East Chicago Avenue, Chicago, Illinois 60611

Angela Manns   Division of Cancer Epidemiology and Genetics, Viral Epidemiology Branch, National Institutes of Health, Bethesda, Maryland 20892

William S. Mason   Fox Chase Cancer Center, 7701 Burholme Avenue, Philadelphia, Pennsylvania 19111

D. J. McCance   Cancer Center and Department of Microbiology, University of Rochester, 601 Elmwood Avenue, Rochester, New York 14642

Patrick S. Moore   Department of Pathology, Columbia University College of Physicians & Surgeons, and Division of Epidemiology, School of Public Health, Columbia University, 630 West 168th Street, New York, New York 10032

Elizabeth Moran   Fels Institute for Cancer Research and Molecular Biology, Temple University School of Medicine, 3307 North Broad Street, Philadelphia, Pennsylvania 19140

M. A. Nead   Department of Microbiology, University of Rochester, 601 Elmwood Avenue, Rochester, New York 14642

Corliss L. Newman   Cancer Center, University of Rochester, Box 704, 601 Elmwood Avenue, Rochester, New York 14642

Joseph D. Rosenblatt   Department of Medicine, Hematology-Oncology Unit, and Department of Microbiology and Immunology, Cancer Center, University of Rochester, Box 704, 601 Elmwood Avenue, Rochester, New York 14642

Thomas Schulz   Division of Genitourinary Medicine, Department of Medical Microbiology, University of Liverpool, Liverpool L69 3GA, United Kingdom

Howard Strickler   Division of Cancer Epidemiology and Genetics, Viral Epidemiology Branch, National Institutes of Health, Bethesda, Maryland 20892

Howard C. Thomas   Department of Medicine, Imperial College School of Medicine at St. Mary's, Norfolk Place, London W2 1NY, United Kingdom

Mark R. Thursz   Department of Medicine, Imperial College School of Medicine at St. Mary's, Norfolk Place, London W2 1NY, United Kingdom

Bruce D. Walker   Infectious Disease Unit, GRJ 504, Department of Medicine, Massachusetts General Hospital, Boston, Massachusetts 02114

Jennifer A. Waters   Department of Medicine, Imperial College School of Medicine at St. Mary's, Norfolk Place, London W2 1NY, United Kingdom

David K. H. Wong   Infectious Disease Unit, GRJ 504, Department of Medicine, Massachusetts General Hospital, Boston, Massachusetts 02114

Brad R. Zerler   CollaGenex Pharmaceuticals, Inc., Newtown, Pennsylvania 18940

PREFACE

It has been 34 years since the isolation of Epstein-Barr virus (EBV), the first virus to be associated with a human tumor. The most recent human tumor virus isolated is another herpesvirus, human herpesvirus 8 (HHV-8), which in 1994 was recognized to be associated with Kaposi's sarcoma. In the intervening years a number of other viruses have been associated with human cancers, although not all have been shown to have a causative role. This book describes the molecular biology and pathogenesis of those viruses where there is overwhelming evidence of a causal association between the virus and human malignancies. In other words, in the absence of infection the cancers described here would be reduced by 95% with a significant reduction in morbidity and mortality, especially in developing countries. Mere infection, however, does not mean cancer will result, since many more people are infected than will develop a malignancy. In fact, one of the major problems in proving that the association is causal or casual in human cancer is the high rate of infection in the general population, given that there are geographical variations in infection rates.

Since the isolation of EBV, extensive studies on the molecular biology of the virus have shown that it codes for several proteins that have proliferative effects on cells, but the sequence of events leading to the malignant state is unclear. The complex nature of EBV or HHV-8 can be compared with the relative molecular simplicity of hepatitis B virus (HBV) or of human papillomaviruses (HPV), which code for only one or three proteins, respectively, which have proliferative properties. While these latter viruses may appear to have simple molecular biological properties, their pathogenesis is complex and involves the activities of viral proteins on multiple host regulatory pathways. The activities of many of these regulatory pathways were elucidated by observations made with small DNA tumor viruses, simian virus 40 (SV40) and adenoviruses; hence this volume includes a discussion of their important contribution to human tumor biology. The immune response of individuals is also important in determining the outcome of infection and is also discussed, in particular with regard to HBV infection, where an ineffective immune response against the virus is a main contributing factor in pathogenesis and disease progression.

It is important to remember the global impact of these viral infections and the resulting cancers on individuals and health care systems. To emphasize this impact, this volume begins with a chapter on the epidemiology of all these viral infections, each discussed subsequently at the molecular level.

The editor hopes that the contents of this book will give the investigator, physician, or student an up-to-date overview of our progress in acquiring some knowledge of the virus/host interactions which lead to cancer, and of the complexity of virus/host interactions generally, most of which have yet to be delineated.

Human Tumor Viruses
Edited by Dennis J. McCance
© 1998 American Society for Microbiology

1

Epidemiology of Viruses Associated with Human Cancer

Paul H. Levine, Howard Strickler, Angela Manns, and Laura Diamondstone

The oncogenic properties of numerous viruses have been well documented in a variety of animals, and it has long been assumed that at least some human malignancies were virus induced. Evidence for viral oncogenicity in humans, however, has necessarily been indirect. The first oncogenic human virus to be documented, Epstein-Barr virus (EBV), appeared to demonstrate a pattern that subsequently identified tumorigenic agents would follow, whereby cancer is a rare outcome of a common infection. The patterns vary widely from virus to virus, however. EBV, for example, is a ubiquitous virus, widely spread throughout the world, whereas human T-cell lymphotropic virus type I (HTLV-I) is far more restricted in its distribution.

Our understanding of viral epidemiology has primarily developed through seroepidemiologic studies; most oncogenic viruses produce lifelong antibodies readily detected in healthy as well as diseased individuals. The continued improvement of serologic assays and the identification of a number of different host responses to viral infection have thus allowed us to readily describe the pattern of infection of several of the better-known oncogenic human viruses, i.e., EBV, HTLV-I, and the hepatitis viruses. In contrast, human papillomavirus (HPV), the widely accepted cause of cervical cancer as well as some other malignancies, has primarily been studied using DNA hybridization methods, as useful serologic assays have only recently become available. The most recently identified potentially oncogenic virus, human herpesvirus 8 (HHV-8), remains relatively unstudied in regard to its worldwide distribution. Despite the different approaches to the evaluation of individual viruses and their associated tumors, a generalized approach to causality was attempted by Evans (77), who modified the Henle-Koch postulates for causation as they might apply to oncogenic viruses. These criteria were subsequently modified (79) but continue to require adaptation as innovative techniques are applied to new candidate oncogenic viruses. This chapter will focus on the current knowledge regarding the transmission of four groups of viruses with well-documented links to human cancer, i.e., the herpesviruses (particularly EBV and HHV-8), the papovaviruses (primarily HPV), the hepatitis viruses, and the retroviruses (HTLV-I), noting the apparent risk factors for the progression of these common infections to cancer.

THE HERPESVIRUSES

EBV (HHV-4)

EBV, a gammaherpesvirus first isolated by Epstein and coworkers from cultured BL cells (74), was shown to be a common infectious agent (104) and the cause of infectious mononucleosis (IM) (107, 203). Subsequent epidemiologic studies indicated that infection usually occurred in childhood; children living in crowded conditions or in developing countries (27, 80, 106) generally become infected at an earlier age than those living in more rural areas and developed countries. As with most herpesviruses, infection of children with EBV usually either is asymptomatic or produces mild disease, but if primary infection is delayed until puberty or afterwards, IM is a frequent outcome, accounting for the observation that IM is predominantly a disease of upper socioeconomic groups. EBV is primarily harbored in the oropharynx and can be transmitted by saliva, although it has also been isolated from the cervix and circulates through the blood contained in B lymphocytes (see reference 80). Because of its ubiquity, EBV being found in every population in which it was sought, it was apparent that the presence of infection was not a reliable predictor of malignancy. Attention then concentrated on the strength of the immune response. The first serologic studies associating EBV with BL (104, 106) and NPC (213) demonstrated higher antibody titers and stronger precipitation bands in cases than in controls, and similar findings in Hodgkin's disease (HD) (128, 157, 158) suggested an etiologic role in that disease as well. Although it subsequently became apparent that elevated antibody titers did not necessarily imply an active EBV-associated disorder, elevated antibody titers being found in cancer families (162) and immunosuppressed individuals (108), the detection of the viral genome in the malignant cells of these cancers (306, 329) confirmed the virus-cancer link in these three malignancies.

Because of its ubiquity, the role of EBV in oncogenesis appears to be determined primarily by cofactors which differ from population to population. EBV-associated BL, for example, primarily occurs in areas of holoendemic malaria, such as sub-Saharan Africa and New Guinea, and the disruption of the immune system by the malarial parasites is speculated to be a predisposing factor leading to EBV oncogenicity. However, the specific biologic mechanisms and other risk factors for the development of BL remain to be determined, as the peak incidence of BL, which occurs in the 10- to 14-year-old age group in sub-Saharan Africa, is only 14/100,000. Genetic predisposition in particular is being investigated as a possible third "Factor X" responsible for this rare tumor.

A far more common EBV-associated malignancy, NPC, also has a peculiar geographic pattern that may provide clues to the cofactors responsible for its occurrence. The group at highest risk for developing NPC is Cantonese Chinese, who begin to show an increased incidence of disease in the fifth decade of life with a peak around age 44, an unusually young age pattern for a carcinoma. Genetic susceptibility appears to play a role in NPC etiology, particularly in Chinese (113, 261), although familial NPC has been reported in non-Chinese as well (166). Ho first suggested that feeding salted fish in infancy was another possible risk factor (69, 118, 119), and subsequent studies have supported his hypothesis. Nitrosamines have been considered as the responsible carcinogenic agent in these studies, but other carcinogens may be implicated in other high-risk populations such as north African adolescents, African-American children, and Inuits (113, 160, 318).

The initial linkage of EBV to specific malignancies was made through serologic studies, as noted above. While the first studies were case-control studies, stronger evidence was provided by retrospective cohort studies which involved the use of serum repositories to investigate predisease antibody patterns.

In addition to being regularly detected in tumor specimens from patients with two human malignancies, endemic Burkitt's lymphoma (BL) and nasopharyngeal carcinoma (NPC), EBV was shown to induce lymphomas in nonhuman primates (75, 259, 274) that resembled EBV-related lymphomas in humans. It is now apparent that for virtually all of the known human tumor viruses, infection is usually asymptomatic and malignancy is a relatively rare outcome.

The initial reports that IM occurred only in individuals who were seronegative (203, 244) suggested that predisease serology could be critical to identifying individuals at risk. The Special Virus Cancer Program, which was charged with developing an antiviral vaccine to protect against cancer, launched an ambitious program under the direction of the International Agency for Research on Cancer to screen approximately 40,000 Ugandan children at risk of developing BL to determine if EBV antibody status could predict the development of BL. Two of the alternative hypotheses were that BL occurred only in those children who escaped EBV infection until a later age (as in IM) or that there was an aberrant immune response with elevated EBV antibody titers. The latter situation appears to be correct (68), since pre-BL sera showed higher antibody titers to immunoglobulin G (IgG) anti-EBV viral capsid antigen (VCA) than age/sex/locality-matched controls. The increased risk of developing BL for children with VCA antibody titers 2 dilutions above the geometric mean titer of the corresponding controls was estimated at 30-fold. However, the study also documented that not all cases of African BL are EBV associated, as cases with low EBV antibody titers also did not have EBV DNA detectable in tumor cells (90). This form of prospective study has subsequently led to similar efforts investigating EBV-related cancers using available serum banks. Such retrospective cohort studies linked EBV more closely to HD (78, 195), non-Hodgkin's lymphoma (NHL) (196), and gastric carcinoma (167).

While serologic studies provided important evidence of EBV oncogenicity in humans, the strongest evidence has been the consistent identification of the virus in the tumor cells. Such studies demonstrated that BL consisted of at least two forms, the "classic" EBV-associated form with chromosome 8 breakpoints upstream of c-*myc*, and "sporadic" BL, which has chromosomal breakpoints within c-*myc* (16). Classic BL is the predominant form in endemic areas (e.g., Africa and New Guinea) and makes up approximately 10 to 20% of cases in nonendemic areas such as the United States (16, 159). BL which is EBV negative occurs sporadically throughout the world; it has been reported to have a different clinical presentation than EBV-associated BL and may be more resistant to chemotherapy (163, 177). The data regarding clinical presentation, however, probably do not reflect biologic differences between EBV-associated and non-EBV-associated cases as much as they do the differences in the age of BL patients.

In comparison to BL, which has a significant proportion of non-EBV-associated cases, a much more consistent association has been observed between EBV and NPC. EBV is found in virtually all cases of undifferentiated NPC, with rare exceptions (204, 234). EBV serology has shown considerable promise in the early detection (324) and monitoring (101, 154, 221) of patients with NPC, and the detection of EBV DNA in the cervical lymph nodes of patients with unknown primary tumors has been reported to result in the successful detection of occult NPC after initial blind biopsies of the nasopharynx were unrewarding. The use of EBV serology as a public health adjunct was attempted in China by Zeng et al., who reported the successful detection of earlier-stage lesions in a series of studies (323–325). The investigation of IgA antibody against EBV was based on the early reports of Wara et al. (302), who recognized that NPC patients had high levels of serum IgA, which the Henles subsequently demonstrated to be the result of high IgA levels to EBV VCA (105). Because of this close association with EBV, NPC has been the target of pilot projects to vaccinate against EBV (222), although the logistic considerations in demonstrating efficacy are considerable (164). A promising report of immunomodulatory treatment of NPC focusing on EBV control (231) also suggests that the consistent association of EBV with this malignancy may provide a useful target for disease control. In the pilot study, a dialyzable lymphocyte extract with specific activity against EBV appeared to provide a significantly longer disease-free interval in treated patients with NPC compared to untreated controls.

Among the many tumors also purported to be linked to EBV, considerable attention has been devoted to HD, gastric cancer, and AIDS-associated lymphomas. As with BL and NPC, HD initially was associated with EBV by serologic studies, but it became clear that high antibody titers could result from immunosuppression. Therefore, the identification of EBV genome in the Reed-Sternberg cells, the large multinucleated cells that are pathognomonic for HD, became the gold standard for defining an EBV-associated case. HD has long been presumed to have an infectious etiology (76), but it is now apparent that HD is heterogeneous epidemiologically and in relation to its association with EBV (127). EBV-associated HD predominates in children (especially in developing countries) and in older individuals (127), suggesting a relationship to immunologic dysfunction seen in other EBV-associated disorders such as BL and immunoblastic lymphomas occurring in immunosuppressed individuals. It is intriguing that environmental factors possibly related to socioeconomics appear to play a major role in childhood HD (64). Recent studies suggest that EBV-associated HD occurs more often in Hispanic children (5, 99), and this may represent a counterpart to EBV-associated BL in sub-Saharan Africa, with another cofactor substituting for malaria as the important cocarcinogen. (For a recent review of patterns of EBV-associated HD in various countries, see reference 94.)

Gastric cancer may also be an important EBV-associated malignancy. This is a potentially important public health issue since gastric carcinoma is so common worldwide, although less than 15% of gastric cancer cases are thought to be EBV associated. A correlation between EBV and histologic subtype has not been demonstrated, although cases with a prominent infiltration of lymphocytes among the epithelial carcinoma cells (resembling EBV-associated lymphoepitheliomas of the nasopharynx and other organs) often contain EBV in the carcinoma cells.

In summary, EBV is detected in several different types of cancer, and it appears that it may have a causal role in some of these tumors. Overall, the economic impact of EBV-associated cancer may be considerable when the incidences of NPC and gastric cancer are taken into account. However, the specific cofactors responsible for turning this ubiquitous, usually harmless virus into a deadly pathogen are poorly understood.

HHV-6

HHV-6, a betaherpesvirus, was initially isolated from patients with lymphoproliferative disorders and AIDS, and serologic studies suggested that it might be etiologically related to several human malignancies, particularly HD (58, 243). In the past decade, as the epidemiology of this virus has been better defined, it is apparent that this virus is the primary cause of roseola (exanthem subitum) (314), and its association with HD has become more tenuous. Torelli et al. first noted the presence of the viral genome in the tissues of three patients with an unusual mediastinal form of HD characterized by a histological pattern of mixed cellularity-nodular sclerosis (285). These authors continue to note the potential importance of this association, but if HHV-6 is etiologically related to some cases of HD, it is an infrequent occurrence. As with EBV, this ubiquitous virus is likely to be reactivated during periods of immunological perturbation, and the serologic association with HD is more likely a reflection of the disease process than an etiologic one (161). HHV-6 has also been reported in the tumor cells in occasional cases of NHL (255), but as with HD, the number of associated cases is apparently small and these studies remain unconfirmed. Potentially more important is the association with oral carcinoma. Detection of the virus in oral tumor cells was noted by Yadav et al. (313) in five of seven cases obtained from the Medical College of Trichor, India, where oral cancer is a major problem for oncologists. With this possible exception, the impact of HHV-6 as a carcinogen is unknown but appears quite minimal, particularly in comparison to EBV.

HHV-8 (KSHV)

Unlike the other human herpesviruses, HHV-8, the putative etiologic agent for Kaposi's sarcoma (KS), was initially detected by representational difference analysis, an innovative procedure which makes this the first herpesvirus isolated by molecular techniques (51, 52). HHV-8 (also called Kaposi's sarcoma-associated herpesvirus [KSHV]) is, like EBV, classified as a gammaherpesvirus. The history of its discovery explains why the early information regarding this herpesvirus was restricted primarily to molecular rather than classic virologic characteristics. Replication of the virus has been achieved using cells obtained from body cavity lymphomas (49). Initial viral isolates were obtained from cell lines from body cavity lymphomas also carrying EBV, thus hampering studies of specificity in clinical populations.

In recent years, purified antigens and newer serologic techniques have been utilized to evaluate antibody patterns in different populations. The diversity of these serologic procedures, however, has made it difficult to interpret the wide-ranging results, which have also varied in regard to the populations under study (Table 1). It is likely that many of the divergent conclusions will be explained by the eventual understanding of the different components of the immune response, similar to the gradual understanding of the complex response to EBV infection where the viral capsid antibody (VCA) proved to be the best marker of previous infection, antibody to the early antigen and soluble complement-fixing antigen correlated with disease activity, and neutralizing antibody as well as other antibodies to membrane-associated antigens detected on living cells proved to be reflective of a good host response and tumor control (220).

Since HHV-8 has been so closely associated with KS, it is reasonable to use KS as a marker for epidemiologic studies of HHV-8, much as adult T-cell leukemia/lymphoma (ATL) has been used as an indicator of HTLV-I infection (see below) and exanthem subitum (roseola) has been used as a marker for studying HHV-6 (155). KS has long been known to be relatively common in older men of Mediterranean descent (52) and, in addition, has been noted to be common in central Africa (282), where aggressive cases of KS were seen in younger patients (66). These observations may partially explain the varying proportions of HHV-8 seropositivity in controls selected for seroepidemiologic studies involving KS patients. For example, Gao et al. (88) and Simpson et al. (262) noted a higher prevalence of antibodies to HHV-8 in Ugandan controls compared to United States controls (Table 1).

Although the mode of HHV-8 transmission in these areas is unknown, studies in non-KS endemic parts of the world suggest that the etiologic agent for this disease, presumably HHV-8, is sexually transmitted. Beral et al. (23) first suggested that the KS agent was transmitted sexually, with expression influenced significantly by immunosuppression. Earlier in the AIDS epidemic it became apparent that KS was most often found in men who had sex with men (70, 224), an observation noted in several countries (224). Antibody to HHV-8 is uncommon in children outside of Africa (156), lending support to the importance of sexual activity as a route of transmission for this virus.

A second possible route of transmission, although less important than sexual contact, is by blood transfusion. As with HTLV-I (see below), whole-cell transfusion rather than serum products may be the source of virus since transfusion recipients with AIDS are three times as likely to acquire KS than hemophiliacs who were infected with HIV by clotting-factor concentrates. Documentation of HHV-8 in the cells of a healthy blood donor (28) has confirmed the epidemiologic data indicating blood transfusion as a potential source of virus. Levy has also suggested that HHV-8 in nasal secretions and saliva could explain the infection of individuals not infected through sexual contact (169).

As with EBV, the diversity of diseases potentially produced by HHV-8 is considerable; this virus is also linked to B-cell body cavity lymphomas (269) and Castleman's disease (264).

PAPOVAVIRUSES

The Family *Papovaviridae*

The family *Papovaviridae* derives its name from its constituent viruses, the papillomaviruses (genus A; HPV) (*pa-*), polyomaviruses (genus B; JC and BK viruses)

Table 1 Serologic studies of HHV-8

Antigen	Assay	Study group	Seropositivity	Reference
HHV-8-infected and uninfected BCP-1 cells	IFA (initial serum dilution 1:160)	AIDS/KS patients (U.S., Italy, Uganda) U.S. blood donors Italian blood donors Ugandan controls	62 (71–88%) 0/122 4/107 (4%) 24/47 (51%)	88
Small viral capsid antigen (SVCA)	Immunoblot	HIV-positive KS patients (U.S.) HIV-positive AIDS patients without KS (U.S.) Children (12 HIV+, 10 HIV−, 25 acutely ill) (U.S.) Hemophilia patients (U.S.) Autoimmune patients (U.S.) Nasopharyngeal carcinoma (China) Healthy adults	42/47 (89%) 11/54 (20%) 0/47 0/25 0/25 0/25 3/28 (11%)	170
ORF65	ELISA	AIDS-KS (U.S., U.K.) Classic Greek KS HIV-infected non-KS homosexual men Hemophilia/HIV infected Hemophilia/HIV uninfected IVDU,[a] HIV infected IVDU, HIV uninfected U.K. blood donors U.S. blood donors Greek controls Ugandan controls (HIV infected) Ugandan controls (HIV uninfected)	60/74 (81%) 17/18 (94%) 5/16 (31%) 0/28 1/56 (2%) 2/38 (5%) 0/25 3/174 (1.7%) 6/117 (5%) 3/26 (12%) 16/34 (47%) 6/17 (35%)	262

BCB-1 cell line latency-associated nuclear antigen (LANA)	IFA	American KS patients
		 African KS patients
		 HIV seropositive
		 General U.S. population
		   ≤15 years old
		   >15 years old
		 "High-risk" non-pregnant women
		 HIV-infected participants
		 Non-HIV-infected participants
			47/91 (52%)
			 28/100 (28%)
			 19/140 (14%)
			
			 0/263 (0%)
			 0/174 (0%)
			 13/386 (3.4%)
			 12/302 (4.0%)
			 1/84 (1.2%)
				133a
				
				 133

Lytic-nuclear antigen	IFA	African countries
		 Haiti, Dominican Republic
		 American KS patients
		 African KS patients
		 HIV seropositive
		 General population, U.S.
		   ≤15 years old
		   >15 years old
			35/219 (16%)
			 0/92 (0%)
			 87/91 (96%)
			 28/28 (100%)
			 97/140 (69%)
			
			 10/263 (3.8%)
			 33/174 (19%)
				153

BCBL-1 cell line, latency-associated nuclear antigen (LANA)	IFA	U.S. pregnant women (HIV+ and HIV−)
		   Haitian women
		   Non-Haitian women
		 Infants
			12/289 (4.2%)
			
			 9/91 (10%)
			 3/198 (1.5%)
			 0/189
				94a

[a] IVDU, intravenous drug users.

(*po*-), and vacuolating virus (simian virus 40 [SV40], a polyomavirus) (*va*-). In general, the *Papovaviridae* are a heterogeneous family. They have in common that they are small viruses (<60 nm) with nonenveloped 72-capsomer icosahedral capsids and a genome made up of double-stranded circular DNA organized into functional regions. The growth cycles of the *Papovaviridae* are slow and involve replication in the nucleus. The viruses may also share important similarities in the manner of their interactions with certain viral and host proteins. For example, polyomavirus T antigen and HPV E6/E7 each interfere with the function of the cellular tumor suppressor genes p53 and Rb. However, papillomaviruses and polyomaviruses do not have common antigenic determinants, they do not share extensive polynucleotide homology, and they differ in many other biologic characteristics.

HPV

General
HPV have the greatest public health relevance of the human *Papovaviridae*. HPV are widely accepted to be the principal causative agents in the development of cervical cancer (36, 248), the second most common malignancy in women worldwide. HPV probably also play a role in several additional lower genital tract tumors, including cancers of the vulva, vagina, anus, and penis. Furthermore, the involvement of HPV in the development of a number of epithelial cancers throughout the body is being investigated, especially aerodigestive tumors such as buccal, laryngeal, and esophageal tumors. Moreover, HPV also cause genital warts and benign skin warts (e.g., plantar and butchers warts), which are common causes of patient complaints (126).

HPV detection
The ability to accurately measure exposure to HPV is essential to research in this field. In general, the presence of HPV is determined by detection of viral DNA in tissue specimens (see Table 2). The development of sensitive and specific laboratory methods for this purpose has enabled a broad understanding of many aspects

Table 2 Summary of common HPV types and their associated lesions[a]

HPV type	Associated lesions
HPV-1, -2, and -4	Benign skin lesions
HPV-5 and -8	Skin cancers in patients with epidermodysplasia verruciformis
HPV-6 and -11	Laryngeal papillomatosis
	Oral and genital condylomata acuminata
	Anogenital neoplasia (low risk of cancer)
HPV-33, -35, -51, -52, and -58	Anogenital neoplasia (intermediate risk of cancer)
	Other squamous carcinomas?
HPV-16, -18, -31, -45, and -56	Anogenital neoplasia (high risk of cancer)
	Other squamous carcinomas?

[a] See text for references.

Table 3 Common methods for detecting HPV DNA[a]

Method	Sensitivity/specificity	Comments
PCR primers		
Consensus primers (e.g., MY09/MY11 or GP5+/GP6+)	Excellent/uncertain	Prone to false-positive results due to contamination
Type-specific primers (e.g., for E6 or E7)	Excellent/uncertain	Prone to false-positive results due to contamination. Typically used to detect limited range of types since each must be amplified separately
Hybrid capture	Good/good	Commercial kit form is available. Assay can be used to measure viral load in semiquantitative manner.
Southern blot	Good/excellent	Extremely labor intensive
Dot blot	Fair-good/fair-good	A simple rapid method
Filter in situ hybridization (FISH)	Poor/poor	The first HPV assay method practical for population-based studies

[a] Adapted from reference 237a.

of HPV biology, natural history, and pathogenicity. Over 70 individual HPV types have been recognized, distinguished from one another by the heterogeneity of their DNA. However, several limitations to these methods and the requirement of tissue specimens for testing have also influenced our understanding of HPV (247).

It is important to appreciate that adequately sensitive and specific methods for HPV DNA detection have only existed for about 10 years at this writing. For years before that, poor laboratory results led to confusion in the field, as false-positive and -negative findings weakened measurements of the association of HPV with cervical cancer and other outcomes in many investigations. Even today's technologies vary considerably in their sensitivity (Table 3). In general, the method of measuring HPV exposure can substantially affect results, and the reader must be keenly aware of the assays being used when comparing findings between HPV studies (246). The additional fact that several novel HPV types which infect the cervix, and a whole group which infect skin, have only recently been reported is evidence of the high degree to which technology continues to affect our understanding of such fundamental issues as whether the viruses are at all present in certain tissues (25, 265).

Tissue preparation methods may also affect the results of HPV DNA assays, as the tissue DNA must be adequate for testing (98). For example, paraffin-embedded tumor blocks are often used for testing. However, the DNA in these materials is often detrimentally affected by the dehydration and fixation process (98). Even frozen materials can sometimes be inadequate, having been obtained from the wrong site (e.g., just to the side of the lesion being tested) or containing too little DNA.

The major limitation of these specimens, however, is their unavailability. The requirement for actual tissue specimens is very restrictive, since except in advanced disease lesions are generally small, tissues can often not be spared for HPV testing since they are needed for histopathologic review, and healthy patients are understandably reluctant to give tissue biopsies for research purposes. Exfoliated tissue specimens scraped or lavaged from tissue surfaces are frequently used instead of biopsies, with good results (299). However, these procedures are limited to exposed or semiexposed tissues. For example, exfoliated cells from the cervix are commonly obtained using a Dacron swab, in a manner similar to taking a Pap smear. A swab for HPV DNA testing is often taken after the Pap smear, making sure to sample both the ecto- and endocervix, and then placed in some type of medium for preservation (299). Even these approaches are limited, though, as testing the cervix does not address HPV in the anus or anywhere else in the body. The issue of obtaining tissue for testing is especially a problem in studying HPV in men, as complete sampling of the genital region is difficult given that the urethra and anus are major sites of infection but uncomfortable sites to sample (15).

Serologic methods for measuring exposure to HPV would seem to be the ideal answer. However, epidemiologically useful serologic tests to measure virus exposure have only recently started to become available, and none of the seroassays to date has yet shown adequate sensitivity and specificity to be used in diagnosing the presence of virus or HPV-related disease in individual patients. The main application for these assays is currently in large population-based studies, in which some misclassification will not be fatal to an investigation. In these contexts HPV serology is beginning to show some usefulness as a measure of HPV exposure independent of DNA results (72, 271).

HPV tissue tropism
Detection of DNA has shown that HPV are highly species and tissue specific, selectively infecting only human epithelial cells and generally only specific types of epithelial tissues. This high level of tropism is further reflected in the location of HPV-related lesions. For example, almost all HPV-related cervical neoplasia, anal neoplasia, and recurrent respiratory papillomatosis occurs at analogous sites within the particular tissues infected, i.e., the site where the squamous and columnar cells of the epithelium come together, called the squamocolumnar junction. In general, each HPV type has a narrow cellular tropism, only infecting the epithelium in selected sites of the body. Overall, HPV types can be roughly grouped according to the epithelial tissues they infect and whether or not they are linked to the development of cancer (126).

HPV of the skin
Benign skin warts, such as plantar warts and warts of the hands, generally contain one of several HPV types: HPV-1, -2, -3, -4, -7, -10, -26, -27, -28, -29, and -41 (see Table 2) (126, 186). Consistent with the narrow cellular tropism of each viral type, these HPV are not typically detected in the lower genital tract, nor do they cause warts in genital epithelium or lesions of the aerodigestive tract (82, 174). Recent evidence has shown that HPV capsid proteins are resistant to desiccation, and

transmission of these benign skin wart HPV types is most likely in the form of fomites (238). In addition, the recent development of degenerate primers for the PCR detection of HPV DNA has resulted in the discovery of a new group of HPV types that may be common in normal skin (37) as well as in skin cancers—though an etiologic connection with any diseases in the general population has not currently been made (265). These same degenerate primers showed that the HPV types (e.g., HPV-5 and -8) which are primarily found in skin cancer specimens from patients with the rare immune disorder epidermodysplasia verruciformis (EV) are also common in normal tissue, possibly ending confusion over the reservoir for these previously difficult-to-detect viruses (37). Skin cancer rates are increased in patients with EV, and HPV-5 and -8 are commonly considered to have a role in the development of these EV-related skin tumors. Notably, skin cancer rates are also increased in individuals who are immune compromised iatrogenically. Evidence of a role for HPV in the development of these skin cancers has been sought in particular, because these patients also frequently develop skin warts and HPV-related anogenital cancers. However, HPV DNA findings in these studies have not presented clear evidence of an etiologic relationship (37, 186). Overall, the data suggest that HPV are ubiquitous viruses of the skin and that certain types may occasionally cause warts or other clinical conditions. The possible role of HPV in skin cancer of immunologically normal individuals is under investigation.

HPV of the aerodigestive tract

The aerodigestive tract is another site known to support infection with HPV. Laryngeal papillomatosis, also called recurrent respiratory papillomatosis, is a well-known but uncommon HPV-related condition characterized by the development of warty benign tumors anywhere in the larynx, although most commonly in the vocal cords (132, 266). The tumors can be solitary and of limited clinical consequence, or can be spread throughout the respiratory tract and be life-threatening (132). Laryngeal papillomas are commonly first recognized in childhood, and recurrences are common following excision, but the condition typically resolves after puberty. There are estimated to be 1,500 new cases of juvenile laryngeal papillomatosis each year. The age distribution is bimodal, however. The onset of tumors can first be noted in adulthood, and juvenile lesions that resolved by puberty can recur in adults (132).

The association of laryngeal papillomatosis with HPV was first indicated by the observation that children with this condition were often born to mothers with a history of genital warts, called condyloma acuminata (13, 82). Estimates suggest that 1:40 to 1:100 births of children born to women with condylomatous acuminata develop respiratory papillomas (132). The mode of transmission is most likely through exposure to the virus during parturition (84). Both types of lesions have been shown to contain predominantly HPV types 6 or 11, and the viruses induce similar morphologic, histologic, and cytologic changes in both locations (82). Interestingly, HPV-11 is the most common form found in juvenile laryngeal papillomas, whereas HPV-6 is the most common in lesions in adults (126). HPV types 6 and 11 are not considered cancer-causing viruses in the anogenital region, nor are they detected in cancers of the larynx. However, there may be an increased risk of laryngeal carcinoma in patients with a history of recurrent respiratory papillomatosis (13), and other types of HPV that are associated with cervical cancer have been detected in laryngeal cancers, according to some but not all reports. Overall, infection of the larynx with HPV types that typically induce benign exophytic lesions in the anogenital region, and their role in laryngeal papillomatosis, are well accepted, whereas a role for HPV types that are linked to the development of cervical cancer has been reported but is controversial.

Oral condyloma acuminatum and squamous papilloma are benign exophytic lesions of the oral pharynx that may occur at any age. HPV DNA has frequently been detected in these lesions, predominantly HPV types 6 and 11, similar to the role of these viruses in the development of benign exophytic lesions in the larynx and in the anogenital regions (215). The lesions may reflect transmission by autoinoculation from another virus body site, or contact with lesions from another individual (200), though this has not been well worked out. These lesions are not generally considered cancer precursor lesions (175). However, a role for HPV in oral cancers has been proposed. Oral leukoplakia is a lesion at high risk of progressing to cancer that occurs most frequently in middle-aged individuals and the elderly, affecting about 4% of adults. Several studies have reported the detection of HPV in these lesions (205). Similarly, several studies have detected HPV, predominantly HPV-16, which is the causative agent in most cervical cancers, in squamous cell carcinomas of the mouth (215, 257). In a recent study, the detection of HPV DNA was significantly more common in mouths of patients with oral cancer than in controls, even after controlling for potential confounders such as smoking and drinking (175).

Overall, however, HPV DNA results in oral cancers have been mixed, and the level of viral DNA expression seems to be low—inconsistent with a causal link with cancer, as it would be expected to find the viral genome in every tumor cell. Furthermore, two recent serologic investigations failed to detect an association between HPV-16 antibodies and the presence of oral cancer (272; Richard Hayes, personal communication). Therefore, as with laryngeal cancer, there is no consensus on the role of HPV in the development of oral cancers.

Two additional aerodigestive cancers have been the subjects of conflicting HPV findings: esophageal cancer and lung cancer. Few investigations have studied the latter. In esophageal cancer, even the serologic data are conflicting. Two studies, one in Shanxii, China, and another in Finland, showed an association with antibodies to HPV-16 (71, 103). However, another study in the United States failed to find a relationship (272).

In general, a subset of HPV types seem to be able to infect both the anogenital region and parts of the aerodigestive tract. There is substantial evidence supporting a role for these viruses in several benign exophytic lesions (particularly HPV-6 and -11), but the data supporting a role for HPV in the development of aerodigestive cancers are less certain.

Genital HPV
Genital HPV may be the most common sexually transmitted viral infection. According to a recent investigation, 30 to 50% of sexually active college-age women are positive for genital HPV DNA at any one time (18). In women, HPV can also

infect the vulva and vagina, in both men and women the anus can be infected, and in men the penis may be infected (297). Many scientists suggest that the lower genital tract should be considered as a whole, as a general site of possible HPV infection, with each component being both a potential reservoir of infection and a potential source of autoinoculation to the rest of the anogenital tract and other parts of the body, as well as a general reservoir for transmission to other individuals (15, 126, 247). Notably, there is no convincing evidence that barrier contraceptive methods reduce the risk of HPV-related cervical cancer (111).

The prevalence of genital HPV DNA has been best studied in specimens taken from the cervix, because of the interest in understanding the role of HPV in cervical cancer and the relative ease of obtaining exfoliated cervical tissue specimens.

The sexual association of cervical HPV infection has been demonstrated in a number of ways. HPV DNA prevalence is directly related to the lifetime number of sexual partners, and it is greater among individuals who initiated sex at an earlier age (17, 112, 308). HPV DNA prevalence is also greater in women whose sexual partners are HPV DNA positive (15). Similarly, several studies have shown that HPV DNA prevalence and sexual experience are greater in the male sex partners of women who develop cervical neoplasia (35, 139, 192). This last point suggests that the sexual mixing pattern of individuals, e.g., with whom they have sex, affects their risk of HPV-related neoplasia. In virgins, HPV DNA is rarely detected. There have been occasional reports suggesting that HPV DNA may be present in cervical specimens from some virgins, and even in infants and children. However, these results have been contested, and, in any case, it is clear that if there are nonsexual routes of transmission of anogenital HPV infection, these are the rare cases (247).

Cervical HPV DNA prevalence peaks in women during their late teenage years or young twenties, soon after many women have initiated sexual intercourse. After that peak, HPV DNA prevalence then decreases incrementally with age, as shown in Fig. 1 (note that the figure is on a log scale) (246). Decreasing prevalence with age can have several interpretations, but most probably reflects clearance of detectable HPV infection over time. Natural history studies of HPV infection in individuals have shown that most HPV infections become undetectable by even the most sensitive DNA detection methods, over a period of months to a few years (114, 247). Clearance of detectable infection is considered the norm (247). Alternatively, a cohort effect could also explain decreasing HPV prevalence with age; i.e., greater sexual experience in today's teenagers and young adults, as compared to older generations, could explain why younger individuals have greater HPV DNA prevalence. However, the same decrease in prevalence with increasing age has been observed in many different populations around the world, arguing against a simple cohort effect due to changes in societal norms of behavior. In any case, such an effect would likely be small relative to the clear tendency for most HPV infections to clear naturally (247). Overall, the age-specific trends in HPV prevalence are thought to reflect a pattern of early lifetime sexual exposure and infections, typically followed by viral clearance.

As compared to the high frequency of exposure to HPV among sexually active individuals, the frequency of HPV-related diseases is much lower (see Fig. 1) (246). Therefore, genital HPV follow the same pattern defined for HPV in many other

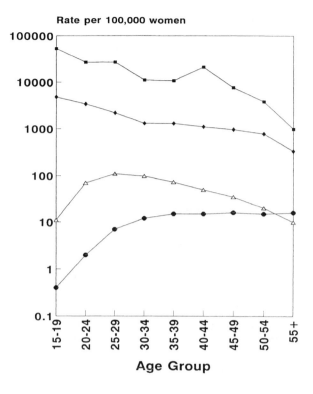

Rate per 100,000 women

Figure 1 Age-specific patterns of HPV infection and cervical neoplasia. (■) Prevalence of HPV DNA in cytologically normal women; (◆) prevalence of koilocytotic atypia (both types of prevalence data were derived from studies of routine cervical cytology screening patients at Kaiser-Permanente, Portland, Oreg.). Also shown are the incidence of CIN-3 (△) and cervical cancer (●), respectively, based on United States national cancer statistics. Note that the figure uses a logarithmic scale. (Reprinted from reference 249a with permission.)

epithelia, i.e., genital HPV are common infections that only occasionally result in clinically recognized pathology.

Over 25 different types of HPV that infect the lower anogenital tract have been described (see Table 2) (126). However, anogenital HPV differ extensively in their individual prevalence, the types of diseases they are associated with, and the level of their association with disease (174, 237). For example, HPV types 6 and 11 are found primarily in benign exophytic lesions of the vulva, called condylomata acuminata or genital warts. In the cervix, HPV-6 and -11 can also cause lesions, but as in the vulva, these lesions tend to be benign in appearance. HPV-6 and -11 are rarely found in cancers of the cervix. For these reasons, these two types are frequently termed "low-risk" HPV types. There are several additional HPV types that are considered low risk for the development of cancer, but HPV-6 and -11 are the most common (174).

High-risk HPV types are defined by their much greater prevalence in cervical cancer specimens than in low-grade lesions or normal cervical tissues. Although there is no strict nomenclature, almost any definition of high-risk HPV types would have to include HPV types 16, 18, 31, 45, and 56 (36, 174, 237). Over 90% of cervical cancers are HPV DNA positive, and these four high-risk HPV types typically account for the vast majority of positive findings (36, 174). About half are specifically HPV-16 positive. Notably, a recent international comparison showed that this pattern was seen in all countries around the world tested, though there were modest variations in the prevalence of the different high-risk HPV types in different regions (36). In Indonesia, HPV-18 rather than HPV-16 was the most common type of HPV DNA detected in cervical cancers, and in cancers in women of African descent HPV-45 was relatively more common than in women from other regions. Other HPV types are less frequently detected in cervical cancers and are often called intermediate risk (36, 174, 237).

HPV-related cervical neoplasia

Cervical cancer tumorigenesis is commonly thought to be a multistage process that occurs over a period of years (188, 228, 249). It is believed to follow a regular pattern, starting with initial HPV infection, followed by persistence of virus expression, and eventually resulting in well-defined cellular changes which worsen over time. This regular pattern of change has resulted in cytologic and histologic nomenclatures that distinguish low grade (early cellular changes) from more severe (later changes) (188, 228). Consistent with the multistage model of cervical cancer tumorigenesis, low-grade cervical neoplasia is most common among older teens and young adults, occurring around the same time that initial HPV infections are most frequent, shortly after many women's sexual debut, whereas high-grade lesions occur mostly among women some 10 years older. Cervical cancer is primarily a disease of women over 35 years of age.

Screening for cervical neoplasia is typically accomplished by cytologic examination of tissue exfoliated from the cervix, stained using modified Papanicolaou staining methods, commonly called Pap smears. Current United States recommendations are that women begin obtaining regular Pap smears as soon as they initiate sexual intercourse or after the age of 18. Treatment of precancerous cervical neoplasia can be accomplished in a number of ways and is usually effective (228).

Since the introduction of routine Pap smears, cervical cancer incidence and mortality rates have dropped dramatically. These decreases have leveled off in recent years, however, and in the United States some possible increases in cervical cancer rates have been noted (141). In much of the developing world, cervical cancer is the most common cancer in women, almost certainly because of the unavailability of Pap smear screening and treatment programs (228). Overall, cervical cancer is the second most common cancer in women worldwide, representing a substantially unnecessary source of morbidity and mortality that disproportionately affects women living in underprivileged conditions.

Like HPV infection, the vast majority of low-grade lesions will spontaneously resolve. Higher-grade lesions often resolve as well, but the risk that they will progress to cancer is much greater (188). For this reason many doctors choose not to treat low-grade lesions and only treat high-grade cervical neoplasia (228). The concerns with this approach are that cytologic detection of a low-grade lesion may sometimes be an indication of severe cervical neoplasia elsewhere in the cervix, and that progression of cervical lesions may not always follow the predicted regular course (this may particularly be an issue among younger women with high-risk HPV infections) (142). Because of the central role of HPV in cervical tumorigenesis, investigators have begun to examine the possible use of HPV typing (e.g., high risk versus low risk) in the triage of patients with abnormal cervical pathology (147).

Other than viral type, the factors determining whether cervical HPV infection will cause cancer have not been well defined. Immunological competence, however, may be among the best-established cofactors. It has been long recognized that iatrogenically immune-compromised individuals such as transplant patients are at increased risk of HPV infection and cervical cancer. Similarly, patients with retrovirus infections may also have increased risk of cervical neoplasia. Specifically, HIV/AIDS patients are more likely to be HPV DNA positive, they are more likely to develop cervical neoplasia, and the risk of cervical lesions in these patients increases with worsening immune status (e.g., a lower CD4$^+$ cell count) (117, 217). In addition, infection with HTLV-I, a type C retrovirus (see below) that may also have some effects on immunity, is associated with higher risk of developing severe cervical neoplasia (270). Among individuals without known reason for immune compromise, a recent study showed diminished production of interleukin-2, a cytokine important in cellular immunity, in responses to HPV epitopes in patients with high-grade cervical neoplasia (291). Overall, an intact immune response to viral antigens seems to be important in controlling cervical HPV infection and its potential progression to cervical cancer.

Viral persistence is another important factor that has often been found to be associated with increased risk of cervical neoplasia, and interestingly the cervical cancer-related HPV types are more likely to persist (114). A number of potential environmental cofactors have also been investigated, including smoking (61), hormonal contraceptive use (40), concomitant infection with other sexually transmitted agents (particularly chlamydia) (41, 142), anogenital hygiene (109), and nutrition (230). However, their associations with the risk of cervical neoplasia have not been firmly established. Now that the central etiologic link between HPV and cervical neoplasia has been made, researchers must reassess most previously identified risk factors for these lesions to determine if they are associated with cervical neoplasia independently of HPV. For example, hormonal contraceptive use may only appear to be associated with cervical neoplasia because HPV infection is more common in the sexually active individuals using these products (248).

Cancers of the anus, vulva, vagina, and penis

HPV DNA, particularly HPV-16, is commonly detected in several extracervical lower genital tract tumors, including cancers of the vulva, vagina, anus, and penis (192, 218, 250, 288). The association of HPV-16 exposure with these tumors has also been confirmed in a large recent HPV serosurvey of cancer patients (272). In addition, the (partial) role of some common etiologic factor in all these tumors is suggested by the repeated observation that individuals who develop one type of anogenital neoplasia are at increased risk of developing others (208, 235). However, many fewer studies of these tumors and their relationship to HPV have been conducted as compared to cervical cancer, and the direct etiologic role for the virus in the development of these tumors has not been as firmly established.

The role of HPV in anal cancers is suggested by several factors. HPV DNA can be detected in most specimens, including cancer metastases, and the biology of anal neoplasia is similar to that of tumors of the cervix. In particular, the lesions tend to be located at the squamocolumnar junction and to show similar histopathology (216). In addition, as discussed above, women with cervical neoplasia are at increased risk of anal neoplasia, suggesting a shared etiology. There are nevertheless a number of important differences in their epidemiology. Anal cancer is an uncommon tumor as compared to cervical cancer, and the median age of patients with anal cancer is about 65 years, 15 to 20 years older than patients with cervical cancer. While anal cancer is primarily a tumor of women, incidence rates are increasing in both men and women. In men, the increase appears to be greatest among homosexuals, a relationship that predates the AIDS epidemic (235).

Vulvar cancer is a rare lesion primarily affecting older women (median age at diagnosis is 71) (Surveillance, Epidemiology, and End Results [SEER]). Two of the three general histologic types of vulvar cancers could be related to HPV, namely baseloid and warty carcinomas, as more than 80% of these tumors may contain HPV. Patients with baseloid and warty vulvar cancers share risk factors with cervical cancer cases, particularly sexual experience, further supporting the notion of a common link to a sexually transmitted agent like HPV (288). Vaginal cancer is also a rare disease of older women, and HPV has been detected in a number of tumors though a link to sexual experience has not been clearly demonstrated (124, 192, 250).

Most studies of penile cancers have found that the majority of tumor specimens contain HPV DNA, and the sexual association of these lesions is supported by the increased risk of penile cancer in the sex partners of patients with cervical cancer and other HPV-related lesions, as well as by a higher number of sex partners reported by penile cancer patients (176, 178). Circumcision, on the other hand, may reduce the risk of penile cancer (176).

Polyomaviruses

An association between human cancers and polyomaviruses has been sought for decades. SV40 was discovered in 1960 as an adventitious contaminant of the early poliovirus vaccines, which were manufactured using Asian macaque kidney cell cultures. The virus was initially not detected because SV40 does not induce cellular changes in its natural host. Because SV40 is more resistant to formalin than is poliovirus, even the formalin-inactivated poliovirus vaccine (IPV) often contained low titers of live SV40. In 1961 SV40 was found to induce sarcomas in hamsters, and in that same year the United States government required that all poliovirus vaccines be free of SV40. However, IPV had a 2-year shelf life, and a diminishing supply of SV40-contaminated vaccines could have been distributed as late as 1963. In the United States, tens of millions of people were probably exposed to low levels of SV40 through IPV. Around the world hundreds of millions of people were probably exposed, but primarily through the oral poliovirus vaccine (OPV). Individuals who received IPV frequently seroconverted for SV40, but the virus has never been isolated from an IPV recipient to demonstrate infection. OPV vaccinees often had virus in their stools, but they never seroconverted, suggesting that the infection was probably localized. Therefore, while it is clear that many people were exposed to SV40 through contaminated poliovirus vaccines, it is unclear whether IPV and OPV were sources of viremia (83, 253).

Epidemiologic studies over the past several decades have examined whether individuals at high probability of exposure to SV40-contaminated poliovirus vaccines, based on year of their birth, had different cancer rates than individuals born just a few years later, who were unexposed. None of the largest studies in the United States or Germany (which used contaminated OPV) showed any associations (83, 89). Most recently, the relation of SV40 with human cancers has become a focus of attention again, as several recent studies using highly sensitive PCR methods have reported detection of the virus in at least four rare human tumors: ependymomas, a type of primary brain cancer that mainly affects infants and children less than 4 years of age (24, 152); choroid plexus papillomas, a primary brain lesion affecting the choroid plexus, most notably in children (24, 152); osteosarcomas, a cancer of bone that particularly affects teenagers (45); and mesotheliomas, a cancer mainly affecting the pleura covering the lungs, which occurs late in life and is highly associated with asbestos exposure (44). Research to understand the relevance of SV40 DNA detection in human tumors is only beginning. Though SV40 DNA has also been found in many normal tissue specimens by some investigators (183), human-to-human transmission of SV40 has never been documented (253), and some researchers failed to detect the virus in cancerous tissues (272). There is currently no serologic evidence of an association between SV40 exposure and human cancers (272).

JCV was discovered in 1971 in the brain tissue of a patient (initials J.C.) with Hodgkin's lymphoma dying of progressive multifocal leukoencephalopathy, a demyelinating disease that also affects AIDS patients. BKV was isolated from the urine of a renal transplant patient (initials B.K.). Both viruses are now recognized as essentially ubiquitous human commensals that can be detected in many normal human tissues, most commonly in specimens from immune-compromised individuals (11). Antibody prevalence to these viruses increases quickly during childhood, and most individuals are seropositive for both agents by young adulthood. JCV and BKV have substantial genomic and molecular biologic similarities to SV40. All three viruses code for T antigen, a protein that can interfere with the function of tumor suppressor genes in vitro. Moreover, like SV40, JCV and BKV inoculations can induce tumors in hamsters and can transform animal and human tissues in vitro. The fact that a virus is ubiquitous does not eliminate the possibility that it may rarely act as a cofactor in the development of human cancers (similar to the role EBV is thought to play in human tumors). However, no strong evidence of an association between JCV and BKV with human cancers has been reported (10).

Conclusions

The *Papovaviridae* are a heterogeneous family of common human DNA viruses that share similar biologic mechanisms for immortalizing human cells in vitro. An etiologic role for HPV in the development of cervical cancer has been well established, and HPV probably play a role in several other lower anogenital tumors. HPV are also known to cause a number of benign lesions of substantial clinical significance. An etiologic role for polyomaviruses in progressive multifocal leukoencephalopathy has been reported. However, the relation of polyomaviruses to human cancers is still uncertain.

HEPATITIS VIRUSES

Introduction

Although the hepatitis A, B, C, D, E, and G viruses are etiologic agents of acute viral hepatitis, only hepatitis B, C, D, and G viruses are known to have a chronic carrier state. Since hepatitis G virus (HGV) has only recently been identified, this chapter will focus on HBV, HCV, and HDV, which have been implicated as etiologic agents of primary hepatocellular carcinoma (HCC). The hepatotropic viruses are quite different from each other with respect to their molecular biology, epidemiology, and pathogenesis. In addition, the heterogeneity of course of infection and clinical outcomes among infected populations is only partly explained by epidemiologic findings such as age at acquisition. Current research suggests that some of this unexplained variation may in part be due to particularly pathogenic genotypes or subtypes of the viruses. It has also been suggested that viral mutations and variants may contribute to the development of primary HCC.

HBV

The hepatitis B Australian antigen was discovered and so named from the observation of a reaction between the serum from an Australian aborigine and that of a United States hemophiliac (32). The HBV gene was cloned and sequenced approximately a decade later (85). The ensuing decade of HBV research resulted in the establishment of viral cultures and mouse lines which facilitated a better understanding of HBV replication and regulation and the role of specific proteins.

HBV is classified as an enveloped DNA virus with some characteristics of a retrovirus, specifically the necessity of reverse transcriptase for replication and the ability to integrate DNA into the host chromosome. The known HBV proteins and gene products include the surface antigen (HBsAg), nucleocapsid antigens (HBeAg and HBcAg), a DNA polymerase product, and an X gene without a well-defined protein. During incubation and acute infection, the serologic markers of HBV infection are HBsAg and HBeAg, as well as HBV DNA. Cleared or resolved HBV infection is indicated by the disappearance of HBsAg and detection of an antibody (anti-HBs). The absence of antibodies to HBcAg and HBsAg reflects susceptibility to HBV infection. The hallmark of chronic infection is serologic evidence of HBsAg for a duration of at least 6 months. Among chronic carriers, approximately 10% will convert to a positive HBeAg state which is a serologic marker of persistent infection and possibly indicative of HBV DNA integration into hepatocytes (38).

Epidemiology

An estimated 300 million people throughout the world have chronic HBV infection (14). In the majority of the developed world, the most common routes of transmission are parenteral and sexual, affecting mostly adolescents and adults. In the United States, HBV infection is considered rare. Estimated from a population-based survey, the prevalence of HBV markers was 4.8%, and only 0.3% had chronic infection determined by HBsAg prevalence (48). In many developing countries, such as sub-Saharan Africa and mainland China where the seroprevalence of HBV is considered endemic, the majority of infections are acquired in infancy or child-hood.

Infection is usually self-limited, but there is a variability in persistence of infection which roughly corresponds to age at acquisition and the geographic distribution of HBV. In the United States, less than 1% of blood donors have chronic infection. In contrast, highly endemic areas such as Southeast Asia have an estimated seroprevalence of HBsAg is as high as 10 to 15% (53). Predictors of chronic carriage include young age at acquisition, male gender, and secretion of HBeAg. Hormonal mechanisms are likely to play a role in male susceptibility to chronic infection, which is supported by experimental evidence of enhanced viral gene expression with steroid hormones (30). Among infants infected from an HBeAg-positive mother, over 80% will become chronic carriers (21, 267). There is also some experimental evidence that genetic variants may persist by evading the immune system (34, 210). HBV clearance is moderately predictable by declining HBsAg titers several weeks after, compared to titers during acute infection (121).

HBV is capable of replicating and producing viral proteins in hepatocytes. A cell-mediated immune response to HBV-infected hepatocytes is involved in both the clearance of and hepatocellular injury from HBV infection. These immune responses are involved in the lysis of HBV-infected hepatocytes, since HBV is not directly cytopathic. It has been demonstrated that cytotoxic T cells in association with major histocompatibility complex group I proteins recognize HBV-infected hepatocytes during acute infection (223), and nucleocapsid antigens are apparent targets for immune-mediated hepatocyte lysis during chronic infection (191, 225). HBV variants may also modify host immune response, pathogenesis, and the course of clinical outcomes (31, 43, 286, 300).

Chronic HBV infection confers a significant risk of HCC, which has been demonstrated from epidemiologic studies with varying degrees of inferential strength. Prevalence studies have correlated geographic areas with a high HCC incidence and high prevalence of chronic HBV infection (33, 122, 232, 275, 276). That is, in regions where HBV is endemic, chronic HBV infection is detected in the majority of HCC cases, whereas only one fourth or fewer of United States HCC cases are considered HBV related. The correlation between high HCC incidence and high HBV prevalence has been observed from dozens of HCC case-control studies conducted throughout the world. These studies have had varying HCC case definitions, referent populations, data to adequately adjust for confounding factors, and HBV serology, but have estimated significant odds ratios associating HCC with chronic HBV infection (134, 151, 171, 233, 287, 316; see reference 125 for a review). Although fewer prospective cohort studies have been conducted, HBV has been found to be a highly significant independent risk factor of incident HCC among populations from endemic HBV regions (19, 214, 232). Cohort studies also indicate that the risk of HBV-related HCC is higher among those infected during infancy or childhood compared to those with HBV infection acquired during adulthood (20, 110, 251). Thus, there is an association between high HBV prevalence and HCC incidence established from epidemiologic findings.

Evidence of HBV infection being directly oncogenic is lacking, and no specific HBV-related oncogenes have been identified. HBV DNA is found integrated into host chromosome, which may promote oncogenesis and unregulated expression. HBV DNA integration has also been detected in human tumor cells from Southern blot analysis studies (39, 73). The body of evidence for viral DNA integration into cellular DNA has grown, yet the sites of integration may be multiple and have not been demonstrated to be proximal to or within selective genes such as onco-genes or tumor suppressor genes (184, 199). The HBx gene has been detected in human HCC, and the protein appears to have *trans*-activating properties (279, 292, 312). Furthermore, HCC has developed in HBx gene transgenic mice (136). In vitro studies have demonstrated that the HBx gene may be involved in altera-tions of p53 function. The tumor suppressor gene p53 is altered in many HCC cases, but this has been observed in regions where both HBV and aflatoxin are prevalent.

HDV

HDV (hepatitis delta virus) is an incomplete virus or viroid-like microorganism with a circular RNA genome that is dependent on the HBV envelope protein for replication. HDV was discovered in the late 1970s (239) and cloned and sequenced approximately a decade later (301). Only one known protein is encoded by HDV, the HDAg, and replication of HDV is likely limited to hepatocytes. At least three HDV genotypes have been identified which are thought to be correlated with severity of associated liver disease (46, 240).

   HDV is endemic in some regions (more than 20% in areas of Mediterranean countries), has had epidemic episodes in other regions, and has a relatively high prevalence among high-risk subgroups such as intravenous drug users. It is esti-mated that 5% of HBsAg carriers are infected with HDV (240).

   Interactions related to both coinfection and superinfection of HDV and HBV have been described. Both are generally associated with a more severe hepatic disease (65). Although HBV is required for HDV replication, evidence of HBV replication is diminished with HDV infection (9, 54). Coinfection with HBV and HDV typically is associated with a severe and often fulminant hepatitis which, if resolved, is followed by clearance of both viruses. In contrast, superinfection is generally followed by chronic carriage of both viruses.

   Predictors of HDV chronic carriage or disease severity are unclear, but per-sons infected in infancy or early childhood generally have a more accelerated and aggressive course of cirrhosis.

HCV

It has been less than a decade since HCV was cloned and sequenced (56). HCV has since been determined to be the etiologic agent of the majority of non-A, non-B posttransfusion-associated hepatitis cases (2). HCV is classified as a distant relative of the *Flaviviridae* family, having a single positive-strand RNA genome and genomic structure related more to pestiviruses than flaviviruses. Most of the viral proteins encoded by HCV have been identified (97, 252). HCV replicates in the liver, and there is recent evidence that it can also replicate in peripheral lymphocytes (327). Like other RNA viruses, many subtypes have been identified and as many as six genotypes are recognized (260). HCV genetic diversity is considered to be related to differing patterns of pathogenesis as well as to immuno-logic responses, which has implications for treatment.

The first evidence linking HCV with HCC came from seroprevalence studies of HCC cases and controls (42, 62). Another link of HCV infection to HCC was suggested by data from Japan describing decreased population-based seroprevalence of HBsAg and an increasing annual incidence of HCC without a parallel increase in HBsAg-associated liver cancers (207). Subsequently, among series of HBsAg-negative HCC cases, over 90% were found to be anti-HCV positive (206). Numerous case-control studies conducted throughout the world consistently reported that a majority of HBsAg-negative HCC cases were infected with HCV (168, 209, 289, 303, 321). A nested case-control study from Taiwan also provided strong evidence of an association of HCV infection with the development of HCC (50).

HCV may be detected in the serum with an enzyme-linked immunosorbent screening assay (ELISA) format using recombinant antigens (146). After HCV infection, antibodies to conserved regions of the HCV genome remain detectable in the serum. Recombinant immunoblot assays are considered confirmatory, and HCV RNA levels detected by reverse transcription and PCR technology have been useful for semiquantification of HCV serum levels. There are no known HCV-specific neutralizing antibodies produced in infected individuals.

The most significant and well-documented route of HCV transmission is parenteral. There has been suggestive but not strong evidence for sexually acquired (1, 3, 172) and community-acquired (4, 281) infections. The prevalence of HCV infection estimated from blood donor serosurveys, using a confirmatory assay, ranges from less than 1% in the U.K. (95), Australia (8), and the United States (268) to between 1 and 3% in Japan (315), the Middle East (26), and some Southeast Asian countries (326). Highest-risk groups, with a seroprevalence above 60%, include subgroups of intravenous drug users and patients with bleeding disorders who received replacement products in which virus inactivation was inadequate.

The majority of HCV infections acquired from transfusions are subclinical. Typically, transfusion-transmitted HCV hepatitis is mild, rarely icteric, and characterized by intermittently elevated liver enzymes after a lag of several weeks to months after infection. Chronic hepatitis develops in approximately half of infected individuals (2), and approximately 20% of those with chronic HCV hepatitis will develop cirrhosis over several decades (63, 138, 187). There are no well-documented factors for progression of liver disease among immunocompetent individuals. There is some evidence that cirrhosis is more likely to develop when liver enzymes are persistently elevated and that disease severity may be correlated with HCV genotypes (278). Mechanisms of HCV immunopathogenesis have not yet been elucidated. Descriptions of potential immune-mediated mechanisms of liver damage with HCV infection are conflicting, although cytotoxic T lymphocytes are considered to play a role in HCV infection remission and pathogenesis (137, 143). In addition to host immune response to infection, disease progression and severity have also been correlated with HCV genetic diversity (278).

Cirrhosis may be a necessary intermediate stage between HCV infection and HCC. In contrast to the evidence linking HBV to HCC, HCV can replicate in hepatocytes but has not been found to be integrated into tumor chromosomes.

No transactivating activity has been demonstrated by any HCV protein, and no HCV-associated oncogenes have been identified. It is likely that HCV-associated HCC may result from proliferation of hepatocytes or malignant transformation as a secondary consequence of hepatic injury.

Virus Interactions

Because of shared transmission routes, coinfection or superinfection with two or more hepatotropic viruses is not a rare occurrence. Modification of the course of infection and liver injury has been recognized with dual (or more) infections with hepatotropic viruses.

There is a clear interaction between HBV and HDV, as the latter needs HBV for its own propagation. Similar to HDV, HCV decreases HBV replication (81, 120, 219, 254, 290). Also similar to HDV, a more severe liver injury may arise from HCV and HBV coinfection (62, 81, 254). Significant independent and additive risks of HBV and HCV coinfection have been estimated from HCC case-control studies (57, 129, 236, 319) and a prospective study of patients with cirrhosis (22). In addition, synergistic effects of HCV and/or HBV infection with other known risk factors of HCC and cirrhosis have been reported, specifically with alcohol (42, 202, 298, 320) and tobacco (293, 320).

Conclusions

These hepatotropic viruses are associated with chronic and progressive liver diseases, including HCC. Definitive oncogenic properties of these viruses have not been elucidated. Mechanisms of virus-related malignant transformation are likely linked to viral integration into host DNA (HBV), severe virus-induced necroinflammatory processes (HDV), and viral replication in hepatocytes resulting in their proliferation and malignant transformation (HCV). Continued research on host immune response and viral genomic diversity is necessary to enhance our understanding of viral clearance and pathogenesis, and for the development of optimal diagnostic markers, treatment modalities, and prevention strategies.

HTLV-I

Background and History

HTLV-I is etiologically linked to ATL, a mature T-cell malignancy. In 1980, HTLV-I was isolated from fresh and cultured lymphocytes of a patient with a cutaneous T-cell lymphoma (227), which at the time was thought to be mycosis fungoides (MF), but in retrospect was ATL. It was the first human type C retrovirus to be found associated with a malignancy, and its isolation was the culmination of many years of investigation on animal retrovirus disease models (87). The discovery was made possible by advances in virological and cell culture techniques, which led to development of assays for the virally encoded reverse transcriptase and made available T-lymphocytic growth factors such as interleukin-2 (86). The clinical observation and description of ATL, as a recognized disease entity, had been made by Takatsuki and colleagues in Japan in 1976 (280, 294). Shortly after that, ATLA (ATL antigen) was identified (115) and found to be molecularly consistent with HTLV-I (317). Through independent and international collaborative efforts, by 1982 the disease was definitively associated with HTLV-I (130).

Epidemiological observations were important in establishing the link between HTLV-I infection and ATL. ATL was initially described as a mature T-cell malignancy with (i) an adult onset, (ii) acute or chronic leukemia with a rapidly progressive terminal course, (iii) leukemic cells with lobulated nuclei resembling "flower" cells, (iv) frequent skin involvement, and (v) frequent lymphadenopathy and hepatosplenomegaly. Early cases were noted to be geographically confined to the southernmost islands of Japan, Kyushu, Shikoku, and Okinawa. From the beginning, a possible viral etiology and genetic susceptibility were thought to be potential causative factors. Reports of ATL associated with HTLV-I among immigrant Caribbean populations in the United Kingdom were soon identified (29, 47). Further evidence was provided that HTLV-I was the causative agent in ATL, even in ethnically distinct and geographically distant populations. The development of serologic assays (245) facilitated the descriptive epidemiology of HTLV-I and revealed that the virus clustered in areas where the disease was found.

Epidemiology of HTLV-I

Geographic distribution

Seroepidemiologic studies have identified that HTLV-I is endemic to populations in southern Japan, the Caribbean, parts of South America, Africa, the Middle East, and the Melanesian Islands. The virus has also been identified in areas within the United States and other parts of North America and Europe, particularly among immigrant populations from endemic areas. Antibodies to the virus can be detected in sera by ELISA. Repeatedly reactive sera are then confirmed with Western blots (immunoblots). The Western blot shows reactivity with viral antigens, and seropositivity is established when *gag* (p19, p24) and *env* (gp46, recombinant gp21) bands to viral genomic structural elements are present. PCR techniques can be used to detect proviral DNA in peripheral blood mononuclear cells from infected HTLV-I carriers or in tumor tissue from patients with ATL (226). Some of the highest seroprevalence rates have been identified in southern Japan and the Caribbean (310). Among populations, the virus increases in seroprevalence with age, and above age 30 to 40 years females have higher prevalence than males.

As the geographic distribution of infection was characterized, with the use of available detection methods, variant HTLV-I strains were identified. For the most part, HTLV-I is genetically very stable with low genetic variability between isolates. The variants were initially discovered through the observation of a high frequency of atypical or indeterminate patterns of reactivity in the standard Western blot. Nucleotide sequencing of virus isolates from Zaire, Papua New Guinea, the Solomon Islands, and aboriginal people of Australia revealed that in Zaire the isolate was 97% similar to that of Japan and the Caribbean, while the Melanesian Islands had sequence homology of only 92% with the other known strains (91, 92). However, the differences observed among these HTLV-I variant strains have not been correlated with different disease manifestations or modes of virus transmission.

Modes of virus transmission

HTLV-I and HIV-1 have similar modes of transmission, although HTLV-I is highly cell associated, resulting in less efficient transmission compared to HIV-1 (179). HTLV-I can be transmitted parenterally, by transfusion, through needle sharing among drug abusers, or via medical injections. It is also transmitted vertically from mother to child, primarily by breast feeding. Sexual transmission from male to female is highly efficient; however, female-to-male and male-to-male contact can also result in transmission. Transmission occurs among 40 to 60% of recipients who receive a seropositive transfusion of cellular blood components, i.e., whole blood, packed red blood cells, or platelets (212). In a prospective study conducted in Jamaica (182), median time to seroconversion following transfusion exposure was 51 days. Donor units stored for less than 7 days were more likely to transmit the virus because of the presence of many viably infected lymphocytes. Recipients receiving immunosuppressive drugs at the time of transfusion were more likely to seroconvert after transfusion exposure. Vertical transmission results in sero-conversion among ~20% of infants born to seropositive mothers, and increased risk of infection is associated with duration of breast feeding greater than 6 months (6). Sexual transmission is more difficult to quantify, although it has been docu-mented in partner studies (273), and among steady partners, longer duration of relationship seems to increase the risk of transmission. Among highly sexually active populations, such as sexually transmitted disease clinic attenders, early age at first sex and higher number of sexual partners heighten risk of transmission (197).

Modes of virus acquisition have been related to increasing the risk of specific HTLV-I-related diseases because of host vulnerability at time of exposure. Vertical transmission has been associated with subsequent development of ATL. It has been hypothesized that early life exposure is critical for later progression to ATL that develops following a 20- to 40-year latency period after infection (198, 311). HTLV-I is also linked to development of HTLV-I-associated myelopathy/tropical spastic paraparesis (HAM/TSP), a chronic, progressive, neurodegenerative condi-tion. Although HAM/TSP can develop following vertical or transfusion transmis-sion (96, 193), due to a preponderance of cases among females, sexual transmission of HTLV-I is thought be a major risk factor for subsequent disease development in certain endemic areas (145). Only rare cases of ATL have been documented after seropositive transfusions (55), and sexually acquired virus resulting in ATL has been difficult to establish.

Nonmalignant Disease Associations

An increasing spectrum of diseases has been associated with HTLV-I. Other non-malignant conditions include infective dermatitis, a childhood skin disorder that may presage development of ATL or HAM/TSP (148). In addition, various benign skin conditions (149), including acquired ichthyosis, scabies, seborrheic eczema, psoriasis, and condylomata acuminata, occur in adults due to the propensity of HTLV-I for skin. Many of these conditions may also precede development of ATL, if observed in HTLV-I carriers. Other conditions suggesting altered immunity in HTLV-I carriers have also been reported. Parasitic infections such as *Strongyloides* sp. commonly occur among HTLV-I carriers (241). *Pneumocystis carinii* infection has been reported among patients with ATL (295). HTLV-I uveitis, arthritis, and Sjogren's syndrome have also been associated with the HTLV-I carrier state (190, 283). However, HAM/TSP is the most common manifestation associated with HTLV-I, other than ATL (123).

HAM/TSP is usually an adult manifestation of HTLV-I infection with a female predominance, although familial, childhood cases have been reported (201). It is characterized by an insidious onset with development of chronic spastic paraparesis, weakness of the lower limbs with hyperreflexia, clonus, and positive Babinski as major presenting symptoms. Laboratory diagnosis should include detection of HTLV-I antibodies in blood and cerebrospinal fluid, although seronegative cases have been reported. In endemic areas, 70 to 100% of cases are HTLV-I seropositive. The virus is polyclonally integrated in host cells, and proviral DNA can also be detected in the central nervous system of patients. Although the immunopathogenesis of this disorder is not completely understood, HAM/TSP patients have a heightened immune response to HTLV-I, which suggests that the host's response to infection may contribute to the disease process. Differences in the host's response to infection may be important in distinguishing among carriers who subsequently develop ATL or HAM/TSP, since distinct mechanisms for disease pathogenesis have been proposed (296).

HTLV-I-Associated Malignancies

ATL
Antibodies to HTLV-I and/or integrated proviral DNA are found in the majority cases of ATL (59). ATL cases have been described in many locales worldwide, mirroring the geographic distribution of HTLV-I. The worldwide prevalence of HTLV-I-associated leukemia/lymphomas is unknown and the incidence in any population depends on the prevalence of viral infection in that region. Highest incidences of ATL are found in southern Japan and the Caribbean basin, where HTLV-I is highly endemic (60, 277). In highly endemic areas such as southern Japan and the Caribbean Islands, the annual incidence of ATL is approximately 3/100,000 per year, accounting for one half of all adult lymphoid malignancies. In the United States, standard mechanisms of cancer surveillance through the Surveillance, Epidemiology and End Results (SEER) Program would suggest that the incidence of ATL is very low. However, through the use of enhanced surveillance in a suspected endemic population in the United States (307), incidence rates resembling those of Japan and the Caribbean have been identified. The lifetime cumulative risk for development of ATL among HTLV-I carriers is 1 to 4.5% (198), and although ATL is a rare outcome, it has an extremely poor prognosis with median survival of less than 1 year.

ATL occurs with nearly equal frequency between men and women, unlike the female predominance among HTLV-I carriers. Occasionally ATL occurs in the pediatric age group with cases reported as young as 5 and 6 years, but these instances are rare (67). The age-specific incidence rate, however, peaks in the sixth or seventh decade of life in Japan and 10 to 20 years younger in Jamaica (60). Between Japanese-American and African-American patients in the United States, similar ethnic differences in the age distribution of patients with ATL have been reported (165), with average age of disease onset being 63 and 39, respectively. It was further observed that the two groups also differed with respect to distribution of clinical features,: African-Americans more commonly presented with lymphoma-type ATL, while Japanese-Americans more frequently presented with the acute type according to criteria established by Shimoyama et al. (256). These differences may be a result of environmental or genetic influences on disease manifestations.

In persons less than age 60, HTLV-I is the primary cause of lymphoma in viral endemic areas. In a study in Jamaica and Trinidad, more than 50% of total lymphoid malignancies were attributable to HTLV-I exposure, with more than 70% of T-cell lymphomas due to HTLV-I (180). Therefore, the interruption of early life infection with HTLV-I would potentially eliminate one-half of the NHL cases in similar viral endemic areas. Other important risk factors in the pathogenesis of ATL that account for the long latent period between HTLV-I infection and disease onset are still unknown. A multistep carcinogenesis model is believed to be important for malignant transformation.

Other factors have been proposed to have potential importance as cofactors with the virus in lymphomagenesis: host immune response associated with specific HLA antigens and haplotypes (296, 309); genetic abnormalities such as p53 mutations and cytogenetic changes (131, 242); farming occupation with possible pesticide exposure (181); and coinfection with EBV (284). Differences in virologic characteristics, such as amount of HTLV-I provirus DNA integrated in host cells or defective provirus, also may be influential in ATL pathogenesis (258). In order to identify the most important determinants of ATL pathogenesis, we must improve our ability to identify HTLV-I carriers at risk for progression. Additional epidemiologic and laboratory investigations are necessary to clarify these relationships.

MF

MF (mycosis fungoides) was initially thought to be associated with HTLV-I, with the original virus isolate identified from a patient whose subsequent clinical course was more consistent with ATL. We now know that ATL has a broad clinical spectrum with several subtypes (acute, lymphoma, chronic, smoldering) which have some clinical overlap with MF (256). MF is a T-cell NHL which also has a predilection for the skin and an erythrodermic leukemic variant, Sezary syndrome (173). The two lymphoma histologic types can be distinguished immunophenotypically (140). MF expresses CD2, CD3, and CD5 but may lack CD7 and does not usually express activation markers such as HLA-DR or CD25. ATL expresses CD2, CD3, and CD5 but may lack CD7 and characteristically expresses activation markers IL-2R (CD25 Tac antigen) and HLA-DR. Both NHL types express the mature, peripheral helper T-cell subset, the CD4$^+$ CD8$^-$ phenotype. In the United States, MF has an incidence of ~1/100,000 population, which reportedly has been increasing since 1973 (305). MF incidence increases with age and has an average age of onset of 50 years. The incidence is higher in males compared to females and is higher in blacks than in whites. Survival (173) is dependent on stage, with the good prognostic risk group having a median survival over 12 years, while the poor prognostic risk group has a median survival of 2.5 years. Despite these distinct features from ATL, following the discovery of HTLV-I, serosurveys of patients with MF were performed and the majority showed that most MF cases were HTLV-I seronegative (229).

Recently, with the advent of PCR, reports of MF cases that are HTLV-I seronegative but DNA positive by PCR have been published (100, 328), although other laboratories have reported no association (135, 194).

Epidemiologic studies of MF, to date, have not revealed many useful etiologic clues (304). Conflicting results have been reported suggesting that MF patients have a propensity for prior and subsequent malignancies, which if true would suggest a possible genetic etiology. Analysis of risk factors associated with possible sexual acquisition of a putative retrovirus has been unrevealing. Additional investigations, using epidemiologic approaches with standardized laboratory methodology, are needed to confirm or refute the possible role of HTLV-I or related retrovirus in the etiology of MF.

Conclusions

Substantial epidemiologic and virologic evidence exists to support the fact that HTLV-I is the etiologic agent important in the pathogenesis of ATL. A role for HTLV-I as the causative agent in other malignancies requires further study. However, HTLV-I through undetermined mechanisms is also associated with various nonmalignant conditions.

CONCLUSIONS

Understanding the relation of viruses to the development of human cancers is important scientifically. However, its greatest relevance is probably in public health, as part of cancer control strategies. That is, as the epidemiology of oncogenic viruses and their central etiologic role in cancer tumorigenicity are better understood, it may be possible to educate people how to avoid infection, to develop vaccines to prevent viral transmission, to identify individuals at high risk of developing disease, and even to target viral therapies against lesions such as virus-induced cancer precursors like HPV-related cervical intraepithelial neoplasia and already established tumors such as NPC, as described in this chapter. In addition, this and other earlier reviews have shown that infections with oncogenic viruses are much more common than the cancers they produce. Therefore, a major focus of research remains on understanding the cofactors that operate in the presence of these viruses increasing the risk of cancer development, so that these too can be avoided. Overall, virus-associated malignancies appear to be associated worldwide with a high tumor burden, particularly in developing nations, and their economic impact, while unknown, is undoubtedly substantial.

REFERENCES

1. **Akahane, Y., T. Aikawa, Y. Sugai, F. Tsuda, H. Okamoto, and S. Mishiro.** 1992. Transmission of HCV between spouses. *Lancet* **339:**1059–1060.
2. **Alter, H. J., R. H. Purcell, J. W. Shih, J. C. Melpolder, M. Houghton, Q. L. Choo, and G. Kuo.** 1989. Detection of antibody to hepatitis C virus in prospectively followed

transfusion recipients with acute and chronic non-A, non-B hepatitis. *N. Engl. J. Med.* **321:**1494–1500.

3. **Alter, M. J., P. J. Coleman, W. J. Alexander, E. Kramer, J. K. Miller, E. Mandel, S. C. Hadler, and H. S. Margolis.** 1989. Importance of heterosexual activity in the transmission of hepatitis B and non-A, non-B hepatitis. *JAMA* **262:**1201–1205.

4. **Alter, M. J., R. J. Gerety, L. A. Smallwood, R. E. Sampliner, E. Tabor, F. Deinhardt, G. Frosner, and G. M. Matanoski.** 1982. Sporadic non-A, non-B hepatitis: frequency and epidemiology in an urban US population. *J. Infect. Dis.* **145:**886–893.

5. **Ambinder, R. F., P. J. Browning, I. Lorenzana, B. G. Leventhal, H. Cosenza, R. B. Mann, E. M. E. MacMahon, R. Medina, V. Cardona, S. Grufferman, A. Olshan, A. Levin, E. A. Petersen, W. Blattner, and P. H. Levine.** 1993. Epstein-Barr virus and childhood Hodgkin's disease in Honduras and the United States. *Blood* **81:**462–467.

6. **Ando, Y., S. Nakano, K. Saito, I. Shimamoto, M. Ichijo, T. Toyama, and Y. Hinuma.** 1987. Transmission of adult T-cell leukemia retrovirus (HTLV-I) from mother to child: comparison of bottle- with breast-fed babies. *Jpn. J. Cancer Res.* **78:**322–324.

7. **Arao, M., S. Kakumu, H. Tahara, K. Yoshioka, and N. Sakamoto.** 1989. Interferon alpha and gamma positive mononuclear cells and HLA antigen expression on hepatocyte membrane in the liver in chronic type B and non-A, non-B hepatitis. *Gastroenterol. Jpn.* **24:**255–261.

8. **Archer, G. T., M. L. Buring, B. Clark, S. L. Ismay, K. G. Kenrick, K. Purusothaman, J. M. Kaldor, W. V. Bolton, and B. R. Wylie.** 1992. Prevalence of hepatitis C virus antibodies in Sydney blood donors. *Med. J. Aust.* **157:**225–227.

9. **Arico, S., M. Aragona, M. Rizzetto, F. Caredda, A. Zanetti, G. Marinucci, S. Diana, P. Farci, M. Arnone, N. Caporaso, A. Ascione, P. Dentico, G. Pastora, G. Raimondo, and A. Craxi.** 1985. Clinical significance of antibody to the hepatitis delta virus in symptomless HBsAg carriers. *Lancet* **ii:**356–358.

10. **Arthur, R. R., S. A. Grossman, B. M. Ronnett, S. H. Bigner, B. Vogelstein, and K. V. Shah.** 1994. Lack of association of human polyomaviruses with human brain tumors. *J. Neurooncol.* **20:**55–58.

11. **Arthur, R. R., and K. V. Shah.** 1989. Occurrence and significance of papovaviruses BK and JC in the urine. *Prog. Med. Virol.* **36:**42–61.

12. **Asou, N., T. Kumagai, S. Uekihara, M. Ishii, M. Sato, K. Sakai, H. Nishimura, K. Yamaguchi, and K. Takatsuki.** 1986. HTLV-I seroprevalence in patients with malignancy. *Cancer* **58:**903–907.

13. **Austen, D. F., P. Reynolds, D. Schottenfeld, and J. F. J. Fraumeni.** 1996. Laryngeal cancer, p. 619–631. *In* D. F. Austen, P. Reynolds, D. Schottenfeld, and J. F. J. Fraumeni (ed.), *Cancer Epidemiology and Prevention*, 2nd ed. Oxford University Press, Inc., New York.

14. **Ayoola, E. A., and Members of the WHO Committee.** 1988. Progress in the control of viral hepatitis: memorandum from a WHO meeting. *Bull. W. H. O.* **66:**443–455.

15. **Baken, L. A., L. A. Koutsky, J. Kuypera, M. R. Kosorok, S. K. Lee, N. B. Kiviat, and K. K. Holmes.** 1995. Genital human papillomavirus infection among male and female sex partners: prevalence and type-specific concordance. *J. Infect. Dis.* **171:** 429–432.

16. **Barriga, F., J. Kiwanuka, M. Alvarez-Mon, B. Shiramizu, B. Huber, P. Levine, and I. Magrath.** 1988. Significance of chromosome 8 breakpoint location in Burkitt's lymphoma: correlation with geographical origin and association with Epstein Barr virus. *Curr. Top. Microbiol. Immunol.* **141:**129–137.

17. **Bauer, H. M., A. Hildesheim, M. H. Schiffman, A. G. Glass, B. B. Rush, D. R. Scott, D. M. Cadell, R. J. Kurman, and M. M. Manos.** 1993. Determinants of genital human

papillomavirus infection in low-risk women in Portland, Oregon. *Sex. Transm. Dis.* **20:**274–278.

18. **Bauer, H. M., Y. Ting, C. E. Greer, J. C. Chambers, C. J. Tashiro, J. Chimera, A. Reingold, and M. M. Manos.** 1991. Genital human papillomavirus infection in female university students as determined by a PCR-based method. *J. Am. Med. Assoc.* **265:** 472–477.

19. **Beasley, R. P., and L. Y. Hwang.** 1984. Hepatocellular carcinoma and hepatitis B virus. *Semin. Liver Dis.* **4:**113–121.

20. **Beasley, R. P., L. Y. Hwang, C. C. Lin, and C. S. Chien.** 1981. Hepatocellular carcinoma and hepatitis B virus: a prospective study of 22707 men in Taiwan. *Lancet* **21:** 1129–1133.

21. **Beasley, R. P., C. Trepo, C. E. Stevens, and W. Smuzness.** 1977. The e antigen and vertical transmission of hepatitis B surface antigen. *Am. J. Epidemiol.***105:**94–98.

22. **Benvegnu, L., G. Fattovich, G. Diodati, F. Noventa, P. Pontisso, F. Tremolada, and A. Alberti.** 1994. Hepatitis C virus infection and replication in patients with hepatocellular carcinoma, p. 706–709. *In* K. Nishioka, H. Suzuki, S. Mishiro, and T. Oda (ed.), *Viral Hepatitis and Liver Disease.* Springer-Verlag, New York.

23. **Beral, V., Peterman, T. A., Berkelman, R. L., Jaffe, H. W.** 1990. Kaposi's sarcoma among persons with AIDS: a sexually transmitted infection? *Lancet* **335:**123–128.

24. **Bergsagal, D. J., M. J. Finegold, J. S. Butel, W. J. Kupsky, and R. L. Garcea.** 1992. DNA sequences similar to those of simian virus 40 in ependymomas and choroid plexus tumors of childhood. *N. Engl. J. Med.* **326:**988–993.

25. **Bernard, H. U., S. K. Chan, M. M. Manos, C. K. Ong, L. L. Villa, H. Delius, C. L. Peyton, H. M. Bauer, and C. M. Wheeler.** 1994. Identification and assessment of known and novel human papillomaviruses by polymerase chain reaction amplification, restriction fragment length polymorphisms, nucleotide sequence, and pylogenetic algorithms. *J. Infect. Dis.* **170:**1077–1085.

26. **Bernvil, S. S., V. J. Andrews, and A. A. Kariem.** 1991. Hepatitis C antibody prevalence in Saudi Arabian blood donors. *Vox Sang.* **61:**279–280.

27. **Biggar, R. J., W. Henle, G. Fleischer, J. Bocker, G. T. Lennette, and G. Henle.** 1978. Primary Epstein-Barr virus infections in African infants. I. Decline of maternal antibody and time of infection. *Int. J. Cancer* **22:**239–243.

28. **Blackbourn, D. J., J. Ambroziak, E. Lennette, M. Adams, B. Ramachandran, and J. A. Levy.** 1997. Infectious human herpesvirus 8 in a healthy North American blood donor. *Lancet* **349:**609–611.

29. **Blattner, W. A., V. S. Kalyanaraman, M. Robert-Guroff, T. A. Lister, D. A. Galton, P. S. Sarin, M. H. Crawford, D. Catovsky, M. Greaves, and R. C. Gallo.** 1982. The human type-C retrovirus, HTLV, in blacks from the Caribbean region, and relationship to adult T-cell leukemia/lymphoma. *Int. J. Cancer* **30:**257–264.

30. **Blum, A. L., R. Stuts, U. P. Haemmerli, P. Schmid, and G. F. Grady.** 1969. A fortuitously controlled study of steroid therapy in acute viral hepatitis. I. Acute disease. *Am. J. Med.* **47:**82–92.

31. **Blum, H. E., E. Galun, T. J. Liang, F. von Weizsacker, and J. R. Wands.** 1991. Naturally occurring missense mutation in the polymerase gene terminating hepatitis B virus replication. *J. Virol.* **65:**1836–1842.

32. **Blumberg, B. S., H. J. Alter, and S. Visnich.** 1965. A "new" antigen in leukemia sera. *JAMA* **191:**541–546.

33. **Blumberg, B. S., and W. T. London.** 1982. Primary hepatocellular carcinoma and hepatitis B virus. *Curr. Probl. Cancer* **6:**1–23.

34. **Bonino, F., A. Demartini, and M. R. Brunetto.** 1991. Diagnostic advances in viral

hepatitis: the discovery of hepatitis C virus and genetic heterogeneity of hepatitis B virus. *Ann. Ist. Super. Sanita.* **27**:547–553.

35. **Bosch, F. X., X. Castellaague, N. Munoz, S. de Sanjose, A. M. Ghaffari, L. C. Gonzalez, M. Gili, I. Izarzugaza, P. Viladiu, and C. Navarro.** 1996. Male sexual behavior and human papillomavirus DNA: key risk factors for cervical cancer in Spain. *J. Natl. Cancer Inst.* **88**:1060–1067.

36. **Bosch, F. X., M. M. Manos, N. Munoz, M. Sherman, A. Jansen, J. Peto, M. H. Schiffman, V. Moreno, R. Kurman, and K. V. Shah.** 1995. Prevalence of HPV DNA in cervical cancer in 22 countries. *J. Natl. Cancer Inst.* **87**:796–802.

37. **Boxman, I. L. A., R. J. M. Berhout, L. H. C. Mulder, M. C. Wolkers, J. N. B. Bavinck, B. J. Vermaer, and J. T. Shegget.** 1996. Detection of human papillomavirus DNA in plucked hairs from renal transplant recipients and healthy volunteers. 15th International Papillomavirus Workshop, Queensland, Australia.

38. **Brechot, C., M. Hadchouel, J. Scotto, F. Degos, P. Charnay, C. Trepo, and P. Tiollais.** 1981. Detection of hepatitis B virus DNA in liver and serum: a direct appraisal of the carrier state. *Lancet* **ii**:765–768.

39. **Brechot, C., C. Pourcel, A. Louise, B. Rain, and P. Tiollsis.** 1980. Presence of integrated hepatitis B virus DNA sequences in cellular DNA of human hepatocellular carcinoma. *Nature* **286**:533–535.

40. **Brinton, L. A.** 1991. Oral contraceptives and cervical neoplasia. *Contraception* **43**: 581–595.

41. **Brinton, L. A., R. Herrero, W. C. Reeves, R. C. de Britton, E. Gaitan, and F. Tenorio.** 1993. Risk factors for cervical cancer by histology [see comments]. *Gynecol. Oncol.* **51**:301–306.

42. **Bruix, J., J. M. Barrera, X. Clavet, G. Ercilla, J. Costa, J. M. Sanchez-Tapias, M. Ventura, M. Vall, M. Bruguera, and C. Bru.** 1989. Prevalence of antibodies to hepatitis C virus in Spanish patients with hepatocellular carcinoma and hepatic cirrhosis. *Lancet* **2**:1004–1006.

43. **Brunetto, M. R., F. Oliveri, G. Rocca, D. Criscuolo, E. Chiaberge, M. Capalbo, E. David, G. Verme, and F. Bonino.** 1989. Natural course and response to interferon of chronic hepatitis B accompanied by antibody to hepatitis B antigen. *Hepatology* **10**:198–202.

44. **Carbone, M., H. I. Pass, P. Rizzo, M. Marinetti, M. Di Muzio, D. J. Mew, A. S. Levine, and A. Procopio.** 1994. Simian virus 40-like DNA sequences in human pleural mesothelioma. *Oncogene* **9**:1781–1790.

45. **Carbone, M., P. Rizzo, A. Procopio, M. Giuliano, H. I. Pass, M. C. Gebhardt, C. Mangham, M. Hansen, D. F. Malkin, and G. Bushart.** 1996. SV40-like sequences in human bone tumors. *Oncogene* **13**:527–535.

46. **Casey, J. L., T. L. Brown, E. J. Colan, F. S. Wignall, and J. L. Gerin.** 1993. A genotype of hepatitis D virus that occurs in northern South America. *Proc. Natl. Acad. Sci. USA* **90**:9016–9020.

47. **Catovsky, D., M. F. Greaves, M. Rose, D. Galton, A. W. Goolden, D. R. McCluskey, J. M. White, I. Lampert, G. Bourikas, R. Ireland, A. I. Brownell, J. M. Bridges, W. A. Blattner, and R. C. Gallo.** 1982. Adult T-cell lymphoma-leukemia in blacks from the West Indies. *Lancet* **1**:639–643.

48. **Centers for Disease Control.** 1989. Racial differences in rates of hepatitis B virus infection-United States, 1976–1980. *Morbid. Mortal. Weekly Rep.* **38**:818–821.

49. **Cesarman, E., P. S. Moore, P. H. Rao, G. Inghirami, D. M. Knowles, and Y. Chang.** 1995. In vitro establishment and characterization of two acquired immunodeficiency syndrome-related lymphoma cell lines (BC-1 and BC-2) containing Kaposi's sarcoma-associated herpesvirus-like (KSHV) DNA sequences. *Blood* **86**:2708–2714.

50. Chang, C. C., M. W. Yu, C. F. Lu, C. S. Yang, and C. J. Chen. 1994. A nested case-control study on association between hepatitis C virus antibodies and primary liver cancer in a cohort of 9,775 men in Taiwan. *J. Med. Virol.* **43**:276–280.

51. Chang, Y., E. Cesarman, M. S. Pessin, F. Lee, J. Culpepper, D. M. Knowles, and P. S. Moore. 1994. Identification of herpesvirus-like DNA sequences in AIDS-associated Kaposi's sarcoma. *Science* **266**:1865–1869.

52. Chang, Y., and P. S. Moore. 1996. Kaposi's sarcoma (KS)-associated herpesvirus and its role in KS. *Infect. Agents Dis.* **5**:215–222.

53. Chen, D. S. 1987. Hepatitis B virus infection, its sequel, and prevention in Taiwan, p. 71–80. *In* K. Okuda and K. G. Ishak (ed.), *Neoplasms of the Liver.* Springer, Tokyo.

54. Chen, P. J., D. S. Chen, C. R. Chen, Y. Y. Chen, H. M. Chen, M. Y. Lai, and J. L. Sung. 1988. Delta infection in asymptomatic carriers of hepatitis B surface antigen: low prevalence of delta activity and effective suppression of hepatitis B virus replication. *Hepatology* **8**:1121–1124.

55. Chen, Y. C., C. H. Wang, I. J. Su, C. Y. Hu, M. J. Chou, T. H. Lee, D. T. Lin, T. Y. Chung, C. H. Liu, and C. S. Yang. 1989. Infection of human T-cell leukemia virus type I and development of human T-cell leukemia/lymphoma in patients with hematologic neoplasms: a possible linkage to blood transfusion. *Blood* **74**:388–394.

56. Choo, Q.-L., G. Kuo, A. J. Weinger, L. R. Overby, D. W. Bradley, and M. Houghton. 1989. Isolation of a cDNA clone derived from a blood-borne non-A, non-B viral hepatitis genome. *Science* **244**:359–362.

57. Chuang, W. L., W. Y. Chang, S. N. Lu, W. P. Su, Z. Y. Lin, S. C. Chen, M. Y. Hsieh, L. Y. Wang, S. L. You, and C. J. Chen. 1992. The role of hepatitis B and C viruses in hepatocellular carcinoma in a hepatitis B endemic area. A case-control study. *Cancer* **69**:2052–2054.

58. Clark, D. A., F. E. Alexander, P. A. McKinney, B. E. Roberts, C. O'Brien, R. F. Jarrett, R. A. Cartwright, and D. E. Onions. 1990. The seroepidemiology of human herpesvirus-6 (HHV-6) from a case-control study of leukaemia and lymphoma. *Int. J. Cancer* **45**:829–833.

59. Clark, J. W., C. Gurgo, G. Franchini, W. N. Gibbs, W. Lofters, C. Neuland, D. Mann, C. Saxinger, R. C. Gallo, and W. A. Blattner. 1988. Molecular epidemiology of HTLV-I-associated non-Hodgkin's lymphomas in Jamaica. *Cancer* **61**:1477–1482.

60. Cleghorn, F. R., A. Manns, R. Falk, P. Hartge, B. Hanchard, N. Jack, E. Williams, E. Jaffe, F. White, and C. Bartholomew. 1995. Effect of HTLV-I infection on non-Hodgkin's lymphoma incidence. *J. Natl. Cancer Inst.* **87**:1009–1014.

61. Coker, A. L., A. J. Rosenberg, M. F. McCann, and B. S. Hulka. 1992. Active and passive cigarette smoke exposure and cervical intraepithelial neoplasia. *Cancer Epidemiol. Biomarkers Prev.* **1**:349–356.

62. Colombo, M., G. Kuo, Q.-L. Choo, M. F. Donato, E. Del Ninno, M. A. Tommasini, N. Dioguardi, and M. Houghton. 1989. Prevalence of antibodies to hepatitis C virus in Italian patients with hepatocellular carcinoma. *Lancet* **2**:1006–1008.

63. Colombo, M., M. G. Rumi, M. F. Donato, M. A. Tommasini, E. Del Ninno, G. Ronchi, G. Kuo, and M. Houghton. 1991. Hepatitis C antibody in patients with chronic liver disease and hepatocellular carcinoma. *Dig. Dis. Sci.* **36**:1130–1133.

64. Correa, F., and G. T. O'Conor. 1971. Epidemiologic patterns of Hodgkin's disease. *Int. J. Cancer* **8**:192–201.

65. Craig, J. R., S. Govindarajan, and K. M. DeCock. 1986. Delta viral hepatitis. Histopathology and course. *Pathol. Ann.* **21**:1–21.

66. Davies, J. N. P. and F. Lothe. 1983. Kaposi's sarcoma in African children, p. 81–86. *In* L. V. Ackerman and J. F. Hurray (ed.), *Symposium on Kaposi's Sarcoma.* S. Karger, Basel.

67. **De Oliveira, M. S., E. Matutes, L. C. Famadas, T. F. Schuls, M. L. Calabro, M. Nucci, M. J. Andrada-Serpa, R. S. Tedder, R. A. Weiss, and D. Catovsky.** 1990. Adult T-cell leukemia/lymphoma in Brazil and its relation to HTLV-I. *Lancet* **336:**987–990.

68. **de-The, G., A. Geser, N. E. Day, P. M. Tukei, E. H. Williams, D. P. Beri, P. G. Smith, A. G. Dean, G. W. Bronkamm, P. Feorino, and W. Henle.** 1978. Epidemiological evidence for causal relationship between Epstein-Barr virus and Burkitt's lymphoma from Ugandan prospective study. *Nature* **274:**756–761.

69. **de-The, G., J. H. C. Ho, and C. Muir.** 1991. Nasopharyngeal carcinoma, p. 737–767. *In* A. S. Evans (ed.), *Viral Infections of Humans: Epidemiology and Control*, 3rd ed. Plenum Publishing Corp, New York.

70. **Des Jarlais, D. C., M. Marmor, P. Thomas, M. Chamberland, S. Zolla-Pazner, and D. J. Sencer.** 1984. Kaposi's sarcoma among four different AIDS risk groups. *N. Engl. J. Med.* **310:**1119. (Letter.)

71. **Dillner, J., P. Knekt, J. T. Schiller, and T. Hakulinen.** 1995. Prospective seroepidemiological evidence that human papillomavirus type 16 is a risk factor for esophageal squamous cell carcinoma. *Br. Med. J.* **311:**1346.

72. **Dillner, J., F. Wiklund, P. Lenner, C. Eklund, V. Frederiksson-Shanazarian, J. T. Schiller, M. Hibma, G. Hallmans, and U. Stendahl.** 1995. Antibodies against linear and conformational epitopes of human papillomavirus type 16 that independently associate with incident cervical cancer. *Int. J. Cancer* **60:**377–382.

73. **Edman, J. C., P. Gray, P. Valenzuela, L. B. Rall, and W. J. Rutter.** 1980. Integration of hepatitis B virus sequences and their expression in a human hepatoma cell. *Nature* **286:**535–538.

74. **Epstein, M. A., B. H. Achong, and Y. M. Barr.** 1964. Virus particles in cultured lymphoblasts from Burkitt's lymphoma. *Lancet* **1:**702–703.

75. **Epstein, M. A., H. Rabin, G. Ball, A. B. Rickinson, J. Jains, and L. V. Melendes.** 1973. Pilot experiments with EB virus in owl monkeys (*Aotus trivirgatus*). II. EB virus in a cell line from an animal with reticuloproliferative disease. *Int. J. Cancer* **12:**319–332.

76. **Evans, A. S.** 1971. The spectrum of infections with Epstein-Barr virus: a hypothesis. *J. Infect. Dis.* **124:**330–337.

77. **Evans, A. S.** 1976. Causation and disease: the Henle-Koch postulates revisited. *Yale J. Biol. Med.* **49:**175–195.

78. **Evans, A. S., and G. W. Comstock.** 1981. Presence of elevated antibody titers to Epstein-Barr virus before Hodgkin's disease. *Lancet* **1:**1183–1186.

79. **Evans, A. S., and N. E. Mueller.** 1990. Viruses and cancer: causal associations. *Ann. Epidemiol.* **1:**71–92.

80. **Evans, A. S., and J. C. Niederman.** 1989. Epstein-Barr virus, p. 265–292. *In* A. S. Evans (ed.), *Viral Infections of Humans: Epidemiology and Control*, 3rd ed. Plenum Publishing Corp., New York.

81. **Fong, T. L., A. M. Di Bisceglie, J. G. Waggoner, S. M. Banks, and J. M. Hoofnagle.** 1991. The significance of antibody to hepatitis C virus in patients with chronic hepatitis B. *Hepatology* **14:**64–67.

82. **Franceshi, S., S. Munos, X. F. Bosch, F. J. F. Snijders, and J. M. M. Walboomers.** 1996. Human papillomavirus and cancers of the upper aerodigestive tract: a review of epidemiological and experimental evidence. *Cancer Epidemiol. Biomarkers Prev.* **5:**567–575.

83. **Fraumeni, J. F., Jr., F. Ederer, and R. W. Miller.** 1963. An evaluation of the carcinogenicity of Simian virus 40 in man. *J. Am. Med. Assoc.* **185:**713–718.

84. **Fredericks, B. D., A. Balkin, H. W. Daniel, J. Schonrock, B. Ward, and I. H. Frazer.**

1993. Transmission of human papillomaviruses from mother to child [see comments]. *Aust. N. Z. J. Obstet. Gynaecol.* **33**:30–32.

85. **Galibert, F., E. Mandart, F. Fitoussi, P. Tiollais, and P. Charnay.** 1979. Nucleotide sequence of the hepatitis B virus genome (subtype ayw) cloned in *E. coli. Nature* **281:** 646–650.

86. **Gallo, R. C., B. J. Poiesz, and F. W. Ruscetti.** 1981. Regulation of human T-cell proliferation: T-cell growth factor and isolation of a new class of type-C retroviruses from human T-cells. *Haematol. Blood Transfus.* **26**:502–513.

87. **Gallo, R. C., and G. J. Todaro.** 1976. Oncogenic RNA viruses. *Semin. Oncol.* **3**:81.

88. **Gao, S. J., L. Kingsley, M. Li, W. Zheng, C. Parravicini, J. Ziegler, R. Newton, C. R. Rinaldo, A. Saah, J. Phair, R. Detels, Y. Chang, and P. S. Moore.** 1996. KSHV antibodies among Americans, Italians and Ugandans with and without Kaposi's sarcoma. *Nat. Med.* **2**:925–928.

89. **Geissler, E.** 1990. SV40 and human brain tumors. *Prog. Med. Virol.* **37**:211–222.

90. **Geser, A., G. de-The, G. Lenoir, N. E. Day, and E. H. Williams.** 1982. Final case reporting from the Uganda prospective study of the relationship between EBV and Burkitt's lymphoma. *Int. J. Cancer* **29**:397–400.

91. **Gessain, A., E. Boeri, R. Yanagihara, R. C. Gallo, and G. Franchini.** 1993. Complete nucleotide sequence of a highly divergent human T-cell leukemia (lymphotropic) virus type I (HTLV-I) variant from Melanesia: genetic and phylogenetic relationship to HTLV-I strains from other geographical regions. *J. Virol.* **67**:1015–1023.

92. **Gessain, A., I. J. Koralnik, J. Fullen, E. Boeri, C. Mora, A. Blank, E. F. Salazar-Grueso, J. Kaplan, W. C. Saxinger, and M. Davidson.** 1994. Phylogenetic study of ten new HTLV-I strains from the Americas. *AIDS Res. Hum. Retroviruses* **10**:103–106.

93. **Gill, P. S., W. Harrington, Jr., M. H. Kaplan, R. C. Ribeiro, J. M. Bennett, H. A. Liebman, M. Bernstein-Singer, B. M. Espina, L. Cabral, and S. Allen.** 1995. Treatment of adult T-cell leukemia-lymphoma with a combination of interferon alfa and zidovudine. *N. Engl. J. Med.* **332**:1744–1748.

94. **Glaser, S. L., J. L. Ruby, S. L. Stewart, R. F. Ambinder, R. F. Jarrett, P. Brousset, G. Pallesen, M. L. Gully, K. Gulfaraz, J. O'Grady, M. Hummel, M. V. Preciado, H. Knecht, K. C. Chan, and A. Claviez.** 1997. Epstein-Barr virus associated Hodgkin's disease: epidemiologic characteristics in international data. *Int. J. Cancer* **70**:375–382.

94a. **Goedert, J. J., D. H. Kedes, and D. Ganem.** 1997. Antibodies to human herpesvirus 8 in women and infants born in Haiti and the USA. *Lancet* **349**:1368.

95. **Goodrick, M. J., N. A. B. Anderson, I. D. Fraser, A. Rouse, and V. Pearson.** 1992. History of previous drug misuse in HCV-positive blood donors. *Lancet* **339**:502.

96. **Gout, O., M. Baulac, A. Gessain, F. Semah, F. Saal, J. Peries, C. Cabrol, C. Foucault-Fretz, D. Laplane, and F. Sigaux.** 1990. Rapid development of myelopathy after HTLV-I infection acquired by transfusion during cardiac transplantation. *N. Engl. J. Med.* **322**:383–388.

97. **Grakoui, A., C. Wychowski, C. Lin, S. M. Feinstone, and C. M. Rice.** 1993. Expression and identification of hepatitis C virus polyprotein cleavage products. *J. Virol.* **67**:1385–1395.

98. **Greer, C. E., J. K. Lund, and M. M. Manos.** 1994. PCR amplification from paraffin-embedded tissues: recommendations on fixatives for long term storage and prospective studies. *PCR Methods Appl.* **3**:S113–S122.

99. **Gulley, M. L., P. A. Eagan, L. Quintanilla-Martinez, A. L. Picado, B. N. Smir, C. Childs, C. D. Dunn, F. E. Craig, J. W. Williams, Jr., and P. M. Banks.** 1994. Epstein-Barr virus DNA is abundant and monoclonal in the Reed-Sternberg cells of Hodgkin's disease: association with mixed cellularity subtype and Hispanic American ethnicity. *Blood* **83**:1595–1602.

100. **Hall, W. W., C. R. Liu, O. Schneewind, H. Takahashi, M. H. Kaplan, G. Roupe, and A. Vahine.** 1991. Deleted HTLV-I provirus in blood and cutaneous lesions of patients with mycosis fungoides. *Science* **253:**317–320.

101. **Halprin, J., A. L. Scott, L. Jacobson, P. H. Levine, H. C. Ho, J. C. Niederman, D. Hayward, and G. Milman.** 1986. Enzyme-linked immunosorbent assay of antibodies to Epstein-Barr virus nuclear and early antigens in patients with infectious mononucleosis and nasopharyngeal carcinoma. *Ann. Intern. Med.* **104:**331–337.

102. **Hamada, T., M. Setoyama, Y. Katshira, T. Furuno, T. Fujiyoshi, S. Sonoda, and M. Tashiro.** 1992. Differences in HTLV-I integration patterns between skin lesions and peripheral blood lymphocytes of HTLV-I seropositive patients with cutaneous lymphoproliferative disorders. *J. Dermatol. Sci.* **4:**76–82.

103. **Han, C., G. Qiao, N. L. Hubbert, L. Li, C. Sun, Y. Wang, D. Xu, Y. Li, D. R. Lowy, and J. T. Schiller.** 1996. Serological association between human papillomavirus 16 infection and esophageal cancer in Shaanxi, China. *J. Natl. Cancer Inst.* **88:**1467–1471.

104. **Henle, G., and W. Henle.** 1966. Immunofluorescence in cells derived from Burkitt's lymphoma. *J. Bacteriol.* **91:**1248–1256.

105. **Henle, G., and W. Henle.** 1976. Epstein-Barr virus-specific IgA serum antibodies as an outstanding feature of nasopharyngeal carcinoma. *Int. J. Cancer* **17:**1–7.

106. **Henle, G., W. Henle, P. Clifford, V. Diehl, G. W. Kafuko, B. G. Kirya, G. Klein, R. H. Morrow, G. M. R. Munuba, P. Pika, P. M. Tukei, and J. L. Ziegler.** 1969. Antibodies to Epstein-Barr virus in Burkitt's lymphoma and control groups. *J. Natl. Cancer Inst.* **43:**1147–1157.

107. **Henle, G., W. Henle, and V. Diehl.** 1968. Relation of Burkitt's tumor-associated herpes type virus to infectious mononucleosis. *Proc. Natl. Acad. Sci. USA* **59:**94–101.

108. **Henle, W., and G. Henle.** 1981. Epstein-Barr virus-specific serology in immunologically compromised individuals. *Cancer Res.* **41:**4222–4225.

109. **Herrero, R., L. A. Brinton, W. C. Reeves, M. M. Brenes, F. Tenorio, R. C. de Britton, E. Gaitan, M. Garcia, and W. E. Rawls.** 1990. Sexual behavior, venereal diseases, hygiene practices, and invasive cervical cancer in a high-risk population. *Cancer* **65:** 380–386.

110. **Heyward, W. L., A. P. Lanier, T. R. Bender, H. H. Hardison, P. H. Dohan, B. J. McMahon, and D. P. Francis.** 1981. Primary hepatocellular carcinoma in Alaskan Natives, 1969–1979. *Int. J. Cancer* **28:**47–50.

111. **Hildesheim, A., L. A. Brinton, K. Mallin, H. F. Lehman, P. Stolley, D. A. Savits, and R. Levine.** 1990. Barrier and spermicidal contraceptive methods and risk of invasive cervical cancer. *Epidemiology* **1:**266–272.

112. **Hildesheim, A., P. Gravitt, M. H. Schiffman, R. J. Kurman, W. Barnes, S. Jones, J. G. Tchabo, L. A. Brinton, C. Copeland, and J. Epp.** 1993. Determinants of genital human papillomavirus infection in low-income women in Washington, D. C. *Sex. Transm. Dis.* **20:**279–285.

113. **Hildesheim, A., and P. H. Levine.** 1993. Etiology of nasopharyngeal carcinoma: a review. *Epidemiol. Rev.* **15:**466–485.

114. **Hildesheim, A., M. H. Schiffman, P. E. Gravitt, A. G. Glass, C. E. Greer, T. Zhang, D. R. Scott, B. B. Rush, P. Lawler, and M. E. Sherman.** 1994. Persistence of type-specific human papillomavirus infection among cytologically normal women [see comments]. *J. Infect. Dis.* **169:**235–240.

115. **Hinuma, Y., K. Nagata, M. Hanaoka, M. Nakai, T. Matsumoto, K. I. Kinoshita, S. Shirakawa, and I. Miyoshi.** 1981. Adult T-cell leukemia: antigen in an ATL cell line and detection of antibodies to the antigen in human sera. *Proc. Natl. Acad. Sci. USA* **78:**6476–6480.

116. **Ho, D. D., L. R. Rota, and M. S. Hirsch.** 1984. Infection of endothelial cells by human T-lymphotropic virus type I. *Proc. Natl. Acad. Sci. USA* **81:**7588–7590.

117. **Ho, G. Y. F., R. D. Burk, I. Fleming, and R. S. Klein.** 1994. Risk of genital human papillomavirus infection in women with human immunodeficiency virus-induced immunosuppression. *Int. J. Cancer* **56:**788–792.

118. **Ho, H. C.** 1971. Incidence of nasopharyngeal cancer in Hong Kong. *UICC Bull. Cancer* **9:**5.

119. **Ho, H. C.** 1975. Epidemiology of nasopharyngeal carcinoma. *J. R. Coll. Surg.* **20:** 223–235.

120. **Hollinger, F. B., G. Dolana, W. Thomas, and F. Gyorkey.** 1984. Reduction in risk of hepatitis transmission by heat-treatment of a human factor VIII concentrate. *J. Infect. Dis.* **150:**250–262.

121. **Hollinger, F. B., and G. R. Dreesman.** 1980. Immunobiology of hepatitis viruses, p. 558–572. *In* N. R. Rose, H. Friedman, and J. L. Fahey (ed.), *Manual of Clinical Laboratory Immunology,* 3rd ed. American Society for Microbiology, Washington, D.C.

122. **Haing, A. W., W. Guo, J. Chen, J. -Y. Li, B. J. Stone, W. J. Blot, and J. F. Fraumeni, Jr.** 1991. Correlates of liver cancer mortality in China. *Int. J. Epidemiol.* **20:**54–59.

123. **Ijichi, S., and M. Osame.** 1995. Human T lymphotropic virus type I (HTLV-I)-associated myelopathy/tropical spastic paraparesis (HAM/TSP): recent perspectives. *Intern. Med.* **34:**713–721.

124. **Ikenberg, H., M. Runge, A. Goppinger, and A. Pfleiderer.** 1990. Human papillomavirus DNA in invasive carcinoma of the vagina. *Obstet. Gynecol.* **76:**432–438.

125. **International Agency for Research on Cancer.** 1994. *IARC Monographs on the Evaluation of Carcinogenic Risks to Humans,* vol. 59, *Hepatitis Viruses.* World Health Organization, Lyon, France.

126. **International Agency for Research on Cancer Working Group.** 1996. *IARC Monographs on the Evaluation of Carcinogenic Risks to Humans,* vol. 64, *Human Papillomaviruses.* IARC, Lyon, France.

127. **Jarrett, R. F., A. Gallagher, D. B. Jones, F. E. Alexander, A. S. Krajewski, A. Kelsey, J. Adams, B. Angus, S. Gledhill, D. H. Wright, R. A. Cartwright, and D. E. Onions.** 1991. Detection of Epstein-Barr virus genomes in Hodgkin's disease: relation to age. *J. Clin. Pathol.* **44:**844–848.

128. **Johannson, B., G. Klein, W. Henle, and G. Henle.** 1970. Epstein-Barr virus (EBV)-associated antibody patterns in malignant lymphoma and leukemia. I. Hodgkin's disease. *Int. J. Cancer* **6:**450–462.

129. **Kaklamani, E., D. Trichopoulos, A. Tzonou, X. Zavitaanos, Y. Koumantaki, A. Hatzakis, C. C. Hsieh, and S. Hatziyannis.** 1991. Hepatitis B and C viruses and their interaction in the origin of hepatocellular carcinoma. *J. Am. Med. Assoc.* **265:** 1974–1976.

130. **Kalyanaraman, V. S., M. G. Sarngadharan, Y. Nakao, Y. Ito, T. Aoki, and R. C. Gallo.** 1982. Natural antibodies to the structural core protein (p24) of the human T-cell leukemia (lymphoma) retrovirus found in sera of leukemia patients in Japan. *Proc. Natl. Acad. Sci. USA* **79:**1653–1657.

131. **Kamada, N., M. Sakurai, K. Miyamoto, I. Sanada, N. Sadamori, S. Fukuhara, S. Abe, Y. Shiraishi, T. Abe, and Y. Kaneko.** 1992. Chromosome abnormalities in adult T-cell leukemia/lymphoma: a karyotype review committee report. *Cancer Res.* **52:** 1481–1493.

132. **Kashima, H., B. Leventhal, P. Mounts, and Papilloma Study Group.** 1985. Scoring system to assess severity and course in recurrent respiratory papillomatosis, p. 125–135. *In* P. M. Howley and T. R. Broker (ed.), *Papillomaviruses: Molecular and Clinical Aspects.* Alan R. Liss, New York.

133. **Kedes, D. H., D. Ganem, N. Ameli, P. Bacchetti, and R. Greenblatt.** 1997. The

prevalence of serum antibody to human herpesvirus 8 (Kaposi sarcoma-associated herpesvirus) among HIV-seropositive and high-risk HIV-seronegative women. *JAMA* **277:**478–481.

133a. **Kedes, D. H., E. Operskalski, M. Busch, R. Kohn, J. Flood, and D. Ganem.** 1996. The seroepidemiology of human herpesvirus 8 (Kaposi's sarcoma-associated herpesvirus): distribution of infection in KS risk groups and evidence for sexual transmission. *Nat. Med.* **2:**918–924.

134. **Kew, M. C., J. Desmyter, A. F. Bradburns, and G. M. Macnab.** 1979. Hepatitis B virus infection in southern African blacks with hepatocellular cancer. *J. Natl. Cancer Inst.* **62:**517–520.

135. **Kikuchi, A., T. Nishikawa, and K. Yamaguchi.** 1997. Absence of human T-cell lymphotropic virus type I in cutaneous T-cell lymphoma. *N. Engl. J. Med.* **336:**296–297.

136. **Kim, C. M., K. Koike, I. Saito, T. Miyamura, and G. Jay.** 1991. HBx gene of hepatitis B virus induces liver cancer in transgenic mice. *Nature* **351:**317–320.

137. **Kita, H., T. Moriyama, T. Kaneko, I. Harase, M. Nomura, H. Miura, I. Nakamura, Y. Yazaki, and M. Imawari.** 1994. Recognition of hepatitis C virus nucleocapsid protein-derived peptides by cytotoxic T lymphocytes, p. 186–189. *In* K. Nishioka, H. Suzuki, S. Mishiro, and T. Oda (ed.), *Viral Hepatitis and Liver Disease.* Springer-Verlag, New York.

138. **Kiyosawa, K., T. Sodeyama, E. Tanaka, Y. Gibo, K. Yoshizawa, Y. Nakano, S. Furuta, Y. Akahane, K. Nishioka, and R. H. Purcell.** 1990. Interrelationship of blood transfusion, non-A, non-B hepatitis and hepatocellular carcinoma: analysis by detection of antibody to hepatitis C virus. *Hepatology* **12:**671–675.

139. **Kjaer, S. K., E. T. De Villiers, C. Dahl, G. Engholm, J. E. Bock, B. F. Vestergaard, E. Lynge, and O. M. Jensen.** 1991. Case-control study of risk factors for cervical neoplasia in Denmark. I. Role of the "male factor" in women with one lifetime sexual partner. *Int. J. Cancer* **48:**39–44.

140. **Knowles, D. M.** 1989. Immunophenotypic and antigen receptor gene rearrangement analysis in T cell neoplasia. *Am. J. Pathol.* **134:**761–785.

141. **Kosary, C. L., L. A. G. Ries, B. A. Miller, B. F. Hankey, A. Harras, and B. K. Edwards.** 1995. SEER cancer statistics review, 1973–1992: tables and graphs. NIH Publication no. 96–2789. National Cancer Institute, Bethesda, Md.

142. **Koutsky, L. A., K. K. Holmes, C. W. Critchlow, C. E. Stevens, P. A. J. Paavonen, A. M. Beckmann, T. A. DeRouen, D. A. Galloway, D. Vernon, and N. B. Kiviat.** 1992. A cohort study of the risk of cervical intraepithelial neoplasia grade 2 or 3 in relation to papillomavirus infection. *N. Engl. J. Med.* **327:**1272–1278.

143. **Koziel, M. J., D. Dudley, J. T. Wong, J. Dienstag, M. Houghton, R. Ralston, and B. D. Walker.** 1992. Intrahepatic cytotoxic T lymphocytes specific for hepatitis C virus in persons with chronic hepatitis. *J. Immunol.* **149:**3339–3344.

144. **Kozuru, M., N. Uike, K. Muta, T. Goto, Y. Suehiro, and M. Nagano.** 1996. High occurrence of primary malignant neoplasms in patients with adult T-cell leukemia/lymphoma, their siblings, and their mothers. *Cancer* **78:**1119–1124.

145. **Kramer, A., E. M. Maloney, O. S. Morgan, P. Rodgers-Johnson, A. Manns, E. L. Murphy, S. Larsen, B. Cranston, J. Murphy, and J. Benichou.** 1995. Risk factors and cofactors for human T-cell lymphotropic virus type I (HTLV-I)-associated myelopathy/tropical spastic paraparesis (HAM/TSP) in Jamaica. *Am. J. Epidemiol.* **142:** 1212–1220.

146. **Kuo, G., Q. L. Choo, H. J. Alter, G. L. Gitnick, A. G. Redeker, R. H. Purcell, T. Miyamura, J. L. Dienstag, M. J. Alter, and C. E. Stevens.** 1989. An assay for circulating antibodies to a major etiologic virus of human non-A, non-B hepatitis. *Science* **244:**362–364.

147. **Kurman, R. J., D. E. Henson, A. L. Herbst, K. L. Noller, and M. H. Schiffman.** 1994. Interim guidelines for management of abnormal cervical cytology. *J. Am. Med. Assoc.* **271:**1866–1869.

148. **La Grenade, L., B. Hanchard, V. Fletcher, B. Cranston, and W. Blattner.** 1990. Infective dermatitis of Jamaican children: a marker for HTLV-I infection. *Lancet* **336:** 1345–1347.

149. **La Grenade, L., S. Richards, W. N. Gibbs, B. Hanchard, E. Williams, N. P. Williams, W. A. Blattner, and A. Manns.** 1995. Cutaneous manifestations for adult T-cell lymphoma/leukemia in Jamaica: a 12 year review. *J. AIDS Hum. Retrovir.* **10:**236.

150. **Lai, M. M., C. M. Lee, F. Y. Bih, and S. Govindarajan.** 1991. The molecular basis of heterogeneity of hepatitis delta virus. *J. Hepatol.* **4:**S121–S124.

151. **Lam, K. C., and M. J. Tong.** 1982. Analytical epidemiology of primary liver carcinoma in the Pacific Basin. *Natl. Cancer Inst. Monogr.* **62:**123–127.

152. **Lednicky, J. A., R. L. Garcea, D. J. Bergsagal, and J. S. Butel.** 1995. Natural simian virus 40 strains are present in human choroid plexus and ependymoma tumors. *Virology* **212:**710–717.

153. **Lennette, E. T., D. J. Blackbourn, and J. A. Levy.** 1996. Antibodies to HHV-8 in the general population and in Kaposi sarcoma patients. *Lancet* **348:**858–861.

154. **Levine, P. H.** 1987. Immunologic markers for Epstein-Barr virus in the control of nasopharyngeal carcinoma and Burkitt's lymphoma. *Cancer Detect. Prev. Suppl.* **1:** 217–223.

155. **Levine, P. H.** 1995. A review of human herpesvirus-6 infections. *Infect. Med.* **12:** 395–402.

156. **Levine, P. H., and D. V. Ablashi.** The new herpesviruses: human herpesvirus-7 and human herpesvirus-8 (KS-virus). *Infect. Med.,* in press.

157. **Levine, P. H., D. V. Ablashi, C. W. Berard, P. P. Carbone, D. E. Waggoner, and L. Malan.** 1970. Elevated antibody titers to herpes-type virus in Hodgkin's disease. *Proc. Am. Assoc. Cancer Res.* **11:**49.

158. **Levine, P. H., D. V. Ablashi, C. W. Berard, P. P. Carbone, D. E. Waggoner, and L. Malan.** 1971. Elevated antibody titers to Epstein-Barr virus in Hodgkin's disease. *Cancer* **27:**416–421.

159. **Levine, P. H., and W. A. Blattner.** 1992. The epidemiology of human virus-associated hematologic malignancies. Leukemia and related diseases. *Leukemia* **6:**54S–59S.

160. **Levine, P. H., and R. R. Connelly.** 1985. Epidemiology of nasopharyngeal cancer. p. 13–34. *In* R. E. Wittes (ed.), *Head and Neck Cancer.* John Wiley & Sons Ltd, Sussex, U.K.

161. **Levine, P. H., P. Ebbesen, D. V. Ablashi, C. W. Saxinger, A. Nordentoft, and R. R. Connelly.** 1992. Antibodies to human herpes virus-6 and clinical course in patients with Hodgkin's disease. *Int. J. Cancer* **51:**53–57.

162. **Levine, P. H., J. F. Fraumeni, Jr., J. I. Reisher, and D. E. Waggoner.** 1974. Antibodies to Epstein-Barr virus-associated antigens in relatives of cancer patients. *J. Natl. Cancer Inst.* **52:**1037–1040.

163. **Levine, P. H., L. S. Kamaraju, R. R. Connelly, C. W. Berard, R. F. Dorfman, I. Magrath, and J. M. Easton.** 1982. The American Burkitt's lymphoma registry: eight years' experience. *Cancer* **49:**1016–1022.

164. **Levine, P. H., J. H. Lubin, and A. S. Evans.** 1994. An Epstein-Barr virus (EBV) vaccine: clinical and epidemiologic considerations, p. 585–591. *In* T. Tursz, J. S. Pagano, D. V. Ablashi, G. de-The, G. Lenoir, and G. R. Pearson (ed.), *Epstein-Barr Virus and Associated Diseases.* John Libbey Eurotext, Ltd., London.

165. **Levine, P. H., A. Manns, E. S. Jaffe, G. Colclough, A. Cavallero, G. Reddy, and W.**

A. Blattner. 1994. The effect of ethnic differences on the patterns of HTLV-I-associated T-cell leukemia/lymphoma (HATL) in the United States. *Int. J. Cancer* **56:**177–181.

166. **Levine, P. H., A. G. Pocinki, P. Madigan, and S. Bale.** 1992. Familial nasopharyngeal carcinoma in patients who are not Chinese. *Cancer* **70:**1024–1029.

167. **Levine, P. H., G. Stemmermann, E. T. Lennette, D. Shibata, and A. M. Y. Nomura.** 1995. Elevated antibody titers to Epstein-Barr virus prior to the diagnosis of Epstein-Barr virus associated gastric adenocarcinoma. *Int. J. Cancer* **60:**642–644.

168. **Levrero, M., A. Tagger, C. Balsano, E. De Marzio, M. L. Avantaggiati, G. Natoli, D. Diop, E. Villa, G. Diodati, and A. Alberti.** 1991. Antibodies to hepatitis C virus in patients with hepatocellular carcinoma. *J. Hepatol.* **12:**60–63.

169. **Levy, J. A.** 1997. Three new human herpesviruses (HHV6, 7, and 8). *Lancet* **349:** 558–563.

170. **Lin, S. -F., R. Sun, L. Heston, L. Gradoville, D. Shedd, K. Haglund, M. Rigsby, and G. Miller.** 1997. Identification, expression, and immunogenicity of Kaposi's sarcoma-associated herpesvirus-encoded small viral capsid antigen. *J. Virol.* **71:**3069–3076.

171. **Lingao, A. L., E. O. Domingo, and K. Nishioka.** 1981. Hepatitis B virus profile of hepatocellular carcinoma in the Philippines. *Cancer* **48:**1590–1595.

172. **Liou, T. C., T. T. Chang, K. C. Young, X. Z. Lin, C. Y. Lin, and H. L. Wu.** 1992. Detection of HCV RNA in saliva, urine, seminal fluid, and ascites. *J. Med. Virol.* **37:** 197–202.

173. **Lorincz, A. L.** 1996. Cutaneous T-cell lymphoma (mycosis fungoides). *Lancet* **347:** 871–876.

174. **Lorincz, A. T., R. Reid, A. B. Jenson, M. D. Greenberg, W. Lancaster, and R. J. Kurman.** 1992. Human papillomavirus infection of the cervix: relative risk associations of 15 common anogenital types. *Obstet. Gynecol.* **79:**328–337.

175. **Maden, C., A. M. Beckmann, D. B. Thomas, B. McKnight, K. J. Sherman, R. L. Ashley, L. Corey, and J. R. Daling.** 1992. Human papillomaviruses, herpes simplex viruses, and the risk of oral cancer in men. *Am. J. Epidemiol.* **135:**1093–1102.

176. **Maden, C., K. J. Sherman, A. M. Beckmann, T. G. Hislop, C. Z. Teh, R. L. Ashley, and J. R. Daling.** 1993. History of circumcision, medical conditions, and sexual activity and risk of penile cancer [see comments]. *J. Natl. Cancer Inst.* **85:**19–24.

177. **Magrath, I. T.** 1982. Malignant lymphomas, p. 473–573. *In* A. S. Levine (ed.), *Cancer in the Young.* Masson, New York.

178. **Malek, R. S., J. R. Goellner, T. F. Smith, M. J. Espy, and M. R. Cupp.** 1993. Human papillomavirus infection and intraepithelial, in situ, and invasive carcinoma of penis. *Urology* **42:**159–170.

179. **Manns, A., and W. A. Blattner.** 1990. Epidemiology of adult T-cell leukemia/lymphoma and the acquired immunodeficiency syndrome, p. 209–239. *In* R. C. Gallo and F. Wong-Staal (ed.), *Retrovirus Biology and Human Disease.* Marcel Dekker, New York.

180. **Manns, A., F. R. Cleghorn, R. T. Falk, B. Hanchard, E. S. Jaffe, C. Bartholomew, P. Hartge, J. Benichou, and W. A. Blattner.** 1993. Role of HTLV-I in development of non-Hodgkin lymphoma in Jamaica and Trinidad and Tobago. The HTLV lymphoma study group. *Lancet* **342:**1447–1450.

181. **Manns, A., R. Falk, E. L. Murphy, W. N. Gibbs, W. S. Lefters, M. Campbell, B. Hanchard, J. Murphy, and W. A. Blattner.** 1990. Risk factors for development of non-Hodgkin's lymphoma in Jamaica. *Proc. Annu. Meet Am. Assoc. Cancer Res.* **31:** A1358. (Abstract.)

182. **Manns, A., R. J. Wilks, E. L. Murphy, G. Haynes, J. P. Figueroa, M. Barnett, B. Hanchard, and W. A. Blattner.** 1992. A prospective study of transmission by transfu-

sion of HTLV-I and risk factors associated with seroconversion. *Int. J. Cancer* **51:** 886–891.

183. **Martini, F., L. Iaccheri, L. Lazzarin, P. Carinci, A. Corallini, M. Gerosa, P. Iuzzolino, G. Barbanti-Brodano, and M. Tognon.** 1996. SV40 early region and large T antigen in human brain tumors, peripheral blood cells, and sperm fluids from healthy individuals. *Cancer Res.* **56:**4820–4825.

184. **Matsubara, K., and T. Tokino.** 1990. Integration of hepatitis B virus DNA and its implications for hepatocarcinogenesis. *Mol. Biol. Med.* **7:**243–260.

185. **Matsuzaki, H., H. Hata, N. Asou, M. Yoshida, F. Matsuno, M. Takeya, K. Yamaguchi, I. Sanada, and K. Takatsuki.** 1992. Human T-cell leukemia virus-1-positive cell line established from a patient with small cell lung cancer. *Jpn. J. Cancer Res.* **83:** 450–457.

186. **McGregor, J. M., and M. H. A. Rustin.** 1994. Human papillomavirus and skin cancer. *Postgrad. Med. J.* **70:**682–685.

187. **Mendenhall, C. L., L. Seaff, A. M. Diehl, S. J. Ghosn, S. W. French, P. S. Gartside, S. D. Rouster, Z. Buskell-Bales, C. J. Grossman, and G. A. Roselle.** 1991. Antibodies to hepatitis B virus and hepatitis C virus in alcoholic hepatitis and cirrhosis: their prevalence and clinical relevance. The VA Cooperative Study Group (No. 119). *Hepatology* **14:**581–589.

188. **Mitchell, M. F., W. N. Hittelman, W. K. Hong, R. Lotan, and D. Schottenfeld.** 1994. The natural history of cervical intraepithelial neoplasia: an argument for intermediate endpoint biomarkers. *Cancer Epidemiol. Biomarkers Prev.* **3:**619–626.

189. **Miyazaki, K., K. Yamaguchi, T. Tohya, T. Ohba, K. Takatsuki, and H. Okamura.** 1991. Human T-cell leukemia virus type I infection as an oncogenic and prognostic risk factor in cervical and vaginal carcinoma. *Obstet. Gynecol.* **77:**107–110.

190. **Mochizuki, M., T. Watanabe, K. Yamaguchi, K. Yoshimura, S. Nakashima, M. Shirao, S. Araki, K. Takatsuki, S. Mori, and N. Miyata.** 1992. Uveitis associated with human T-cell lymphotropic virus type I. *Am. J. Ophthalmol.* **114:**123–129.

191. **Mondelli, M., G. M. Vergani, A. Alberti, D. Vergani, B. Portmann, A. L. Eddleston, and R. Williams.** 1982. Specificity of T lymphocytes cytotoxicity to autologous hepatocytes in chronic hepatitis B virus infection: evidence that T cells are directed against HBV core antigen expressed on hepatocytes. *J. Immunol.* **129:**2773–2778.

192. **Monsonego, J., L. Zerat, F. Catalan, and Y. Coscas.** 1993. Genital human papillomavirus infections: correlation of cytological, colposcopic and histological features with viral types in women and their male partners. *Int. J. STD AIDS* **4:**13–20.

193. **Mori, M., N. Ban, and K. Kinoshita.** 1988. Familial occurrence of HTLV-I-associated myelopathy. *Ann. Neurol.* **23:**100.

194. **Moudgil, T., A. Molina, G. L. Norman, A. M. Levine, and P. S. Gill.** 1994. Absence of HTLV-I or HTLV-II proviral DNA in PBMC's of patients with mycosis fungoides (MF) with serum TH$_2$ cytokine profile. *Blood* **84:**478a.

195. **Mueller, N., A. Evans, and N. L. Harris.** 1989. Hodgkin's disease and Epstein-Barr virus: altered antibody pattern before diagnosis. *N. Engl. J. Med.* **320:**696–701.

196. **Mueller, N., A. Mohar, A. Evans, N. L. Harris, G. W. Comstock, E. Jellum, K. Magnus, N. Orentreich, B. F. Polk, and J. Vogelman.** 1991. Epstein-Barr virus antibody patterns preceding the diagnosis of non-Hodgkin's lymphoma. *Int. J. Cancer* **49:**387–393.

197. **Murphy, E. L., J. P. Figueroa, W. N. Gibbs, A. Brathwaite, M. Holding-Cobham, D. Waters, B. Cranston, B. Hanchard, and W. A. Blattner.** 1989. Sexual transmission of human T-lymphotropic virus type I (HTLV-I). *Ann. Intern. Med.* **111:**555–560.

198. **Murphy, E. L., B. Hanchard, J. P. Figueroa, W. N. Gibbs, W. S. Lofters, M. Campbell, J. J. Goedert, and W. A. Blattner.** 1989. Modelling the risk of adult T-cell

leukemia/lymphoma in persons infected with human T-lymphotropic virus type I. *Int. J. Cancer* **43**:250–253.

199. **Nagaya, T., T. Nakamura, T. Tokino, T. Tsurimoto, M. Imai, T. Mayumi, K. Kamino, K. Yamamura, and K. Matsubara.** 1987. The mode of hepatitis B virus DNA integration in chromosomes of human hepatocellular carcinoma. *Genes Dev.* **1**:773–782.

200. **Naghashfar, Z., M. J. Kutcher, E. Sawada, and K. V. Shah.** 1985. Infection of the oral cavity with genital tract papillomaviruses. *UCLA Symp. Mol. Cell. Biol. N. Ser.* **32**:155–163.

201. **Nakagawa, M., S. Izumo, S. Ijichi, H. Kubota, K. Arimura, M. Kawabata, and M. Osama.** 1995. HTLV-I-associated myelopathy: analysis of 213 patients based on clinical features and laboratory findings. *J. Neurovirol.* **1**:50–61.

202. **Nalpas, B., F. Driss, S. Pol, B. Hamelin, C. Housset, C. Brechot, and P. Berthelot.** 1991. Hepatitis C infection in alcoholics. *Alcohol Alcohol. Suppl.* **1**:321–322.

203. **Niederman, J. C., R. W. McCollum, G. Henle, and W. Henle.** 1968. Infectious mononucleosis: clinical manifestations in relation to EB virus antibodies. *JAMA* **203**:205–209.

204. **Niedobitek, G., M. L. Hansmann, H. Berbst, L. S. Young, D. Dienemann, C. A. Hartmann, T. Finn, S. Pitteroff, A. Welt, I. Anagnostopoulos, R. Friedrich, H. Lobeck, C. K. Sam, I. Araujo, A. B. Rickinson, and H. Stein.** 1991. Epstein-Barr virus and carcinomas: undifferentiated carcinomas but not squamous cell carcinomas of the nasopharynx are regularly associated with the virus. *J. Pathol.* **165**:17–24.

205. **Nielson, H., B. Norrild, P. Vedtofte, F. Praetorius, J. Reibel, and P. Holmstrup.** 1996. Human papillomavirus in oral premalignant lesions. *Eur. J. Cancer* **32B**:264–270.

206. **Nishioka, K., J. Watanabe, S. Furuta, E. Tanaka, S. Iino, H. Suzuki, T. Tsuji, M. Yano, G. Kuo, Q. L. Choo, M. Houghton, and T. Oda.** 1991. A high prevalence of antibody to the hepatitis C virus in patients with hepatocellular carcinoma in Japan. *Cancer* **67**:429–433.

207. **Nishioka, K., J. Watanabe, S. Furuta, E. Tanaka, H. Suzuki, S. Iino, T. Tsuji, M. Yano, G. Kuo, Q. L. Choo, M. Houghton, and T. Oda.** 1991. Antibody to the hepatitis C virus in acute hepatitis and chronic liver diseases in Japan. *Liver* **11**:65–70.

208. **Ogunbiyi, O. A., J. H. Scholefield, G. Robertson, J. H. Smith, F. Sharp, and K. Rogers.** 1994. Anal human papillomavirus infection and squamous neoplasia in patients with invasive vulvar cancer. *Obstet. Gynecol.* **83**:212–216.

209. **Ohkoshi, S., H. Kojina, H. Tawaraya, T. Miyajima, T. Kamimura, H. Asakura, A. Satoh, S. Hirose, M. Jiijikata, N. Kato, and K. Shimotohno.** 1990. Prevalence of antibody against non-A, non-B hepatitis in Japanese patients with hepatocellular carcinoma. *Jpn. J. Cancer Res.* **81**:550–553.

210. **Okamoto, H., S. Yotsumoto, Y. Akahane, T. Yamanaka, Y. Miyazaki, Y. Sugai, F. Tsuda, T. Tanaka, Y. Miyakawa, and M. Mayumi.** 1990. Hepatitis B viruses with precore region defects prevail in persistently infected hosts along with seroconversion to the antibody against e antigen. *J. Virol.* **64**:1298–1303.

211. **Okayama, A., T. Maruyama, N. Tachibana, K. Hayashi, T. Kitamura, N. Mueller, and H. Tsubouchi.** 1995. Increased prevalence of HTLV-I infection in patients with hepatocellular carcinoma associated with hepatitis C virus. *Jpn. J. Cancer Res.* **86**:1–4.

212. **Okochi, K., H. Sato, and Y. Hunuma.** 1984. A retrospective study on transmission of adult T cell leukemia virus by blood transfusion: seroconversion in recipients. *Vox Sang.* **46**:245–253.

213. **Old, L. J., E. A. Boyse, H. F. Oettgen, E. de Harven, G. Geering, E. Williamson, and P. Clifford.** 1966. Precipitation antibody in human serum to an antigen present in cultured Burkitt's lymphoma cells. *Proc. Natl. Acad. Sci. USA* **56**:1699–1704.

214. **Oshima, A., H. Tsukuma, T. Hiyama, I. Fujimoto, H. Yamano, and M. Tanaka.**

1984. Follow-up study of HBs Ag-positive blood donors with special reference to effect of drinking and smoking on development of liver cancer. *Int. J. Cancer* **34:** 775–779.

215. **Ostwald, C., P. Muller, M. Barten, K. Rutsatz, M. Sonnenburg, K. Milde-Langosch, and T. Loning.** 1994. Human papillomavirus DNA in oral squamous cell carcinomas and normal mucosa. *J. Oral Pathol. Med.* **23:**220–225.

216. **Palefsky, J. M.** 1994. Anal human papillomavirus infection and anal cancer in HIV-positive individuals: an emerging problem. *AIDS* **8:**283–295.

217. **Palefsky, J. M.** 1995. Human papillomavirus-associated malignancies in HIV-positive men and women. *Curr. Opin. Oncol.* **7:**437–441.

218. **Palefsky, J. M., E. A. Holly, J. Gonzalez, J. Berline, D. K. Ahn, and J. S. Greenspan.** 1991. Detection of human papillomavirus DNA in anal intraepithelial neoplasia and anal cancer. *Cancer Res.* **51:**1014–1019.

219. **Pastore, G., L. Monno, T. Santantonio, G. Angarano, M. Milella, A. Giannelli, and J. R. Fiore.** 1990. Hepatitis B virus clearance from serum and liver after acute hepatitis delta virus superinfection in chronic HBsAG carriers. *J. Med. Virol.* **31:**284–290.

220. **Pearson, G. R.** 1980. Epstein-Barr virus: immunology, p. 739–767. *In* G. Klein (ed.), *Viral Oncology.* Raven Press, New York.

221. **Pearson, G. R., B. Johansson, and G. Klein.** 1978. Antibody-dependent cellular cytotoxicity against Epstein-Barr virus-associated antigens in African patients with nasopharyngeal carcinoma. *Int. J. Cancer* **22:**120–125.

222. **Pearson, G. R., and P. H. Levine.** 1993. Epstein-Barr virus vaccine—the time to proceed is now!, p. 349–356. *In* B. Roizman, R. Whitley, and C. Lopez (ed.), *Human Herpesvirus: Biology, Pathogenesis and Treatment.* Raven Press, New York.

223. **Penna, A., F. V. Chisari, A. Bertoletti, G. Missale, P. Fowler, T. Giuberti, F. Fiaccadori, and C. Ferrari.** 1991. Cytotoxic T lymphocytes recognize an HLA-A2-restricted epitope within the hepatitis B virus nucleocapsid antigen. *J. Exp. Med.* **174:**1565–1570.

224. **Peterman, T. A., H. W. Jaffe, and V. Beral.** 1993. Epidemiologic clues to the etiology of Kaposi's sarcoma. *AIDS* **7:**605–611.

225. **Pignatelli, M., J. Waters, A. Lever, S. Iwarson, R. Gerety, and H. C. Thomas.** 1987. Cytotoxic T-cell responses to the nucleocapsid proteins of HBV in chronic hepatitis. Evidence that antibody modulation may cause protracted infection. *J. Hepatol.* **4:** 15–21.

226. **Poiesz, B. J., G. D. Ehrlich, B. C. Byrns, K. Wells, S. Kwok, and J. Sninsky.** 1990. The use of the polymerase chain reaction in the detection, quantification and characterization of human retroviruses. *Med. Virol.* **9:**47–75.

227. **Poiesz, B. J., F. W. Ruscetti, A. F. Gazdar, P. A. Bunn, J. D. Minna, and R. C. Gallo.** 1980. Detection and isolation of type C retrovirus particles from fresh and cultured lymphocytes of a patient with cutaneous T-cell lymphoma. *Proc. Natl. Acad. Sci. USA* **77:**7415–7419.

228. **Ponten, J., H. O. Adami, R. Bergstrom, J. Dillner, L. G. Friberg, L. Gustafsson, A. B. Miller, D. M. Parkin, P. Sparen, and D. Trichopoulos.** 1995. Strategies for global control of cervical cancer [see comments]. *Int. J. Cancer* **60:**1–26.

229. **Posner, L. E., M. Robert-Guroff, V. S. Kalyanaraman, B. J. Poiesz, F. W. Ruscetti, B. Fossieck, P. A. Bunn, Jr., J. D. Minna, and R. C. Gallo.** 1981. Natural antibodies to the human T cell lymphoma virus in patients with cutaneous T cell lymphomas. *J. Exp. Med.* **154:**333–346.

230. **Potischman, N., and L. A. Brinton.** 1996. Nutrition and cervical neoplasia. *Cancer Causes Control* **7:**113–126.

231. **Prasad, U., M. A. Jalaludin, P. Rajadurai, G. Pizza, C. DeVinci, D. Viza, and P.**

H. Levine. 1996. Transfer factor with anti-EBV activity as an adjuvant therapy for nasopharyngeal carcinoma: a pilot study. *Biotherapy* **9:**109–115.

232. Prince, A. M., and P. Alcabes. 1982. The risk of development of hepatocellular carcinoma in hepatitis B virus carriers in New York. A preliminary estimate using death-records matching. *Hepatology* **2:**15S–20S.

233. Prince, A. M., W. Szmuness, J. Michon, J. Demaille, G. Diebolt, J. Linhard, C. Quenum, and M. Sankale. 1975. A case/control study of the association between primary liver cancer and hepatitis B infection in Senegal. *Int. J. Cancer* **16:**376–383.

234. Raab-Traub, N., K. Flynn, G. Pearson, A. Huang, P. H. Levine, A. Lanier, and J. Pagano. 1987. The differentiated form of nasopharyngeal carcinoma contains Epstein-Barr Virus DNA. *Int. J. Cancer* **39:**25–29.

235. Rabkin, C. S., R. J. Biggar, M. Melbye, and R. E. Curtis. 1992. Second primary cancers following anal and cervical carcinoma: evidence of shared etiologic factors. *Am. J. Epidemiol.* **136:**54–58.

236. Rapicetta, M., T. Stroffolini, M. Charamonte, C. Tiribelli, E. Villa, R. G. Simonetti, M. A. Stazi, P. Chionne, T. Bertin, L. Croce, P. Trande, and A. Magliocci. 1994. Antibody pattern of HCV infection and hepatocellular carcinoma in Italy: a case control study, p. 703–705. *In* K. Nishioka, H. Suzuki, S. Mishiro, and T. Oda (ed.), *Viral Hepatitis and Liver Disease.* Springer-Verlag, Tokyo.

237. Rattray, C., H. D. Strickler, C. Escoffery, A. Manns, M. H. Schiffman, J. Palefsky, B. Hanchard, and W. A. Blattner. 1996. Type specific prevalence of HPV DNA among Jamaican colposcopy patients. *J. Infect. Dis.* **173:**718–721.

237a. Reid, A. E., and A. T. Lorincz. 1995. Human papillomavirus tests. *Bailliere's Clin. Obstet. Gynecol.* **9:**65–103.

238. Richard, B. S., D. R. Lowy, and J. T. Schiller. 1996. Pseudotype HPV16 virions are resistant to desiccation: implications for non-sexual transmission by fomites. 15th International Papillomavirus Workshop, Queensland, Australia.

239. Rizzetto, M., A. Ponzetto, and I. Forzani. 1977. Epidemiology of hepatitis delta virus: overview, p. 1–20. *In* J. L. Gerin, R. H. Purcell, and M. Rizzetto (ed.), *The Hepatitis Delta Virus.* Wiley-Liss, New York.

240. Rizzetto, M., A. Ponzetto, and I. Forzani. 1991. Epidemiology of hepatitis delta virus: overview. *Prog. Clin. Biol. Res.* **364:**1–20.

241. Robinson, R. D., J. F. Lindo, F. A. Neva, A. A. Gam, P. Vogel, S. I. Terry, and E. S. Cooper. 1994. Immunoepidemiologic studies of *Strongyloides stercoralis* and human T lymphotropic virus type I infections in Jamaica. *J. Infect. Dis.* **169:**692–696.

242. Sakashita, A., T. Hattori, and C. Miller. 1992. Mutations of the p53 gene in adult T-cell leukemia. *Blood* **79:**477–80.

243. Salahuddin, S. Z., D. V. Ablashi, P. D. Markham, S. F. Josephs, S. Sturzenegger, M. Kaplan, G. Halligan, P. Biberfeld, F. Wong-Staal, B. Kramarsky, and R. C. Gallo. 1986. Isolation of a new virus, HBLV, in patients with lymphoproliferative disorders. *Science* **234:**596–601.

244. Sawyer, R. W., A. S. Evans, J. C. Niederman, and R. W. McCollum. 1971. Prospective studies of a group of Yale University freshmen. I. Occurrence of infectious mononucleosis. *J. Infect. Dis.* **123:**263–269.

245. Saxinger, C., and R. C. Gallo. 1982. Methods in laboratory investigation: application of the indirect enzyme-linked immunosorbent assay microtest to the detection and surveillance of human T cell leukemia-lymphoma virus. *Lab. Invest.* **49:**371–377.

246. Schiffman, M. H. 1992. Recent progress in defining the epidemiology of human papillomavirus infection and cervical neoplasia. *J. Natl. Cancer Inst.* **84:**394–398.

247. Schiffman, M. H. 1994. Epidemiology of cervical human papillomavirus infections. *Curr. Top Microbiol. Immunol.* **186:**55–81.

248. Schiffman, M. H., H. M. Bauer, R. N. Hoover, A. G. Glass, D. M. Cadell, B. B. Rush, D. R. Scott, M. E. Sherman, R. J. Kurman, and S. Wacholder. 1993. Epidemiologic evidence showing that human papillomavirus infection causes most cervical intraepithelial neoplasia. *J. Natl. Cancer Inst.* **85:**958–964.

249. Schiffman, M. H., and L. A. Brinton. 1995. The epidemiology of cervical carcinogenesis. *Cancer* **76:**1888–1901.

249a. Schiffman, M. H., and M. E. Sherman. 1994. HPV testing to improve cervical cancer screening, p. 265–277. *In* S. Srivastava, S. M. Lippman, W. K. Hong, and J. L. Mulshine (ed.), *Early Detection of Cancer: Molecular Markers.* Futura Publishing Co., Inc., Armonk, N.Y.

250. Schneider, A., G. Meinhardt, R. Kirchmayr, and V. Schneider. 1991. Prevalence of human papillomavirus genomes in tissues from the lower genital tract as detected by molecular in situ hybridization. *Int. J. Gynecol. Pathol.* **10:**1–14.

251. Seeff, L. B., G. W. Beebe, J. H. Hoofnagle, J. E. Norman, Z. Buskell-Bales, J. G. Waggoner, N. Kaplowitz, R. S. Koff, J. L. Petrini, Jr., and E. R. Schiff. 1987. A serologic follow-up of the 1942 epidemic of postvaccination hepatitis in the United States Army. *N. Engl. J. Med.* **316:**965–970.

252. Selby, M. J., Q. L. Choo, K. Berger, G. Kuo, E. Glazer, M. Eckart, C. Lea, D. Chien, C. Kuo, and M. Houghton. 1993. Expression, identification and subcellular localization of the proteins encoded by the hepatitis C viral genome. *J. Gen. Virol.* **74:** 1103–1113.

253. Shah, K., and N. Nathanson. 1976. Human exposure to SV40: review and comment. *Am. J. Epidemiol.* **103:**1–12.

254. Sheen, I. S., Y. F. Liaw, C. M. Chu, and C. C. Pao. 1992. Role of hepatitis C virus infection in spontaneous hepatitis B virus surface antigen clearance during chronic hepatitis B virus infection. *J. Infect. Dis.* **165:**831–834.

255. Shen, Y. Y., H. A. Ming, N. Jahan, M. Manak, E. S. Jaffe, and P. H. Levine. 1993. In situ hybridization detection of human herpesvirus 6 in biopsy specimens from Chinese patients with non-Hodgkin's lymphoma. *Arch. Pathol. Lab. Med.* **117:**502–506.

256. Shimoyama, M., and Members of the Lymphoma Study Group (1984–1987). 1991. Diagnostic criteria and classification of clinical subtypes of adult T-cell leukaemia-lymphoma. *Br. J. Haematol.* **79:**428–437.

257. Shindoh, M., I. Chiba, M. Yasuda, T. Saito, K. Funaoka, T. Kohgo, A. Amemiya, Y. Sawada, and K. Fujinaga. 1995. Detection of human papillomavirus DNA sequences in oral squamous cell carcinomas and their relation to p53 and proliferating cell nuclear antigen expression. *Cancer* **76:**1513–1521.

258. Shinzato, O., S. Ikeda, S. Momita, Y. Nagata, S. Kamihira, E. Nakayama, and H. Shiku. 1991. Semiquantitative analysis of integrated genomes of human T-lymphotropic virus type I in asymptomatic virus carriers. *Blood* **78:**2082–2088.

259. Shope, T., D. Dechairo, and G. Miller. 1973. Malignant lymphoma in cotton-top marmosets following inoculation of Epstein-Barr virus. *Proc. Natl. Acad. Sci. USA* **70:** 2487–2489.

260. Simmonds, P., E. C. Holmes, T. A. Cha, S. W. Chan, F. McOmish, B. Irvine, E. Beall, P. L. Yap, J. Kolberg, and M. S. Urdea. 1993. Classification of hepatitis C virus into six major genotypes and a series of subtypes by phylogenetic analysis of the NS-5 region. *J. Gen. Virol.* **74:**2391–2399.

261. Simons, M. J., G. B. Wee, E. H. Goh, S. H. Chan, K. Shanmugaratham, N. E. Day, and G. de-The. 1976. Immunogenetic aspects of nasopharyngeal carcinoma. IV. Increased risk in Chinese of nasopharyngeal carcinoma associated with a Chinese-related HLA profile (A2, Singapore 2). *J. Natl. Cancer Inst.* **57:**977–980.

262. Simpson, G. R., T. F. Schulz, D. Whitby, P. M. Cook, C. Boshoff, L. Rainbow, M.

R. Howard, S. -J. Gao., R. A. Bohenzky, P. Simmonds, C. Lee, A. de Ruiter, A. Hatzakis, R. S. Tedder, I. V. D. Weller, R. A. Weiss, and P. S. Moore. 1996. Prevalence of Kaposi's sarcoma associated herpesvirus infection measured by antibodies to recombinant capsid protein and latent immunofluorescence antigen. *Lancet* **349:** 1133–1138.

263. Smedile, A., F. Rosina, E. Chiaberge, V. Lattore, G. Saracco, M. R. Brunetto, F. Bonino, G. Verme, and M. Rizzetto. 1991. Presence and significance of hepatitis B virus replication in chronic type D hepatitis, p. 185–195. *In* J. L. Gerin, R. H. Purcell, and M. Rizzetto (ed.), *The Hepatitis Delta Virus.* Wiley-Liss, New York.

264. Soulier, J., L. Grollet, E. Oksenhendler, P. Cacoub, D. Cazals-Hatem, P. Babinet, M. F. d'Agay, J. P. Clauvel, M. Raphael, and L. Degos. 1995. Kaposi's sarcoma-associated herpesvirus-like DNA sequences in multicentric Castleman's disease. *Blood* **86:**1276–1280.

265. Stancu, A., A. Schmitt, B. Moschik, B. Schmidt, S. P. Wilezynski, and T. Iftner. 1996. Prevalence of human papillomaviruses in malignant skin lesions of unselected non-immunosuppressed individuals. 15th International Papillomavirus Workshop, Queensland, Australia.

266. Stern, R. C., R. E. Behrman, R. Kliegman, and W. E. Nelson. 1996. Neoplasms of the larynx, p. 1209–1210. *In* R. C. Stern, R. E. Behrman, R. Kliegman, and W. E. Nelson (ed.), *Nelson Textbook of Pediatrics,* 15th ed. W.B. Saunders Company, Philadelphia.

267. Stevens, C. E., R. P. Beasley, J. Tsui, and W. -C. Lee. 1975. Vertical transmission of hepatitis B antigen in Taiwan. *N. Engl. J. Med.* **292:**771–774.

268. Stevens, C. E., P. E. Taylor, J. Pindyck, Q. -L. Choo, D. W. Bradley, G. Kuo, and M. Houghton. 1990. Epidemiology of Hepatitis C virus. A preliminary study in volunteer blood donors. *JAMA* **263:**49–53.

269. Strauchen, J. A., A. D. Hauser, D. Burstein, R. Jimenez, P. S. Moore, and Y. Chang. 1996. Body cavity-based malignant lymphoma containing Kaposi sarcoma-associated herpesvirus in an HIV-negative man with previous Kaposi sarcoma. *Ann. Intern. Med.* **125:**822–825.

270. Strickler, H. D., C. Rattray, C. Escoffary, A. Manns, M. H. Schiffman, C. Brown, B. Cranston, E. Hanchard, J. M. Palefsky, and W. A. Blattner. 1995. Human T-cell lymphotropic virus type I and severe cervical neoplasia of the cervix in Jamaica. *Int. J. Cancer* **61:**23–26.

271. Strickler, H. D., M. H. Schiffman, C. Eklund, A. G. Glass, D. R. Scott, M. E. Sherman, S. Wacholder, R. J. Kurman, M. M. Manos, and J. Schiller. 1997. Evidence for at least two distinct groups of humoral immune reactions to papillomavirus antigens in women with SIL. *Cancer Epidemiol. Biomarkers and Prev.* **6:**183–188.

272. Strickler, H. D., M. H. Schiffman, K. V. Shah, C. S. Rabkin, J. T. Schiller, S. Wacholder, B. Clayman, and R. P. Viscidi. 1996. A survey of human papillomavirus 16 antibodies in patients with various cancers. 15th International Papillomavirus Workshop, Queensland, Australia.

273. Stuver, S. O., N. Tachibana, A. Okayama, S. Shioiri, Y. Tsunetoshi, K. Tsuda, and N. E. Mueller. 1993. Heterosexual transmission of human T cell leukemia/lymphoma virus type I among married couples in southwestern Japan: an initial report from the Miyazaki cohort study. *J. Infect. Dis.* **167:**57–65.

274. Sundar, S. K., P. H. Levine, D. V. Ablashi, S. A. Leiseca, G. R. Armstrong, J. L. Cicmanec, G. A. Parker, and M. Nonoyama. 1981. Epstein-Barr virus-induced malignant lymphoma in a white-lipped marmoset. *Int. J. Cancer* **27:**107–111.

275. Szmuness, W. 1978. Hepatocellular carcinoma and the hepatitis B virus: evidence for a causal association. *Prog. Med. Virol.* **24:**40–69.

276. Tabor, E., R. J. Gerety, C. L. Vogel, A. C. Bayley, P. P. Anthony, C. H. Chan, and

L. F. Barker. 1977. Hepatitis B virus infection and primary hepatocellular carcinoma. *J. Natl. Cancer Inst.* **58**: 1197–1200.

277. Tajima, K., and T. Kuroishi. 1985. Estimation of rate of incidence of ATL among ATLV (HTLV-I) carriers in Kyushu, Japan. *Jpn. J. Clin. Oncol.* **15**:423–430.

278. Takada, N., S. Takase, N. Enomoto, A. Takada, and T. Date. 1992. Clinical backgrounds of the patients having different types of hepatitis C virus genomes. *J. Hepatol.* **14**:35–40.

279. Takada, S., and K. Koike. 1990. X protein of hepatitis B virus resembles a serine protease inhibitor. *Jpn. J. Cancer Res.* **12**: 1191–1194.

280. Takatsuki, K., T. Uchiyama, K. Sagwa, and J. Yodoi. 1976. Adult T cell leukemia in Japan, p. 73–78. *In* S. Seno, R. Takaku, and S. Irino (ed.), *Topics in Hematology.* Excerpta Medica Amsterdam, Amsterdam.

281. Tassopoulos, N. C., M. Alikiotis, F. Limotirkis, P. Nikolakakis, H. Mela, and M. Paraloglou-Ionnides. 1988. Acute sporadic non-A, non-B hepatitis in Greece. *J. Med. Virol.* **26**:71–77.

282. Templeton, A. C. 1972. Studies in Kaposi's sarcoma. *Cancer* **30**:884–867.

283. Terada, K., S. Katamine, K. Eguchi, R. Moriuchi, M. Kita, H. Shimada, I. Yamashita, K. Iwata, Y. Tsuji, and S. Nagataki. 1994. Prevalence of serum and salivary antibodies to HTLV-1 in Sjogren's syndrome. *Lancet* **344**:1116–1119. [See comments.]

284. Tokunaga, M., S. Imai, Y. Uemura, T. Tokudome, T. Osato, and E. Sato. 1993. Epstein-Barr virus in adult T-cell leukemia/lymphoma. *Am. J. Pathol.* **143**:1263–1269.

285. Torelli, G., R. Marasca, M. Luppi, L. Selleri, S. Ferrari, F. Narni, M. T. Mariano, M. Federico, L. Ceccherini-Nelli, M. Bondinelli, G. Montagnani, M. Montorsi, and T. Artusi. 1991. Human herpes virus-6 in human lymphomas: identification of specific sequences in Hodgkin's lymphoma by polymerase chain reaction. *Blood* **77**: 2251–2258.

286. Tran, A., D. Kremsdorf, F. Capel, C. Houssett, C. Dauguet, M. A. Petit, and C. Brechot. 1991. Emergence of and takeover by hepatitis B virus (HBV) with rearrangements in the pre-S/S and pre-C/C genes during chronic HBV infection. *J. Virol.* **65**: 3566–3674.

287. Trichopoulos, D., N. E. Day, E. Kaklamani, A. Tsonou, N. Munoz, X. Zavitsanos, Y. Koumantaki, and A. Trichopoulou. 1987. Hepatitis B virus, tobacco smoking and ethanol consumption in the etiology of hepatocellular carcinoma. *Int. J. Cancer* **39**: 45–49.

288. Trimble, C. L., A. Hildesheim, L. A. Brinton, K. V. Shah, and R. J. Kurman. 1996. Heterogeneous etiology of squamous carcinoma of the vulva. *Obstet. Gynecol.* **87**: 59–64.

289. Tsai, J. F., J. E. Jeng, M. S. Ho, W. Y. Chang, Z. Y. Lin, and J. H. Tsai. 1994. Hepatitis B and C virus infection as risk factors for hepatocellular carcinoma in Chinese: a case-control study. *Int. J. Cancer* **56**:619–621.

290. Tsiquaye, K. N., B. Portmann, G. Tovey, H. Kessler, S. Hu, X. Z. Lu, A. J. Zuckerman, J. Craske, and R. Williams. 1983. Non-A, non-B hepatitis in persistent carriers of hepatitis B virus. *J. Med. Virol.* **11**:179–189.

291. Tsukui, T., A. Hildesheim, M. H. Schiffman, J. Lucci, 3rd, D. Contois, P. Lawler, B. B. Rush, A. T. Lorincz, A. Corrigan, R. D. Burk, W. Qu, M. A. Marshall, D. Mann, M. Carrington, M. Clerici, G. M. Shearer, D. P. Carbone, D. R. Scott, R. A. Houghten, and J. A. Berzofsky. 1996. Interleukin 2 production in vitro by peripheral lymphocytes in response to human papillomavirus-derived peptides: correlation with cervical pathology. *Cancer Res.* **56**:3967–3974.

292. Twu, J. S., and R. H. Schloemer. 1987. Transcriptional trans-activating function of hepatitis B virus. *J. Virol.* **61**:3448–3453.

293. Tzonou, A., D. Trichopoulos, E. Kaklamani, X. Zavitsanos, Y. Koumantaki, and C. C. Haieb. 1991. Epidemiologic assessment of interactions of hepatitis-C virus with seromarkers of hepatitis-B and -D viruses, cirrhosis and tobacco smoking in hepatocellular carcinoma. *Int. J. Cancer* **49**:377–380.

294. Uchiyama, T., J. Yodoi, K. Sagawa, K. Takatsuki, and H. Uchino. 1977. Adult T-cell leukemia: clinical and hematologic features of 16 cases. *Blood* **50**:481–492.

295. Ueda, N., K. Iwata, H. Tokuoka, T. Akagi, J. Ito, and M. Misushima. 1979. Adult T-cell leukemia with generalized cytomegalic inclusion disease and pneumocystis carinii pneumonia. *Acta Pathol. Jpn.* **29**:221–232.

296. Usuku, K., S. Sonoda, M. Osame, S. Yashiki, K. Takahashi, M. Matsumoto, T. Sawada, K. Tsuji, M. Tara, and A. Igata. 1988. HLA haplotype-linked high immune responsiveness against HTLV-I in HTLV-I-associated myelopathy: comparison with adult T-cell leukemia/lymphoma. *Ann. Neurol.* **23**(Suppl.):S143–S150.

297. Van Doornum, G. J., M. Prins, L. H. Juffermans, C. Hooykaas, J. A. van den Hoek, R. A. Coutinho, and W. G. Quint. 1994. Regional distribution and incidence of human papillomavirus infections among heterosexual men and women with multiple sexual partners: a prospective study. *Genitourin. Med.* **70**:240–246.

298. Vargas, V., L. Castella, and J. I. Esteban. 1990. High frequency of antibodies to the hepatitis C virus among patients with hepatocellular carcinoma. *Ann. Intern. Med.* **112**:232–233.

299. Vermund, S. H., M. H. Schiffman, G. L. Goldberg, D. B. Ritter, A. Weltman, and R. D. Burk. 1989. Molecular diagnosis of genital human papillomavirus infection: comparison of two methods used to collect exfoliated cervical cells. *Am. J. Obstet. Gynecol.* **160**:304–308.

300. Wanda, J. R., Y. K. Fujita, K. J. Isselbacher, C. Degott, H. Schellekens, M. C. Dazza, V. Thiers, P. Tiollais, and C. Brechot. 1986. Identification and transmission of hepatitis B virus-related variants. *Proc. Natl. Acad. Sci. USA* **83**:6608–6612.

301. Wang, K. S., Q. L. Choo, A. J. Weiner, J. H. Ou, R. C. Najarian, R. M. Thayer, G. T. Mullenbach, K. J. Denniston, J. L. Gerin, and M. Houghton. 1986. Structure, sequence and expression of the hepatitis delta (delta) viral genome. *Nature* **323**: 508–514.

302. Wara, W. M., D. W. Wara, T. L. Phillips, and A. Ammahh. 1975. Elevated IgA in carcinoma of the nasopharynx. *Cancer* **35**:1313–1315.

303. Watanabe, Y., S. Harada, I. Saito, and T. Miyamura. 1991. Prevalence of antibody against the core protein of hepatitis C virus in patients with hepatocellular carcinoma. *Int. J. Cancer* **48**:340–343.

304. Weinstock, M. A. 1991. A registry-based case-control study of mycosis fungoides. *Ann. Epidemiol.* **1**:533–539.

305. Weinstock, M. A., and J. W. Horm. 1988. Mycosis fungoides in the United States: increasing incidence and descriptive epidemiology. *JAMA* **260**:42–46.

306. Weiss, L. M., L. A. Movahed, and R. A. Estmir. 1989. Detection of Epstein-Barr viral genomes in Reed-Sternberg cells of Hodgkin's disease. *N. Engl. J. Med.* **320**:502–506.

307. Welles, S. L., P. H. Levine, E. M. Joseph, L. J. Goberdhan, S. Lee, A. Miotti, J. Cervantes, M. Bertoni, E. Jaffe, and H. Dosik. 1994. An enhanced surveillance program for adult T-cell leukemia in central Brooklyn. *Leukemia* **8**(Suppl. 1):S111–S115.

308. Wheeler, C. M., C. A. Parmenter, W. C. Hunt, T. M. Becker, C. E. Greer, A. Hildesheim, and M. M. Manos. 1993. Determinants of genital human papillomavirus infection among cytologically normal women attending the University of New Mexico student health center. *Sex. Transm. Dis.* **20**:286–289.

309. White, J. D., J. A. Johnson, J. M. Nam, B. Cranston, B. Hanchard, T. A. Waldmann, and A. Manns. 1996. Distribution of human leukocyte antigens in a population of

black patients with human T-cell lymphotropic virus type I-associated adult T-cell leukemia/lymphoma. *Cancer Epidemiol. Biomarkers Prev.* **5**:873–877.

310. **Wiktor, S. Z., and W. A. Blattner.** 1991. Epidemiology of human T-cell leukemia virus type I (HTLV-I), p. 171–192. *In* R. C. Gallo and G. Jay (ed.), *The Human Retroviruses.* Academic Press, Inc., San Diego.

311. **Wilks, R., B. Hanchard, O. Morgan, E. Williams, B. Cranston, M. L. Smith, P. Rodgers-Johnson, and A. Manna.** 1996. Patterns of HTLV-I infection among family members of patients with adult T-cell leukemia/lymphoma and HTLV-I associated myelopathy/tropical spastic paraparesis. *Int. J. Cancer* **65**:272–273.

312. **Wollersheim, M., U. Debelka, and P. H. Hofschneider.** 1988. A transactivating function encoded in the hepatitis B virus X gene is conserved in the integrated state. *Oncogene* **3**:545–552.

313. **Yadav, M., A. Chandrashekran, D. M. Vasudevan, and D. V. Ablashi.** 1994. Frequent detection of human herpesvirus 6 in oral carcinoma. *J. Natl. Cancer Inst.* **86**:1792–1794.

314. **Yamanishi, K., T. Okuno, K. Shiraki, M. Takahashi, T. Kondo, Y. Asano, and T. Kurata.** 1988. Identification of human herpes virus-6 as a causal agent for exanthem subitum. *Lancet* **1**:1065–1067.

315. **Yano, M., H. Yatauhashi, O. Inoue, and M. Koga.** 1991. Epidemiology of hepatitis C virus in Japan: role in chronic liver disease and hepatocellular carcinoma. *J. Gastroenterol. Hepatol.* **1**(Suppl.):31–35.

316. **Yah, F. S., C. C. Mo, S. Luo, B. E. Henderson, M. J. Tong, and M. C. Yu.** 1985. A serological case-control study of primary hepatocellular carcinoma in Guangxi, China. *Cancer Res.* **45**:872–873.

317. **Yoshida, M., I. Miyoshi, and Y. Hinuma.** 1982. Isolation and characterization of retrovirus from cell lines of human adult T-cell leukemia and its implication in the disease. *Proc. Natl. Acad. Sci. USA* **79**:2031–2035.

318. **Yu, M. C., and B. E. Henderson.** 1996. Nasopharyngeal cancer, p. 893–919. *In* D. Schottenfeld and J. F. Fraumeni, Jr. (ed.), *Cancer Epidemiology and Prevention.* Oxford University Press, New York.

319. **Yu, M. C., M. J. Tong, P. Coursaget, R. K. Ross, S. Govindarajan, and B. E. Henderson.** 1990. Prevalence of hepatitis B and C viral markers in black and white patients with hepatocellular carcinoma in the United States. *J. Natl. Cancer Inst.* **82**:1038–1041.

320. **Yu, M. W., S. L. You, A. S. Chang, S. N. Lu, Y. F. Liaw, and C. J. Chen.** 1991. Association between hepatitis C virus antibodies and hepatocellular carcinoma in Taiwan. *Cancer Res.* **51**:5621–5625.

321. **Yuki, N., N. Hayashi, A. Kasahara, H. Hagiwara, K. Katayama, H. Fusamoto, and T. Kamada.** 1992. Hepatitis B virus markers and antibodies to hepatitis C virus in Japanese patients with hepatocellular carcinoma. *Dig. Dis. Sci.* **37**:65–72.

322. **Zachoval, R., J. Abb, V. Zachoval, J. Eisenburg, G. R. Pape, G. Paumgartner, and F. Deinhardt.** 1988. Interferon alpha in hepatitis type B and non-A, non-B. Defective production by peripheral blood mononuclear cells in chronic infection and development of serum interferon in acute disease. *J. Hepatol.* **6**:364–368.

323. **Zeng, Y.** 1985. Sero-epidemiological studies on nasopharyngeal carcinoma in China. *Adv. Cancer Res.* **44**:121–138.

324. **Zeng, Y., L. G. Zhang, H. Y. Li, M. G. Jan, Q. Zhang, Y. C. Wu, Y. S. Wang, and G. R. Su.** 1982. Serological mass survey for early detection of nasopharyngeal carcinoma in Wuzhou City, China. *Int. J. Cancer* **29**:139–141.

325. **Zeng, Y., L. G. Zhang, Y. C. Wu, Y. S. Huang, N. Q. Huang, J. Y. Li, Y. B. Wang, M. K. Jiang, Z. Fang, and N. N. Meng.** 1985. Prospective studies on nasopharyngeal carcinoma in Epstein-Barr virus IgA/VCA antibody-positive persons in Wuzhou City, China. *Int. J. Cancer* **36**:545–547.

326. **Zhang, Y. -Y., B. G. Hansson, A. Widell, and E. Nordenfelt.** 1992. Hepatitis C antibodies and hepatitis C virus RNA in Chinese blood donors determined by ELISA, recombinant immunoblot assay and polymerase chain reaction. *Acta Pathol. Microbiol. Immunol. Scand.* **100:**851–855.

327. **Zignego, A. L., D. Macchia, M. Monti, V. Thiers, M. Mazzetti, M. Foschi, E. Maggi, S. Romagnani, P. Gentilini, and C. Brachot.** 1992. Infection of peripheral mononuclear blood cells by hepatitis C virus. *J. Hepatol.* **15:**382–386.

328. **Zucker-Franklin, D., E. E. Coutavas, M. G. Rush, and D. C. Zouzias.** 1991. Detection of human T-lymphotropic virus-like particles in cultures of peripheral blood lymphocytes from patients with mycosis fungoides. *Proc. Natl. Acad. Sci. USA* **88:**7630–7634.

329. **Zur Hausen, R., H. Schulte-Holthausen, G. Klein, W. Henle, G. Henle, P. Clifford, and L. Santesson.** 1970. EB-virus DNA in biopsies of Burkitt tumors and anaplastic carcinomas of the nasopharynx. *Nature* **228:**1056–1057.

Human Tumor Viruses
Edited by Dennis J. McCance
© 1998 American Society for Microbiology

2 Introduction to DNA Tumor Viruses: Adenovirus, Simian Virus 40, and Polyomavirus

George R. Beck, Jr., Brad R. Zerler, and Elizabeth Moran

Human adenovirus, simian virus 40 (SV40), and murine polyomavirus are small DNA viruses, classified as tumor viruses because they can induce tumors in rodents and immortalize primary cells in vitro. These small DNA tumor viruses are not usually associated with human cancer or with tumors in their natural adult hosts. However, due to their ease of culture they have served since the 1950s as important laboratory models to probe the molecular mechanisms underlying carcinogenesis.

Adenovirus, SV40, and polyomavirus have each served as valuable experimental systems for probing cellular pathways involved in DNA replication and the regulation of gene expression and cell proliferation. These viruses have small genomes, encoding only the virus structural proteins and a limited number of regulatory proteins. The viruses rely heavily on host cell products for viral gene expression and replication. These viruses are each adapted to infect and reproduce in differentiated, generally quiescent, epithelial cells. Thus the viruses encode proteins able to compromise normal cell growth restriction mechanisms, thereby activating gene expression and replication systems specific for cycling cells. In their natural hosts, this activation of cell growth functions is tempered by an interplay of virus and host factors such that tumorigenesis does not result. However, the viruses provide extremely useful tumor models in heterologous hosts.

The value of these viruses was realized in a very particular way in the mid-1980s when years of study on adenovirus and SV40 converged to show that these relatively unrelated viruses transform cells by similar molecular mechanisms. This breakthrough led rapidly to the wider realization that these mechanisms are common to other DNA tumor viruses, including ones such as human papillomavirus (HPV) that are more closely associated with human cancer.

Work on adenovirus and SV40 in particular has been of profound importance in revealing the central roles of the retinoblastoma gene product (pRb) and p53 in the control of cell growth. Loss or altered function of these two products is a contributing factor in the large majority of human tumors. Our current understanding of the role of these two proteins in the course of human carcinogenesis, even in the absence of viruses, has become much clearer as a result of these virus

51

studies. This chapter reviews the historical role of adenovirus and SV40 in illuminating the molecular basis of pRb and p53 function. More recently recognized interactions between the transforming proteins of adenovirus, SV40, and polyomavirus and cell growth control products are also considered here.

HISTORICAL PERSPECTIVE

Isolation of Adenovirus, SV40, and Polyomavirus

Adenovirus was first isolated in the 1950s from throat swabs of military recruits. The virus proved easy to propagate in human cells in culture, and so was introduced to laboratory study. Dozens of strains of adenovirus have been isolated from humans, other mammals, and even birds. The value of human adenovirus for carcinogenesis studies was realized soon after its isolation when the virus was found to be able to transform rodent cells in vitro. In 1962, human adenovirus type 12 (Ad12) was shown to be able to induce malignant tumors in newborn hamsters (176). Adenoviruses have since been divided into three groups on the basis of their oncogenicity. Adenoviruses of subgroups A and B can cause tumors in rodents, but those of subgroup C, including the common laboratory isolates Ad2 and Ad5, generally do not cause tumors in animals although they retain the ability to transform cells in culture.

SV40 was first detected and isolated in the late 1950s. This monkey virus was discovered when preparations of human poliovirus, cultured in primary simian cells and treated for use as vaccines, were found to be contaminated with an agent that could induce tumors in the newborn hamsters routinely used to test the safety of the vaccine. Because these tumors did not develop until months after the vaccine had seemed to pass toxicity tests, many individuals received tumor virus-contaminated vaccine doses. SV40 was identified as the oncogenic agent from the vaccine preparation (49, 64) and was studied intensively in culture. Despite early fears, SV40 has not been associated with human cancer (88, 132). More recently, two SV40-related viruses, JC virus and BK virus, have been identified in humans (62, 138). A majority of people become infected with these viruses in childhood, and the infection can persist with no overt disease during a normal lifetime. However, immunosuppressing conditions in some individuals can lead to virus proliferation and subsequent pathology. In particular, the incidence of JC-related progressive multifocal leukoencephalopathy has risen markedly as a consequence of the AIDS epidemic, heightening clinical interest in this relatively little-studied virus. Laboratory studies of SV40 have led the way in understanding the biological effects of JC and BK viruses.

Murine polyomavirus has a genetic organization similar to that of SV40. It was discovered as a result of studies on murine leukemia virus. The presence of a second tumor virus in tissue extracts from leukemia virus-infected mice was inferred from the occasional induction of salivary gland tumors instead of the expected leukemia. The tumorigenic agent was isolated in 1953 (71) and subsequently named mouse polyomavirus because of its ability to cause a variety of different tumors in newborn or immunodeficient mice (165). Infection with polyomavirus rarely causes tumors in normal adult mice.

Identification of the Virus Transforming Genes

Molecular studies of virus transformation became intense in the mid-1970s when restriction enzymes and mammalian cell transfection techniques became available. An early use of these enzymes was to cleave isolated adenovirus DNA into recognizable, reproducible, easily isolated fragments that could be expressed individually in mammalian cells. From such studies the minimal sequences required for the transforming properties of adenovirus were identified (70). Similar studies defined the transforming sequences of SV40 (69) and, later, polyomavirus (172, 174, 175).

In general, DNA tumor virus gene expression can be divided into early and late phases. The early genes encode proteins that prepare the host cell to replicate DNA. The late genes encode the virus structural proteins. The transforming proteins are part of early gene expression. In the approximately 36,000-bp linear adenovirus genome, the transforming fragments encompass two transcription units, the products of the genes designated early region 1A (E1A) and early region 1B (E1B) (Fig. 1A). By differential splicing, each of these regions gives rise to two or more mRNA sequences (10). The most important messages encode the 13S and 12S proteins from E1A (named for the sedimentation coefficient of their respective mRNAs) and the 55K and 19K proteins from E1B (Fig. 1B). In the 5,243-bp circular SV40 genome, the array of transforming products is simpler (Fig. 2). A transcription unit giving rise to two distinctly sized proteins, the large tumor antigen (large T antigen) and the small tumor antigen (small t antigen), is sufficient for SV40 transforming functions. SV40 large T antigen has transforming activity independent of small t antigen, but the efficiency of transformation, especially in primary cells, is strongly enhanced by expression of small t antigen (11, 14). Polyomavirus has a genetic structure similar to that of SV40, but expresses a middle T antigen and a recently recognized "tiny t" antigen (150) in addition to large T and small t antigens (Fig. 3). Middle T antigen is the major transforming protein of the polyomavirus tumor antigens. Similar studies were also done on HPV, the major clinically relevant human DNA tumor virus under study at the time. Papillomaviruses have been recognized since the 1930s as a family of viruses associated with cutaneous warts in mammals (161). Initially they were less studied than adenovirus and SV40 because they were more difficult to propagate in cell culture. However, by the early 1980s, it was clear that the HPV subfamilies fall into two groups. The first is associated with benign warts that rarely progress to cancer. The second is associated with anogenital lesions at high risk for neoplastic progression and the majority of cervical carcinomas. HPVs have small circular genomes like SV40 and polyomavirus, but they express a series of early genes more like adenovirus. In cells derived from human cervical carcinomas, it is the HPV E6 and E7 genes that are consistently retained and expressed. HPV is reviewed in detail later in this volume (see chapters 6 and 7). Excellent reviews of the basic biological features of adenovirus, SV40, and polyomavirus are presented in references 30 and 160.

Interactions between Viral Transforming Genes and Cellular Proteins

While the SV40 and adenovirus studies (as well as HPV studies) identified transforming genes which were soon sequenced, no relationship was immediately ap-

A

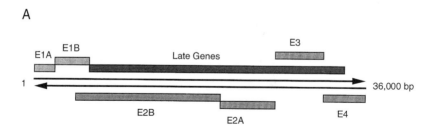

B

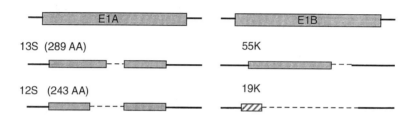

Figure 1 Schematic representation of adenovirus genome transcription products and the translation products of the E1A and E1B transforming genes. (A) The Ad2 or Ad5 genome consists of approximately 36,000 bp of linear double-stranded DNA. Adenovirus transcribes from both strands (the direction is indicated by the arrows). The early (E) and major late transcription units are represented as bars. Each transcription unit is further processed to produce a variety of products. The early transcription units produce the transforming proteins and proteins involved in virus replication, gene expression, and host interactions. The late proteins are mostly virus structural proteins. (B) The E1A and E1B genes produce a series of mRNAs through alternate splicing. The splice junctions are represented by dashed lines. The most prominent E1A splice products early in infection are the 13S and 12S mRNAs, named for their sedimentation coefficients. The corresponding 13S and 12S proteins are translated in the same reading frame and differ only by an internal region of 46 amino acids unique to the 13S product. The 13S unique region is required for efficient transcription of the remaining viral transcription units. However, all of the functions required for transformation of rodent cells in culture are contained within the common sequence represented by the 12S product. E1B produces two distinct proteins which initiate from different reading frames and have no sequence in common. As indicated in the text, the E1B 55K product targets p53. The 19K product is a homolog of the cellular protein Bcl-2, which restricts the apoptosis response. AA, amino acids.

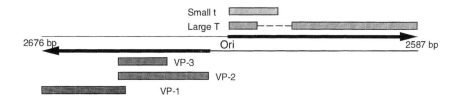

Figure 2 Linear representation of the SV40 genome and transcription products. The SV40 genome consists of 5,243 bp of circular double-stranded DNA. SV40 transcribes from sequences near the single origin of replication in the clockwise direction to produce the early mRNAs (direction of transcription represented by the rightward arrow), and in the counterclockwise direction to produce the late mRNAs (direction of transcription represented by the leftward arrow). The late transcription unit encodes the virus structural proteins (VP-1, VP-2, and VP-3). The early transcription unit is differentially spliced (the splice site is represented by the dashed line) to produce mRNAs encoding the two SV40 transforming proteins, large T and small t antigens. Large T and small t antigens share an amino-terminal region of 82 residues, followed by areas of sequence unique to each.

parent among the DNA or deduced protein sequences linking the transforming activity in the different viruses. Moreover, the sequences themselves provided little insight into the molecular mechanisms used by the transforming fragments. An early clue to SV40 large-T-antigen mechanisms of action came from the development of T-antigen-specific antibodies. Coimmunoprecipitation with these antibodies revealed a 53-kDa cellular protein (designated p53) tightly bound to T antigen (107, 114). The stable association between T antigen and p53 suggested that p53 was an important cellular target for the virus transforming activity. Immune complex association of p53 with large T antigen can be seen in Fig. 4. Comparison with adenovirus showed that p53 is targeted by an adenovirus tumor antigen as well, the 55K product of the adenovirus E1B gene (152). p53 is also targeted by the HPV E6 protein (36, 182). These findings provided early hints that these tumor virus families act through similar molecular targets and encouraged a number of laboratories to clone and study p53. The role of the tumor viruses in elucidating the functions of p53 is considered further below.

An important technique being developed through the rapidly progressing field of molecular biology was the ability to do site-specific mutagenesis. E1A was an attractive target for this powerful technique for several reasons. E1A is smaller than T antigen (289 amino acid residues compared with 708). Also, while both E1A and E1B are required for induction of a fully transformed phenotype, E1A alone is sufficient to activate the cell cycle and induce some degree of proliferation (86, 151). Thus, some separation of function can be achieved by looking at E1A alone. The dozens of isolated adenovirus strains made E1A sequence available from several different groups, such that conserved and presumably functionally important amino acid residues could be identified. Another advantage was the ease with which E1A could be shuttled between plasmid and virus constructs. This property made it feasible to create multiple mutants and transfer them to

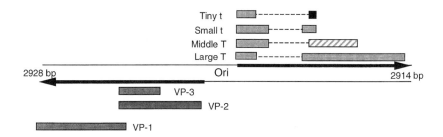

Figure 3 Linear representation of the murine polyomavirus genome and transcription products. The polyomavirus genome consists of 5,295 bp of circular double-stranded DNA with a genetic organization similar to that of SV40. Polyomavirus transcribes from sequences near the single origin of replication in the clockwise direction to produce the early mRNAs (direction of transcription represented by the rightward arrow), and in the counterclockwise direction to produce the late mRNAs (direction of transcription represented by the leftward arrow). The late transcription unit encodes the virus structural proteins (VP-1, VP-2, and VP-3). The early transcription unit is differentially spliced (the splice site junctions are represented by the dashed lines) to produce four messages encoding the early proteins: large T, middle T, small t, and tiny t antigens. All of the polyomavirus T antigens share amino-terminal protein sequence, and the carboxy-terminal sequence of small t antigen is contained within large T antigen. Small t antigen has an internal region that is not present in large T or tiny t antigens, but is common to middle T antigen. The carboxy-terminal regions of middle T antigen and tiny t antigen are unique. Despite the overall similarity of the polyomavirus and SV40 genome structures, the T antigens differ significantly between the two viruses. The large T antigens of both contain J domains and functional pRb family binding motifs. However, polyomavirus large T antigen does not show p53 binding activity and does not transform efficiently in the absence of middle T antigen. Middle T antigen is the major transforming agent of the polyomavirus genome. Its transforming activity is largely localized to its unique sequence, which encodes binding sites for a number of cytoplasmic cell proteins involved in growth factor signaling (see also Fig. 9).

virus vectors, which can enter essentially 100% of a targeted cell population in contrast to the much lower efficiency of plasmid transfection. Another advantage which was unknown at the start, but became clear as the mutagenesis studies proceeded, is that the E1A proteins comprise several discrete and independent domains that are relatively insensitive to distal changes in secondary structure. The stability of the E1A proteins to rather extensive deletions permitted the active regions to be mapped to within very narrow limits.

Studies on virus transforming protein interactions with cellular products benefited from the development of specific monoclonal antibodies. Monoclonal antibodies are advantageous in protein studies because these hybridoma cultures produce a single antibody which is reactive with a single epitope. Because they do not cover the whole surface of the target protein, monoclonal antibodies are more likely than polyclonal antibodies to reveal proteins that are in stable association with the target

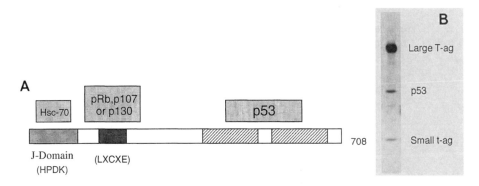

Figure 4 Schematic of the 708-residue SV40 large T antigen gene product and major protein binding sites linked with transforming activities. (A) SV40 large T antigen contains an amino-terminal J domain of approximately 80 residues which mediates interaction with cellular hsc70. This interaction promotes the degradation of p130 and posttranslational modifications of p107, effects that correlate with increased cell cycle activity. This interaction may have consequences for other targets of large T function as well, including components of replication complexes. The J domain is present in both the large T and small t antigens. Sequences downstream of the J domain are not common to small t antigen. These include the pRb family binding site, which is specified by the conserved Leu-X-Cys-X-Glu (LXCXE) motif. p53 is bound via two noncontiguous regions near the C terminus of the protein. The potential Bcl-2 homology domain falls between the p53 binding regions. SV40 large T antigen has a variety of other functions not indicated here. It participates directly in transcription of the SV40 late genes and interacts with the basal transcription machinery. It also participates directly in replication complexes and encodes ATPase and helicase active regions which are required for its role in replication. The murine polyomavirus large T antigen has regions of both homology and dissimilarity with SV40 large T antigen. Polyomavirus large T antigen contains an amino-terminal J domain that binds hsc70 and an LXCXE motif that mediates binding to the pRb family. However, polyomavirus large T antigen does not include a recognizable p53 binding function, nor does it transform well in the absence of middle T antigen, which has no direct homolog in SV40. The properties of middle T antigen are illustrated in Fig. 9. (B) p53 is present in immune complexes isolated from infected cells using a monoclonal antibody reactive with large T antigen. In the autoradiogram shown, isotopically labeled cells were lysed and the lysate was immunoprecipitated with an antibody that reacts with the N-terminal region common to large T and small t antigens. Visualization of immune complexes such as this one first called attention to p53 and were the impetus for the cloning and characterization of this important cell growth regulatory molecule. Not all important associations are readily detectable in this manner. Detection of pRb association with large T antigen requires more sensitive assays, although this binding activity is readily apparent with the E1A proteins (shown in Fig. 5).

protein. Monoclonal antibodies became a particularly important tool in E1A studies. Monoclonal antibodies, as well as specific antipeptide antibodies, revealed a series of cellular proteins tightly associated with E1A, but p53 was not among them (77, 190). The most prominent were a group of three proteins ranging in size from 105 to 130 kDa, plus a larger species of 300 kDa (Fig. 5). The site-specific mutagenesis

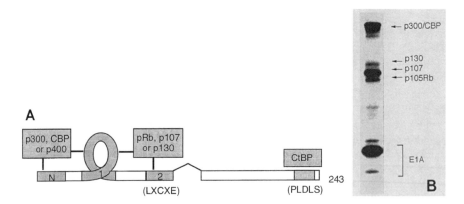

Figure 5 Representation of the 243-residue E1A 12S product and protein binding activities. (A) The 12S protein is represented by the bar. The break in the bar indicates the splice site, where the first and second exons are joined. Shaded regions are those that have been linked with binding activities and transforming functions. Conserved regions 1 and 2 are areas that are extensively conserved among adenoviruses of different serotypes. Conserved region 1 contributes to the nonoverlapping binding sites of both the p300 and pRb families and so is represented as a loop. The key residues required for pRb family binding are the Leu-X-Cys-X-Glu (LXCXE) motif in conserved region 2. The same motif mediates pRb binding activity in SV40 and polyomavirus large T antigens. A key residue required for p300 binding is a conserved Arg (R) at position 2 in the amino-terminal region (N). The E1A second exon contains an activity that restricts the tumorigenic potential of cells immortalized by E1A. This function is linked with binding of a C-terminal binding protein (CtBP) through a conserved five-residue motif, PLDLS (154). (B) Immune complexes isolated from infected cells using a monoclonal antibody with specificity for the carboxy-terminal region of the E1A proteins show stable association with p300, CBP, and the pRb family (p105-Rb, p107, and p130). This autoradiogram shows an electrophoretic separation of the proteins associated with the wild-type 12S E1A product. (p300 and CBP do not resolve distinguishably in the conditions of this gel.) Various E1A mutants have been generated that abrogate binding to either p300/CBP or the pRb family. Such mutants have been very useful in distinguishing the respective roles of the p300 and pRb families in pathways regulating cell growth and differentiation. Closer examination of the E1A complexes revealed that they also contain cyclin A and cyclin E, with their associated cyclin-dependent kinases, products which are not as prominent in the conditions of this gel. The presence of the cyclins in the E1A complexes is dependent on the presence of the pRb-related proteins, which led to the understanding that they do not contact E1A directly, but are associated through p130 and p107.

studies revealed that a very small region, requiring no more than 21 amino acid residues, plays a critical role in E1A transforming functions (112, 131). This region is conserved to about 50% identity among adenoviruses of different serotypes. Most critical for transforming activity is a shared motif spanning five residues: LXCXE (where X is any amino acid). Narrowing one of the E1A transforming regions down to such a small stretch of sequence revealed what was not apparent before, that SV40 and polyomavirus large T antigens contain a region of very similar structure. Indeed, the corresponding region from SV40 large T antigen can substitute functionally for the E1A sequence (129). Analysis of E1A-associated proteins showed that their contact residues in E1A correspond down to the single-amino-acid level with the E1A residues most critical for transforming functions (50, 186). The 300-kDa product binds to E1A at the extreme amino terminus in a region that shows no clear homology to T antigen, but is required for the E1A-mediated transforming function. The group of three proteins ranging from 105 to 130 kDa all interact with the LXCXE motif, which suggests that they might be structurally related. The identification of the 105-kDa E1A binding product as the product of the retinoblastoma tumor susceptibility gene (pRb) (185) led to a series of breakthroughs that forged clear links between DNA tumor virus transforming functions and cellular proliferation controls. The existence of similar-sized proteins with E1A binding properties much like p105-Rb prompted the cloning of p107 (55) and p130 (75, 109, 124), pRb family members that also play fundamental roles in cell growth control. The E1A mutant studies that led to the association of p300 and the pRb family with E1A transformation functions have been the subject of several reviews (see for example references 8, 45, and 130).

The interactions of the various DNA tumor virus transforming gene products with p53 and pRb suggested that these virus proteins have evolved a strategy of binding directly to key regulators of cell growth and modifying their functions. This contrasts with the general mechanism of tumorigenesis seen among the RNA tumor viruses, which have usually acquired a cellular proto-oncogene that is altered by integration into the viral genome such that it acts to promote cellular proliferation. Further work on the DNA tumor virus transforming products has revealed a fascinating number of direct protein interactions, many of them conserved among the different virus families. The contribution of E1A and large-T-antigen studies to the current understanding of the molecular roles of p53 and the pRb family in the control of cell growth is discussed further below. Additional interactions of the adenovirus, SV40, and polyomavirus transforming proteins are also considered in the following sections.

p53

Targeting of pRb and p53 appears to be nearly universal among the DNA tumor viruses, consistent with the finding that loss of these two tumor suppressors is characteristic of a wide spectrum of tumors. In SV40, p53 and pRb are both bound by large T antigen (reviewed in reference 117). In adenovirus, separate protein products target pRb and p53, a situation which has made this virus in some ways more useful for elucidating the roles of these cellular products.

E1A and E1B are the only adenovirus genes consistently found integrated and expressed in adenovirus-transformed cultured rodent cells (42, 61, 159). In general, E1A by itself is sufficient to activate the cell cycle, inducing genes required for DNA synthesis and repeated rounds of cell division. Although E1A expression readily induces cell proliferation, most colonies of primary cells proliferating in response to E1A fail to become established as indefinitely dividing lines. The E1B products on their own show no ability to induce proliferation, but with E1B coexpression the number of primary cell colonies able to proliferate indefinitely in response to E1A is greatly enhanced (148). Such studies ultimately suggested that E1B blocks the apoptotic response of cells to unregulated DNA synthesis and proliferation. The binding of p53 by the 55-kDa product of the E1B gene is consistent with the view, developed from p53 studies directly, that p53 plays a critical role in regulating apoptotic pathways. These kinds of studies also contributed to the current understanding that tumor development depends as much on loss of apoptotic mechanisms in a cell as on the activation of pathways that promote proliferation. For reviews of the role of p53 in regulating apoptosis and cell growth, see references 68, 100, and 108.

Interestingly, p53 was originally thought to be an oncogene, i.e., an activator of cell growth, because early isolates of the p53 cDNA could cooperate with other oncogenes, including E1A, to establish primary cells (53, 94, 142). However, further insight into p53 structure and function revealed that the early isolates were actually mutants acting dominantly to suppress wild-type p53 function, and that wild-type p53 acts as a tumor suppressor (52, 59). The extremely widespread occurrence of p53 mutations in naturally occurring tumors implies that alteration of p53 function is a required event in the formation of most tumors (reviewed in references 84 and 178). These studies also emphasized the virtue of isolating regulators of cell growth from primary cells rather than from the tumor-derived cells typically used in laboratory culture.

The DNA tumor viruses played a direct and pivotal role in focusing attention on p53, which is now the subject of numerous studies, and have continued to contribute to the current understanding of the role of p53 in cell growth. Two principal, related effects of p53 are to restrict the activity of the cyclin-dependent kinases which drive the cell cycle, and to initiate the apoptotic pathway in cells that have escaped normal cell cycle control. The key cell cycle regulation pathways and the points of intervention of the DNA tumor virus transforming proteins are shown schematically in Fig. 6. Functional domains identified within p53 include a sequence-specific DNA binding domain, an amino-terminal transcriptional activation domain, a carboxy-proximal oligomerization domain, and a carboxy-terminal domain that regulates the activity of the DNA binding domain (Fig. 7). Mutant forms of p53 act as dominant suppressors of the wild-type p53 transformation suppression and transcription activities, presumably because of the oligomerization activity. p53 transactivation function is modulated by association with a 90-kDa cellular protein, MDM-2. MDM-2 can block the transcription-stimulating activity of p53 (128) and is amplified at high frequency in tumors that contain wild-type p53 genes (136), consistent with the interpretation that p53 transcription-stimulating activity is required for its tumor suppressor activity and that increased MDM-2 levels can contribute to transformation by inactivating this function of p53.

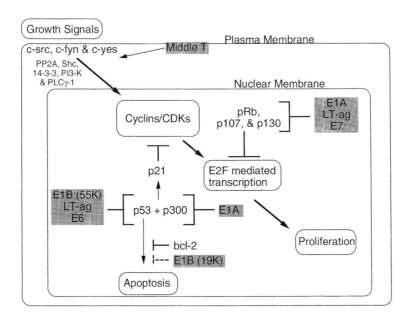

Figure 6 Cellular targets of the DNA tumor virus transforming proteins. The tumor viruses have evolved similar strategies for targeting host cell pathways important for cellular proliferation and apoptosis. This diagram illustrates the roles of the major host proteins targeted by the viral protein products (gray boxes). A common target among tumor viruses is the pRb family of proteins. This protein family functions, in part, by regulating the E2F family of transcription factors. The E2F family of proteins is responsible for the transcription of many key cell cycle regulatory genes. The adenovirus E1A proteins, the SV40 and polyomavirus large T antigens (LT-ag), and the HPV E7 protein all bind directly to the pRb family of proteins. This, in turn, abrogates control of E2F transcription, allowing the cell to enter the cell cycle. Another key regulator of cell growth is p53. p53 can stop the cell cycle by upregulating, among other targets, the cyclin-dependent kinase inhibitor p21. Depending on other signals, p53 can also initiate the cell death machinery (apoptosis). Adenovirus E1B 55K, SV40 large T antigen, and the HPV E6 protein all target p53 directly. Adenovirus E1A targets p53 indirectly by binding p300, which normally serves as a transcriptional coactivator of p53. Still another way in which adenovirus targets apoptotic pathways is through its E1B 19K protein product. This protein mimics the growth survival protein Bcl-2. Polyomavirus middle T antigen has evolved a different method for circumventing cell growth control. It is a plasma membrane-bound protein that targets the growth factor signal transduction cascade. Middle T antigen usurps this signal cascade by binding key components of the signaling pathway, including the Src family of tyrosine kinases (c-*src*, c-*fyn*, and c-*yes*), protein phosphatase 2A (PP2A), the Shc-Grb2 complex, 14-3-3 proteins, phosphatidylinositol 3-kinase (PI3-K), and phospholipase C (PLCγ-1).

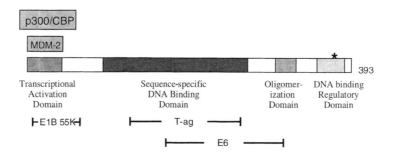

Figure 7 Functional regions of human p53. The bar represents the 393-residue open reading frame of human p53. Major functional regions are indicated by shading. The binding sites for the Ad5 E1B 55K protein, SV40 large T antigen, and HPV E6 are indicated by lines below the bar. p300 and CBP interact with p53 in the amino-terminal activation domain, a region also involved in MDM2 (mouse double minute 2) binding. p300/CBP-mediated acetylation at a site near the carboxy terminus (represented by the asterisk) relieves the repressive effect that the carboxy-terminal regulatory domain exerts on p53 site-specific DNA binding. This permits p53 to bind to DNA target sequences with greater affinity and increases transcription of p53 targets such as the cyclin-dependent kinase inhibitor, p21, which inhibits cell cycle progression (see also Fig. 6).

One important characteristic revealed by DNA tumor virus transforming protein probes is the variety of ways in which p53 function can be compromised. The DNA tumor virus transforming proteins that bind pRb generally do so through a shared amino acid motif that contacts a single pRb site. However, the DNA tumor viruses show a diversity of approaches in targeting p53. No consensus binding site for p53 has been identified in the DNA tumor virus transforming products, and binding studies indicate that adenovirus E1B 55K, SV40 large T antigen, and HPV E6 each target the p53 protein through different p53 regions (Fig. 7). T antigen binds in the DNA binding region, where most naturally occurring p53 mutations are found, and E1B 55K binds in the amino-terminal activation domain (191). HPV E6 is unique among these virus proteins in that it promotes ubiquitin-mediated degradation of p53 and has helped to focus attention on protein stability as an important means of regulating p53 function, even in the absence of viruses (155). E6 studies have also proved useful in identifying other protein components of the ubiquitin pathway. An E6-associated protein (E6-AP) mediates E6 binding to an N-terminal domain of p53 and acts together with E6 to target p53 for destruction (89–91, 121). Despite the diversity of approaches to compromising p53 function, the phenotypes of the DNA tumor virus probes support the general conclusion that the ability of p53 to suppress unregulated cell growth is closely linked with its transcription-regulating function (e.g., references 126 and 191).

One of the most profound insights into p53 mechanisms of action has been assisted by a DNA tumor virus transforming protein that does not contact p53 directly. It has been known for some years that E1A can impede p53-associated transcriptional activation of a major target, the cyclin-dependent kinase inhibitor p21$^{cip-1/waf-1}$. More recently, experiments linking this activity with the integrity of the p300 binding site in E1A have contributed to the realization that p300 serves as a transcriptional cofactor for p53. These results are discussed further below (E1A-Targeted p300).

OTHER CELLULAR TARGETS INVOLVED IN APOPTOTIC PATHWAYS

The single E1B gene uses differential splicing and alternative reading frames to produce two different protein products of nonoverlapping sequence, the 55K and 19K proteins (Fig. 1). In vitro manipulation of the E1B gene to produce 55K-only and 19K-only cDNA sequences allowed their functions to be probed individually. While their structures are completely unique with respect to each other, both express a biological activity that opposes the host cell apoptosis response. The lack of any evidence that 19K was directly targeting p53 suggested that 19K might act at a different point in cell apoptotic pathways. Another key regulator of cellular apoptosis is the product of the *bcl-2* gene, first identified as part of the most common translocation in human B-cell follicular lymphoma (177) and as an onco-gene with the unusual property of extending cell survival rather than stimulating cell proliferation. An extensive family of Bcl-2-like proteins with shared regions of homology is now recognized (reviewed in reference 79). Members of the Bcl-2 family include proteins such as Bax, a promoter of apoptosis, as well as apoptosis inhibitors such as Bcl-2. Bax was cloned as a Bcl-2 binding protein. Heterodimeri-zation occurs between different protein members, and the decision of a cell to live or die depends on the relative levels of these proteins (137). The molecular mechanism by which apoptosis is actually induced is not yet clear, but Bcl-2 fam-ily-mediated signals for cell death correlate with increased levels of a group of cellular cysteine proteases, now designated caspases, which cleave a variety of cellular substrates including nuclear lamins and the retinoblastoma protein (re-viewed in reference 146).

Comparison of the 19K sequence with the *bcl-2* sequence, combined with targeted mutagenesis studies, established that 19K itself is a member of the Bcl-2 protein family (reviewed in references 170 and 183). 19K shares homology with each of the three short regions of conserved sequence, the Bcl-2 homology domains BH1, BH2, and BH3, that currently define Bcl-2 family members. 19K expression and increased Bcl-2 expression are functionally interchangeable in adenovirus infection and transformation (148, 171), and both 19K and Bcl-2 can block E1A-induced caspase activity in adenovirus-infected cells (15). E1B 19K has served as the basis for identifying previously unknown cellular products that are now recognized as part of Bcl-2-related apoptotic pathways (16, 58, 74). The ability to mimic Bcl-2 may turn out to be a more general aspect of DNA tumor virus func-tion. Epstein-Barr virus (EBV), a human DNA virus that establishes a persistent latent infection in the majority of the human population and which is associated with Burkitt's lymphoma, encodes a viral homolog of Bcl-2, BHRF-1 (83, 143). Another human tumor-associated DNA virus, human herpesvirus 8 (HHV-8), which is linked with Kaposi's sarcoma, encodes a Bcl-2 homolog as well (25). EBV and HHV-8 are reviewed in detail later in this volume (see chapters 3, 4, and 5). The BH3 region is poorly conserved in the EBV and HHV-8 homologs, suggesting that it is not essential for anti-apoptotic function. Continued analysis of the viral products is likely to shed significant light on important structural and functional aspects of the cellular Bcl-2 family. A recent analysis of SV40 large T antigen indicates that it contains a region able to prevent apoptosis independently of binding and inactivation of p53 (31). This function has not been linked experimentally with Bcl-2-mediated pathways, but it maps to a region of large T antigen that shares sequences characteristic of BH1.

pRb

The Rb gene product is a suppressor of tumorigenesis. Its allelic loss confers an increased susceptibility to specific kinds of tumors, most notably childhood retinoblastomas and osteosarcomas in early adulthood (reviewed in references 99 and 181). Loss of pRb in individual cells appears to be an important step in carcinogenesis of almost every type, as Rb mutations are found with high frequency in many kinds of tumors.

The central role of pRb in regulating S-phase-specific gene expression was revealed largely through E1A genetics. The most important products encoded by the E1A gene are the differentially spliced 12S and 13S mRNAs, which produce similar proteins distinguished only by an internal region of 46 amino acids specific to the 13S product. E1A is required for transactivation of the remainder of the viral genome and for activation of DNA synthesis in quiescent host cells to support viral replication. Cloning of the individual 12S and 13S cDNAs revealed that the 12S product is sufficient to stimulate proliferation of quiescent primary cells (192) and provided the means to study this activity apart from the general transcriptional activation effected by the 13S product. Site-specific mutagenesis of the 12S product revealed two distinct regions required for cell growth-stimulating activity. One region contains the conserved LXCXE motif. Deletion mutagenesis and single-residue substitutions demonstrated that this motif is highly important for transforming function (reviewed in references 8 and 130). The very narrow definition of the active site made possible by the fortunate arrangement of E1A structure led to the realization that the large T antigens of SV40 and polyomavirus share this functional motif with E1A. In quick succession, a conserved LXCXE motif was identified in the E7 gene products of the various serotypes of HPV and pRb was found to interact with the large T antigens of SV40 and polyomavirus and the E7 gene products of specific HPV serotypes (38, 46). pRb is bound by each of these proteins through the consensus binding site, and the affinity of HPV E7 for pRb correlates with oncogenicity among the genital HPV types (80).

E1A genetics played a further specific role in revealing the mode of action of the Rb protein. As indicated above, the 13S E1A product is a general activator of host cell transcription, while the 12S product does not activate most of the cell and virus transcription elements that are responsive to the 13S product. However, transcription of cell-cycle-specific genes is a central feature of cell cycle activation in quiescent cells, so there was considerable interest in promoters that could be activated in response to the 12S product. A key component to the successful identification of a 12S-responsive promoter element was to study transactivation in quiescent cells rather than the constitutively active laboratory strains, such as

HeLa cells, that were most commonly employed. Indeed, HeLa cells have proved to be a most disadvantageous line for studying pRb function as they are now understood to harbor HPV E6 and E7 genes, consistent with their origin from a cervical carcinoma.

The adenovirus E2 promoter proved to be the key to a first understanding of pRb transcription function. The adenovirus E2 region encodes the products that adenovirus contributes to DNA replication, and its expression is linked with activation of host cell DNA synthesis (reviewed in reference 35). E2 reporter constructs are activated by the E1A 12S product in normal quiescent cells. E2 promoter studies had shown that a promoter binding protein factor, designated E2F, exists in HeLa cells in a free form, but exists in most normally regulated cells in protein complexes that can be dissociated by E1A in a region 2-dependent manner (4). The same single-amino-acid residues in E1A that are critical for E2F activation and dissociation of E2F complexes to the active form were those found critical for pRb binding (149). This suggested that pRb might be the factor being dissociated from E2F. pRb-specific antibodies confirmed the presence of pRb in E2F complexes in the absence of E1A (6, 22), and parallel studies showed that immobilized pRb in cell lysates shows a DNA binding activity that selects the E2F promoter element sequence specifically from a random pool of oligonucleotides (27). This series of observations on E2F promoter activation was generally reproduced using the T antigens and HPV E7 (23).

One further observation critical to the basic elucidation of pRb function was derived specifically from studies with SV40 T antigen. It was clear from immune precipitation of pRb that it exists in a mixed cell population as a series of phosphorylated forms. The realization that T antigen binds only to the most hypophosphorylated forms suggested that this form was targeted preferentially because it is the principal form active in cell growth suppression (118). Cell cycle studies following this observation showed that pRb undergoes cyclical phosphorylation (17, 24, 39, 119, 127), suggesting a mechanism for how pRb, while continuously present in dividing cells, could permit successive rounds of cell division.

Since these historic studies, the field has progressed to the understanding that pRb is one of a family of related proteins, pRb, p107, and p130, each of which is bound by E1A and SV40 T antigen (Fig. 4 and 5). p130 appears to regulate progression from G0 to G1 in a manner analogous to the role exerted by pRb at the transition from G1 to S phase. E2F was cloned through its pRb binding activity (82, 95, 158). The properties of the first cloned E2F cDNA product led quickly to the realization that E2F is also part of a protein family. At least five different E2F genes express products that bind in different combinations with the members of the pRb family to regulate gene expression in successive waves throughout the cell cycle. Each of the pRb family members shows distinct regulatory functions at different stages of the cell cycle (reviewed in reference 106).

E1A had at least one more important contribution to make to the understanding of pRb function and cell cycle regulation. While p300 and the pRb family are the most prominent of the proteins seen associated with E1A in immune complexes (Fig. 5), others are present as well. Most notably, cellular products independently linked with cell cycle regulation, cyclin A and cyclin E, are present, along with the associated cyclin-dependent kinase CDK2 (57, 145). The mapping of the cyclin A/E binding site to essentially the same residues required for binding the pRb family helped to show that the cyclins do not bind E1A directly but are present in E1A complexes as a consequence of their association with p107 and p130 (54, 56, 63, 87). From these clues, the cyclins were also recognized as components of E2F complexes (5, 133). Thus, the properties of the relatively simple E1A protein revealed an intimate series of connections between multiple families of seemingly unrelated proteins, each of which had been isolated as a critical regulator of cell cycle activity. These include the E2F family of transcription factors, the pRb family of growth suppressors, and the families of cyclins and cyclin-dependent kinases. The interaction of these cellular factors and the mechanisms by which they regulate cell-cycle-specific events have been the subject of several excellent reviews (9, 93, 123, 162, 163). The initial studies on the pRb family revealed how these proteins can physically transfer signals from the cyclin-dependent kinases to the transcription apparatus in the form of E2F.

Since these discoveries, it has become clear that pRb interacts with a variety of other cellular proteins. These include other transcription factors and also proteins with diverse functions such as nuclear matrix attachment. In most cells expression of tissue-specific genes is linked with restriction of cell cycle activity. pRb (and p300, as discussed below) appears to participate significantly in the regulation of both processes. While neither pRb nor p300 is a tissue-specific factor in its own right, each has a role in coordinating tissue-specific gene expression with cell cycle activity in a variety of tissues. The regulatory role of pRb in tissue-specific gene expression is reflected in interactions with factors such as EII-1 and MyoD, tissue-specific factors involved in the development of lymphoid lineages and muscle cells, respectively (reviewed in reference 181).

One approach to a more complete understanding of pRb function and its normal cellular partners involves the identification of cellular proteins that contain the LXCXE motif. Such a diverse assortment of proteins contain this motif, and have the potential to bind pRb at least in vitro, that it is not yet possible to judge from the simple presence of the motif whether a protein has a physiological role in association with pRb. However, one such protein has been implicated as a cellular partner of pRb in vivo. This is BRG-1, a human homolog of the yeast Swi2 protein, which is essential for control of the yeast mating type switch (reviewed in reference 144). BRG-1 and Swi2 each exist in vivo in stable complex with a series of interacting proteins, known as the SWI/SNF complex. Human SWI/SNF complexes are implicated in the normal control of cell development in response to nuclear hormone receptors. BRG-1 is altered or lost in various human tumor lines, and the ability of exogenously introduced BRG-1 to suppress aspects of the transformed phenotype in such cells requires the integrity of the LXCXE motif (44). The potential ability of E1A or T antigen to modulate the activity of SWI complexes has not yet been studied in depth, but it is intriguing that at least one other class of DNA tumor viruses has been found to target the human SWI complexes. EBV nuclear antigen 2 (EBNA2), a product required for EBV-mediated immortalization of B cells, interacts in vivo with the BRG-1-associated protein hSNF5 (188).

OTHER PROTEINS IMPLICATED BY THE VIRAL ONCOGENES AS REGULATORS OF CELL GROWTH

E1A-Targeted p300

A major advantage E1A has offered to the study of cell growth-regulating mechanisms is a structure comprising several distinct functional sites that are relatively stable and active when expressed as deletion fragments. These mutants show clearly that E1A contains at least two distinct active sites that target cell cycle regulation. These correspond to the binding site for the pRb family and to the binding site for the cellular p300 protein (Fig. 5). In primary rodent cells either binding site is sufficient to induce S phase and replication of the host cell genome, but neither alone is sufficient to induce normal rounds of cell division; cells expressing single-site mutants stall in S phase or arrest in G2.

Targeting of p300 does not activate E2F, so its molecular functions are largely distinct from those of the pRb family (180). Cloning and sequencing of p300 revealed that it is highly homologous to a cyclic AMP response element binding protein (CREB) coactivator, CBP (CREB binding protein) (1, 28, 48). A 400-kDa band has also been observed in E1A immune complexes with a binding profile similar to that of p300, and this is likely to be another member of the p300 family (7). p300 and CBP are so similar that their functions have not yet been clearly distinguished in laboratory studies. Both are targeted by E1A and both can bind CREB, at least in vitro. Both p300 and CBP are present in intracellular complexes with the TATA binding protein, TBP (37), and both can bind TFIIB in vitro (103). p300 and CBP have been shown to bind to a number of important upstream activators, including several of the nuclear hormone receptors (21, 76, 96; reviewed in reference 47) (Fig. 8). p300 and CBP are also now known to serve as cofactors for p53-mediated transcriptional activation (3, 73, 110, 156). The binding properties of p300 and CBP allow them to form a bridge between site-specific upstream activating factors and the basal transcription machinery. Thus they are integral components of both basic and universal pathways regulating cell growth responses.

p300 and CBP link two important aspects of regulated gene expression. They help to restrict cell cycle activity while promoting tissue-specific gene expression. As coactivators of p53-mediated transactivation, p300/CBP enhance the ability of p53 to activate p21 expression, which restricts cell cycle progression by inhibiting the action of the cyclin-dependent kinases. As cofactors for numerous specific upstream activators such as the nuclear hormone receptors, p300/CBP are also part of tissue-specific gene expression complexes in a wide variety of cells. Indeed, it appears that p300/CBP and the pRb family each play pivotal roles in coordinating tissue-specific gene expression with cell cycle activity. Studies with large T antigen and the E1A mutants that segregate the p300 and pRb family binding functions are contributing to the elucidation of the separate roles of each of these proteins in the differentiation of various tissues.

p300 and CBP play another unique and important function in gene expression. Besides linking upstream activators with the basal transcription machinery, they link transcription complexes with chromatin remodeling activities (Fig. 8). p300 and CBP contain both intrinsic and associated histone acetyltransferase activity (135, 189), suggesting that chromatin modification is an essential part of their role in regulating transcription. The targets of the p300 acetylase activity are not limited to histones. The activity of a number of cellular proteins may be modulated by p300-mediated acetylation. In particular, recent studies indicate that p300 can acetylate p53 and that this modification strongly enhances the DNA binding activity of p53 (72). Continued study of p300 and CBP should be a key approach to the question of how selective gene activation is accomplished in the context of intact chromosomes.

Targeting p300 and CBP is clearly important in E1A-mediated cell cycle activation and concomitant repression of tissue-specific gene expression. It is not yet clear whether or how other DNA tumor viruses target p300-mediated pathways. What is clear from genetic studies of E1A, the large T antigens, and HPV E7 is that targeting of the pRb family is not sufficient for the cell proliferation-inducing activity of these viral oncogenes in primary cells. Each of these products apparently performs an additional function somewhat equivalent to E1A targeting of p300. SV40 large T antigen can associate with p300, apparently via the T-antigen interaction with p53 (111). E1A binding to p300, and T antigen binding to p53, have a similar effect on the transcriptional activity of p300/p53 complexes, which

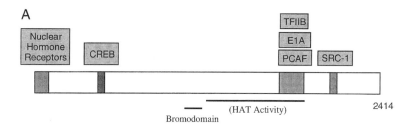

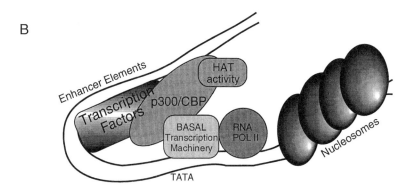

Figure 8 Functions of p300. (A) The bar represents the 2,414-residue open reading frame of p300. At least four distinguishable protein binding sites have been identified in p300 and are indicated by shading. An amino-terminal region binds various nuclear hormone receptors including the retinoic acid receptor, the estrogen receptor, and the glucocorticoid receptor. p300 contains a binding site for the cyclic AMP-responsive element binding protein (CREB). The CREB

may be sufficient to account for their shared biological activities. However, experimental evidence also suggests that T antigen has important transforming functions in its amino terminus, a region independent of p53 binding. The activities of the T-antigen amino terminus are discussed below.

The T-Antigen Amino Terminus and hsc70

Efficient SV40 large T antigen-mediated transforming activity requires an intact amino terminus in addition to the binding sites for p53 and the pRB family. The amino terminus of T antigen has recently been shown to contain a functional J domain (Fig. 4), a motif of approximately 70 amino acid residues originally described in the *Escherichia coli* DnaJ protein, a member of a highly conserved class of molecular chaperones. J domains are an interaction site for members of the 70-kDa heat shock protein (hsp) family. DNAJ and hsp70 act together to perform a variety of functions related to the folding, transport, and degradation of various protein targets. The amino-terminal sequences of T antigen comprise a novel J domain that mediates a specific interaction with the constitutively expressed hsp70 family member hsc70 (19, 153, 164). This interaction is sufficient to activate an ATPase activity in hsc70 (164).

Cellular targets of the T antigen/hsc70 complex are likely to include a variety of proteins, some more related to virus replication and assembly than to transforming activity. However, specific effects on the pRb-related proteins p107 and p130 correlate with the integrity of the J domain motif in T antigen, and these proteins are likely to be critical transformation targets. The integrity of the hsc70 binding site correlates with alterations in the phosphorylation state of p107 and p130 and

◀──

binding region also serves as a binding site for various other transcription factors including c-*myb* and c-*jun*. p300 contains a bromodomain, but no specific function has yet been linked with this motif. p300 has both intrinsic and associated histone acetyltransferase (HAT) activities. Associated acetylases include P/CAF (p300/CBP-associated factor). p300 also contains an intrinsic HAT activity, which maps to the region indicated by the line below the bar. p300 has at least one other protein binding region, a region near the carboxy terminus which binds the nuclear hormone receptor cofactor SRC-1. Sequences in this region also mediate binding to other transcription factors including YY-1. The binding site for p53 association has not yet been localized definitively on p300. In addition to binding various upstream activators of transcription, p300 interacts with the basal transcription machinery. p300 and CBP are stably associated with TBP (TATA binding protein) complexes in vivo, and p300 binds directly in vitro to TFIIB (transcription factor IIB) at a region similar to that used by P/CAF. (B). The combination of properties, binding to upstream activators, association with the basal transcription machinery, and the presence of HAT activity suggests that p300 integrates signals from each of these sources. p300 can bind multiple upstream activators and/or cofactors simultaneously and thus has the potential to synergize the effects of these molecules. Similarly, p300 can connect upstream activation signals to the basal transcription machinery. The HAT activity may contribute to transcriptional activation through multiple channels. p300 can acetylate histones, suggesting that it contributes to chromatin remodeling. p300 can also mediate acetylation of at least one upstream activator, p53. This acetylation enhances p53 site-specific DNA binding activity and consequent transactivation function. Acetylation of components of the basal transcription machinery may also be an aspect of p300 function.

with ubiquitin-mediated degradation of p130, in cells expressing large T antigen (167). These effects depend on both the J domain and the LXCXE motif, suggesting that T antigen must bind p130 and p107 together with hsc70 in order for hsc70 to direct the posttranslational modification of these proteins. The presence of the J domain also has implications for the role of small t antigen. The J domain is within the region of sequence that is shared by small t and large T antigens (although the LXCXE motif is not). The J domain sequence is present in all the T antigens of polyomavirus. This domain comprises almost the entire sequence of polyomavirus tiny t antigen, where it is also able to activate the ATPase activity of hsc70 (150).

The Targets of Middle T Antigen

The most notable feature of polyomavirus relative to SV40 is the production of additional alternative splice products from the transforming gene. Polyomavirus large T antigen shares some SV40 large T transforming functions. It binds pRb through an analogous LXCXE motif, and its amino terminus has J domain structure and activity. However, polyomavirus large T antigen has not been found to bind p53. Polyomavirus small t antigen is similar to SV40 small t antigen and shares an amino-terminal J domain with its respective large T antigen. A product unique to polyomavirus is middle T antigen which, like small t antigen, is formed through alternative splicing. Middle T antigen is critical to polyomavirus transforming activity (172, 174, 175). The adenovirus and SV40 transforming proteins target predominantly nuclear products that are involved in transcription complexes. Middle T antigen diverges from this pattern by targeting cellular proteins generally concerned with signal transduction (Fig. 6).

Middle T antigen is a plasma membrane-bound protein present in infected cells in two forms of about 56 and 58 kDa, differing in the extent of C-terminal serine phosphorylation (157). Middle T antigen mediates its transforming function through association with and modulation of, the activities of normal cell proteins that transmit cell proliferation signals (Fig. 6 and 9). These include three *src* kinase family members, c-*src* (12, 32, 33), c-*yes* (101), and c-*fyn* (26, 85, 104). Binding to middle T activates the tyrosine kinase activity of these proto-oncogenes, which then phosphorylate various substrates including middle T antigen. The tyrosine-phosphate-containing sites in middle T antigen serve in turn as binding sites for a number of other cellular proteins. Middle T binding proteins include the 85-kDa regulatory subunit of phosphatidylinositol 3-kinase (PI 3-kinase) (2), phospholipase C-γ-1 (PLCγ-1) (168), the adaptor protein complex of Shc and Grb2 (20, 40), and 14-3-3 (140) proteins. As indicated above, middle T antigen also contains an N-terminal J domain.

Another activity common to polyomavirus middle T and small t antigens, and also to SV40 small t antigen, is binding of the catalytic (36-kDa) and regulatory (63-kDa) subunits of protein phosphatase 2A (PP2A) (141, 179). The PP2A binding site is not part of the large T antigens. The small t antigens are able to activate transcription of various targets, including E2F-responsive promoters (115). Whether the PP2A interaction contributes to the transactivation- and/or transformation-enhancing activity of small t antigen is not yet clear. PP2A binding may be another function that is conserved among the DNA tumor virus families,

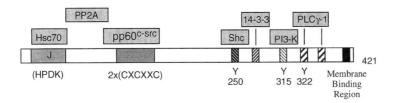

Figure 9 Binding activities of murine polyomavirus middle T antigen. The bar represents the 421-residue open reading frame of murine polyomavirus middle T antigen. Major functional regions are indicated by shading. The amino-terminal 79 residues comprise a J domain motif, which mediates interaction with the cellular hsc70 protein. The approximate binding sites for PP2A and various cell signaling molecules bound by middle T antigen are also indicated. Middle T antigen binds c-*src* (and other *src* family members including c-*yes* and c-*fyn*). c-*src* binding is mediated by two Cys-X-Cys-X-X-Cys motifs (2X-CXCXXC). Binding activates the tyrosine kinase activity of the *src* family members and promotes tyrosine phosphorylation at sites indicated by "Y." Phosphorylation at these sites in turn promotes the binding of other signal transfer proteins, including the Shc-Grb2 complex, phosphatidylinositol 3-kinase (PI3-K), and PLCγ-1. Middle T antigen also binds 14-3-3 proteins. Binding of these proteins promotes cell growth signals in a manner analogous to the binding of growth factors to plasma membrane receptors (see also Fig. 6). The polyomavirus small t antigen shares the J domain sequence with the other T antigens. Additional sequence shared with middle T antigen includes the PP2A binding region, but does not include functions downstream of PP2A binding (see also Fig. 3). The more recently discovered tiny t antigen also shares the J domain sequence and associated hsc70 binding function.

though, as one of the adenovirus early region gene products also binds PP2A. Adenovirus early region 4 is differentially spliced to produce several open reading frames encoding proteins that are not required for the known transforming activities of adenovirus, but do modulate E1A effects on host cell transcription activity. E4 open reading frame 4 (E4-ORF4) binds to and activates the serine/threonine-specific phosphatase activity of PP2A (98) and can inhibit E1A-mediated activation of certain target genes (13, 98).

The identification and analysis of the protein interactions of middle T antigen has led to an increased understanding of the flow of regulatory signals within the cell. From its position in the plasma membrane, middle T antigen stimulates cell signals much like an oncogenic growth factor (reviewed in reference 125). These signals proceed through *ras*- and MAP kinase-mediated pathways, culminating in the downstream activation of specific transcription factors. SV40 and adenovirus are not known to encode proteins that interact with the cellular family of Src tyrosine kinase signaling molecules. However, lymphocytes persistently infected with the Epstein-Barr human DNA tumor virus (EBV) express a viral protein, LMP2A, that associates with Src family kinases (18, 116). It is another curious distinction that polyomavirus does not appear to target p53 directly. It is likely that the pleiotropic mitogenic signals triggered by middle T antigen activate com-

pensating pathways. Notably, coexpression of adenovirus E1A with either activated *ras* or polyoma middle T antigen has an effect in primary baby rat kidney
cells similar to that of coexpression with E1B, i.e., to promote the outgrowth of
fully established cells (151).

E1A C-Terminal Binding Protein

E1A has offered another view of cell growth controls through the activity of its
C-terminal region. E1A mRNAs are derived from two exons. Early studies had
shown that the protein region encoded by the first exon is sufficient for the ability
of E1A to induce proliferation in primary cells. The binding sites for the pRb and
p300 families are entirely contained within the first exon. A closer look at the
colonies of proliferating cells induced by E1A constructs expressing only the first
exon, however, revealed that they grow more aggressively than those expressing
wild-type E1A. They are also tumorigenic in rodents even though cells transformed by the E1A gene of adenovirus serotype 5 are not ordinarily tumorigenic
in rodents (43, 113, 169). These properties suggested that wild-type E1A normally
restrains development of a fully transformed phenotype, and that this restraint
is dependent on second exon sequences. This interpretation is supported by the
finding that E1A can reverse aspects of the transformed phenotype in cells transformed by other oncogenes, or in human tumor cell lines (60, 147). Elevated levels
of metalloprotease gene expression have been found in cells transformed with E1A
constructs lacking the second exon (113). The tumorigenesis modulating activity of
E1A has been mapped to a conserved amino acid motif near the C terminus and
has been correlated with E1A binding to a cellular protein designated C-terminal
binding protein (CtBP) (Fig. 5) (154). The molecular function of the CtBP/E1A
complex is not yet known, but it is an intriguing possibility that this association
controls expression of products that determine aspects of cellular attachment and
metastasis. Presumably the maintenance of a latent infection is favored if the virus
restrains infected cells from detaching and developing metastatic properties.

VIRUS/HOST INTERACTIONS

The human adenoviruses, SV40, and murine polyomavirus have each contributed
importantly to our understanding of the molecular events that underlie carcinogenesis. However, cancer as it occurs in the organism involves the interplay of
more than the changes that affect cell cycle regulatory molecules such as p53 and
pRb. It is a striking property of the DNA tumor viruses that some are tumorigenic
in their natural hosts, while others with similar molecular effects are not. Among
the viruses considered here, adenovirus is the only one that naturally infects humans. Adenovirus interactions with its human hosts are of considerable importance for at least two reasons. One is that adenovirus targets the major tumor
suppressors pRb and p53 very effectively, yet, in contrast to viruses such as HPV,
has never been associated with human cancer (e.g., reference 120). The second
reason for interest in adenovirus/human interactions is that the biological properties of adenovirus have made it an attractive gene therapy vector (reviewed in
reference 134). It is likely that increasingly more engineered virus strains will be

introduced into patients, including persons with cancer and/or immune deficiencies.

Human adenoviruses of a variety of serotypes infect the population with high frequency and can establish a persistent latent infection. Some human adenovirus strains are tumorigenic in rodents, but none has been associated with human cancer, even in immunocompromised hosts. The human counterparts of SV40 virus, JC and BK viruses, are widespread in the population but are not associated with human cancer although they do have disease manifestations in immunocompromised individuals. A notable few human DNA viruses, in particular HPV, EBV, and hepatitis B virus, are associated with human carcinogenesis with measurable frequencies. HPV includes strains that range from low risk to high risk for progression to cervical carcinoma. The degree of risk of cancer resulting from infection with the genital isolates correlates well with affinity for pRb association (80). However, the E1A gene products from most adenovirus serotypes bind pRb as stably as any of the HPV E7 gene products, so this property alone does not explain the difference in carcinogenesis potential. In addition, the E7 protein of HPV-1, a virus causing benign plantar warts, binds pRb as well as the E7 proteins from oncogenic viruses (92). At present we simply do not know why some viruses with the ability to drive the cell cycle contribute to carcinogenesis while others do not. The tissue tropism of the individual virus may be one contributing factor. A clearer understanding of the interplay between viral and host factors is essential to an eventual understanding of the general role of viruses in human cancer. Recent work on adenovirus has begun to establish some of the ways that the virus eludes host responses to maintain a persistent infection.

The Host Immune Response

Adenovirus encodes products capable of protecting infected cells from lysis mediated by tumor necrosis factor (TNF) and cytotoxic T lymphocytes (CTLs). The virus also encodes products that counteract the antiviral effects of interferons. The host mechanisms that attack cells infected with a foreign agent such as a virus are similar to mechanisms that guard against the emergence of cells with tumorigenic potential. Thus, virus studies offer insight into the means by which tumor cells can evade host responses as well as the means by which potentially tumorigenic viral agents can persist. In addition, attempts to use adenovirus as a vector for gene therapy against diseases including malignancies must consider the immune responses active at the tumor site and aspects of the host response that may affect the ability of the virus to deliver exogenous gene products to their target sites.

Much of the ability of adenovirus to guard against the TNF- and CTL-mediated response resides in the products encoded by the E3 region. Because E3 function is largely related to the host response at the organism level, it is dispensable to the virus life cycle in tissue culture and has been deleted in many common laboratory isolates. Whether to include it in vectors designed for gene therapy is an important strategic consideration. One mechanism by which the E3 region may protect infected cells from lysis by CTLs is by reducing surface expression of major histocompatibility complex (MHC) class I antigens. Recognition of infected cells by CTLs requires that viral peptide antigens be displayed on the cell surface in a complex with an MHC class I antigen. The E3-encoded 19-kDa glycoprotein (E3 19K), the most abundant of the E3-encoded products, localizes to the membrane of the endoplasmic reticulum and binds to the peptide binding domain of MHC class I antigens, preventing the transport of MHC antigens to the cell surface (31). Reduced expression of class I antigens on the cell surface is expected to inhibit the ability of CTLs to recognize infected cells. Experiments to test this mechanism in the commonly studied human viruses have relied mostly on infection of rodents. It is not yet clear how significant reduced surface expression of MHC class I antigens is to viral persistence in the natural host. Adenovirus does not reduce surface expression of MHC class I antigens in all laboratory cell lines tested, and mouse adenovirus, which establishes a persistent infection in its natural host, does not appear to reduce surface levels of MHC antigens (102).

Adenovirus can also counteract the cytolytic effects of TNF-α. This activity appears to be mediated by E3 and E1B gene products. Several E3 proteins (14.7K, 10.4K, and 14.5K) are involved in this pathway. The 14.7K protein (66), or a heterodimeric complex between the 14.5K protein and the 10.4K protein (67), can protect infected cells against TNF-α-mediated cytolysis. Although the mechanism of action needs to be fully determined, the anticytolytic effects of the E3 proteins may be a consequence of their ability to inhibit the release of arachidonic acid following the stimulation of phospholipase A2 by TNF-α (41). Inhibition of arachidonic acid release correlates with reduced cytokine production and inhibition of the TNF-α-induced inflammatory response as well as reduced cytolysis. The E1B gene products can also protect against some effects of TNF-α (78), which kills some cells by necrosis and others by apoptosis. Some of these effects can be countered by the E1B 19K protein, which can protect human cells in culture from TNF-α-induced cytolysis (65, 184). These various virus products do not all prevent TNF-α-mediated cytolysis in all cell types examined, which perhaps explains the need for partially redundant mechanisms. The ability of the virus to counteract the apoptosis response may be essential to the maintenance of persistent infection. The E3 transcription unit encodes at least three additional protein products: 11.6K, 6.7K, and 12.5K. The 11.6K product accumulates to high levels at very late stages of infection and promotes cell lysis at the appropriate time in the virus replication cycle, allowing mature virus to be released from infected cells (173). The roles of the E3 6.7K and 12.5K proteins are not yet known. Their significance may become more apparent with increasing use of adenovirus vectors in vaccine and gene therapy. The functions of the adenovirus E3 transcription region are reviewed in reference 187.

Yet another adenovirus defense against the host immune response is encoded by the VA (virus-associated) RNA transcription unit, which antagonizes the host interferon response. The VA RNAs are not translated products. They comprise two extremely abundant short RNA species, VA RNA$_I$ and VA RNA$_{II}$, transcribed from the major late region of the virus genome. Interferons induce a protein kinase that is activated by double-stranded RNA (which is often formed during virus infections due to transcription off both strands of the viral genome). The kinase shuts down protein synthesis in the infected cell by phosphorylating an elongation factor, eIF-2, leading to inhibition of protein synthesis. Activation of the kinase depends on its binding to double-stranded RNA as a dimer and undergoing autophosphorylation. The VA RNAs are short molecules that form double-stranded hairpin structures that bind the kinase but prevent autophosphorylation. In this situation, the kinase is not activated and protein synthesis continues in the infected cell (reviewed in reference 122).

EBV uses a very similar strategy, encoding abundantly expressed small RNA molecules that bind and inhibit activation of the interferon-induced RNA-dependent protein kinase (51; reviewed in reference 29).

Adenovirus Applications in Gene Therapy

The adenovirus genome can be manipulated readily using recombinant DNA techniques. The virus is also able to infect a wide variety of tissues and to infect nondividing cells. These properties have led to the use of adenoviruses as vectors for expression of heterologous genes, as delivery systems for gene therapeutics, and for recombinant vaccines. Adenovirus is not associated with serious illness in immunocompetent individuals. For more than two decades unattenuated adenoviruses have been used successfully as oral vaccines, offering evidence of the safety of such vectors as potential gene therapy tools. Since the initial construction and in vitro expression of recombinant adenovirus vectors containing the SV40 T antigen and herpes simplex virus thymidine kinase genes and, in 1990, the ground-breaking use of adenoviruses to deliver the gene for ornithine transcarbamylase (OTC) in mice harboring a genetic OTC defect (166), adenovirus has been used to deliver numerous genes in in vivo preclinical models.

In various anticancer strategies now being tested, adenovirus will be used to restore expression of tumor suppressors such as p53, pRb, and p21 to tumor cells. Other strategies that may prove useful in anticancer therapy rely on virus products alone. The observation that the E1A gene products have tumor suppression activity in numerous transformed lines (discussed under E1A C-Terminal Binding Protein, above) raises the possibility of the use of E1A expression in cancer therapy. Indeed this property, studied so far only in the E1A products of human Ad2/5 strains, may be part of the reason that adenovirus is not normally tumorigenic in its human host.

Another strategy, now being explored in phase I/II human trials, uses recombinant adenovirus to kill tumor cells that lack functional p53. Adenovirus needs to inactivate p53 to replicate efficiently in culture. The basis of the antitumor strategy is the prediction that a mutant virus lacking E1B would be unable to grow in normal cells, but would replicate in, and subsequently lyse, tumor cells lacking p53. In this scenario the virus would replicate and spread through the tumor, killing cells by lytic infection. This could constitute a very specific antitumor agent and one that alleviates one of the most difficult problems in gene therapy strategies, i.e., how to deliver the active gene to all the cells of a tumor. The virus has performed effectively in tissue culture and mouse tumor models. Normal human cells have proved highly resistant to killing by the virus, which efficiently kills human tumor cells lacking functional p53. The virus also shows antitumor activity in nude mouse-human tumor xenografts, an effect that can be increased when virus treatment is combined with standard chemotherapeutic agents such as cisplatin (81).

CONCLUSIONS

It would be hard to overstate the importance of the study of the small DNA tumor viruses to the understanding of human cancer. These viruses provided the first clear view of the functional links between multiple disparate families of cell cycle-regulating molecules. They pointed directly to p53 and pRb as critical determinants of cellular ability to resist transformation to a tumorigenic phenotype. They offered a basis for an understanding of the contribution of clinically relevant viruses to human cancer and have revealed potential molecular bases for the distinction between related virus strains that differ in their oncogenicity.

Acknowledgments
We thank Dennis McCance for his encouragement during the writing of this review. This work was supported by Public Health Service grants CA 53592 (E.M.) and CA09214 (G.R.B., Jr.).

REFERENCES

1. **Arany, Z., W. R. Sellers, D. M. Livingston, and R. Eckner.** 1994. E1A-associated p300 and CREB-associated CBP belong to a conserved family of coactivators. *Cell* **77**:799–800.
2. **Auger, K. R., C. L. Carpenter, S. E. Shoelson, H. Piwnica-Worms, and L. C. Cantley.** 1992. Polyomavirus middle T antigen-pp60c-src complex associates with purified phosphatidylinositol 3-kinase in vitro. *J. Biol. Chem.* **267**:5408–5415.
3. **Avantagiatti, M. L., V. Ogryzko, K. Gardner, A. Giordano, A. S. Levine, and K. Kelly.** 1997. Recruitment of p300/CBP in p53 dependent pathways. *Cell* **89**:1175–1184.
4. **Bagchi, S., P. Raychaudhuri, and J. R. Nevins.** 1990. Adenovirus E1A proteins can dissociate heteromeric complexes involving the E2F transcription factor: a novel mechanism for E1A-transactivation. *Cell* **62**:659–669.
5. **Bandara, L. R., J. P. Adamczewski, T. Hunt, and N. B. La Thangue.** 1991. Cyclin A and the retinoblastoma gene product complex with a common transcription factor. *Nature* **352**:249–251.
6. **Bandara, L. R., and N. B. La Thangue.** 1991. Adenovirus E1A prevents the retinoblastoma gene product from complexing with a cellular transcription factor. *Nature* **351**: 494–507.
7. **Barbeau, D., R. Charbonneau, S. G. Whalen, S. T. Bayley, and P. E. Branton.** 1994. Functional interactions within adenovirus E1A protein complexes. *Oncogene* **9**: 359–373.
8. **Bayley, S. T., and J. S. Mymryk.** 1994. Adenovirus E1A proteins and transformation (review). *Internat. J. Oncol.* **5**:425–444.
9. **Beijersbergen, R. L., and R. Bernards.** 1996. Cell cycle regulation by the retinoblastoma family of growth inhibitory proteins. *Biochim. Biophys. Acta* **1287**:103–120.
10. **Berk, A. J., and P. A. Sharp.** 1977. Sizing and mapping of early adenovirus mRNAs by gel electrophoresis of S1 endonuclease-digested hybrids. *Cell* **12**:721–732.
11. **Bikel, I., X. Montano, M. E. Agha, M. Brown, M. McCormack, J. Boltax, and D. M. Livingston.** 1987. SV40 small t antigen enhances the transformation activity of limiting concentrations of SV40 large T antigen. *Cell* **48**:321–330.
12. **Bolen, J. B., C. J. Thiele, M. A. Israel, W. Yonemoto, L. A. Lipsich, and J. S. Brugge.** 1984. Enhancement of cellular src gene product associated tyrosyl kinase activity following polyomavirus infection and transformation. *Cell* **38**:767–777.
13. **Bondesson, M., K. Ohman, M. Manervik, S. Fan, and G. Akusjarvi.** 1996. Adenovi-

rus E4 open reading frame 4 protein autoregulates E4 transcription by inhibiting E1A transactivation of the E4 promoter. *J. Virol.* **70**:3844–3851.

14. **Bouck, N., N. Beales, T. Shenk, P. Berg, and G. di Mayorca.** 1978. New region of the simian virus 40 genome required for efficient viral transformation. *Proc. Natl. Acad. Sci. USA* **75**:2473–2477.

15. **Boulakia, C. A., G. Chen, F. W. Ng, J. G. Teodoro, P. E. Branton, D. W. Nicholson, G. G. Poirier, and G. C. Shore.** 1996. Bcl-2 and adenovirus E1B 19 kDA protein prevent E1A-induced processing of CPP32 and cleavage of poly(ADP-ribose) polymerase. *Oncogene* **12**(3):529–535.

16. **Boyd, J. M., S. Malstrom, T. Subramanian, L. K. Venkatesh, U. Schaeper, B. Elangovan, C. D'Sa-Eipper, and G. Chinnadurai.** 1994. Adenovirus E1B 19 kDa and Bcl-2 proteins interact with a common set of cellular proteins. *Cell* **79**:341–351.

17. **Buchkovich, K., L. A. Duffy, and E. Harlow.** 1989. The retinoblastoma protein is phosphorylated during specific phases of the cell cycle. *Cell* **58**:1097–1105.

18. **Burkhardt, A. L., J. B. Bolen, E. Kieff, and R. Longnecker.** 1992. An Epstein-Barr virus transformation-associated membrane protein interacts with *src* family tyrosine kinases. *J. Virol.* **66**:5161–5167.

19. **Campbell, K. S., K. P. Mullane, I. A. Aksoy, H. Stubdal, J. Zalvide, J. M. Pipas, P. A. Silver, T. M. Roberts, B. S. Schaffhausen, and J. A. DeCaprio.** 1997. DnaJ/hsp40 chaperone domain of SV40 large T antigen promotes efficient viral DNA replication. *Genes Dev.* **11**:1098–1110.

20. **Campbell, K. S., E. Ogris, B. Burke, W. Su, K. R. Auger, B. J. Druker, B. S. Schaffhausen, T. M. Roberts, and D. C. Pallas.** 1994. Polyoma middle tumor antigen interacts with SHC protein via the NPTY (Asn-Pro-Thr-Tyr) motif in middle tumor antigen. *Proc. Natl. Acad. Sci. USA* **91**:6344–6348.

21. **Chakravarti, D., V. La Morte, M. Nelson, T. Nakajima, I. G. Schulman, H. Juguilon, M. Montminy, and R. Evans.** 1996. Role of CBP/p300 in nuclear receptor signalling. *Nature* **383**:99–103.

22. **Chellappan, S. P., S. Hiebert, M. Mudryj, J. R. Horowitz, and J. R. Nevins.** 1991. The E2F transcription factor is a cellular target for the RB protein. *Cell* **65**:1053–1061.

23. **Chellappan, S., V. B. Kraus, B. Kroger, K. Munger, P. M. Howley, W. C. Phelps, and J. R. Nevins.** 1991. Adenovirus E1A, simian virus 40 tumor antigen, and human papillomavirus E7 protein share the capacity to disrupt the interaction between transcription factor E2F and the retinoblastoma gene product. *Proc. Natl. Acad. Sci USA* **89**:4549–4553.

24. **Chen, P. L., P. Scully, J. Y. Shew, J. Y. Wang, and W. H. Lee.** 1989. Phosphorylation of the retinoblastoma gene product is modulated during the cell cycle and cellular differentiation. *Cell* **58**:1193–1198.

25. **Cheng, E. H., J. Nicholas, D. S. Bellows, G. S. Hayward, H. G. Guo, M. S. Reitz, and J. M. Hardwick.** 1997. A Bcl-2 homolog encoded by Kaposi sarcoma-associated virus, human herpesvirus 8, inhibits apoptosis but does not heterodimerize with Bax or Bak. *Proc. Natl. Acad. Sci. USA* **94**:690–694.

26. **Cheng, S. H., R. Harvey, P. C. Espino, K. Samba, T. Yamamoto, K. Toyoshima, and A. E. Smith.** 1988. Peptide antibodies to the human c-fyn gene product demonstrate pp59$^{c\text{-fyn}}$ is capable of complex formation with the middle-T antigen of polyomavirus. *EMBO J.* **7**:3845–3855.

27. **Chittenden, T., D. M. Livingston, and W. G. Kaelin, Jr.** 1991. The T/E1A-binding domain of the retinoblastoma product can interact selectively with a sequence-specific binding protein. *Cell* **65**:1073–1082.

28. **Chrivia, J. C., R. P. S. Kwok, N. Lamb, M. Hagiwara, M. R. Montminy, and R. H.**

Goodman. 1993. Phosphorylated CREB binds specifically to the nuclear protein CBP. *Nature* **365:**855–859.

29. **Clemens, M. J., K. G. Laing, I. W. Jeffrey, A. Schofield, T. V. Sharp, A. Edia, V. Matys, M. C. James, and V. J. Tilleray.** 1994. Regulation of the interferon-inducible eIF-2 alpha protein kinase by small RNAs. *Biochimir* **76:**770–778.

30. **Cole, C.** 1996. *Polyomavirinae:* the viruses and their replication, p. 917–946. *In* B. N. Fields, D. M. Knipe, and P. M. Howley (ed.), *Fundamental Virology,* 3rd ed. Lippincott-Raven, Philadelphia.

31. **Conzen, S. D., C. A. Snay, and C. N. Cole.** 1997. Identification of a novel antiapoptotic functional domain in simian virus 40 large T antigen. *J. Virol.* **71:**4536–4543.

32. **Courtneidge, S. A., and A. E. Smith.** 1983. Polyoma virus transforming protein associates with the product of the c-src cellular gene. *Nature* **303:**435–439.

33. **Courtneidge, S. A., and A. E. Smith.** 1984. The complex of polyoma virus middle-T antigen and pp60^{c-src}. *EMBO J.* **3:**585–591.

34. **Cox, J. H., J. R. Bennink, and J. W. Yewdell.** 1991. Retention of adenovirus E19 glycoprotein in the endoplasmic reticulum is essential to its ability to block antigen presentation. *J. Exp. Med.* **174:**1629–1637.

35. **Cress, W. D., and J. R. Nevins.** 1996. Use of the E2F transcription factor by DNA tumor virus regulatory proteins. *Curr. Top. Microbiol. Immunol.* **208:**63–78.

36. **Crook, T., J. A. Tidy, and K. H. Vousden.** 1991. Degradation of p53 can be targeted by HPV E6 sequences distinct from those required for p53 binding and trans-activation. *Cell* **67:**547–556.

37. **Dallas, P. B., P. Yaciuk, and E. Moran.** 1997. Characterization of monoclonal antibodies raised against p300: both p300 and CBP are present in intracellular TBP complexes. *J. Virol.* **71:**1726–1731.

38. **DeCaprio, J. A., J. W. Ludlow, J. Figge, J. Y. Shew, C. M. Huang, W.-H. Lee, E. Marsilio, E. Paucha, and D. M. Livingston.** 1988. SV40 large tumor antigen forms a specific complex with the product of the retinoblastoma-susceptibility gene. *Cell* **54:**275–283.

39. **DeCaprio, J. A., J. W. Ludlow, D. Lynch, Y. Furukawa, J. Griffin, H. Piwnica-Worms, C. M. Huang, and D. M. Livingston.** 1989. The product of the retinoblastoma susceptibility gene has properties of a cell cycle regulatory element. *Cell* **58:**1085–1095.

40. **Dilworth, S. M., C. E. Brewster, M. D. Jones, L. Lanfrancone, G. Pelicci, and P. G. Pelicci.** 1994. Transformation by polyoma virus middle T-antigen involves the binding and tyrosine phosphorylation of Shc. *Nature* **367:**87–90.

41. **Dimitrov, T., P. Krajcsi, T. W. Hermiston, A. E. Tollefson, M. Hannink, and W. S. Wold.** 1997. Adenovirus E3-10.4K/14.5K protein complex inhibits tumor necrosis factor-induced translocation of cytosolic phospholipase A2 to membranes. *J. Virol.* **71:**2830–2837.

42. **Doerfler, W.** 1968. The fate of the DNA of adenovirus type 12 in baby hamster kidney cells. *Proc. Natl. Acad. Sci. USA* **60:**636–643.

43. **Douglas, J. L., S. Gopalakrishnan, and M. P. Quinlan.** 1991. Modulation of transformation of primary epithelial cells by the second exon of the Ad5 E1A12S gene. *Oncogene* **6:**2093–2103.

44. **Dunaief, J. L., B. E. Strober, S. Guha, P. A. Khavari, K. Alin, J. Luban, M. Begemann, G. R. Crabtree, and S. P. Goff.** 1994. The retinoblastoma protein and BRG-1 form a complex and cooperate to induce cell cycle arrest. *Cell* **79:**119–130.

45. **Dyson, N., and E. Harlow.** 1992. Adenovirus E1A targets key regulators of cell proliferation. *Cancer Surv.* **12:**161–195.

46. **Dyson, N., P. M. Howley, K. Munger, and E. Harlow.** 1989. The human papillomavi-

rus-16 E7 oncoprotein is able to bind to the retinoblastoma gene product. *Science* **243:** 934–937.

47. **Eckner, R.** 1996. p300 and CBP as transcriptional regulators and targets of oncogenic events. *Biol. Chem.* **377:**685–688.

48. **Eckner, R., M. E. Ewen, D. Newsome, J. A. DeCaprio, J. B. Lawrence, and D. M. Livingston.** 1994. Molecular cloning and functional analysis of the adenovirus E1A-associated 300-kd protein (p300) reveals a protein with properties of a transcriptional adaptor. *Genes Dev.* **8:**869–884.

49. **Eddy, B. E., G. S. Borman, G. E. Grubbs, and R. D. Young.** 1962. Identification of the oncogenic substance in rhesus monkey cell kidney cell cultures as simian virus 40. *Virology* **17:**65–75.

50. **Egan, C., T. N. Jelsma, J. A. Howe, S. T. Bayley, B. Ferguson, and P. E. Branton.** 1988. Mapping of cellular protein-binding sites on the products of early region 1A of human adenovirus type 5. *Mol. Cell. Biol.* **8:**3955–3959.

51. **Elia, A., K. G. Laing, A. Schofield, V. J. Tilleray, and M. J. Clemens.** 1996. Regulation of the double-stranded RNA-dependent protein kinase PKR by RNAs encoded by a repeated sequence in the Epstein-Barr virus genome. *Nucleic Acids Res.* **24:**4471–4478.

52. **Eliyahu, D., N. Goldfinger, O. Pinhasi-Kimhi, G. Shaulsky, Y. Shurnik, N. Aral, V. Rotter, and M. Oren.** 1988. Meth A fibrosarcoma cells express two transforming mutant p53 species. *Oncogene* **3:**313–321.

53. **Eliyahu, D., A. Raz, P. Gruss, D. Givol, and M. Oren.** 1984. Participation of p53 cellular tumor antigen in transformation of normal embryonic cells. *Nature* **312:** 646–649.

54. **Ewen, M. E., B. Faha, E. Harlow, and D. M. Livingston.** 1992. Interaction of p107 with cyclin A independent of complex formation with viral oncoproteins. *Science* **255:**85–87.

55. **Ewen, M. E., Y. G. Xing, J. B. Lawrence, and D. M. Livingston.** 1991. Molecular cloning, chromosomal mapping, and expression of the cDNA for p107, a retinoblastoma gene product-related protein. *Cell* **66:**1155–1164.

56. **Faha, B., M. E. Ewen, L.-H. Tsal, D. M. Livingston, and E. Harlow.** 1992. Interaction between human cyclin A and adenovirus E1A-associated p107 protein. *Science* **255:** 87–90.

57. **Faha, B., E. Harlow, and E. Lees.** 1993. The adenovirus E1A-associated kinase consists of cyclin E-p33^{cdk2} and cyclin A-p33^{cdk2}. *J. Virol.* **67:**2456–2465.

58. **Farrow, S. N., J. H. White, I. Martinou, T. Raven, K. T. Pun, C. J. Grinham, J. C. Martinou, and R. Brown.** 1995. Cloning of a bcl-2 homologue by interaction with adenovirus E1B 19K. *Nature* **374:**731–733.

59. **Finlay, C. A., P. W. Hinds, and A. J. Levine.** 1989. The p53 proto-oncogene can act as a suppressor of transformation. *Cell* **57:**1083–1093.

60. **Frisch, S. M.** 1994. E1A induces the expression of epithelial characteristics. *J. Cell. Biol.* **127:**1085–1096.

61. **Gallimore, P. H., P. A. Sharp, and J. Sambrook.** 1974. Viral DNA in transformed cells. II. A study of the sequences of adenovirus 2 DNA in nine lines of transformed rat cells using specific fragments of the viral genome. *J. Mol. Biol.* **89:**49–72.

62. **Gardner, S. D., A. M. Field, D. V. Coleman, and B. Hulme.** 1971. New human papovavirus (B. K.) isolated from urine after renal transplantation. *Lancet* **1:** 1253–1257.

63. **Giordano, A., C. McCall, P. Whyte, and B. R. Franza.** 1991. Human cyclin A and the retinoblastoma protein interact with similar but distinguishable sequences in the adenovirus E1A gene product. *Oncogene* **6:**481–486.

64. **Girardi, A. J., B. H. Sweet, V. B. Slotnick, and M. R. Hilleman.** 1962. Development

of tumors in hamsters inoculated in the neo-natal period with vacuolating virus, SV40. *Proc. Soc. Exp. Biol. Med.* **109:**649–660.

65. **Gooding, L. R., L. Aquino, P. J. Duerksen-Hughes, D. Day, T. M. Horton, S. Yei, and W. S. M. Wold.** 1991. The E1B-19K protein of group C adenoviruses prevents cytolysis by tumor necrosis factor of human cells but not mouse cells. *J. Virol.* **65:** 3083–3094.

66. **Gooding, L. R., L. W. Elmore, A. E. Tollefson, H. A. Brady, and W. S. M. Wold.** 1988. A 14,700 MW protein from the E3 region of adenovirus inhibits cytolysis by tumor necrosis factor. *Cell* **53:**341–346.

67. **Gooding, L. R., R. Ranheim, A. E. Tollefson, L. Aquino, P. Duerksen-Hughes, T. M. Horton, and W. S. M. Wold.** 1991. The 10,400- and 14,500-dalton proteins encoded by region E3 of adenovirus function together to protect many but not all mouse cell lines against lysis by tumor necrosis factor. *J. Virol.* **65:**4114–4123.

68. **Gottlieb, T. M., and M. Oren.** 1996. p53 in growth and control of neoplasia. *Biochim. Biophys. Acta* **1287:**77–102.

69. **Graham, F. L., P. J. Abrahams, C. Mulder, H. L. Heijneker, S. O. Warnaar, F. A. De Vries, W. Fiers, and A. J. van der Eb.** 1975. Studies on in vitro transformation by DNA and DNA fragments of human adenoviruses and simian virus 40. *Cold Spring Harbor Symp. Quant. Biol.* **39:**637–650.

70. **Graham, F. L., A. J. van der Eb, and H. L. Heijneker.** 1974. Size and location of the transforming region in human adenovirus type 5 DNA. *Nature* **251:**687–691.

71. **Gross, L.** 1953. A filterable agent, recovered from Ak leukemic extracts, causing salivary gland carcinomas in C3H mice. *Proc. Soc. Exp. Biol. Med.* **83:**414–421.

72. **Gu, W., and R. G. Roeder.** 1997. Activation of p53 sequence-specific DNA binding by acetylation of the p53 C-terminal domain. *Cell* **90:**595–606.

73. **Gu, W., X.-L. Shi, and R. G. Roeder.** 1997. Synergistic activation of transcription by CBP and p53. *Nature* **387:**819–822.

74. **Han, J., P. Sabbatini, and E. White.** 1996. Induction of apoptosis by human Nbk/Bik, a BH3-containing protein that interacts with E1B 19K. *Mol. Cell. Biol.* **16:**5857–5864.

75. **Hannon, G. J., D. Demetrick, and D. Beach.** 1993. Isolation of the Rb-related p130 through its interaction with CDK2 and cyclins. *Genes Dev.* **7:**2378–2391.

76. **Hanstein, B., R. Eckner, J. DiRenzo, S. Halachmi, H. Liu, B. Searcy, R. Kurokawa, and M. Brown.** 1996. p300 is a component of an estrogen receptor coactivator complex. *Proc. Natl. Acad. Sci. USA* **93:**11540–11545.

77. **Harlow, E., P. Whyte, B. R. Franza, Jr., and C. Schley.** 1986. Association of adenovirus early region 1 proteins with cellular polypeptides. *Mol. Cell. Biol.* **6:**1579–1589.

78. **Hashimoto, S., A. Ishii, and S. Yonehara.** 1991. The E1B oncogene of adenovirus confers cellular resistance to cytotoxicity of tumor necrosis factor and monoclonal anti-Fas antibody. *Int. Immunol.* **3:**343–351.

79. **Hawkins, C. J., and D. L. Vaux.** 1997. The role of the Bcl-2 family of apoptosis regulatory proteins in the immune system. (Review.) *Semin. Immunol.* **9:**25–33.

80. **Heck, D. V., C. L. Yee, P. M. Howley, and K. Munger.** 1992. Efficiency of binding the retinoblastoma protein correlates with the transforming capacity of the E7 oncoproteins of the human papillomaviruses. *Proc. Natl. Acad. Sci. USA.* **89:**4442–4446.

81. **Heise, C., A. Sampson-Johannes, A. Williams, F. McCormick, D. D. Von Hoff, and D. H. Kirn.** 1997. ONYX-015, an E1B-attenuated adenovirus, causes tumor-specific cytolysis and antitumoral efficacy that can be augmented by standard chemotherapeutic agents. *Nat. Med.* **3:**639–645.

82. **Helin, K., J. A. Lees, M. Vidal, N. Dyson, E. Harlow, and A. Fatteay.** 1992. A cDNA encoding a pRB binding protein with the properties of the transcription factor E2F. *Cell* **70:**337–350.

83. **Henderson, S., D. Huen, M. Rowe, C. Dawson, G. Johnson, and A. Rickinson.** 1993. Epstein-Barr virus-coded BHRF1 protein, a viral homologue of Bcl-2, protects human B cells from programmed cell death. *Proc. Natl. Acad. Sci. USA* **90:**8479–8483.

84. **Hollstein, M., D. Sidransky, B. Vogelstein, and C. C. Harris.** 1991. p53 mutations in human cancers. *Science* **253:**49–53.

85. **Horak, I. D., T. Kawakami, F. Gregory, K. C. Robbins, and J. B. Bolen.** 1989. Association of p60^{fyn} with middle tumor antigen in murine polyomavirus-transformed rat cells. *J. Virol.* **63:**2343–2347.

86. **Houweling, A., P. J. van den Elsen, and A. J. van der Eb.** 1980. Partial transformation of primary rat cells by the leftmost 4.5% fragment of adenovirus 5 DNA. *Virology* **105:**537–550.

87. **Howe, J. A., and S. T. Bayley.** 1992. Effects of Ad5 E1A mutant viruses on the cell cycle in relation to binding of cellular proteins including the retinoblastoma protein and cyclin A. *Virology* **186:**15–24.

88. **Howley, P. M., A. J. Levine, F. P. Li, D. M. Livingston, and A. S. Rabson.** 1991. Lack of SV40 DNA in tumors from scientists working with SV40 virus. *N. Engl. J. Med.* **324:**494.

89. **Huibregtse, J. M., M. Scheffner, and P. M. Howley.** 1991. A cellular protein mediates association of p53 with the E6 oncoprotein of human papillomavirus type 16 or 18. *EMBO J.* **10:**4129–4135.

90. **Huibregtse, J. M., M. Scheffner, and P. M. Howley.** 1993. Cloning and expression of the cDNA for E6-AP, a protein that mediates the interaction of the human papillomavirus E6 oncoprotein with p53. *Mol. Cell. Biol.* **13:**775–784.

91. **Huibregtse, J. M., M. Scheffner, and P. M. Howley.** 1993. Localization of the E6-AP regions that direct human papillomavirus E6 binding, association with p53, and ubiquitination of associated proteins. *Mol. Cell Biol.* **13:**4918–4927.

92. **Ibaraki, T., M. Satake, N. Kurai, M. Ichijo, and Y. Ito.** 1993. Transactivating activities of the E7 genes of several types of human papillomavirus. *Virus Genes* **7(2):**187–192.

93. **Jansen-Durr, P.** 1996. How viral oncogenes make the cell cycle. *Trends Genet.* **12:** 270–275.

94. **Jenkins, J. R., K. Rudge, and G. A. Currie.** 1984. Cellular immortalization by a cDNA clone encoding the transformation associated phosphoprotein p53. *Nature* **312:** 651–654.

95. **Kaelin, W. G., Jr., W. Krek, W. R. Sellers, J. A. DeCaprio, F. Ajchenbaum, C. S. Fuchs, T. Chittenden, Y. Li, P. J. Farnham, and M. A. Blanar.** 1992. Expression cloning of a cDNA encoding a retinoblastoma-binding protein with E2F-like properties. *Cell* **70:**351–364.

96. **Kamei, Y. L. Xu, T. Heinzel, J. Torchia, R. Kurokawa, B. Gloss, S.-C. Lin, R. A. Heyman, D. W. Rose, C. K. Glass, and M. Rosenfeld.** 1996. A CBP integrator complex mediates transcriptional activation and AP-1 inhibition by nuclear receptors. *Cell* **85:** 403–414.

97. **Kelley, W. L., and C. Georgopoulos.** 1997. The T/t common exon of simian virus 40, JC, and BK polyomavirus T antigens can functionally replace the J-domain of the *Escherichia coli* DnaJ molecular chaperone. *Proc. Natl. Acad. Sci. USA* **94:**3679–3684.

98. **Kleinberger, T., and T. Shenk.** 1993. Adenovirus E4orf4 protein binds to protein phosphatase 2A, and the complex down regulates E1A-enhanced *junB* transcription. *J. Virol.* **67:**7556–7560.

99. **Knudson, A. G.** 1996. Hereditary cancer: two hits revisited. *J. Cancer Res. Clin. Oncol.* **122:**135–140.

100. **Ko, L. J., and C. Prives.** 1996. p53: puzzle and paradigm. *Genes Dev.* **10:**1054–1072.

101. **Kornbluth, S., M. Sudol, and H. Hanafusa.** 1987. Association of the polyomavirus middle-T antigen with c-yes protein. *Nature* **325:**171–173.

102. **Kring, S. C., and K. R. Spindler.** 1996. Lack of effect of mouse adenovirus type I infection on cell surface expression of major histocompatibility complex class I antigens. *J. Virol.* **70:**5495–5502.

103. **Kwok, R. P. S., J. R. Lundblad, J. C. Chrivia, J. P. Richards, H. P. Bachinger, R. G. Brennan, S. G. E. Roberts, M. R. Green, and R. H. Goodman.** 1994. Nuclear protein CBP is a coactivator for the transcription factor CREB. *Nature* **370:**223–226.

104. **Kypta, R. M., A. Hemming, and S. A. Courtneidge.** 1988. Identification and characterization of p59^{fyn} (a src-like protein tyrosine kinase) in normal and polyomavirus transformed cells. *EMBO J.* **7:**3837–3844.

105. **La Thangue, N. B.** 1994. DRTF1/E2F1: an expanding family of heteromeric transcription factors implicated in cell cycle control. *Trends Biol. Sci.* **19:**108–114.

106. **La Thangue, N. B.** 1996. E2F and the molecular mechanisms of early cell-cycle control. *Bioch. Soc. Trans.* **24(1):**54–59.

107. **Lane, D. P., and L. V. Crawford.** 1979. T antigen is bound to a host protein in SV40-transformed cells. *Nature* **278:**261–263.

108. **Levine, A. J.** 1997. p53, the cellular gatekeeper for growth and division. *Cell* **88:** 323–331.

109. **Li, Y., C. Graham, S. Lacy, A. M. V. Duncan, and P. Whyte.** 1993. The adenovirus E1A-associated 130-Kd protein is encoded by a member of the retinoblastoma gene family and physically interacts with cyclins A and E. *Genes Dev.* **7:**2366–2377.

110. **Lill, N. L., S. R. Grossman, D. Ginsburg, J. DeCaprio, and D. M. Livingston.** 1997. Binding and modulation of p53 by p300/CBP coactivators. *Nature* **387:**823–827.

111. **Lill, N. L., M. J. Tevethia, R. Eckner, D. M. Livingston, and N. Modjtahedi.** 1997. p300 family members associate with the carboxyl terminus of simian virus 40 large tumor antigen. *J. Virol.* **71:**129–137.

112. **Lillie, J. W., M. Green, and M. R. Green.** 1986. An adenovirus E1A protein region required for transformation and transcriptional repression. *Cell* **46:**1043–1051.

113. **Linder, S., P. Popowicz, C. Svensson, H. Marshall, M. Bondesson, and G. Akusjarvi.** 1992. Enhance invasive properties of rat embryo fibroblasts transformed by adenovirus E1A mutants with deletions in the carboxy-terminal exon. *Oncogene* **7:**439–443.

114. **Linzer, D. I., and A. J. Levine.** 1979. Characterization of a 54K dalton cellular SV40 tumor antigen present in SV40-transformed cells and uninfected embryonal carcinoma cells. *Cell* **17:**43–52.

115. **Loeken, M. R.** 1992. Simian virus 40 small t antigen transactivates the adenovirus E2A promoter by using mechanisms distinct from those used by adenovirus E1A. *J. Virol.* **66:**2551–2555.

116. **Longnecker, R., B. Druker, T. M. Roberts, and E. Kieff.** 1991. An Epstein-Barr virus protein associated with cell growth transformation interacts with a tyrosine kinase. *J. Virol.* **65:**3681–3692.

117. **Ludlow, J. W.** 1993. Interactions between SV40 large-tumor antigen and the growth suppressor proteins, pRB and p53. *FASEB J.* **7:**866–871.

118. **Ludlow, J. W., J. A. DeCaprio, C.-M. Huang, W.-H. Lee, E. Paucha, and D. M. Livingston.** 1989. SV40 large T antigen binds preferentially to an underphosphorylated member of the retinoblastoma susceptibility gene product family. *Cell* **56:**57–65.

119. **Ludlow, J. W., J. Shon, J. M. Pipas, D. M. Livingston, and J. A. DeCaprio.** 1990. The retinoblastoma susceptibility gene product undergoes cell cycle-dependent dephosphorylation and binding to and release from SV40 large T. *Cell* **60:**387–396.

120. **Mackey, J. K., P. M. Rigden, and M. Green.** 1976. Do highly oncogenic group A

human adenoviruses cause human cancer? Analysis of human tumors for adenovirus 12 transforming sequences. *Proc. Natl. Acad. Sci. USA* **73:**4657–4661.

121. **Mansur, C. P., B. Marcus, S. Dalal, and E. J. Androphy.** 1995. The domain of p53 required for binding HPV E6 is separable from the degradation domain. *Oncogene* **10:**457–465.

122. **Mathews, M. B.** 1995. Structure, function, and evolution of adenovirus virus-associated RNAs. *Curr. Top. Microbiol. Immunol.* **199:**173–187.

123. **Mayol, X., and X. Grana.** 1997. pRB, p107 and p130 as transcriptional regulators: role in cell growth and differentiation, p. 157–169. *In* L. Meijier, S. Guidet, and M. Phillipe (ed.), *Progress in Cell Cycle Research*, vol. 3. Plenum Press, New York.

124. **Mayol, X., X. Grana, A. Baldi, N. Sang, Q. Hu, and A. Giordano.** 1993. Cloning of a new member of the retinoblastoma gene family (pRb2) which binds to the E1A transforming domain. *Oncogene* **8:**2561–2566.

125. **Messerschmitt, A. S., N. Dunant, and K. Ballmer-Hofer.** 1997. DNA tumor viruses and Src family tyrosine kinases, an intimate relationship. (Minireview.) *Virology* **227:**271–280.

126. **Mietz, J. A., T. Unger, J. M. Huibregtse, and P. M. Howley.** 1992. The transcriptional transactivation function of wild-type p53 is inhibited by SV40 large T-antigen and by HPV-16 E6 oncoprotein. *EMBO J.* **11:**5013–5020.

127. **Mihara, K., X. R. Cao, A. Yen, S. Chandler, B. Driscoll, A. L. Murphree, A. T'Ang, and Y. K. Fung.** 1989. Cell cycle-dependent regulation of phosphorylation of the human retinoblastoma gene product. *Science* **246:**1300–1303.

128. **Momand, J., G. P. Zambetti, D. C. Olson, D. George, and A. J. Levine.** 1992. The mdm-2 oncogene product forms a complex with the p53 protein and inhibits p53-mediated transactivation. *Cell* **69:**1237–1245.

129. **Moran, E.** 1988. A region of SV40 large T antigen can substitute for a transforming domain of the adenovirus E1A gene products. *Nature* **334:**168–170.

130. **Moran, E.** 1994. Cell growth control mechanisms reflected through protein interactions with the adenovirus E1A gene products. *Semin. Virol.* **5:**327–340.

131. **Moran, E., B. Zerler, T. M. Harrison, and M. B. Mathews.** 1986. Identification of separate domains in the adenovirus E1A gene for immortalization activity and the activation of virus early genes. *Mol. Cell. Biol.* **6:**3470–3480.

132. **Mortimer, E. A., Jr., M. L. Lepow, E. Gold, F. C. Robbins, G. J. Burton, and J. F. Fraumeni, Jr.** 1981. Long-term follow-up of persons inadvertently inoculated with SV40 as neonates. *N. Engl. J. Med.* **305:**1517–1518.

133. **Mudryj, M., S. H. Devoto, S. W. Hiebert, T. Hunter, J. Pines, and J. R. Nevins.** 1991. Cell cycle regulation of the E2F transcription factor involves an interaction with cyclin A. *Cell* **65:**1243–1253.

134. **Munaf, A., N. R. Lemoine, and C. J. A. Ring.** 1994. The use of DNA viruses and vectors for gene therapy. *Gene Ther.* **1:**367–384.

135. **Ogryzko, V. V., R. L. Schlitz, V. Russanova, B. H. Howard, and Y. Nakatani.** 1996. The transcriptional coactivators p300 and CBP are histone acetyltransferases. *Cell* **87:**953–959.

136. **Oliner, J. D., K. W. Kinzler, P. S. Meltzer, D. L. George, and B. Vogelstein.** 1992. Amplification of a gene encoding a p53-associated protein in human sarcomas. *Nature* **358:**80–83.

137. **Oltvai, Z. N., C. L. Milliman, and S. J. Korsmeyer.** 1993. Bcl-2 heterodimerizes in vivo with a conserved homolog, Bax, that accelerates programmed cell death. *Cell* **74:**609–619.

138. **Padgett, B. L., D. L. Walker, G. M. ZuRhein, R. J. Eckroade, and B. H. Dessel.** 1971.

Cultivation of a papova-like virus from human brain with progressive multifocal leukoencephalopathy. *Lancet* **1:**1257–1260.

139. **Paggi, M. G., A. Baldi, F. Bonetto, and A. Giordano.** 1996. Retinoblastoma protein family in cell cycle and cancer: a review. *J. Cell. Biochem.* **62:**418–430.

140. **Pallas, D. C., H. Fu, L. C. Haehnel, W. Weller, R. J. Collier, and T. M. Roberts.** 1994. Association of polomavirus middle tumor antigen with 14-3-3 proteins. *Science* **265:**535–537.

141. **Pallas, D. C., L. K. Shahrik, B. L. Martin, S. Jaspers, T. B. Miller, D. L. Brautigan, and T. M. Roberts.** 1990. Polyoma small and middle T antigens and SV40 small t antigen form stable complexes with protein phosphatase 2A. *Cell* **60:**167–176.

142. **Parada, L. F., H. Land, R. A. Weinberg, D. Wolf, and W. Rotter.** 1984. Cooperation between gene encoding p53 tumor antigen and ras in cellular transformation. *Nature* **312:**649–651.

143. **Pearson, G. R., J. Luka, L. Petti, J. Sample, M. Birkenbach, D. Braun, and E. Kieff.** 1987. Identification of an Epstein-Barr virus early gene encoding a second component of the restricted early antigen complex. *Virology* **160:**151–161.

144. **Peterson, C. L., and J. W. Tamkun.** 1995. The SWI-SNF complex: a chromatin remodelling machine. *Trends Biochem. Sci.* **20:**143–146.

145. **Pines, J., and T. Hunter.** 1990. Human cyclin A is adenovirus E1A-associated protein p60 and behaves differently from cyclin B. *Nature* **346:**760–763.

146. **Porter, A. G., P. Ng, and R. U. Janicke.** 1997. Death substrates come alive. *Bioessays* **19:**501–507.

147. **Pozzatti, R., M. McCormick, M. A. Thompson, and G. Khoury.** 1988. The E1a gene of adenovirus type 2 reduces the metastatic potential of ras-transformed rat embryo cells. *Mol. Cell. Biol.* **8:**2984–2988.

148. **Rao, L., M. Debbas, P. Sabbatini, D. Hockenbery, S. Korsmeyer, and E. White.** 1992. The adenovirus E1A proteins induce apoptosis, which is inhibited by the E1B 19-kDa and Bcl-2 proteins. *Proc. Natl. Acad. Sci. USA* **89:**7742–7746.

149. **Raychaudhuri, P., S. Bagchi, S. H. Devoto, V. B. Kraus, E. Moran, and J. R. Nevins.** 1991. Domains of the adenovirus E1A protein required for oncogenic activity are also required for dissociation of E2F transcription factor complexes. *Genes Dev.* **5:** 1200–1211.

150. **Riley, M. I., W. Yoo, N. Y. Mda, and W. R. Folk.** 1997. Tiny T antigen: an autonomous polyomavirus T antigen amino-terminal domain. *J. Virol.* **71:**6068–6074.

151. **Ruley, H. E.** 1983. Adenovirus early region 1A enables viral and cellular transforming genes to transform primary cells in culture. *Nature* **304:**602–606.

152. **Sarnow, P., Y. S. Ho, J. Williams, and A. J. Levine.** 1982. Adenovirus E1B-58kd tumor antigen and SV40 large tumor antigen are physically associated with the same 54kd cellular protein. *Cell* **28:**387–394.

153. **Sawal, E. T., and J. S. Butel.** 1989. Association of a cellular heat shock protein with simian virus 40 large T antigen in transformed cells. *J. Virol.* **63:**3961–3973.

154. **Schaeper, U., J. M. Boyd, S. Verma, E. Uhlmann, T. Subramanian, and G. Chinnadural.** 1995. Molecular cloning and characterization of a cellular phosphoprotein that interacts with a conserved C-terminal domain of adenovirus E1A involved in negative modulation of oncogenic transformation. *Proc. Natl. Acad. Sci. USA* **92:** 10467–10471.

155. **Scheffner, M., B. A. Werness, J. M. Huibregtse, A. J. Levine, and P. M. Howley.** 1990. The E6 oncoprotein encoded by human papillomavirus 16 and 18 promotes the degradation of p53. *Cell* **63:**1129–1136.

156. **Scolnick, D. M., N. Chehab, E. S. Stavridi, M. Lien, L. Caruso, E. Moran, S. L. Berger, and T. D. Halazonetis.** 1997. CREB-binding protein and p300/CBP-associated

factor are transcriptional coactivators of the p53 tumor suppressor protein. *Cancer Res.* **57:**3693–3696.

157. **Segawa, K., and Y. Ito.** 1982. Differential subcellular localization of in vivo phosphorylated and non-phosphorylated middle-sized tumor antigen of polyomavirus and its relationship to middle-sized tumor antigen phosphorylating activity in vitro. *Proc. Natl. Acad. Sci. USA* **79:**6812–6816.

158. **Shan, B., X. Zhu, P.-L. Chen, T. Durfee, Y. Yang, D. Sharp, and W.-H. Lee.** 1992. Molecular cloning of cellular genes encoding retinoblastoma-associated proteins: identification of a gene with properties of the transcription factor E2F. *Mol. Cell. Biol.* **12:**5620–5631.

159. **Sharp, P. A., U. Pettersson, and J. Sambrook.** 1974. Viral DNA in transformed cells. I. A study of the sequences of adenovirus 2 DNA in a line of transformed rat cells using specific fragments of the viral genome. *J. Mol. Biol.* **86:**709–726.

160. **Shenk, T.** 1996. *Adenoviridae:* the viruses and their replication, p. 979–1016. *In* B. N. Fields, D. M. Knipe, and P. M. Howley (ed.), *Fundamental Virology,* 3rd ed. Lippincott-Raven, Philadelphia.

161. **Shope, R. E., and E. W. Hurst.** 1993. Infectious papillomatosis of rabbits; with a note on the histopathology. *J. Exp. Med.* **58:**607–624.

162. **Sidle, A., C. Palaty, P. Dirks, O. Wiggan, M. Kiess, R. M. Gill, A. K. Wong, and P. Hamel.** 1996. Activity of the retinoblastoma family of proteins, PRB, p107, and p130, during cellular proliferation and differentiation. *Crit. Rev. Biochem. Mol. Biol.* **31:**237–271.

163. **Slansky, J. E., and P. J. Farnham.** 1996. Introduction to the E2F family: protein structure and gene regulation. *Curr. Top. Microbiol. Immunol.* **208:**1–30.

164. **Srinivasan, A., A. J. McClellan, J. Vartikar, I. Marks, P. Cantalupo, Y. Li, P. Whyte, K. Rundell, J. L. Brodsky, and J. M. Pipas.** 1997. The amino-terminal transforming region of simian virus large T and small t antigens functions as a J domain. *Mol. Cell. Biol.* **17:**4761–4773.

165. **Stewart, S. E., B. E. Eddy, and N. G. Borgese.** 1958. Neoplasms in mice inoculated with a tumor agent carried in tissue culture. *J. Natl. Cancer Inst.* **20:**1223–1243.

166. **Stratford-Perricaudet, L. D., J. F. Levrero, M. Chasse, M. Perricaudet, and P. Briand.** 1990. Evaluation of the transfer and expression in mice of an enzyme-encoding gene using a human adenovirus vector. *Human Gene Ther.* **1:**241–256.

167. **Stubdal, H., J. Zalvide, K. S. Campbell, C. Schweitzer, T. M. Roberts, and J. A. DeCaprio.** 1997. Inactivation of pRB-related proteins p130 and p107 mediated by the J domain of simian virus 40 large T antigen. *Mol. Cell. Biol.* **17:**4979–4990.

168. **Su, W., W. Liu, B. S. Schaffhausen, and T. M. Roberts.** 1995. Association of polyomavirus middle tumor antigen with phospholipase C-gamma 1. *J. Biol. Chem.* **270:** 12331–12334.

169. **Subramanian, T., M. La Regina, and G. Chinnadurai.** 1989. Enhanced ras oncogene mediated cell transformation and tumorigenesis by adenovirus 2 mutants lacking the C-terminal region of E1a protein. *Oncogene* **4:**415–420.

170. **Subramanian, T., B. Tarodi, and G. Chinnadurai.** 1995. Functional similarity between adenovirus E1B 19-kDa protein and proteins encoded by Bcl-2 proto-oncogene and Epstein-Barr virus BHRF1 gene. *Curr. Top. Microbiol. Immunol.* **199**(Part 1): 153–161.

171. **Tarodi, B., T. Subramanian, and G. Chinnadurai.** 1993. Functional similarity between adenovirus E1B 19K gene and bcl2 oncogene: mutant complementation and suppression of cell death induced by DNA damaging agents. *Int. J. Oncol.* **3:**467–472.

172. **Templeton, D., and W. Eckhart.** 1982. Mutation causing premature termination of the polyoma virus medium T antigen blocks cell transformation. *J. Virol.* **41:**1014–1024.

173. **Tollefson, A. E., A. Scaria, T. W. Hermiston, J. S. Ryers, L. J. Wold, and W. S. Wold.** 1996. The adenovirus death protein (E3-11.6K) is required at very late stages of infection for efficient cell lysis and release of virus from infected cells. *J. Virol.* **70:** 2296–2306.

174. **Treisman, R., A. Cowie, J. Favaloro, P. Jat, and R. Kamen.** 1981. The structures of the spliced mRNAs encoding polyoma virus early region proteins. *J. Mol. Appl. Genet.* **1:** 83–92.

175. **Treisman, R., U. Novak, J. Favaloro, and R. Kamen.** 1981. Transformation of rat cells by an altered polyoma virus genome expressing only the middle-T protein. *Nature* **292:** 595–600.

176. **Trentin, J. J., Y. Yabe, and G. Taylor.** 1962. The quest for human cancer viruses. *Science* **137:** 835–849.

177. **Tsujimoto, Y., and C. M. Croce.** 1986. Analysis of the structure, transcripts, and protein products of bcl-2, the gene involved in human follicular lymphoma. *Proc. Natl. Acad. Sci. USA* **83:** 5214–5418.

178. **Vogelstein, B., and K. W. Kinzler.** 1992. p53 function and dysfunction. *Cell* **70:** 523–526.

179. **Walter, G., R. Ruediger, C. Slaughter, and M. Mumby.** 1990. Association of protein phosphatase 2A with polyoma virus medium tumor antigen. *Proc. Natl. Acad. Sci. USA* **87:** 2521–2525.

180. **Wang, H.-G. H., E. Moran, and P. Yaciuk.** 1995. E1A promotes association between p300 and pRB in multimeric complexes that are required for normal biological activity. *J. Virol.* **69:** 7917–7924.

181. **Weinberg, R. A.** 1995. The retinoblastoma protein and cell cycle control. *Cell* **81:** 323–330.

182. **Werness, B. A., A. J. Levine, and P. M. Howley.** 1990. Association of human papillomavirus types 16 and 18 E6 proteins with p53. *Science* **248:** 76–79.

183. **White, E.** 1996. Life, death, and the pursuit of apoptosis. *Genes Dev.* **10:** 1–15.

184. **White, E., P. Sabbatini, M. Debbas, W. S. Wold, D. I. Kusher, and L. R. Gooding.** 1992. The 19-kilodalton adenovirus E1B transforming protein inhibits programmed cell death and prevents cytolysis by tumor necrosis factor alpha. *Mol. Cell. Biol.* **12:** 2570–2580.

185. **Whyte, P., K. Buchkovich, J. M. Horowitz, S. H. Friend, M. Raybuck, R. A. Weinberg, and E. Harlow.** 1988. Association between an oncogene and an anti-oncogene: the adenovirus E1A proteins bind to the retinoblastoma gene product. *Nature* **334:** 124–129.

186. **Whyte, P., N. M. Williamson, and E. Harlow.** 1989. Cellular targets for transformation by the E1A proteins. *Cell* **56:** 67–75.

187. **Wold, W. S., A. E. Tollefson, and T. W. Hermiston.** 1995. E3 transcription unit of adenovirus. *Curr. Top. Microbiol. Immunol.* **199**(Part 1):237–274.

188. **Wu, D. Y., G. V. Kalpana, S. P. Goff, and W. H. Schubach.** 1996. Epstein-Barr virus nuclear protein 2 (EBNA2) binds to a component of the human SNF-SWI complex, hSNF5/Ini1. *J. Virol.* **70:** 6020–6028.

189. **Yang, X.-J., V. V. Ogryzko, J.-I. Nishikawa, B. H. Howard, and Y. Nakatani.** 1996. A p300/CBP-associated factor that competes with the adenoviral oncoprotein E1A. *Nature* **382:** 319–324.

190. **Yee, S., and P. Branton.** 1985. Detection of cellular proteins associated with human adenovirus type 5 early region 1A polypeptides. *Virology* **147:** 142–153.

191. **Yew, P. R., and A. J. Berk.** 1992. Inhibition of p53 transactivation required for transformation by adenovirus early 1B protein. *Nature* **357:** 82–85.

192. **Zerler, B., E. Moran, K. Maruyama, J. Moomaw, T. Grodzicker, and H. E. Ruley.** 1986. Analysis of adenovirus E1A coding sequences which enable *ras* and *pmt* oncogenes to transform cultured primary cells. *Mol. Cell. Biol.* **6:** 887–899.

Human Tumor Viruses
Edited by Dennis J. McCance
© 1998 American Society for Microbiology

3

Kaposi's Sarcoma-Associated Herpesvirus (Human Herpesvirus 8)

Thomas F. Schulz, Yuan Chang, and Patrick S. Moore

The discovery of a novel herpesvirus in Kaposi's sarcoma (KS) tumors in 1994 has opened new research avenues to investigate this intriguing neoplasm. As epidemiologic evidence accumulates, it is becoming increasingly clear that KS-associated herpesvirus (KSHV) (or human herpesvirus 8 [HHV-8]) is the likely infectious cause of KS and related neoplastic disorders. However, little is currently known about how the virus is transmitted or which additional factors are required for KS tumorigenesis.

Despite its recent discovery, the virus is already providing important insights into how DNA tumor viruses behave and misbehave in the human host. KS has long been suspected to be caused by an infectious agent, presumably a tumor virus (68, 69, 155). Prior to the AIDS epidemic, the tumor had a restricted geographic distribution, with the highest rates of disease being found in Central African and Mediterranean countries. With the advent of the AIDS epidemic and its coincident epidemic of AIDS-KS, surveillance studies have consistently demonstrated that KS is predominantly a tumor of homosexual/bisexual male AIDS patients compared to other male AIDS risk groups. A variety of sexual behaviors (e.g., greater numbers of sex partners) increase the risk for KS, and even among AIDS patients the tumor shows geographic clustering. These observations are consistent with a sexually transmitted agent (14, 16, 163, 208). Immunosuppression is an important cofactor in KS pathogenesis, as shown by the occurrence of KS in posttransplant patients and its occasional regression after the reduction of iatrogenic immunosuppression. Whether KS is a hyperplastic process driven by cytokine dysregulation or represents the oligo- or monoclonal expansion of virus-transformed endothelial cells remains controversial.

EPIDEMIOLOGY OF KSHV

Detection of KSHV in Various Diseases

KS

After its initial discovery in AIDS-KS by representational difference analysis (37), KSHV was found to be consistently detectable using PCR in nearly all biopsies

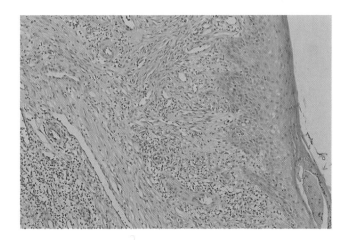

Figure 1 Photomicrograph of a KS lesion showing characteristic spindle cells interwoven into the dermis and forming intersecting bands of tumor cells with irregular capillary channels underneath an intact epidermis. A prominent inflammatory infiltrate of plasma cells, lymphocytes, and macrophages is frequently present, suggesting an immunologic response to tumor-specific antigens (magnification, ×23; hematoxylin and eosin stain).

of all epidemiological forms of KS, but was rarely found in tissues from persons without KS. In addition to KS in AIDS patients (37), these forms include KS in human immunodeficiency virus (HIV)-seronegative individuals both from Mediterranean countries ("classic KS") and from Africa ("endemic KS") (25, 30, 41, 55, 137, 154, 191, 211), KS in iatrogenically immunosuppressed transplant recipients (25, 30, 154), and KS in HIV-negative homosexual men (25, 137). Near-universal detection of KSHV DNA in all of the histologically indistinguishable forms of KS (e.g., classic, epidemic, endemic, and posttransplantation) suggests that they share a common etiology (Fig. 1). The amount of viral DNA in KS biopsies can be variable, with the signal intensity obtained by Southern blot hybridization ranging from undetectable to an estimated 10 to 20 viral genome copies per cell. Nonaffected tissues proximal to KS lesions in KS patients are more likely to have detectable virus genome than more distant tissues (23, 37, 55, 137), suggesting that the virus is localized primarily to KS lesions but can disseminate to nondiseased tissues as well. Geographic variation in prevalence between highly endemic and nonendemic regions may account for higher detection rates in some control populations (18, 38a, 226). The overall detection rate of KSHV DNA in all forms of KS is approximately 95% for over 500 KS lesions that have been tested by PCR by numerous groups (for review see reference 156).

PCR using 35 to 40 amplification cycles will nearly always yield a specific signal when DNA is extracted from fresh or frozen KS tissue samples. The specificity and sensitivity of PCR detection can be enhanced by Southern blotting for the PCR product, but this does not reduce the likelihood of contamination giving a false-positive result. DNA extracted from formaldehyde-fixed, paraffin-embedded

KS tissue often requires nested PCR to obtain a positive signal, which also dramatically increases the likelihood for intraexperimental contamination. The PCR detection rate also depends on the histological stage of a KS lesion and is higher in plaque and fully developed nodular lesions than in early patch lesions (154). It has also been suggested that a decrease in viral DNA may precede the regression of iatrogenic KS and AIDS-KS lesions, but these findings need to be confirmed in larger case series (6).

Within KS lesions, KSHV DNA is localized to endothelial and spindle cells as demonstrated by PCR-in situ hybridization, in situ hybridization, microdissection, and immunohistochemistry (23, 59a, 113, 172, 207, 210). In situ hybridization for expression of the ORFK12/T0.7 KSHV gene (207, 210) has been particularly helpful in demonstrating virus infection in most, if not all, KS tumor cells. In keeping with the results obtained by conventional PCR, in situ techniques also detect KSHV signals more readily in nodular than in early patch KS lesions (59a). However, this may be related to measured levels of gene expression in early lesions as opposed to viral DNA copy number. Immunohistochemical detection of a latency-expressed protein (LANA/ORF73) has confirmed the results of nucleic acid localization studies (172). Taken together, localization studies provide convincing evidence that KS tumor cells are the primary host cells of KSHV in KS lesions and that the majority, if not all, of tumor cells are infected with virus. Several independent lines of evidence therefore suggest that KSHV infects and persists in KS spindle cells in vivo. In contrast, primary cell cultures derived from KS lesions lose detectable KSHV immediately after being established (6, 7, 59, 108) and only a few cultures have been reported to maintain detectable KSHV for several passages (6). Two permanent cell lines derived from KS lesions have been shown to be tumorigenic and to contain an identical chromosomal abnormality (118). These two cell lines both lack detectable KSHV (59). While this observation was initially considered an argument against KSHV being required for the development of KS lesions, the weight of localization studies, cell biology experiments, and epidemiological data now suggests that these in vitro cell lines either are derived from uninfected cells or lose virus infection rapidly after explantation.

Lymphoproliferative diseases

Body cavity-based primary effusion lymphomas (BCBL/PEL). Pre-AIDS KS has been epidemiologically linked to a higher than expected rate of secondary lymphoproliferative disorders by Safai et al. in a hospital registry review (182). Further, several uncommon B-cell disorders, such as Castleman's disease (CD) and angioimmunoblastic lymphadenopathy with dysproteinemia (AILD), frequently are associated with coincident KS (164). This suggests that KS may be part of a larger set of neoplastic manifestations resulting from KSHV infection that also includes lymphoproliferative disorders.

During the initial investigation of KSHV, several non-Hodgkin's lymphomas with an unusual phenotype (104) were found to be positive for KSHV (37). Subsequent evaluations by Cesarman et al. (33) found that only "body cavity-based lymphomas" (BCBL) (96) harbored viral DNA in a survey of 193 lymphomas from patients with and without AIDS. Previously, BCBL were noted to have a strong predilection for body cavities, to be associated with EBV infection, and to lack c-

myc rearrangement (96). It is now well established that KSHV is consistently found in this rare B-cell lymphoma variant, which is alternatively referred to as primary effusion lymphomas (PEL) (9, 32, 33, 62, 66, 106, 145, 159, 160, 209). The KSHV viral load (approximately 50 to 100 virus copies per cell) in BCBL/PEL is nearly always higher than that found in KS tissues and can be detected by Southern hybridization of lesion DNA (34). The virus genome is episomal in BCBL/PEL; however, evidence suggests that genomic integration of the virus may also occur (172).

BCBL/PEL have distinctive molecular genetic and morphological features which distinguish them from other large-cell immunoblastic lymphomas (145). While BCBL/PEL were first described in AIDS patients (96, 104), they may also rarely occur in HIV-seronegative persons (35, 209). Most BCBL/PEL which have been described are coinfected with Epstein-Barr virus (EBV) and KSHV (33), although BCBL/PEL have been identified which are infected with KSHV alone, particularly in patients without AIDS (32, 35, 174, 209). Terminal repeat analysis demonstrates that the EBV strains infecting BCBL/PEL are monoclonal, suggesting that EBV may play an initiating role in a portion of BCBL/PEL. BCBL/PEL cells lack most B-cell-specific antigens, such as CD19 and CD20, but exhibit clonal immunoglobulin gene rearrangements consistent with a monoclonal B-cell origin (21, 34). Most other lymphocyte surface expression markers and adhesion molecules required for cell homing and aggregation are not expressed, possibly accounting for the effusion phenotype of the lymphomas (21).

PEL cell lines infected with KSHV have been established and at present are the only cell culture system that can maintain the virus in vitro (11, 34, 62, 65, 174, 183, 184). Like the parental lymphomas, some PEL cell lines are dually infected with EBV and KSHV whereas others are positive for KSHV alone. KSHV is generally latent in these cells, but a fraction of BCBL/PEL cells in some cell lines can be induced into lytic virus replication by treatment with phorbol esters or butyrate, which is analogous to EBV producer cell lines (174). Other cell lines, such as BC-1/HBL-6, are relatively refractory to producing whole virions, although structural genes characteristic for lytic virus replication can be induced by chemical treatment (128, 129, 139).

Other than BCBL/PEL, KSHV has not been reproducibly found in other B- and T-cell lymphomas (33, 62, 160, 161, 177). The possible exception to this is primary central nervous system lymphomas in AIDS patients, where KSHV has been variably detected by PCR by some groups (119). Quantitative studies suggest that detection of viral DNA in some of these lymphomas may be due to "spillover" of infected circulating B cells rather than direct infection of lymphoma cells (61). One report on the detection of KSHV in cutaneous T-cell lymphomas (185) has not been confirmed by another group (161).

CD and related lymphoproliferative disorders. Castleman's disease (CD) is an atypical lymphoproliferative disorder which may present as generalized lymphadenopathy (multicentric CD [MCD]) or as a solitary enlarged lymph node, usually in the mediastinum. Both animal models (27, 28) and human studies (31, 90, 111) suggest that CD occurs in the setting of interleukin-6 (IL-6) overexpression. CD occurs in two histological variants, the hyaline-vascular type and the plasma cell type, the latter frequently being multicentric and associated with both KS and malignant lymphomas. While both hyaline-vascular and plasma cell types can be positive for KSHV (13, 67, 119, 206), KSHV infection is primarily associated with multicentric disease, particularly of the plasma cell variant. Biopsies from HIV-infected patients with CD-like features are frequently positive for KSHV by both PCR and Southern hybridization, whereas in HIV-uninfected individuals only a minority of samples have detectable KSHV DNA (13, 67, 206). The variable detection of KSHV in these forms of CD suggests that this may be a heterogeneous disorder due to overexpression of IL-6-like activity. Support for this comes from the finding that among those lesions infected with KSHV, expression of vIL-6, a functional cytokine with IL-6-like activity (136, 153), is readily detected by immunohistochemical staining (159a). Thus, some MCD cases may be related to KSHV vIL-6 overexpression while others are due to endogenous huIL-6 overexpression.

Another KS-related disorder, AILD (angioimmunoblastic lymphadenopathy with dysproteinemia), is not consistently found to be infected with KSHV despite the frequent co-occurrence of the two disorders. In an Italian study (119), KSHV was detected in lesions from 3 of 15 cases of HIV-negative AILD and 4 of 23 reactive lymphadenopathies from HIV-seronegative subjects by PCR. This may be related to the general prevalence of KSHV infection in Italy, since other groups have failed to detect KSHV in AILD tissues of patients from geographic areas with a low KSHV prevalence (53). Detection of KSHV in follicular hyperplasia (type IA follicular hyperplasia) has also been reported (67), and KSHV was detected in angiolymphoid hyperplasia of the skin of four HIV-seronegative Hungarian patients by Gyulai and colleagues (77). The latter condition is an uncommon benign disorder characterized by the proliferation of atypical endothelial cells accompanied by an infiltration of lymphocytes and eosinophils. While KSHV has been reported to occur in dendritic cells of multiple myeloma lesions (175) (in which vIL-6 is proposed to indirectly maintain the expansion of tumor cells), these findings were not confirmed by others (119a, 120a, 157). Therefore, aside from BCBL/PEL and MCD, detection of KSHV is not always a consistent feature for other lymphoproliferative disorders, and a role for KSHV in their pathogenesis appears unlikely.

Other pathologies

Many controversial reports exist on the PCR detection of KSHV in a variety of tumors. Some of these discrepancies may be due to the marked regional differences in KSHV prevalence which are now becoming apparent. As previously indicated, studies based on PCR alone (particularly those using nested PCR) are extremely susceptible to intraexperimental contamination that is difficult to detect and occurs even in experienced laboratories. Sequencing of PCR products alone will not exclude contamination unless care is taken to independently amplify and clone multiple products to take into account Taq incorporation errors and to achieve highly redundant bidirectional sequencing. KSHV DNA was detected by PCR in squamous cell carcinomas from four immunosuppressed individuals in the United States (171), but this could not be replicated by others (3, 24, 53, 224). A specific association of KSHV with squamous cell carcinoma is therefore unlikely. Two groups reported the presence of KSHV in angiosarcomas (78, 124), while other studies failed to find any association between KSHV and angiosarcomas (25, 37, 53, 94, 218). Individual case reports and case series are likely to continue to be reported relating KSHV to various pathologies. In evaluating these studies, it is important to keep in mind possible accidental associations due to geographic variations in KSHV prevalence and artifactual associations due to PCR contamination. Whenever possible, PCR detection of viral DNA in an unusual setting should be confirmed by non-PCR-based techniques.

Detection of KSHV in tissues and body fluids

Like other gammaherpesviruses, KSHV infects and may persist in a latent form in lymphocytes and lymphoid tissues (7, 18, 37, 44, 51, 110, 139, 231). Cell separation studies suggest that $CD19^+$ B cells are the primary circulating reservoir for the virus (7, 19); however, circulating peripheral blood mononuclear cells (PBMC) will be positive only in about 50% of patients with clinical KS under standard amplification conditions using nested PCR (140, 231). Despite this low detection rate, several studies have found a strong predictive association between the KSHV positivity in PBMC by nested PCR and the eventual risk of developing KS among HIV-positive patients (110, 140, 231). In these studies, low-risk control patients (blood donors, oncology patients) generally have low KSHV DNA detection rates. In Italy, a country of intermediate prevalence, KSHV has been detected in approximately 9% of PBMC and lymphoid tissues of HIV-uninfected individuals (18, 226). Even higher positivity rates can be expected in hyperendemic African countries where the prevalence of infection may exceed 50%. The viral load in PBMC estimated by quantitative PCR generally correlates with the presence and extent of KS. Thus the viral load is higher in HIV-negative patients with advanced or disseminated disease than in those with nonprogressing indolent disease (26, 108). Important considerations for PCR-based studies using PBMC are that the technique is at best semiquantitative and that the amount of DNA used in standard PCR reactions may not sample a large enough cell volume to accurately determine the KSHV infection status of infected individuals (51).

Only limited data are available on the persistence of KSHV in PBMC, of which B cells and perhaps macrophages may harbor KSHV genomes (7, 19c). One study reported the presence of linear KSHV genomes in PBMC, reflecting the presence of productively infected cells (51). Additionally, $CD34^+$ spindle cells can be isolated from the peripheral blood of KS patients as well as from control patients without KS (29). When isolated from KS patients, these cells may be positive for KSHV DNA by PCR and have an identical immunophenotype to spindle cells found in KS lesions (202). Thus, they may represent circulating precursor cells of KS lesions or shedding of KS lesion spindle cells into the circulation. Since similar cell cultures which are KSHV negative can be established from the peripheral blood of patients without KS, the significance of this finding remains unclear.

There is wide variation in the detection of KSHV shedding into semen, and early studies demonstrating high semen detection rates from healthy donors have led some authors to conclude that this virus is ubiquitous in North American and European populations (116, 135). These early studies, however, have not proven to be reproducible or generalizable (7, 76, 89, 121, 134). Among semen samples from patients with clinically overt KS, KSHV is detectable for only a minority of patients (0 to 33%). It is unknown whether some KS patients continually shed virus into their semen whereas others do not, or whether all KS patients have periodically detectable virus in semen. This is an important question to be answered given the likely sexual mode of virus transmission. In one study of 18 KS patients, the only 2 patients found to have detectable virus in semen were positive at two separate time points, consistent with the first possibility (76).

Semen samples from 30% of United States semen donors (116), and initially from 90% of Italian semen donors (135), were reported to contain KSHV. The results of the latter study were adjusted down to 23% in an extended analysis (134). In contrast, no positive samples were found among 115 donors in the United Kingdom (89), 22 in France (109), or 20 in northwest Italy (45). Higher detection rates (13%) were found in healthy Italian donors from southern Italy, where classic KS is endemic (226). There is no readily available explanation for the discrepancies between these studies, although the current weight of evidence suggests that KSHV is not ubiquitous and is not present in the vast majority of donor semen from Europe or the United States (for review, see reference 19b). Possible explanations for the high rates initially detected include sampling of hyperendemic populations and PCR contamination during nested PCR.

A potential sexual route of transmission has also focused attention on the prostate gland as a reservoir for KSHV. KSHV is preferentially detected in semen plasma rather than spermatocytes, suggesting secretion into seminal fluids (47, 89, 135). However, it cannot be excluded that KSHV-infected mononuclear cells may occasionally represent the source of KSHV in semen. A survey of tissues from AIDS-KS patients found prostate tissues to consistently harbor virus genome (47). In situ hybridization using the T0.7 probe indicated expression of a latent viral gene in prostatic glandular epithelium (207). This study also detected KSHV in prostate biopsies from men without KS but reserved judgment as to the epidemiological implications of this observation. High rates of KSHV detection by nested PCR in prostate tissue of Italian men have been reported (135). However, other studies have failed to find prostatic tissues from most men without KS to be infected (109, 180, 214). It is possible that widespread detection of viral nucleic acids in prostate, like that in semen, may be due to geographic or technical differences in KSHV detection.

Several herpesviruses (e.g., EBV, HHV6, HHV7) are shed into saliva, providing a ready means of transmission through oral secretions. KSHV DNA is also frequently detected in salivary secretions (19a, 20, 105) or in oropharyngeal tissues (52, 229) of patients with HIV and AIDS-KS. These findings may represent exfoliation of infected cells from preclinical KS lesions, since KS commonly occurs in the oropharynx (93) and KSHV was not found in non-KS lymphoid tissues from patients with AIDS in one study (229). If transmission from oral secretions is a significant means of human-human transmission, sexual activity and/or specific sexual behaviors (e.g., oral-genital or oral-anal contact) might contribute to or be correlated with the epidemiologic patterns of KS (14, 16). Since it is not known whether infectious virus can be transmitted through the oropharyngeal epithelium, the implications of these findings for KSHV transmission remain unknown but warrant further study.

In addition to studies suggesting persistence of KSHV in prostate tissues from KS patients described above (47, 135, 207), one study has found that dorsal root ganglia may harbor viral DNA in patients with AIDS-KS (46). Although gammaherpesviruses are not known to be neurotropic, this observation, if confirmed, may help to explain the unusual dermatomal distribution of KS lesions which frequently occurs among AIDS-KS patients.

Serologic Studies

Serological assays

Estimates of the KSHV infection prevalence in various populations have benefited from the development of specific first-generation serologic assays. While these assays all require technical improvements, results from studies using various assays are qualitatively and, frequently, quantitatively similar. Immunofluorescence (IFA), Western blot (WB) (immunoblot), and enzyme-linked immunoadsorbent (ELISA) assays to detect antibodies against latent and lytic antigens of KSHV have been developed (summarized in Table 1).

Most currently available serological assays for KSHV are based on KSHV-infected BCBL/PEL cell line antigens, although recombinant antigen- and peptide-based assays have also been developed. The first identification of a KSHV-specific antigen was the detection of a latency-associated nuclear antigen by immunofluorescence using BCBL/PEL cells (HBL-6, BC-1) dually infected with KSHV and EBV

Table 1 Serologic assays for detection of KSHV antibodies

		Antibody detection rate (%)						
		Study populations		Control populations (without KS)				
Antigen	Assay	HIV-seronegative KS	AIDS-KS	Homo-sexual men	Hemo-philiac men	Blood donors	Other control groups	References
Latent antigen								
LANA	IFA	85–100	83–88	18–30	0–3 (U.K./U.S.)	0–3 (U.K./U.S.)	12–30 (Greece)	65, 101, 112, 139, 199
						4–21 (Italy)	51 (Uganda)	
	WB	100	80	18		0 (U.S.)	62 (Uganda)	64
Lytic antigens								
40-kDa protein	WB		67	13				129
VP19 (ORF65)	ELISA/WB	94	75–80	31	2 (U.K.)	2–5 (U.K./U.S.)	12 (Greece)	199
						9–22 (Italy)	35 (Uganda)	
VP23 (ORF26)	ELISA		ca. 40			ca. 20 (Germany)		8
			ca. 60			ca. 20 (U.S.)		49
Whole cell	IFA		96–100	90–100		0–25 (U.S.)	32–100 (Africa)	112, 204

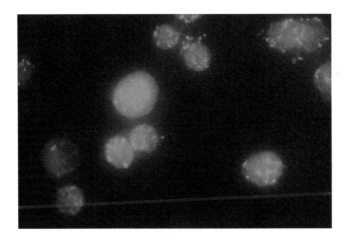

Figure 2 Characteristic immunofluorescence staining of the KSHV-infected BCP-1 cell line with serum at 1:160 dilution from a patient with KS. The KSHV LANA encoded by ORF73 forms a speckled nuclear pattern, suggesting subnuclear localization, which is distinguishable from the diffuse, nonspecific cytoplasmic seroreactivity that is commonly seen with BCBL/PEL cell lines (65, 99).

(139). The use of these cell lines in IFA required laborious preadsorption of EBV-specific antibodies to avoid cross-reactivity. The antigen localization pattern has a characteristic specular immunofluorescence localized to the nucleus (Fig. 2). With the identification of EBV-negative BCBL/PEL cell lines, IFA testing for KSHV latent antigen antibodies has allowed rapid serologic testing. WB assay of BC-1 nuclei also identified a high-molecular-weight doublet antigen (p226/234) expressed during virus latency (64). Subsequent studies (172) established that both the high-molecular-weight WB nuclear antigen and the IFA nuclear antigen are encoded by the ORF73-protein, now called LANA (latency-associated nuclear antigen) (101). A highly acidic internal repeat (the moi repeat region) may account for the retarded electrophoretic mobility of LANA compared to its predicted protein size (130 kDa), and posttranslational modification is likely to account for the doublet pattern seen by WB (172).

The first identification of a lytic viral antigen was achieved by the differential induction of KSHV and EBV using butyrate, which allowed detection of a specific 40-kDa WB band (129). Although dually infected cell lines are of limited use for serologic studies because of the potential for cross-reactivity to EBV antigens (176), results obtained with these early assays demonstrated that specific antibodies were present in serum of KS patients and that the virus is unlikely to be ubiquitous in North American populations (64, 129, 139).

Widespread screening of KS risk groups and the general population became possible when the first BCBL/PEL cell lines infected with KSHV alone were established (174). These cell lines express LANA and can be used to screen large numbers of serum samples without preadsorption (65, 101). Examination of panels of sera from populations at high and low risk for KS indicated that antibodies to the ORF73/LANA antigen predict the likelihood of KS developing among AIDS patients. This, as well as the observation that only 0 to 3% of United States and United Kingdom blood donors have antibodies to this latent protein (65, 101, 199), makes it the most specific KSHV diagnostic antigen so far identified. Assays based on LANA also show reasonable (80 to 95%) sensitivity among AIDS-KS patients (although lower sensitivity rates have been seen in some studies [75] and the antigen detection rate may be susceptible to serum freeze-thaw cycles [64]). Sensitivity may even be higher among KS patients without severe immunosuppression (65, 101, 199). Nonspecific cross-reactive antibodies to cytoplasmic antigens interfere with the LANA IFA at low serum dilutions, requiring either the isolation of whole nuclei (101) or use of dilute serum (65). This has been a particular problem for sera collected from African countries, which frequently have high levels of nonspecific reactivity for reasons which are poorly understood.

Lytic cycle (structural) KSHV antigens have also been found to react with sera from KS patients. In addition to the 40-kDa antigen discussed above, which is recognized by 67% of KS sera (129), some patients have antibodies to other lytic-phase KSHV antigens of approximately 27 kDa and 60 kDa (128). There is so far no indication of cross-reactivity of these proteins with corresponding EBV proteins, but their specificity requires further investigation. A 19-kDa capsid-related protein encoded by ORF65 has been used as a recombinant protein in ELISA and is recognized by approximately 80% of AIDS KS sera and 85 to 90% of classic KS sera (199). About 3 to 5% of United Kingdom and United States blood donors show reactivity with this protein. Its immunogenic determinants are located within the carboxy-terminal 80 amino acids, and this region is 21% identical to the corresponding region in its EBV homolog BFRF3 (199). Although vp19/ORF65 is not recognized by most sera from EBV-positive individuals and does not react with a high-titered EBV-positive serum (199), the question of whether there may be occasional cross-reactivity with EBV is not yet completely resolved. Concordance between the truncated recombinant ORF65 antigen and LANA IFA antigen is high (around 80 to 85%) in sera from KS patients, but lower in blood donor sera. Similar results have also been reported by another group using the full-length recombinant protein (117).

Other recombinant lytic-phase proteins are actively being investigated as potential serologic antigens. The minor capsid protein vp23, encoded by ORF26, has been expressed as a recombinant protein and used as a serological antigen (8). Significantly more AIDS-KS sera than HIV-negative control sera react with this antigen, and no clear-cut evidence for EBV cross-reactivity was found despite relatively high homology (49% amino acid identity) with its EBV homolog, BDLF1. However, this antigen detects only approximately 40% of sera from AIDS-KS patients (8). A synthetic peptide derived from ORF26 has been reported to react with 60% of AIDS-KS sera and 20% of blood donor sera (49). Whether ORF26-derived recombinant proteins or peptides can be of use in a cocktail of defined KSHV antigens requires further investigation. A recombinant carboxy-terminal fragment of the major capsid protein (MCP), encoded by ORF25, has also been investigated and shown to be cross-reactive with high-titered EBV antibodies (8). This observation is in line with the high homology (56% amino acid identity) of the ORF25 protein with its EBV homolog, BcLF1 (139), and indicates that care should be taken to exclude cross-reactive KSHV antigens from an antigen cocktail to be used for diagnostic purposes.

In addition to these defined structural proteins, other assays involve the detection of antibodies to unknown lytic antigens. An IFA assay on a BCBL/PEL cell line treated with phorbol ester to induce the lytic replication cycle (112, 204) has been described. This assay detects antibodies in nearly 100% of KS patients. The reported antibody detection rate for United States blood donors is higher (20%) than with other latent (64, 65) or recombinant lytic antigen assays (199). In contrast, Smith et al. (204), using a similar assay format with a serum dilution of 1:40 and Evans blue counterstaining to reduce nonspecific background fluorescence, found none of 52 United States blood donors positive. That cytoplasmic seroreactivity to KSHV-infected BCBL/PEL cells can be cross-adsorbed by EBV-containing cell lines (65) suggests that nonspecific reactivity may be a major problem with these assays, especially at low serum dilutions. This is consistent with the finding of Smith et al. (204) that sera reactive by lytic IFA at a serum dilution of less than 1:40 could not be confirmed by radioimmunoprecipitation.

Given the problem of specifically detecting antibodies to KSHV in individuals with a 90% or greater background prevalence of EBV, we shall most likely be experimenting with several combinations of antigens and assay formats before the best solution is found. LANA IFA screening is frequently used in seroepidemiologic studies, which can be confirmed or supplemented with vp19/ORF65 antigen assays. In the near future, competitive assays, new antigens, or cocktails of antigens may provide more reliable, sensitive, and specific methods for KSHV antibody detection.

Seroprevalence and transmission of KSHV

While different serological assays for KSHV-specific antibodies give varying estimates of the seroprevalence of this virus, all studies are relatively concordant in demonstrating that KSHV, unlike the majority of other herpesviruses, is not a ubiquitous human infection. Antibodies to LANA are present in about 85% of AIDS-KS patients and in greater than 90% of individuals with classic KS (64, 65, 101, 112, 199). Among persons studied in the United States, the United Kingdom, and Denmark, about a third of HIV-infected homosexual men without KS are KSHV LANA seropositive compared to 8% of HIV-uninfected sexually transmitted disease (STD) clinic attenders, 0 to 3% of HIV-uninfected blood donors, 0 to 3% of patients with hemophilia, and 0% of intravenous drug users (64, 65, 100, 101, 112, 126, 199). American women, who are at low risk for AIDS-KS, have correspondingly low LANA antibody titers regardless of HIV status (100). Antibody positivity to vp19/ORF65 shows a very similar distribution (81% of AIDS KS, 94% of classic KS, 31% of HIV-infected homosexual men without KS, 2 to 5% of HIV-negative English and American blood donors, 1% of patients with hemophilia, 3% of intravenous drug users) (199). Therefore, the distribution of antibodies to both of these antigens suggests that KSHV is an uncommon infection in the general population of these countries where it is likely to be sexually transmitted.

The phorbol ester-induced lytic antigen IFA has higher detection rates (>95%) than either the ORF65 or LANA assays among AIDS-KS patients. Using this assay,

Lennette and colleagues (112) found that prevalence in adult North American populations ranged from 18 to 28% and as high as 90% among HIV-infected homosexual men without KS. As indicated above, Smith et al. (204), using a similar assay but counterstaining with Evans blue, found a generally lower antibody detection rate for several sentinel populations, including North American blood donors. The question of sensitivity versus specificity for these assays is not yet resolved; nonetheless, these results broadly concur in that prevalence of KSHV mirrors the risk of KS among various populations in North America and Europe (16), albeit at a much higher baseline positivity rate than when using the LANA or ORF65 assays.

Seroconversion to KSHV positivity has been examined (64, 65) using follow-up sera from HIV-positive homosexual men enrolled in the Multicenter AIDS Cohort Study (MACS). In these studies, 40 men who eventually developed AIDS-KS were compared with 40 control men who were matched by $CD4^+$ counts and who developed AIDS but not KS. LANA positivity as measured by both WB and IFA formats found a 80 to 90% seropositivity rate for AIDS-KS patients but only 18 to 30% positivity for controls who did not develop KS. In examining banked sera from patients destined to develop KS, 50% of the AIDS-KS group had seroconverted by 34 to 46 months prior to clinical diagnosis of KS (Fig. 3B). Using serial endpoint dilution titers to quantitate antibody responses, these authors found that LANA antibody levels remained consistent and elevated for up to 8 years after seroconversion until KS development (Fig. 3A). This pattern of seroreactivity does not appear to reflect reactivation of virus with immunosuppression and instead argues strongly for seroconversion reflecting initial infection with KSHV. Although LANA seroconversion appears to reflect adult KSHV infection (consistent with a sexually transmitted agent), no studies have been performed to determine whether or not LANA seropositivity is a late response. It remains possible that infection occurs a number of months prior to the first appearance of LANA antibodies.

Low positivity rates for ORF65 and LANA antibodies are found in blood donors of Northern Europe and North America, but much higher rates occur in Mediterranean and African populations. In Greece, preliminary studies suggest that 10 to 20% of HIV-negative surgical patients without KS have antibodies to ORF65 or LANA (199; unpublished data). In Italy, there is evidence for significant regional differences in seroprevalence measured by these two antigens, ranging from 4% LANA positivity in Milan (northwestern Italy [65]) to about 35% LANA and ORF65 positivity in Apulia (southeastern Italy [31a]). Even within northern Italy there may be considerable geographic variation in KSHV prevalence rates (31a, 65, 231a). Thus, in European populations, KSHV appears to be more common in regions with higher endemic incidences of HIV-seronegative KS. A careful comparison of KS incidence with KSHV seroprevalence in Southern Europe is required before definitive conclusions can be reached. These marked differences in seroprevalence within Europe may also help to explain reports of frequent KSHV genome detection by PCR in tissues from non-KS patients in highly endemic countries such as Italy. Seroprevalence to LANA and ORF65 is much higher (>50%) in countries of East and Central Africa (9a, 65, 112, 199). This suggests that infection with KSHV may approach near-universal levels in some African populations and is almost certainly more common than in northern Europe and North America.

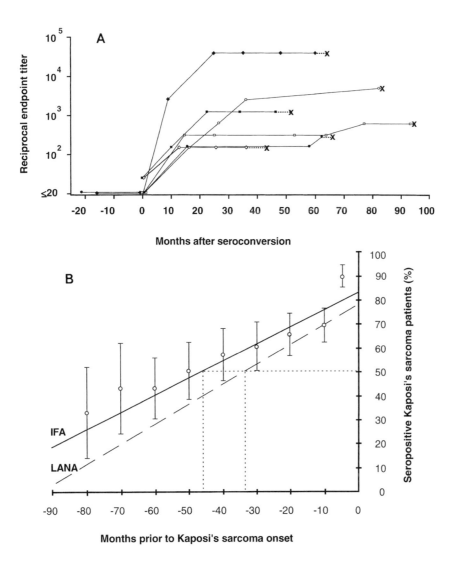

Figure 3 (A) Immunoglobulin G (IgG) antibody kinetics of seroconversion with the LANA IFA for six homosexual, HIV-positive men followed by the MACS study who developed KS. All six patients seroconverted from a LANA IFA endpoint titer of 1:40 or less to greater than 1:160 at time 0 and were subsequently followed for 43 to 96 months until they developed KS (marked by X). Antibody titers did not significantly vary after seroconversion, suggesting that seroconversion is caused by initial KSHV infection rather than reactivation associated with immunosuppression. (B) Cumulative LANA seropositivity rate by IFA (solid line) and WB (dashed line) for 40 homosexual, HIV-positive men from entry into the MACS study until development of KS at time 0. Seroconversion rates were linear over time for both assays, with the median seropositivity rate being 32 months by WB and 46 months by IFA. Estimated sensitivities of these latent antigen assays at KS onset (0 timepoint) are 80 to 90%. Reprinted with permission from reference 65.

Serological studies demonstrate that, irrespective of the type of antigen used, KSHV infection is more common among STD clinic attendees than among blood donors (101, 112, 199). A detailed analysis of behavioral risk factors among Danish homosexual men revealed that behavioral variables such as high numbers of sex partners and unprotected anal receptive intercourse increased the risk for KSHV infection (126). Having had sexual contact with American homosexual men in the early 1980s also markedly enhanced the likelihood among Danish homosexual men of having, or acquiring, antibodies to KSHV, suggesting that KSHV was probably introduced into this community in the late 1970s or early 1980s (126). Oral-anal contact was not a risk factor for KSHV seropositivity among the Danish men, but was found to be a significant risk factor for Australian HIV-positive STD clinic attenders (74). Further detailed risk factor studies may be needed to delineate the precise mechanisms by which KSHV is transmitted among homosexual men and other high-risk populations.

Taken together, these serologic studies strongly suggest that KSHV is sexually transmitted in countries with low KS prevalence. Risk factors which have previously been shown to increase the risk for KS (16) are also associated with increased likelihood of KSHV infection, providing further evidence that KSHV is indeed the postulated KS agent. For homosexual men, initial infection is most likely to occur after sexual debut during adulthood, consistent with a sexually transmitted agent. However, little is known about specific sexual behaviors which place homosexual men or others at risk for infection. The exceptionally high seroprevalence seen in some African countries, even among preadolescent children, indicates that nonsexual modes of transmission may also play a role in virus dissemination in these countries. In this regard, the epidemiology of KSHV may resemble that of hepatitis B virus. The syndrome of KS occurring among young children in highly endemic African countries (a condition which is exceedingly rare in developed countries) suggests the possibility of vertical virus transmission in some settings (242).

Sequence variation among different KSHV isolates

KSHV obtained from various sources has a high degree of sequence conservation in most portions of the genome. Comparison of a 20-kb region sequenced from both a KS lesion and a BCBL/PEL cell line demonstrated only 0.1% nucleotide variation between the two isolates (139, 181). Over 90% of the genome was recently sequenced from a KS lesion and shows a high level of conservation similar to the prototype BC-1 strain sequence (148, 181).

As one of the two RDA fragments originally used to identify the virus (37), the ORF26 gene has been examined by a number of groups for sequencing studies. This region has limited sequence variation, rendering it unsuitable for molecular epidemiologic studies (25, 44, 121, 137, 243). Through a combined analysis of several genomic sites, one group has found up to 1.5% overall nucleotide variation between isolates, allowing strains to be grouped into three different main variants, provisionally termed A, B, and C (243). Most KSHV sequences amplified from United States AIDS patients or HIV-negative Mediterranean KS patients have the A or B pattern, whereas sequences derived from Central or Eastern African biopsies belonged mainly to the B or C group. The majority of American AIDS-KS patients were found to have nearly identical KSHV sequences, indicating that they may be derived from a single variant spread during the AIDS epidemic (243).

Care will be needed to confirm these findings since there is a low degree of overall phylogenetic discrimination available from these sites. This finding is in keeping, however, with seroepidemiologic studies indicating that KSHV may have been introduced into the homosexual community in Denmark in the early 1980s through contact with homosexual men from the United States (126).

The low degree of sequence variability for genomic regions thus far examined prevents detailed molecular epidemiologic studies on the origin and transmission patterns of KSHV. In herpesvirus saimiri (HVS) the highest degree of sequence variation between different isolates is found at the left end of the genome in an area featuring the saimiri transforming protein (STP) and Tip genes, which are essential for the transformation of T cells by this virus. A similar highly variable region (ORFK1) of KSHV is present in the portion of the genome corresponding to the HVS STP/Tip genes. A comparison of sequences (107a, 148, 181) from different isolates for the ORFK1 gene shows high levels of sequence variability. This gene may thus serve as a phylogenetically rich sequence for molecular epidemiologic studies.

VIROLOGY OF KSHV

In Vitro Culture, Transmission, and Biology

A number of cell lines derived from BCBL/PEL have been established allowing high-titered in vitro culture of KSHV. These cell lines are the primary reagents for characterizing the behavior of KSHV since in vitro transmission to uninfected cell lines is not currently practical for biologic studies. Some BCBL/PEL cells are coinfected with EBV, while others are infected with KSHV alone. Several of the dually infected BCBL/PEL-derived cells lines appear to be under tight latent control and do not spontaneously express late lytic cycle genes required for virus packaging (128). Lytic cycle gene expression in these cell lines can be demonstrated through the use of inducing agents such as phorbol esters or butyrate, but whole virion assembly has not been demonstrated by electron microscopy. Because of the tight latency of KSHV in EBV-coinfected BCBL/PEL cell lines, a major advance in the study of KSHV was the description by Renne and colleagues of the EBV-negative BCBL/PEL cell line BCBL-1, which is readily inducible to produce virions with phorbol ester treatment (174).

Cultivation of the virus in cells uninfected with EBV has led to the development of serologic tests (64, 65, 101; see detailed discussion above) and studies of KSHV gene expression during latency and lytic replication (240). Typical 100- to 120-nm herpes virion structures can be readily seen by electron microscopy in BCBL-1 after 12-*O*-tetradecanoyl phorbol-13-acetate (TPA) treatment (Fig. 4), and virus particles can be banded by centrifugation (240), allowing accurate estimation of the virus genome size from encapsidated virions (173). Tumors induced in immunodeficient mice injected with the EBV-negative KS-1 strain BCBL/PEL cells produce typical herpes-type virions in those cells undergoing lytic replication as shown by electron microscopy (183).

Techniques allowing efficient in vitro KSHV transmission, however, have not yet been established. This has hampered studies of the virus's replication and biology. Initial transmission experiments using the phorbol-induced BC-1 cell line (139) demonstrated filterable genome transmission to Raji, OMK, Molt-3, and cord

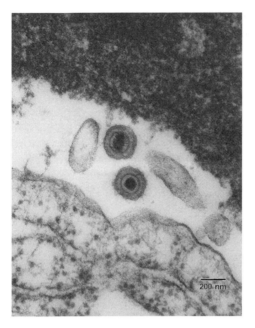

Figure 4 Extracellular, mature KSHV particles, found among cellular debris of lysed KS-1 cells which were derived from a body cavity-based lymphoma chronically infected with KSHV. These virus particles are seen exhibiting two different planes of sectioning through the center of the virus particles. Both of these mature herpesvirus particles exhibit an electron-dense viral DNA core region, surrounded by spherical capsid structure. The capsids of the virions are enclosed by an electron-lucent layer surrounded by the viral envelopes, which are seen with attached protein spikes. (Courtesy D. Ablashi.)

blood lymphocytic cells (139), which can be inhibited by UV light and antiherpesvirus drugs (127), but the level of virus replication in recipient cells was only detectable by PCR or in situ hybridization. Replication competence and virion production in BC-1 have not been unequivocally established (128, 139) since this cell line is coinfected with EBV and the KSHV strain has a large genomic duplication. However, similar low levels of unstable transmission occur using BCBL/PEL cell lines infected with KSHV alone. Paradoxically, most recipient cell lines undergo cytolysis within 24 h of exposure to virus-containing supernatants, despite the lack of substantial virus replication (60, 139).

The most suitable cell line thus far described for transmission studies is 293, which has been used to directly culture KSHV from KS tissues (60). These authors first described the direct transmission and serial propagation of KSHV from explanted tissue to 293 cells, which is enhanced by the addition of IL-2 to the culture medium. These experiments provide direct evidence that a portion of infected KS cells spontaneously undergo lytic replication, a finding supported by electron microscopic and in situ hybridization studies (92, 158, 207, 228). Serial passage demonstrates that the virus is replication competent, but, similar to other cell lines, 293 cells do not appreciably amplify KSHV in vitro. Nonetheless, refinement of 293 culture techniques may eventually overcome these difficulties.

It is currently unknown why KSHV is not readily cultivable in vitro in non-BCBL/PEL cell lines. It is possible that BCBL/PEL cells have an undescribed mutation or are infected at a precursor stage allowing initial infection and carriage of the virus at high copy number. Despite early speculation, evidence suggests that KSHV is not a defective virus requiring EBV for its propagation (139). The resistance of KSHV-infected KS spindle cells to in vitro culture is reminiscent of EBV-infected nasopharyngeal cancer cell cultures. The inability to transmit KSHV in vitro remains a major obstacle for cellular transformation studies and is an area of active research.

Effect of Antiviral Drugs on KSHV

Several inhibitors of herpesviral DNA polymerases, active against lytic but not latent herpesvirus infection, have been examined for their activity against KSHV in vitro and KS in vivo. Phosphonoacetic acid (Foscarnet) has been found to induce the regression of KS lesions in one small study (141) and to reduce the frequency of KS lesions in three large follow-up studies (70, 95, 132; although also see reference 48). A similar effect was reported for ganciclovir in two studies (70, 132), but not in another (95), and no evidence for acyclovir activity has been reported (95, 132). Despite this reported effect on the tumor, no effect on the detectability of virus DNA in PBMC was reported for patients receiving Foscarnet and/or ganciclovir in one study (91). The in vivo effect of antiherpesvirus drugs on the development of KS lesions is mirrored by in vitro studies on KSHV replication in lytically induced BCBL-1 cells. Foscarnet, ganciclovir, and cidofivir showed in vitro inhibitory activity at achievable pharmacological concentrations, whereas acyclovir did not (99, 124a, 149a).

At present it seems likely that Foscarnet and possibly ganciclovir, but not acyclovir, have some activity in preventing KS among AIDS patients. Potential toxicity and comorbidity from administration of these drugs needs to be carefully weighed against their potential benefit before antiviral therapies are instituted in clinical settings. No well-controlled studies are available to determine the effects of these drugs on established KS lesions, and currently available data are from retrospective analyses, not clinical trials. These studies point to the possibility that potent, low-toxicity oral antiherpesvirus drugs may eventually play a role in treatment of KSHV-related disorders. Of fundamental importance is the fact that these drugs are DNA polymerase inhibitors that are likely to be active only against lytically replicating virus. Thus, it appears likely the lytic replication plays a critical role in either the initiation or maintenance of the KS lesion (99).

GENOMIC STUDIES

The gene orientation and sequence of KSHV genome derived both from PEL (153, 181) and from KS lesions (139, 148) are fundamentally the same, indicating that the KSHV strains inducing these two malignancies are likely to be identical. The genomic structure of the virus (Fig. 5) is similar to that of HVS (5), with a single contiguous 140.5-kb long unique region (LUR) containing all identified coding regions (181). The LUR is flanked on either side by a variable-length terminal repeat (TR) region, composed of 30 to 40 801-bp repeat units having a high (85%) G:C content, which acts as the presumed site for viral genome circularization and linearization (181). The size of wild-type KSHV is approximately 165 kb based on studies of genome banded from productive BCBL/PEL cells (173), which has been confirmed through Gardella gel electrophoresis (51) and whole genome mapping (181).

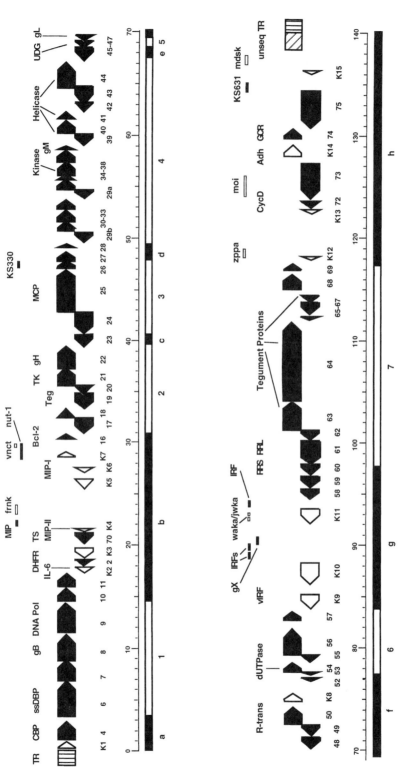

Figure 5 The KSHV ~140.5-kb LUR encodes at least 81 ORFs and is flanked by the TR region composed of multiple TR units. Additional genes are likely to be delineated through experimental studies. Genes with homology to other herpesviruses (solid black) fall into regions which are well characterized gene blocks (1 through 7; open segments) conserved among herpesviruses. Genes unique to KSHV and related rhadinoviruses, some of which are homologs to cell signaling and regulatory genes, are designated with a K prefix (open segments) and lie in intervening nonconserved gene blocks. Terminal repeat sequences are composed

The TR Region

The TR region is a conserved feature of herpesviruses which serves as the site for circularization of the linear genome during the transition to a latent plasmid genome. The two KSHV strains studied thus far (BC-1 and BCP-1) have 30 to 40 TR units in the TR region. In the BC-1 strain, the leftmost TR unit is partially deleted and the rightmost portion of the TR has not been determined due to difficulties in sequencing through a repeat region abutting the right TR (181). Similar difficulties have been found in cloning right-hand portions of the genome from a KS lesion library (148). Recent efforts to clone and sequence the right end of the genome from some samples have apparently been successful and suggest potential strain variation between different isolates (153a). No expressed open reading frames (ORFs) have so far been identified within the TR region.

Replication of KSHV is likely to be similar to that of other herpesviruses, in which the genome replicates by a rolling-circle mechanism (178). In this model, viral genomes are replicated as a continuous long concatemer as the DNA replication fork proceeds around the circular genome. Individual virus genomes are then cleaved within the TR region and packaged into virions. KSHV has conserved packaging and cleavage signal sequences within each TR unit consistent with this model (181). For EBV, this process of linearization and recircularization results in variable loss or gain of individual TR units during each lytic replication cycle. Differences in TR region size can thus be used to demonstrate monoclonal expansion of latently infected cell populations in tumor tissues (168). If a virus population increases by clonal cellular expansion (through latent viral replication without lytic replication), a homogeneous TR length will be present in the virus population, resulting in a single TR DNA band on Southern hybridization with a TR probe. If lytic replication and de novo infection occur within the diseased tissue, a variety of TR lengths will be present, resulting in a laddering DNA pattern on Southern hybridization.

Southern hybridization using a KSHV TR probe on DNA from KS lesions and BCBL/PEL cell lines may therefore be useful to examine issues of clonality (181). BC-1, which is under tight latent replication control, gives a uniform TR banding pattern when digested with *Taq*I, a restriction enzyme which does not cut within the TR region. Two BC-1 TR bands are present (35 kb and 7 kb) resulting from the duplication and insertion of a portion of the LUR into the TR region, and the same banding pattern is present in HBL-6, a cell line independently derived from the same tumor as BC-1, suggesting that the duplication occurred in the parental human tumor rather than in tissue culture. It is likely that this duplication is the cause for the apparent 270-kb size of the BC-1 genome seen by pulsed-field gel electrophoresis (127, 139). The presence of the same polymorphic KSHV TR pattern in both cell lines, together with studies showing EBV and cellular monoclonality of the tumor, suggests that a single progenitor cell became infected with both EBV and KSHV prior to monoclonal expansion. While the replication competence of BC-1 has not been unambiguously demonstrated, the viral genome can be induced into replication by butyrate treatment (128) and late lytic cycle genes are expressed by either phorbol ester or butyrate treatment (128, 187).

BCP-1, which undergoes spontaneous lytic replication, has a major band at 35 kb and lower-molecular-weight laddering at approximately 800-bp intervals.

This may represent TR length polymorphisms due to lytic virus replication, similar to the TR laddering pattern seen with lytic EBV replication. In contrast to the BCP-1 pattern, KS DNA has a monoclonal or oligoclonal TR pattern (181). These studies support HUMERA X-linked chromosome inactivation studies, performed on KS lesions from women, which also suggest that KS has a monoclonal cellular component (169, 170). Caution is needed to interpret these data until TR region polymorphism studies similar to those performed for EBV (168, 181) can be completed. Conflicting HUMERA studies on KS lesions have been reported (51a).

Coding Regions of the Virus

The 140.5-kb KSHV LUR is larger than corresponding regions of HVS and EBV, encoding at least 81 predicted reading frames (181) (Fig. 5). ORFs were named according to their corresponding HVS homologs (5) by sequence homology (genes without significant HVS homology are given a K prefix). It is likely that additional genes will be identified experimentally. The LUR contains several gene blocks which are conserved among all subfamilies of herpesviruses (39). Several blocks (designated 1 through 7) contain genes conserved among herpesvirus subfamilies while others (designated a through h) are only shared with other rhadinoviruses or are unique to KSHV alone.

In the conserved gene blocks, genes are found encoding major structural proteins (e.g., ORF25, the major capsid protein), DNA synthetic enzymes (e.g., ORF21, the thymidine kinase, and ORF9, the DNA polymerase), and glycoproteins (e.g., gB [ORF8], gH [ORF22], gM [ORF38]) present in other families of herpesviruses. The original DNA BamHI fragments used to identify the virus (KS330Bam and KS631Bam) are located within the ORF26 and ORF75 genes, respectively. The gene block regions a through h contain genes which distinguish rhadinoviruses from other herpesviruses (5). The right end of the KSHV genome (ORF72 through ORF74), for example, contains rhadinovirus genes not found in EBV. However, EHV-2, also a gamma-2 herpesvirus, lacks ORF72 and ORF73 homologs (215). ORF2 and ORF70, encoding dihydrofolate reductase and thymidylate synthase genes, respectively, are found in HVS but are rearranged in KSHV relative to their positions in HVS (181).

KSHV does not encode identifiable sequence homologs to EBV or HVS genes known to play roles in viral transformation. The high degree of synteny between KSHV and HVS genomes indicates a close evolutionary relationship between the two viruses. This is also seen by phylogenetic analyses using the amino acid sequences of highly conserved proteins (139). KSHV is the only rhadinovirus (gamma-2 herpesvirus) within the *Gammaherpesvirinae* subfamily to naturally infect humans. Gammaherpesviruses were originally biologically defined as a distinct group by their tropism for cells of hematopoeitic origin (178); this assignment has been confirmed through sequence-based phylogenetic analyses.

PCR amplification using consensus primers from gammaherpesvirus DNA polymerase genes have identified two other new herpesviruses (RFHVMn, RFHVMm) in captive *Macaca nemestrina* and *Macaca mulatta* monkeys (179). These viruses may have a closer evolutionary relationship to KSHV than HVS (179) and may be associated with an epidemic of retroperitoneal fibromatosis affecting immunosuppressed monkeys at primate centers in the northwestern United States in the 1970s and 1980s (222). An additional new gammaherpesvirus has been found in rhesus monkeys; its relationship to the other macaque viruses remains to be determined (51b). Sequencing studies of RFHVMm and RFHVMn genomes show 83 to 84% amino acid identity at the DNA polymerase fragment examined compared to 67% for the HVS DNA polymerase; however, extended sequencing studies are needed to resolve the precise phylogenetic relationships between these nonhuman and human viruses.

Virus Gene Expression

Like other herpesviruses, KSHV appears to have a latent replication cycle, in which virus plasmid DNA replicates in synchrony with host cell replication, and a lytic replication cycle in which new viral progeny are produced. So far, both latent and lytic replication gene expression have been studied mainly in BCBL/PEL cell lines (128, 139, 174, 187, 212, 240). During latent replication in BCBL/PEL cell lines, viral gene expression is limited to relatively few strongly expressed transcripts (187, 240). Lytic replication is associated with expression of many structural proteins and replication enzymes required for virion packaging and release from the cell (128, 240) which can be induced with butyrate (129) or phorbol esters (139, 174). Initial analysis suggests that there may be tissue-specific expression patterns for some KSHV genes, and thus determination of lytic and latent gene expression in BCBL/PEL may not be applicable to KS tissues (136).

KSHV gene expression in BCBL/PEL

Three major patterns of KSHV gene expression have been identified which are similar to those found for other herpesviruses. Since de novo transmission systems are not available, classification of KSHV gene expression into latent, immediate-early, and late replication cycle patterns cannot be rigorously evaluated. Gene expression patterns are therefore delineated in BCBL/PEL cells by whether expression is constitutive (latent, class I), requires induction with phorbol esters or butyrate (lytic, class III), or is expressed constitutively but is enhanced by induction agent treatment (class II) (187). Latent genes (class I) are expressed in uninduced cells, and transcriptional activity is not affected by inducing agents such as phorbol esters (187). These genes include ORFK13, ORF72, and ORF73 (36, 64, 174). Two polycistronic latently expressed transcripts have been identified from the region including genes ORFK13, ORF72, and ORF73. A 6.6-kb transcript includes all these three latently expressed genes, while an alternatively spliced 2.2-kb latent transcript originating at the same promoter includes ORFK13 and ORF72 alone (172, 189). ORF72 encodes a homolog of mammalian D-type cyclins (36) shown to be functionally active in phosphorylating the retinoblastoma tumor suppressor protein in association with cyclin-dependent kinases (38, 71, 114). ORF73 encodes the 223–234-kDa LANA (172). ORFK13 may have antiapoptotic properties (217).

A second group of genes (class II) are expressed in uninduced BCBL/PEL cells, but, unlike structural late gene products which are not expressed in the absence of phorbol esters or butyrate (117, 128), their expression is increased by phorbol ester treatment (140, 187). The ORFK12 transcript is expressed at high levels in BCBL/PEL cells, but is further induced by phorbol ester treatment, and has been used as a sensitive in situ hybridization probe (207, 210, 240, 241).

ORFK12 encodes T0.7, an abundantly expressed polyadenylated transcript which may be translated into a small hydrophobic protein (241). T1.1 and T0.7 are the most abundant class II transcripts expressed in BCBL/PEL cells (239, 240), but other class II transcripts also include the KSHV-encoded cytokines such as vIL-6 (ORFK2) (136). While vIL-6 is expressed in uninduced BCBL/PEL-cell lines and KSHV-positive lymphoma cells in vivo, vIL-6 protein is rarely found in KS spindle cells (136; Y. Chang and M.T. Weisse, unpublished observations). T0.7 is characteristically classified as a latent transcript although it, like T1.1, is inducible with phorbol ester treatment (187) and therefore is considered a class II transcript (187).

Class III transcripts are only expressed after chemical induction in BCBL/PEL cells under tight latency control and presumably reflect lytic replication (187). Class III genes include most of the genes encoding structural proteins, such as ORF25 (major capsid protein), and DNA replication enzymes. Evidence for class III lytic gene expression can also be obtained by use of DNA polymerase antagonists which inhibit the KSHV lytic gene expression program. A transcriptional map of KSHV gene expression in BC-1 has recently been prepared by Sarid et al. (187). While most genes clearly fit into these expression patterns, future studies are likely to clarify the expression programs used by KSHV throughout its life cycle.

KSHV gene expression in KS tissues

KSHV establishes a persistent infection in most KS spindle cells as demonstrated by expression studies of transcripts for K12, K13, ORF72, and ORF73 (172, 207, 210). A subpopulation (approximately 10%) of KS spindle cells also express the polyadenylated nuclear T1.1 RNA (207) in a pattern similar to that of a lytic phase structural gene (ORF26). Expression of T1.1 may therefore be indicative of lytic replication within KS lesions, suggesting that a subpopulation of KS spindle cells produce KSHV virions. Intranuclear herpesvirus-like particles or intranuclear inclusions characteristic of herpesviruses can be identified in KS tissues (92, 158, 228), and DNA polymerase inhibitors may have clinical activity against KS tumors. Expression of ORF74, encoding a functional chemokine receptor (10), is also detected by reverse transcriptase PCR (36). Because of the presence of both latently and lytically infected spindle cells in KS tissues, it is not clear whether this gene is expressed during the latent or the lytic viral expression programs.

Molecular Piracy and KSHV

With whole genome sequencing of large DNA viruses (5, 39, 122, 181, 193, 195, 225), it has become increasingly clear that piracy of cellular regulatory and signaling protein genes is a common phenomenon (4, 143). KSHV has acquired a far greater number and diversity of homologs to cellular genes than other human herpesviruses (136, 181). These homologs include a member of the family of complement-binding proteins similar to CR2, three macrophage inflammatory protein (MIP) chemokines, IL-6, Bcl-2, an interferon regulatory factor (IRF) protein, a D-type cyclin, an IL-8 receptor-like G protein-coupled receptor (GCR), and an NCAM-like adhesin molecule (36, 40, 136, 148, 149, 152, 153, 181, 188). A number of these genes are also found in other viruses, where they may serve a similar function (5, 12, 122, 195). For example, complement-binding protein homologs are found in poxviruses as well as HVS and MHV68 (5, 122, 193, 195, 225); cyclin and

GCR homologs are present in HVS and MHV68 (151, 225); both EBV and HVS encode functional Bcl-2 genes (42, 203); and an NCAM-like homolog is present in the genomes of HHV-6 and HHV-7 (72, 150). The sequence determination of KSHV has proven very useful in delineating common requirements for herpesvirus interactions with cell regulatory and signaling pathways (181). There is a remarkable correspondence between cellular homologs encoded by KSHV to cellular genes known to be induced after EBV infection, suggesting that both viruses have conserved mechanisms for modifying the same regulatory pathways. The "transparency" (138) of the KSHV genome, resulting from the many regulatory factors that have been pirated by the virus, points to specific regulatory pathways that are important in maintaining herpesvirus infection.

In addition to regulatory and signaling protein genes, homologs to DNA synthetic enzymes are encoded by KSHV, including a DNA polymerase, a dihydrofolate reductase, a thymidylate synthase, a thymidine kinase, and the ribonucleotide reductase subunit genes (139, 181). Some of all of these genes are also shared with other herpesviruses. An intriguing feature of the cellular counterparts of these genes is that they are under transcriptional control of the E2F family of transcriptional regulators (1). It is therefore unlikely that KSHV has pirated these genes solely to supplement cellular DNA synthetic enzymes during lytic replication. Instead, it appears likely that KSHV encodes key checkpoint proteins controlling DNA synthesis, allowing viral DNA replication outside of the S phase of the cell cycle or when circumvention of pRb-imposed shutdown of DNA synthesis is required (138).

POTENTIAL KSHV ONCOGENES

Although no homologs to EBV or HVS transforming genes have been identified by sequence analysis, KSHV encodes recognizable viral homologs to at least five genes that either are established human proto-oncogenes or contribute to maintenance of a transformed cell phenotype (181). All of these genes are encoded by regions of the genome which are nonconserved and thus may contribute to the particular pathologic characteristics associated with KSHV infection. However, with the exception of vIRF, in vitro transformation studies are only now being initiated, and whether or not these genes play roles in KSHV-related neoplasia remains to be determined. Piracy of cellular proto-oncogenes by KSHV has been hypothesized to inhibit intrinsic antiviral activities of tumor suppressor pathways (138a).

v-Bcl-2 (ORF16)

KSHV ORF16 possesses significant homology to cellular Bcl-2 and other members of the Bcl-2 family. This family of genes is involved in the regulation of apoptosis, the process of programmed cell death. Although their mechanism of action is poorly understood, conserved sequence motifs such as the BH1 and BH2 domains are necessary for members of this family to achieve their functional effect by binding to each other in homodimeric or heterodimeric interactions. Cellular Bcl-2-like proteins appear to form ion pores which are intrinsic to their activity in the mitochondrial membrane (130). While some members of the Bcl-2 family of protein prevent apoptosis, other members such as Bax act to promote apoptosis (for review see reference 82). Some of the proapoptotic members of this family, such as Bax, may be constituent components for final common apoptotic pathways resulting from diverse signaling pathways, such as p53 or CD95/Fas activation.

A striking number of RNA and DNA viruses encode proteins which are either are sequence homologs to Bcl-2 or mimic the antiapoptotic activity of Bcl-2 (196, 216). The conservation of functional homologs to Bcl-2 among both tumor and non-tumor viruses suggests that inhibition of apoptosis is critical for successful virus replication. Viral Bcl-2 homologs are commonly thought to be primarily active during lytic replication, during which time the cell may be particularly prone to apoptotic responses (216). In this case, virus-encoded Bcl-2 homologs may keep the cell alive long enough to achieve maximum virion replication. Alternatively, v-Bcl-2 may contribute to neoplastic cell expansion by functioning as an antiapoptotic factor in enhancing survival of infected cells. Dysregulated expression of the human Bcl-2 gene has been associated with neoplasia and overexpression of Bcl-2 mRNA and protein as a result of chromosomal translocation t(14:18) as seen in follicular cell lymphomas (42, 223).

The KSHV Bcl-2 (ORF16) shares only 16% amino acid identity with its cellular counterpart yet retains recognizable BH1 and BH2 domains which are sites of homo/heterodimerization (40, 188). Functional studies indicate that Bcl-2 prevents Bax-mediated toxicity or apoptosis in yeast, transfected fibroblasts, and Sindbis virus-infected cells. There is contradictory evidence about the ability of v-Bcl-2 to heterodimerize with human Bcl-2/Bax members. Using a yeast two-hybrid system, Sarid and colleagues (188) have found evidence that v-Bcl-2/hu-Bcl-2 heterodimerize, which is similar to the interactions of the HVS homolog (146); Cheng and colleagues (40), however, did not find specific interactions between the KSHV protein and other Bcl-2-like proteins by using coimmunoprecipitation. These differences could be due to the sensitivities or specificities of the two different assays or due to differences in the v-Bcl-2 constructs. Additional studies will be needed to clarify likely interactions between v-Bcl-2 and human Bcl-2 family members; however, both studies demonstrate that the v-Bcl-2 has functional antiapoptotic activity. v-Bcl-2, like its EBV homolog BHRF1 (144), appears to be primarily expressed during lytic virus replication (M. Sturzl, personal communication). Transcription of this gene can be detected in KS lesions by reverse transcriptase PCR (188), but this is not unexpected even if v-Bcl-2 is not expressed during latency, given the high sensitivity of this technique and the proportion of KS tumor cells undergoing lytic virus replication in a tumor.

vIL-6 (ORFK2)

IL-6 overexpression has been implicated in a number of human malignancies, especially tumors of B-cell origin. This cytokine has been found to act in a paracrine or autocrine manner to stimulate growth of multiple myeloma (98, 103), acute myeloblastic leukemia (86), and some lymphomas (236, 237). Myeloma and plasmacytoma cell lines have been established which are paracrine dependent on IL-6 (84, 85, 102).

There is also strong clinicopathologic and experimental evidence that IL-6 dysregulation may also play a role in nonneoplastic lymphoid disease such as CD (90, 111, 120, 238). Patients with CD frequently present with systemic findings and symptoms of generalized lymphadenopathy with plasmacytosis, fever, hyper-gammaglobulinemia, anemia, and increased levels of plasma acute-phase proteins. These symptoms have been correlated with elevated IL-6 serum levels and IL-6 protein expression in lymphoid tissues. Further, a mouse model for CD has been developed by overexpression of IL-6 in hematolymphoid cells (27, 28). CD has been epidemiologically linked to KS, and a subset of these cases have been found to contain KSHV.

A unique gene (ORFK2) encoded by KSHV is 25% identical at the amino acid level to hu-IL-6 (136, 149, 153). The virus protein is a secreted cytokine capable of preventing apoptosis of the IL-6-dependent mouse myeloma cell line B9 (136, 153). vIL-6 is secreted by BCBL/PEL cells infected by KSHV but not appreciably expressed in KS tissues infected with the virus, suggesting that vIL-6 has a tissue-specific role in the two tumors (136). Recently, a lymphoproliferative disorder occurring in KSHV-infected transplant patients has been recognized in which vIL-6 expression is likely to induce a hyperplastic B-cell response (123) (Fig. 6).

In addition to functional studies demonstrating an ability to maintain B9 proliferation (136, 153), vIL-6 activates STAT1, STAT3, and Jak1 phosphorylation in HepG2 hepatoma cells (133). This pattern of Jak-STAT signal pathway activation is similar to that caused by huIL-6 (133). huIL-6 and vIL-6, however, differ markedly in their cellular IL-6 receptor interactions. IL-6 signal transduction is mediated by binding to the specific IL-6Rα (gp80) protein which associates with the transduction unit of the complex, gp130, which is shared among various members of the IL-6 cytokine family (56). Since gp130 is expressed on a number of cell types, it is largely the expression of the IL-6Rα subunit which determines the tissue

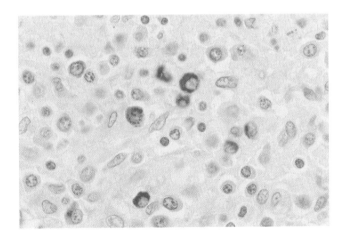

Figure 6   vIL-6 expression (Vector Red chromagen) in KSHV-infected plasmacytoid infiltrate found in the spleen of a patient with posttransplantation KS and generalized lymphadenopathy (123). Cells show typical cytoplasmic immunostaining for vIL-6 expression (red) with nuclear exclusion (magnification, ×230; Mayer's hematoxylin counterstain).

target specificity for IL-6. Whereas huIL-6 requires both the IL-6Rα and the gp130 protein for signal transduction, vIL-6 is able to induce STAT1/3 activation by binding to gp130 alone (133). It is unknown whether or not this results in fundamental differences in cell types activated by vIL-6 which could contribute to KSHV-related pathogenesis.

vIL-6 may thus contribute to cellular proliferation, particularly in B-cell lineages, by inhibition of normal B-cell apoptosis signaling (80, 115). Studies in myeloma cell lines demonstrate that huIL-6 induces Bcl-xL and that the effects of exogenous IL-6 on IL-6-dependent cells can be mimicked by transfection and overexpression of Bcl-2 (192). While similar studies have not been performed with KSHV vIL-6, its ability to activate cell signaling through the shared gp130 subunit indicates that it is likely to have a comparable function. KSHV infection of stromal cells of multiple myeloma lesions has been described (175), and vIL-6 expression has been implicated in maintaining tumor cells. This work has not been confirmed by detailed serologic and PCR-based studies (119a, 120a, 157), and thus future studies will be required to determine what role, if any, KSHV vIL-6 plays in multiple myeloma.

The close correspondence between signal pathways activated by EBV infection and those activated by virus-encoded homologs during KSHV infection (181) is seen for IL-6 signal transduction (136). EBV-infected B cells have increased cellular expression of huIL-6, and EBV-derived lymphoblastoid cell lines (LCL) become autocrine dependent on IL-6 (221). While EBV-infected LCL are immortalized but not tumorigenic when injected into nude mice, overexpression of huIL-6 in LCL allows these cells to generate lymphomas (190). EBV-related posttransplantation lymphoproliferative disease is also associated with huIL-6 overexpression (220).

v-Cyclin (ORF 72)

Cyclins are a family of proteins, largely defined by sequence homology, involved in the regulation of the cell cycle and DNA replication (for review see references 197 and 198). Subtypes of cyclins are defined according to where they act in the cell cycle. The D-type cyclins, to which the KSHV v-cyclin shows strongest amino acid similarity, function during G1. Hypophosphorylated pRb inhibits entry into S phase by binding E2F transcriptional factors, preventing transcription of DNA synthesis enzymes. D-type cyclins directly regulate cell passage through the G1 cell cycle checkpoint (166) by binding to cyclin-dependent kinases, resulting in hyperphosphorylation of the pRb tumor suppressor protein by the cyclin-dependent kinase-cyclin complex (97).

Cyclin D1 is an oncogene in humans overexpressed in mantle cell lymphomas and parathyroid tumors (142). Recent studies suggest that most if not all tumors require inhibition of the pRb-mediated checkpoint control for successful proliferation (198). This can be achieved in a variety of ways including homozygous deletion of pRb or cyclin-dependent kinase inhibitor genes (CDKIs, p21, p16, and p27) or by cyclin overexpression.

Several rhadinoviruses including KSHV possess homologs to D-type cyclins (36, 151, 225). The KSHV v-Cyc (ORF72), expressed in both KS lesions and latently infected B cells, has 27% amino acid identity with human cyclin D2 (36). The typical "cyclin box" sequence of cellular cyclins, responsible for interaction with cyclin-dependent kinases and required for pRb phosphorylation, is conserved in KSHV v-Cyc (38). While EBV does not possess a cyclin homolog, cooperative interaction between the EBV proteins EBNA-LP and EBNA2 may induce overexpression of cyclin D2 (200, 201), possibly achieving a functional effect similar to KSHV v-Cyc expression.

The activity of v-Cyc has been evaluated in a bioassay using SAOS-2 cells, which have homozygous deletions of both pRb and p53 (58). When wild-type pRb is transfected into SAOS-2, the cells stop replicating and develop a characteristic senescent phenotype. Cotransfection of the KSHV v-Cyc with wild-type pRb, however, prevents SAOS-2 entry into senescence and induces continuous proliferation (38). v-Cyc orchestrates phosphorylation of pRb at authentic sites in vitro (38) and coimmunoprecipitates with Cdk6 (71, 114), demonstrating functional similarity to cellular D-type cyclins. v-Cyc has a broader range of activity than D-type cyclins, however, and can induce phosphorylation of H1 histone, indicating that it may be active in other phases of the cell cycle in addition to G1 (71, 114). Cellular cyclin-dependent kinase inhibitors do not inhibit v-Cyc (212a), which may allow the virus to escape DNA replication control as a cellular antiviral mechanism (138a).

Transfection of the v-Cyc into NIH 3T3 cells paradoxically does not lead to cell proliferation. Instead, transfectants appear to die of apoptosis (C. Boshoff, R. Sarid, P. S. Moore, and Y. Chang, unpublished observations). One explanation for this paradoxical result is that E2F dysregulation triggers p53 activation, leading to cellular apoptosis (83, 107, 194, 235). This is similar to the effect seen with cellular D-type cyclin overexpression (205) and is likely due to E2F activation of p53 (2, 198). Indeed, p53 has been referred to as the "guardian" of pRb (232) in that dysregulation of the G1 checkpoint results in p53 activation. Activation of p53 can induce cyclin-dependent kinase inhibitors, such as p21, which restores cell cycle arrest. However, if this fails to stop unscheduled cell entry, it is likely that p53 activation leads to apoptosis and cell death. p53 activation may therefore act as a fail-safe mechanism to prevent uncontrolled cell proliferation resulting from defects in the pRb tumor suppressor pathway.

v-GCR (ORF74)

A G-protein coupled receptor (GCR) gene (ORF74) with high sequence homology to the IL-8R is also found in both KSHV and HVS (5, 36, 75a). Analogous to the other proto-oncogenes already discussed, the GCR is phylogenetically similar to an orphan receptor induced by EBV infection, EBI-1. Transient expression studies of the KSHV v-GCR indicate that it is constitutively active and either does not require ligand binding or is activated by a ligand commonly expressed in cell culture medium (10). Despite its constitutive activation state, the receptor also binds chemokines belonging to both the CXC and CC families, a property shared with the Duffy antigen receptor. Transfection of the KSHV GCR into rat kidney fibroblasts (NRK-49F) enhances their proliferation, suggesting that it may contribute to virus-induced tumor formation (10). The angiogenic activity of IL-8 also suggests that endothelial cells may be particularly responsive to v-GCR expression, contributing to KS tumor formation. Evidence for this was demonstrated by transfection experiments showing that the v-GCR can transform NIH 3T3 cells and induce elaboration of angiogenic factors (12a). This gene is not expressed in latently infected B cells, and further studies are needed to determine whether it is expressed in, and responsible for, proliferative responses in endothelial cells.

vIRF (ORFK9)

A unique gene which is not found in EBV or HVS is the KSHV interferon regulatory factor (vIRF, ORFK9) (136, 181, 242a). This gene has low but significant homology to genes encoding the IRF family of proteins responsible for interferon signal transduction and regulation. IRFs are either transcriptional activators or repressors induced through the Jak-STAT pathway by interferon receptor signaling (for review see references 213 and 213a). Two members of this pathway, IRF-1 and IRF-2, are antagonistic in their effects to each other. IRF-1 positively amplifies activated transcription of interferon-stimulated genes, resulting in the characteristic cellular changes including transcription of p21 and major histocompatibility complex (MHC) antigens (87, 131, 186). p21 induction by interferon signaling leads to cell cycle arrest through dephosphorylation of pRb, and MHC upregulation leads to enhanced antigen presentation for cell-mediated immunity. Thus, interferon signaling may prevent viral replication by inhibiting viral DNA synthesis and by making the infected cell a better immunologic target. IRF-2 is synthesized late in the interferon signaling cascade and negatively regulates IRF-1 (213).

Interferon signal transduction may play an important role in preventing cellular transformation as well as inducing cellular antiviral responses. IRF-1 expression, like that of D-type cyclins, varies with the state of the cell cycle, being elevated during the resting G0 phase (79). Transfection of IRF-2 can fully transform NIH 3T3 cells to growth in low serum conditions and cause tumor formation in nude mice. As expected for the antagonistic activity of IRF-1 and IRF-2, this can be prevented by cotransfection of IRF-1 with the transformed phenotype of NIH 3T3 cells, depending on the relative ratios of IRF-1 and IRF-2 concentrations (79).

vIRF appears to act as a negative regulator of interferon signaling, analogous to IRF-2. Transfection of vIRF into 293, HeLa, and microvascular endothelial cells inhibits class I interferon-mediated induction of reporter plasmids containing interferon-stimulated response elements (ISRE) (63, 242a). Like IRF-2, stable transfection of the vIRF gene into NIH 3T3 cells results in cell transformation as measured by in vitro transformation assays. Nude mice injected with the vIRF-transformed NIH 3T3 cells rapidly develop fibrosarcomas expressing the viral gene. vIRF transformants have down-regulation of p21 which may contribute to cell cycle dysregulation and transformation. Surface expression of MHC class I may also be down-regulated by vIRF, suggesting that virus-infected cells may express this gene to escape from host immune surveillance. Currently, the mechanism by which vIRF interacts with the interferon signal cascade is still unclear since direct binding to ISRE sequences or to members of the interferon signal cascade has not been found (63, 242a). The KSHV vIRF has only a partially conserved DNA binding domain thought to be critical for direct interactions of IRF-like transcriptional regulators with ISRE (136). Like vIL-6, vIRF does not appear to be expressed in KS tissues to a significant degree, but is expressed and inducible in BCBL/PEL (136). Several other sequences with low degrees of homology to IRF-like genes are present in KSHV, but it is not known whether or not they represent actual ORFs (149, 181).

Other Potential Oncogenes

LANA (ORF73)

The LANA antigen encoded by ORF73 is positionally conserved in the HVS genome but has not been described as a major latency-associated antigen for HVS. The sequence homology between the HVS and the KSHV proteins is low, with LANA having a long acidic repeat region and a leucine zipper motif at the carboxyl-terminal portion of the repeat region (181). The acidic repeat domain is likely to account for its aberrantly slow electrophoretic migration on polyacrylamide gel electrophoresis (64, 172). There is no current evidence indicating that LANA is involved in cellular transformation, although it is expressed on a polycistronic 6-kb transcript that includes ORFK13 (v-FLIP) and ORF72 (v-Cyc) (172, 187, 189). By analogy to EBV, in which restricted subsets of genes are expressed during latency, LANA remains as a candidate protein involved in KSHV tumorigenesis. It may primarily be involved in controlling transcription of other viral genes, rather than having a direct effect on cell immortalization.

v-FLIP (ORFK13)

ORFK13 is transcribed during latency along with v-Cyc (ORF72) (172, 187, 189). No functional studies of this gene in KSHV have yet been reported; however, the corresponding homolog in equine herpesvirus 2 (E8) protein acts as a dominant negative inhibitor of Fas/CD95-mediated apoptotic signaling, and hence has been given the name FLICE-inhibitor protein (v-FLIP) (17, 217). The corresponding HVS homolog (ORF71) is expressed during lytic reactivation and, like v-Bcl-2, has been hypothesized to inhibit terminal lytic-phase apoptosis (217). While there is no significant BLAST homology between the HVS and the KSHV proteins (hence the K13 designation for the KSHV protein), the KSHV v-FLIP retains a conserved DEDD motif characteristic for FLICE inhibitors and thus is likely to also abrogate CD95-mediated apoptosis. However, since the KSHV K13 is transcribed during latency, it may be expected to exhibit an antiapoptotic activity that could contribute to tumorigenesis if it functions in a manner analogous to the HVS ORF71 protein.

Viral MIPs (ORFK4, -K4.1, and -K6)

KSHV is also unique in encoding homologs to MIP CC chemokines. Like vIL-6 and vIRF, these genes are expressed as class II transcripts (136). Two of these genes (ORFK4 and ORFK6) encode proteins that have high homology to huMIP-1α and RANTES, which have been shown to have in vitro activity in inhibiting HIV-1 entry into cells by binding to the CCR5 coreceptor (43, 54). vMIP-I also inhibits HIV entry into cells expressing CCR5; recent studies suggest that these chemokines also may have activity in binding to other chemokine receptors, such as CCR3 (22, 101a, 136). Unlike the cellular chemokines, both vMIP-I and vMIP-II are highly angiogenic in chick chorioallantoic membrane assays, leading to speculation that they may contribute to KS through paracrine hyperplastic proliferation of spindle cells (22). vMIP-II, but not vMIP-I, induces eosinophil chemotaxis, presumably through CCR3 activation (22). The vMIP-III (ORFK4.1) shows low homology to human β-chemokines; it is unknown whether this is an expressed gene and, if so, whether it has similar functional activity.

Functional dissection of the KSHV genome is still in its infancy, and it is likely that additional viral genes will be found which can alter cell cycle control and enhance cellular proliferation. While several genes have been found in isolation to be candidates for induction of KSHV-related tumorigenesis, their roles may be dramatically different in the context of whole viral genome expression. Caution is warranted in ascribing in vivo tumorigenic activity to any of these viral genes, but their identification and initial characterization provide an important basis for a more sophisticated understanding of the interactions of KSHV with the infected cell.

KSHV AND ITS RELATIONSHIP TO OTHER DNA TUMOR VIRUSES

Rapid progress has been made over the past few years in delineating the cellular pathways leading to neoplasia (for review see references 57 and 198). Deletions of positive regulators of pRb or overexpression of pRb inhibitors are a frequent occurrence in human cancers, as are mutations in p53. It is becoming increasingly evident that neoplastic proliferation may require, at minimum, simultaneous inhibition of both pRb and p53 tumor suppressor pathways (198), as well as other less well-defined tumor suppressor pathways. pRb and p53 pathways interact with each other such that dysregulated transit through the pRb checkpoint leads to either p53-mediated growth arrest or apoptosis (50). Activation of p53 in this well-characterized feedback loop prevents dysregulated cell cycling alone from leading to neoplasia (232).

DNA tumor viruses have been essential tools in dissecting out transformation pathways, and KSHV promises to provide a unique model for investigating virus-induced transformation. Tumor viruses which are distantly related to each other have coevolved specific proteins to inhibit both pRb and p53, suggesting that these may be conserved features for a wide variety of viruses (for review, see references 138a, 196, 219, 227, and 232). This convergent evolution suggests that tumor suppressor pathways serve the dual function of defending the cell from virus infection as well as dysregulated proliferation (138, 138a, 147, 230, 234). Both viral infection and uncontrolled cell growth have in common dysregulated DNA replication, and thus common approaches to the control of dysregulated replication are not unexpected. Antiviral responses to infection include shutdown of cell cycle, induction of apoptosis, and enhancement of immunologic surveillance (136, 138a), which are also mechanisms to prevent tumor cell proliferation. It is apparent that a virus which successfully inhibits these antiviral defenses may contribute to uncontrolled cell proliferation.

Like other DNA tumor viruses, KSHV encodes a specific inhibitor of the pRb tumor suppressor pathway in the form of v-Cyc (ORF72). This protein, expressed during latency, is likely to be required to maintain cell cycling, even under conditions which promote cell cycle shutdown. In a similar fashion to the adenovirus E1A protein, which directly inhibits pRb (233), unopposed expression of KSHV v-Cyc appears likely to lead to apoptotic cell death. Identification of apoptosis inhibitors to prevent apoptotic activation by v-Cyc or other virus products may be integral to understanding KSHV-induced transformation.

The best-characterized KSHV apoptosis inhibitor is v-Bcl-2, which prevents Bax-mediated cell death in yeast and apoptosis in mammalian cells. Both HVS and EBV have similar proteins which are functionally conserved. These genes, however, appear to be largely induced during the lytic cycle of replication. While v-Bcl-2 is expressed as a class II transcript in BCBL/PEL (188), it remains to be seen whether this protein can prevent the infected cell from progressing into apoptosis. v-IL-6 may serve a similar function in preventing apoptosis and initiating an autocrine loop within the cell, although it is expressed in lymphoid tissues, not KS lesions, and is unlikely to play a role in the latter tumor (136). v-FLIP is expressed during latency in both BCBL/PEL and KS lesions and could contribute to prevention of apoptosis. It is not currently clear that p53 induces apoptosis through FLICE-mediated pathways or that v-FLIP can overcome the p53-mediated apoptotic responses induced by the v-Cyc on pRb. Finally, vIRF may contribute to tumorigenesis both through inhibitory effects on p21 induction and by inhibiting tumor-specific immune recognition. This KSHV gene provides evidence for an additional parallel with other DNA tumor virus proteins. The adenovirus E1A protein inhibits interferon-induced transcription through direct binding to the p300/CREB transcriptional adapter protein. It is unknown whether these KSHV proteins or other proteins yet to be examined play a direct role in KSHV tumorigenesis.

Tumor formation by DNA tumor viruses is likely to be an accidental event. Most tumor viruses are not tumorigenic in their natural hosts under normal circumstances, and there is no apparent evolutionary advantage in causing a tumor since it does not enhance lytic progeny production and is deleterious to the host (138). Most DNA tumor viruses are well adapted to replicate latently in their natural host without inducing cancers. A good example of this is human adenoviruses, which only cause tumors when crossing into new species. Similarly, rhadinoviruses, with the exception of KSHV, generally do not cause tumors in their natural hosts (73, 125). In the native host, immunologic surveillance also plays an important role in preventing virus-induced tumorigenesis. EBV, a normally benign viral infection in persons with intact immunity, induces lymphomas and soft-tissue sarcomas primarily in the setting of transplant-related or HIV-induced immunosuppression. Thus it is reasonable to assume that large numbers of healthy persons can be infected by KSHV and not manifest tumor development unless they are confronted by severe immunosuppression subsequent to transplantation or AIDS.

Humans are probably the natural host for KSHV, although thorough evaluation of other primate species has not yet been performed. KSHV-related tumors, however, primarily occur only in severely immunocompromised and infected persons. Evidence that immunologic surveillance is the primary mechanism for suppressing KSHV-induced tumorigenesis comes from surveillance statistics of the AIDS epidemic. Up to 50% of North American homosexual male AIDS patients develop KS over the course of their lifetimes (88), a rate that is tens of thousands-fold higher than for the general HIV-negative United States population (15). Further, the rate of posttransplant KS is proportional to the geographic prevalence of KSHV (see for example references 81, 162, and 167), suggesting that immunosuppression directly determines the risk of tumor formation. Among HIV-negative patients, KS is primarily a disease of the elderly which is consistent with nonspecific age-related decline in immune surveillance among those persons infected with KSHV.

There remains reasonable disagreement over whether or not KS is a mono- or polyclonal expansion of tumor cells driven by direct viral infection or a hyperplastic proliferation related to exogenous cytokines. HUMERA and KSHV TR studies lend support to the former possibility for KS, and current evidence makes clear that PEL are fully transformed lymphomas which in part result from KSHV gene expression. Because of the high rate of KS among homosexual male AIDS patients, the requirement for a specific oncogenic cellular mutation in addition to KSHV infection for initiation of KS tumorigenesis seems unlikely. Two transformed, aneuploid in vitro KS spindle cell lines have been reported to have in common 3(p14) translocations (165), but neither is infected with KSHV (59). No consistent cell mutations are found in KS tumor cells in vitro, and the rare nature of these transformed cell lines brings into question their generalizability. However, if KS tumor cells have a clonal origin (169, 170) this would suggest that KSHV infection alone is insufficient to generate an autonomously proliferating KS tumor cell, and that other host cell mutations may be required for clonal proliferation. PEL, on the other hand, may represent a cell line in which KSHV provides part of the genetic support for cellular transformation, which is supplemented by EBV coinfection or endogenous cell mutations.

CONCLUSION

In the short time since the first report of its discovery in 1994, research on KSHV has matured to the point where it can begin to provide important and generalizable insights into viral tumorigenesis. The transparency of its genome makes it an excellent system for studying the effects of isolated viral genes on cellular growth control and regulation pathways. At the same time, its close functional correspondence to EBV and other DNA tumor viruses allows important comparisons to be made on basic features of oncogenic viruses. The unique epidemiology of KSHV and its public health importance, especially to parts of sub-Saharan African, suggest that this virus be accorded an important priority in the development of techniques for its control and treatment.

Acknowledgments
T.F.S. is supported by the Medical Council of Great Britain, the Cancer-Research Campaign, and the European Concerted Action on KS Pathogenesis. Y.C. and P.S.M. are supported by the National Institutes of Health and the Centers for Disease Control and Prevention. This chapter is dedicated to the late American composer Frank Zappa.

Note in Proof

ORF K1

The ORF K1 gene is on the far left-hand end of the genome on the KSHV standard map (Fig. 5) at a position equivalent to that of the TIP and STP oncogenes of herpesvirus saimiri (HVS). While it has no apparent sequence similarity to known cellular or viral genes, it is predicted to encode a small transmembrane protein. Although ORF K1 is generally not expressed in BCBL/PEL cells in culture without TPA induction (107a, 187), it may have an important role in cell growth deregulation. Lee et al. (in press) substituted the KSHV ORF K1 gene for the HSV STP gene in an infectious recombinant HVS viral construct and showed that infection of marmoset T cells with this ORF K1-expressing HVS virus was able to generate IL-2-independent, immortalized T cell lines and induce lymphoproliferative disease in common marmosets (H. Lee, R. Veazey, K. Williams, M. Li, J. Guo, F. Neipel,

B. Fleckenstein, A. Lackner, R. C. Desrosiers, and J. U. Jung, "Deregulation of cell growth by the K1 gene of Kaposi's sarcoma-associated herpesvirus," *Nat. Med.*, in press). Thus, in the context of HVS gene expression, the ORF K1 can in some ways functionally substitute for the HVS STP. Sequence variation at the K1 locus between different KSHV isolates is predicted to affect the extracellular portion of this protein, while the intracellular amino acid sequence is highly conserved between the few isolates thus far examined.

REFERENCES

1. **Adams, P. D., and W. J. Kaelin.** 1995. Transcriptional control by E2F. *Semin. Cancer Biol.* **6:**99–108.
2. **Adams, P. D., and W. J. Kaelin.** 1996. The cellular effects of E2F overexpression. *Curr. Top. Microbiol. Immunol.* **208:**79–93...
3. **Adams, V., W. Kempf, M. Schmid, B. Muller, J. Briner, and G. Burg.** 1995. Absence of herpesvirus-like DNA sequences in skin cancers of non-immunosuppressed patients. *Lancet* **346:**1715.
4. **Ahuja, S. K., J. L. Gao, and P. M. Murphy.** 1994. Chemokine receptors and molecular mimicry. *Immunol. Today* **15:**281–287.
5. **Albrecht, J.-C., J. Nicholas, D. Biller, K. R. Cameron, B. Biesinger, C. Newman, S. Wittmann, M. A. Craxton, H. Coleman, B. Fleckenstein, and R. W. Honess.** 1992. Primary structure of the Herpesvirus saimiri genome. *J. Virol.* **66:**5047–5058.
6. **Aluigi, M. G., A. Albini, S. Carlone, L. Repetto, R. De Marchi, A. Icardi, M. Moro, D. Noonan, and R. Benelli.** 1996. KSHV sequences in biopsies and cultured spindle cells of epidemic, iatrogenic and Mediterranean forms of Kaposi's sarcoma. *Res. Virol.* **147:**267–275.
7. **Ambroziak, J. A., D. J. Blackbourn, B. G. Herndier, R. G. Glogau, J. H. Gullett, A. R. McDonald, E. T. Lennette, and J. A. Levy.** 1995. Herpes-like sequences in HIV-infected and uninfected Kaposi's sarcoma patients. *Science* **268:**582–583.
8. **Andre, S., O. Schatz, J. R. Bogner, H. Zeichhardt, M. M. Stoffler, H. U. Jahn, R. Ullrich, A. K. Sonntag, R. Kehrn, and J. Haas.** 1997. Detection of antibodies against viral capsid proteins of human herpesvirus 8 in AIDS-associated Kaposi's sarcoma. *J. Mol. Med.* **75:**145–152.
9. **Ansari, M. Q., D. B. Dawson, R. Nador, C. Rutherford, N. R. Schneider, M. J. Latimer, L. Picker, D. M. Knowles, and R. W. McKenna.** 1996. Primary body cavity-based AIDS-related lymphomas. *Am. J. Clin. Pathol.* **105:**221–229.
9a. **Ariyoshi, K., M. Schim van der Loeff, T. Corrah, F. Cham, P. M. Cook, D. Whitby, R. A. Weiss, and T. F. Schulz.** Kaposi's sarcoma and human herpesvirus 8 (HHV8) in HIV-1 and HIV-2 infection in The Gambia. *J. Hum. Virol.*, in press.
10. **Arvanitakis, L., R. E. Geras, A. Varma, M. C. Gershengorn, and E. Cesarman.** 1997. Human herpesvirus KSHV encodes a constitutively active G-protein-coupled receptor linked to cell proliferation. *Nature* **385:**347–350.
11. **Arvanitakis, L., E. A. Mesri, R. G. Nador, J. W. Said, A. S. Asch, D. M. Knowles, and E. Cesarman.** 1996. Establishment and characterization of a primary effusion (body cavity-based) lymphoma cell line (BC-3) harboring Kaposi's sarcoma-associated herpesvirus (KSHV/HHV-8) in the absence of Epstein-Barr virus. *Blood* **88:** 2648–2654.
12. **Baer, R., A. T. Bankier, P. L. Biggin, P. L. Deininger, P. J. Farrell, T. J. Gibson, G. Hatfull, G. S. Hudson, S. C. Satchwell, C. Seguin, P. S. Tuffnell, and B. G. Barrell.** 1984. DNA sequence and expression of the B95-8 Epstein-Barr virus genome. *Nature* **310:**207–211.
12a. **Bais, C., B. Santomasso, O. Coso, L. Arvanitakis, E. Geras-Raaka, J. S. Gutkind, A. S. Asch, E. Cesarman, M. C. Gerhengorn, and E. A. Mesri.** 1998. G-protein-coupled receptor of Kaposi's sarcoma-assocated herpesvirus is a viral oncogene and angiogenesis activator. *Nature* **391:**86–89.

13. **Barozzi, P., M. Luppi, L. Masini, R. Marasca, M. Savarino, M. Morselli, M. G. Ferrari, M. Bevini, G. Bonacorsi, and G. Torelli.** 1996. Lymphotrophic herpesvirus (EBV, HHV-6, HHV-8) DNA sequences in HIV negative Castleman's disease. *J. Clin. Pathol. Mol. Pathol.* **49:**M232–M235.

14. **Beral, V.** 1991. Epidemiology of Kaposi's sarcoma. *Cancer Surveys* **10:**5–22.

15. **Beral, V., D. Bull, S. Darby, I. Weller, C. Carne, M. Beecham, and H. Jaffe.** 1992. Risk of Kaposi's sarcoma and sexual practices associated with faecal contact in homosexual or bisexual men with AIDS. *Lancet* **339:**632–635.

16. **Beral, V., T. A. Peterman, R. L. Berkelman, and H. W. Jaffe.** 1990. Kaposi's sarcoma among persons with AIDS: a sexually transmitted infection? *Lancet* **335:**123–128.

17. **Bertin, J., R. C. Armstrong, S. Ottilie, D. A. Martin, Y. Wang, S. Banks, G. H. Wang, T. G. Senkevich, E. S. Alnemri, B. Moss, M. J. Lenardo, K. J. Tomaselli, and J. I. Cohen.** 1997. Death effector domain-containing herpesvirus and poxvirus proteins inhibit both Fas- and TNFR1-induced apoptosis. *Proc. Natl. Acad. Sci. USA* **94:** 1172–1176.

18. **Bigoni, B., R. Dolcetti, L. De Lellis, A. Carbone, M. Boiocchi, E. Cassai, and D. Di Luca.** 1996. Human herpesvirus 8 is present in the lymphoid system of healthy persons and can reactivate in the course of AIDS. *J. Infect. Dis.* **173:**542–549.

19. **Blackbourn, D. J., J. Ambroziak, E. Lennette, M. Adams, B. Ramachandran, and J. A. Levy.** 1997. Infectious human herpesvirus 8 in a healthy North American blood donor. *Lancet* **349:**609–611.

19a. **Blackbourn, D. J., E. T. Lennette, J. Ambroziak, D. V. Mourich, and J. A. Levy.** 1998. Human herpesvirus 8 detection in nasal secretions and saliva. *J. Infect. Dis.* **177:** 213–216.

19b. **Blackbourn, D., and J. A. Levy.** 1997. Human herpesvirus 8 in semen and prostate. *AIDS* **11:**249–250.

19c. **Blasis, C., C. Zietz, B. Haar, F. Neipel, S. Esser, N. H. Brockmeyer, E. Tsachler, S. Colombini, B. Ensoli, and M. Sturzl.** 1997. Monocytes in Kaposi's sarcoma lesions are productively infected by human herpesvirus 8. *J. Virol.* **71:**7963–7968.

20. **Boldough, I., P. Szaniszlo, W. A. Bresnahan, C. M. Flaitz, M. C. Nichols, and T. Albrecht.** 1996. Kaposi's sarcoma herpesvirus-like DNA sequences in the saliva of individuals infected with human immunodeficiency virus. *Clin. Infect. Dis.* **23:** 406–407.

21. **Boshoff, C., S.-J. Gao, L. Healy, S. Matthews, A. Thomas, R. Warnke, J. Strauchen, E. Matutes, R. A. Weiss, P. S. Moore, O. W. Kamel, and Y. Chang.** In vivo characterization of KSHV positive primary effusion lymphoma (PEL) cells. *Blood,* in press.

22. **Boshoff, C., Y. Endo, P. D. Collins, Y. Takeuchi, J. D. Reeves, V. L. Schweikert, M. Siani, T. Sasaki, T. J. Williams, P. W. Gray, P. S. Moore, Y. Chang, and R. A. Weiss.** 1997. Angiogenic and HIV inhibitory functions of KSHV-encoded chemokines. *Science* **278:**290–294.

23. **Boshoff, C., T. F. Schulz, M. M. Kennedy, A. K. Graham, C. Fisher, A. Thomas, J. O. McGee, R. A. Weiss, and J. J. O'Leary.** 1995. Kaposi's sarcoma-associated herpesvirus infects endothelial and spindle cells. *Nat. Med.* **1:**1274–1278.

24. **Boshoff, C., S. Talbot, M. Kennedy, J. O'Leary, T. Schulz, and Y. Chang.** 1996. HHV8 and skin cancers in immunosuppressed patients. *Lancet* **347:**338–339. (Letter.)

25. **Boshoff, C., D. Whitby, T. Hatziionnou, C. Fisher, J. van der Walt, A. Hatzakis, R. Weiss, and T. Schulz.** 1995. Kaposi's sarcoma-associated herpesvirus in HIV-negative Kaposi's sarcoma. *Lancet* **345:**1043–1044.

26. **Brambilla, L., V. Boneschi, E. Berti, M. Corbellino, and C. Parravicini.** 1996. HHV8 cell-associated viraemia and clinical presentation of Mediterranean Kaposi's sarcoma. *Lancet* **347:**1338.

27. **Brandt, S. J., D. M. Bodine, C. E. Dunbar, and A. W. Nienhuis.** 1990. Dysregulated interleukin 6 expression produces a syndrome resembling Castleman's disease in mice. *J. Clin. Invest.* **86:**592–599.

28. **Brandt, S. J., D. M. Bodine, C. E. Dunbar, and A. W. Nienhuis.** 1990. Retroviral-mediated transfer of interleukin-6 into hematopoietic cells of mice results in a syndrome resembling Castleman's disease. *Curr. Top. Microbiol. Immunol.* **166:**37–41.

29. **Browning, P. J., J. M. Sechler, M. Kaplan, R. H. Washington, R. Gendelman, R. Yarchoan, B. Ensoli, and R. C. Gallo.** 1994. Identification and culture of Kaposi's sarcoma-like spindle cells from the peripheral blood of human immunodeficiency virus-1-infected individuals and normal controls. *Blood* **84:**2711–2720.

30. **Buonaguro, F. M., M. L. Tornesello, E. Beth-Giraldo, A. Hatzakis, N. Mueller, R. Downing, B. Biryamwaho, S. D. K. Sempala, and G. Giraldo.** 1996. Herpesvirus-like DNA sequences detected in endemic, classic, iatrogenic and epidemic Kaposi's sarcoma (KS) biopsies. *Int. J. Cancer* **65:**25–28.

31. **Burger, R., J. Wendler, K. Antoni, G. Helm, J. R. Kalden, and M. Gramatzki.** 1994. Interleukin-6 production in B-cell neoplasias and Castleman's disease: evidence for an additional paracrine loop. *Ann. Hematol.* **69:**25–31.

31a. **Calabro, L., J. Sheldon, A. Favero, G. R. Simpson, J. R. Fiore, E. Gomez, G. Angarano, L. Chieco-Bianchi, and T. F. Schulz.** Seroprevalence of Kaposi's sarcoma-associated herpesvirus (KSHV/HHV8) in different regions of Italy. *J. Hum. Virol.*, in press.

32. **Carbone, A., A. Gloghini, E. Vaccher, V. Zagonel, C. Pastore, P. Dalla Palma, F. Branz, G. Saglio, R. Volpe, V. Tirelli, and G. Gaidano.** 1996. Kaposi's sarcoma-associated herpesvirus DNA sequences in AIDS-related and AIDS-unrelated lymphomatous effusions. *Br. J. Haematol.* **94:**533–543.

33. **Cesarman, E., Y. Chang, P. S. Moore, J. W. Said, and D. M. Knowles.** 1995. Kaposi's sarcoma-associated herpesvirus-like DNA sequences are present in AIDS-related body cavity based lymphomas. *N. Engl. J. Med.* **332:**1186–1191.

34. **Cesarman, E., P. S. Moore, P. H. Rao, G. Inghirami, D. M. Knowles, and Y. Chang.** 1995. *In vitro* establishment and characterization of two AIDS-related lymphoma cell lines containing Kaposi's sarcoma-associated herpesvirus-like (KSHV) DNA sequences. *Blood* **86:**2708–2714.

35. **Cesarman, E., R. Nador, and D. M. Knowles.** 1996. Body-cavity-based lymphoma in an HIV-seronegative patient without Kaposi's sarcoma-associated herpesvirus-like DNA sequences. *N. Engl. J. Med.* **334:**272–273. (Reply.)

36. **Cesarman, E., R. G. Nador, F. Bai, R. A. Bohenzky, J. J. Russo, P. S. Moore, Y. Chang, and D. M. Knowles.** 1996. Kaposi's sarcoma-associated herpesvirus contains G protein-coupled receptor and cyclin D homologs which are expressed in Kaposi's sarcoma and malignant lymphoma. *J. Virol.* **70:**8218–8223.

37. **Chang, Y., E. Cesarman, M. S. Pessin, F. Lee, J. Culpepper, D. M. Knowles, and P. S. Moore.** 1994. Identification of herpesvirus-like DNA sequences in AIDS-associated Kaposi's sarcoma. *Science* **265:**1865–1869.

38. **Chang, Y., P. S. Moore, S. J. Talbot, C. H. Boshoff, T. Zarkowska, D. Godden-Kent, H. Paterson, R. A. Weiss, and S. Mittnacht.** 1996. Cyclin encoded by KS herpesvirus. *Nature* **382:**410.

38a. **Chang, Y., J. L. Ziegler, H. Wabinga, E. Katongle-Mbidde, C. Boshoff, D. Whitby, T. F. Schulz, R. A. Weiss, and P. S. Moore.** 1996. Kaposi's sarcoma-associated herpesvirus and Kaposi's sarcoma in Africa. *Arch. Intern. Med.* **156:**202–204.

39. **Chee, M. S., S. B. Bankier, C. M. Bohni, R. C. Brown, T. Horsnell, C. A. Hutchison, T. Kouzarides, J. A. Martignetti, E. Preddie, S. C. Satchwell, P. Tomlinson, K. M. Weston, and B. G. Barrell.** 1990. Analysis of the protein coding content of the sequence of cytomegalovirus strain AD169. *Curr. Top. Microbiol. Immunol.* **154:**125–169.

40. **Cheng, E. H., J. Nicholas, D. S. Bellows, G. S. Hayward, H. G. Guo, M. S. Reitz, and J. M. Hardwick.** 1997. A Bcl-2 homolog encoded by Kaposi sarcoma-associated virus, human herpesvirus 8, inhibits apoptosis but does not heterodimerize with Bax or Bak. *Proc. Natl. Acad. Sci. USA* **94:**690–694.

41. **Chuck, S., R. M. Grant, E. Katongole-Mbidde, M. Conant, and D. Ganem.** 1996.

Frequent presence of a novel herpesvirus genome in lesions of human immunodeficiency virus-negative Kaposi's sarcoma. *J. Infect. Dis.* **173**:248–251.

42. **Cleary, M. L., S. D. Smith, and J. Sklar.** 1986. Cloning and structural analysis of cDNAs for bcl-2 and a hybrid bcl-2/immunoglobulin transcript resulting from the t(14;18) translocation. *Cell* **47**:19–28.

43. **Cocchi, F., A. L. DeVico, D. A. Garzino, S. K. Arya, R. C. Gallo, and P. Lusso.** 1995. Identification of RANTES, MIP-1 alpha, and MIP-1 beta as the major HIV-suppressive factors produced by CD8+ T cells. *Science* **270**:1811–1815.

44. **Collandre, H., S. Ferris, O. Grau, L. Montagnier, and A. Blanchard.** 1995. Kaposi's sarcoma and new herpesvirus. *Lancet* **345**:1043.

45. **Corbellino, M., G. Bestetti, M. Galli, and C. Parravicini.** 1996. Absence of HHV-8 in prostate and semen. *N. Engl. J. Med.* **335**:1238–1239. (Letter.)

46. **Corbellino, M., C. Parravicini, J. T. Aubin, and E. Berti.** 1996. Kaposi's sarcoma and herpesvirus-like DNA sequences in sensory ganglia. *N. Engl. J. Med.* **334**:1341–1342. (Letter.)

47. **Corbellino, M., L. Poirel, G. Bestetti, M. Pizzuto, J. T. Aubin, M. Capra, C. Bifulco, E. Berti, H. Agut, G. Rizzardini, M. Galli, and C. Parravicini.** 1996. Restricted tissue distribution of extralesional Kaposi's sarcoma-associated herpesvirus-like DNA sequences in AIDS patients with Kaposi's sarcoma. *AIDS Res. Hum. Retroviruses* **12**: 651–657.

48. **Costagliola, D., and M. Mary-Krause.** 1995. Clinical Epidemiology Group from Centres d'Information et de Soins de l'Immunodeficience Humaine. 1995. Can antiviral agents decrease the occurrence of Kaposi's sarcoma? *Lancet* **346**:578. (Letter.)

49. **Davis, D. A., R. W. Humphrey, F. M. Newcomb, T. R. O'Brien, J. J. Goedert, S. E. Straus, and R. Yarchoan.** 1997. Detection of serum antibodies to a Kaposi's sarcoma-associated herpesvirus-specific peptide. *J. Infect. Dis.* **175**:1071–1079.

50. **Debbas, M., and E. White.** 1993. Wild-type p53 mediates apoptosis by E1A, which is inhibited by E1B. *Genes Dev.* **7**:546–554.

51. **Decker, L. L., P. Shankar, G. Khan, R. B. Freeman, B. J. Dezube, J. Lieberman, and L. D. Thorley.** 1996. The Kaposi sarcoma-associated herpesvirus (KSHV) is present as an intact latent genome in KS tissue but replicates in the peripheral blood mononuclear cells of KS patients. *J. Exp. Med.* **184**:283–288.

51a. **Delabesse, E., E. Oksenhendel, C. Lebbe, O. Verola, and B. Varet.** 1997. Molecular analysis of clonality in Kaposi's sarcoma. *J. Clin. Pathol.* **50**:664–668.

51b. **Desrosiers, R. C., V. G. Sasseville, S. C. Czajak, X. Zhang, K. G. Mansfield, A. Kaur, R. P. Johnson, A. A. Lackner, and J. U. Jung.** 1997. A herpesvirus of rhesus monkeys related to the human Kaposi's sarcoma-associated herpesvirus. *J. Virol.* **71**: 9764–9769.

52. **Di Alberti, L., S. L. Ngui, S. R. Porter, P. M. Speight, C. M. Scully, J. M. Zakrewska, I. G. Williams, L. Artese, A. Piattelli, and C. G. Teo.** 1997. Presence of human herpesvirus 8 variants in the oral tissues of human immunodeficiency virus-infected persons. *J. Infect. Dis.* **175**:703–707.

53. **Dictor, M., E. Rambech, D. Way, M. Witte, and N. Bendsoe.** 1996. Human herpesvirus 8 (Kaposi's sarcoma-associated herpesvirus) DNA in Kaposi's sarcoma lesions, AIDS Kaposi's sarcoma cell lines, endothelial Kaposi's sarcoma simulators, and the skin of immunosuppressed patients. *Am. J. Pathol.* **148**:2009–2016.

54. **Dragic, T., V. Litwin, G. P. Allaway, S. R. Martin, Y. Huang, K. A. Nagashima, C. Cayanan, P. J. Maddon, R. A. Koup, J. P. Moore, and W. A. Paxton.** 1996. HIV-1 entry into CD4+ cells is mediated by the chemokine receptor CC-CKR-5. *Nature* **381**: 667–673.

55. **Dupin, N., M. Grandadam, V. Calvez, I. Gorin, J. T. Aubin, S. Harvard, F. Lamy, M. Leibowitch, J. M. Huraux, J. P. Escande, and H. Agut.** 1995. Herpesvirus-like DNA in patients with Mediterranean Kaposi's sarcoma. *Lancet* **345**:761–762.

56. **Economides, A. N., J. V. Ravetch, G. D. Yancopoulos, and N. Stahl.** 1995. Designer cytokines: targeting actions to cells of choice. *Science* 270:1351–1353.

57. **Elledge, S. J.** 1996. Cell cycle checkpoints: preventing an identity crisis. *Science* 274: 1664–1671.

58. **Ewen, M. E., H. K. Sluss, C. J. Sherr, H. Matsushime, J. Kato, and D. M. Livingston.** 1993. Functional interactions of the retinoblastoma protein with mammalian D-type cyclins. *Cell* 73:487–497.

59. **Flamand, L., R. A. Zeman, J. L. Bryant, I. Y. Lunardi, and R. C. Gallo.** 1996. Absence of human herpesvirus 8 DNA sequences in neoplastic Kaposi's sarcoma cell lines. *J. AIDS Hum. Retrovir.* 13:194–197.

59a. **Foreman, K. E., P. E. Bacon, E. D. Hsi, and B. J. Nickoloff.** 1997. In situ polymerase chain-reaction-based localization studies support role of human herpesvirus 8 as the cause of two AIDS-related neoplasms. *J. Clin. Invest.* 99:2971–2978.

60. **Foreman, K. E., J. J. Friborg, W. P. Kong, C. Woffendin, P. J. Polverini, B. J. Nickoloff, and G. J. Nabel.** 1997. Propagation of a human herpesvirus from AIDS-associated Kaposi's sarcoma. *N. Engl. J. Med.* 336:163–171.

61. **Gaidano, G., D. Capello, C. Pastore, A. Antinori, A. Gloghini, A. Carbone, L. M. Larocca, and G. Saglio.** 1997. Analysis of human herpesvirus type 8 infection in AIDS-related and AIDS-unrelated central nervous system lymphoma. *J. Infect. Dis.* 175:1193–1197.

62. **Gaidano, G., K. Cechova, Y. Chang, P. S. Moore, D. M. Knowles, and F. R. Dalla.** 1996. Establishment of AIDS-related lymphoma cell lines from lymphomatous effusions. *Leukemia* 10:1237–1240.

63. **Gao, S.-J., C. Boshoff, S. Jayachandra, R. A. Weiss, Y. Chang, and P. S. Moore.** 1997. KSHV ORF K9 (vIRF) is an oncogene that inhibits the interferon signaling pathway. *Oncogene* 15:1979–1986.

64. **Gao, S.-J., L. Kingsley, D. R. Hoover, T. J. Spira, C. R. Rinaldo, A. Saah, J. Phair, R. Detels, P. Parry, Y. Chang, and P. S. Moore.** 1996. Seroconversion of antibodies to Kaposi's sarcoma-associated herpesvirus-related latent nuclear antigens prior to onset of Kaposi's sarcoma. *N. Engl. J. Med.* 335:233–241.

65. **Gao, S. J., L. Kingsley, M. Li, W. Zheng, C. Parravicini, J. Ziegler, R. Newton, C. R. Rinaldo, A. Saah, J. Phair, R. Detels, Y. Chang, and P. S. Moore.** 1996. KSHV antibodies among Americans, Italians and Ugandans with and without Kaposi's sarcoma. *Nat. Med.* 2:925–928.

66. **Gessain, A., J. Briere, D. C. Angelin, F. Valensi, H. M. Beral, F. Davi, M. A. Nicola, A. Sudaka, N. Fouchard, J. Gabarre, X. Troussard, E. Dulmet, J. Audouin, J. Diebold, and G. de The.** 1997. Human herpes virus 8 (Kaposi's sarcoma herpes virus) and malignant lymphoproliferations in France: a molecular study of 250 cases including two AIDS-associated body cavity based lymphomas. *Leukemia* 11:266–272.

67. **Gessain, A., A. Sudaka, J. Briere, N. Fouchard, M.-A. Nicola, B. Rio, M. Arborio, X. Troussard, J. Audouin, J. Diebold, and G. de The.** 1996. Kaposi sarcoma-associated herpes-like virus (human herpesvirus type 8) DNA sequences in multicentric Castleman's disease: is there any relevant association in non-human immunodeficiency virus-infected patients? *Blood* 87:414–416.

68. **Giraldo, G., E. Beth, and F. M. Buonaguro.** 1984. Kaposi's sarcoma: a natural model of interrelationships between viruses, immunologic responses, genetics and oncogenesis. *Antibiot. Chemother.* 32:1–11.

69. **Giraldo, G., E. Beth, and F. Haguenau.** 1972. Herpes-type particles in tissue culture of Kaposi's sarcoma from different geographic regions. *J. Natl. Cancer Inst.* 49:1509.

70. **Glesby, M. J., D. R. Hoover, S. Weng, N. M. H. Graham, J. P. Phair, R. Detels, M. Ho, and A. Saah.** 1996. Use of antiherpes drugs and the risk of Kaposi's sarcoma: data from the Multicenter AIDS Cohort Study. *J. Infect. Dis.* 173:1477–1480.

71. **Godden-Kent, D., S. J. Talbot, C. Boshoff, Y. Chang, P. S. Moore, R. A. Weiss, and

S. Mittnacht. 1997. The cyclin encoded by Kaposi's sarcoma associated herpesvirus (KSHV) stimulates cdk6 to phosphorylate the retinoblastoma protein and histone H1. *J. Virol.* **71:**4193–4198.

72. **Gompels, U. A., J. Nicholas, G. Lawrence, M. Jones, B. J. Thomson, M. E. Martin, S. Efstathiou, M. Craxton, and H. A. Macaulay.** 1995. The DNA sequence of human herpesvirus-6: structure, coding content, and genome evolution. *Virology* **209:**29–51.

73. **Grassmann, R., B. Fleckenstein, and R. C. Desrosiers.** 1994. Viral transformation of human T lymphocytes. *Adv. Cancer Res.* **63:**211–244.

74. **Grulich, A. E., J. M. Kaldor, O. Hendry, K. H. Luo, N. J. Bodsworth, and D. A. Cooper.** 1997. Risk of Kaposi's sarcoma and oroanal sexual contact. *Am. J. Epidemiol.* **145:**673–679.

75. **Grulich, A. E., S. J. Olsen, K. Luo, O. Hendry, P. Cunningham, D. Cooper, S. J. Gao, Y. Chang, P. S. Moore, and J. Kaldor.** Kaposi's sarcoma-associated herpesvirus: a sexually transmissible infection. Submitted for publication.

75a. **Guo, H. G., P. Browning, J. Nicholas, G. S. Hayward, E. Tschachler, J. W. Jiang, M. Sadowska, M. Raffeld, S. Colombini, R. C. Gallo, and M. J. Reitz.** 1997. Characterization of a chemoline receptor-related gene in human herpesvirus 8 and its expression in Kaposi's sarcoma. *Virology* **228:**371–378.

76. **Gupta, P., M. K. Singh, C. Rinaldo, M. Ding, H. Farzadegan, A. Saah, D. Hoover, P. Moore, and L. Kingsley.** 1996. Detection of Kaposi's sarcoma herpesvirus DNA in semen of homosexual men with Kaposi's sarcoma. *AIDS* **10:**1596–1598.

77. **Gyulai, R., L. Kemeny, E. Adam, F. Nagy, and A. Dobozy.** 1996. HHV8 DNA in angiolymphoid hyperplasia of the skin. *Lancet* **347:**1837. (Letter.)

78. **Gyulai, R., L. Kemeny, M. Kiss, E. Adam, F. Nagy, and A. Dobozy.** 1996. Herpesvirus-like DNA sequence in angiosarcoma in a patient without HIV infection. *N. Engl. J. Med.* **334:**540–541.

79. **Harada, H., M. Kitagawa, N. Tanaka, H. Yamamoto, K. Harada, M. Ishihara, and T. Taniguchi.** 1993. Anti-oncogenic and oncogenic potentials of interferon regulatory factors-1 and -2. *Science* **259:**971–974.

80. **Hardin, J., S. MacLeod, I. Grigorieva, R. Chang, B. Barlogie, H. Xiao, and J. Epstein.** 1994. Interleukin-6 prevents dexamethasone-induced myeloma cell death. *Blood* **84:**3063–3070.

81. **Harwood, A. R., D. Osoba, and S. L. Hofstader.** 1979. Kaposi's sarcoma in recipients of renal transplants. *Am. J. Med.* **67:**759–765.

82. **Hawkins, C. J., and D. L. Vaux.** 1994. Analysis of the role of bcl-2 in apoptosis. *Immunol. Rev.* **142:**127–139.

83. **Hiebert, S. W., G. Packham, D. K. Strom, R. Haffner, M. Oren, G. Zambetti, and J. L. Cleveland.** 1995. E2F-1:DP-1 induces p53 and overrides survival factors to trigger apoptosis. *Mol. Cell. Biol.* **15:**6864–6874.

84. **Hilbert, D. M., M. Kopf, B. A. Mock, G. Kohler, and S. Rudikoff.** 1995. Interleukin 6 is essential for in vivo development of B lineage neoplasms. *J. Exp. Med.* **182:**243–248.

85. **Hirano, T., S. Suematsu, T. Matsusaka, T. Matsuda, and T. Kishimoto.** 1992. The role of interleukin 6 in plasmacytomagenesis. *CIBA Found. Symp.* **167:**188–196.

86. **Hoang, T., A. Haman, O. Goncalves, G. G. Wong, and S. C. Clark.** 1988. Interleukin-6 enhances growth factor-dependent proliferation of the blast cells of acute myeloblastic leukemia. *Blood* **72:**823–826.

87. **Hobeika, A. C., P. S. Subramaniam, and H. M. Johnson.** 1997. IFNα induces the expression of the cyclin-dependent kinase inhibitor p21 in human prostate cancer cells. *Oncogene* **14:** 1165–1170.

88. **Holmberg, S., S. Buchbinder, L. Conley, L. Wong, M. Katz, K. Penley, R. Hershow, and F. Judson.** 1995. The spectrum of medical conditions and symptoms before acquired immunodeficiency syndrome in homosexual and bisexual men infected with the human immunodeficiency virus. *Am. J. Epidemiol.* **141:**395–404.

89. Howard, M. R., D. Whitby, G. Bahadur, F. Suggett, C. Boshoff, F. M. Tenant, T. F. Schulz, S. Kirk, S. Matthews, I. Weller, R. S. Tedder, and R. A. Weiss. 1997. Detection of human herpesvirus 8 DNA in semen from HIV-infected individuals but not healthy semen donors. *AIDS* **11**:F15–F19.

90. Hsu, S. M., J. A. Waldron, S. S. Xie, and B. Barlogie. 1993. Expression of interleukin-6 in Castleman's disease. *Hum. Pathol.* **24**:833–839.

91. Humphrey, R. W., T. R. O'Brien, F. M. Newcomb, H. Nishihara, K. M. Wyvill, G. A. Ramos, M. W. Saville, J. J. Goedert, S. E. Straus, and R. Yarchoan. 1996. Kaposi's sarcoma (KS)-associated herpesvirus-like DNA sequences in peripheral blood mononuclear cells: association with KS and persistence in patients receiving anti-herpesvirus drugs. *Blood* **88**:297–301.

92. Ioachim, H. L. 1995. Kaposi's sarcoma and KSHV. *Lancet* **346**:1360. (Letter.)

93. Jin, Y.-T., S.-T. Tsai, J.-J. Yan, F.-F. Chen, W.-Y. Lee, W.-Y. Li, H. Chiang, and I.-J. Su. 1996. Presence of human herpesvirus-like DNA sequences in oral Kaposi's sarcoma. *Oral Surg. Oral Med. Oral Pathol.* **81**:442–444.

94. Jin, Y.-T., S.-T. Tsai, J.-J. Yan, J.-H. Hsiao, Y.-Y. Lee, and L.-J. Su. 1996. Detection of Kaposi's sarcoma-associated herpesvirus-like DNA sequence in vascular lesions: a reliable diagnostic marker for Kaposi's sarcoma. *Am. J. Clin. Pathol.* **105**:360–363.

95. Jones, J., T. Peterman, S. Chu, and H. Jaffe. 1995. AIDS-associated Kaposi's sarcoma. *Science* **267**:1078–1079.

96. Karcher, D. S., F. Dawkins, and C. T. Garrett. 1992. Body cavity-based non-Hodgkin's lymphoma (NHL) in HIV-infected patients: B-cell lymphoma with unusual clinical, immunophenotypic, and genotypic features. *Lab. Invest.* **92**:80a.

97. Kato, J., H. Matsushime, S. W. Hiebert, M. E. Ewen, and C. J. Sherr. 1993. Direct binding of cyclin D to the retinoblastoma gene product (pRb) and pRb phosphorylation by the cyclin D-dependent kinase CDK4. *Genes Dev.* **7**:331–342.

98. Kawano, M., H. Tanaka, H. Ishikawa, M. Nobuyoshi, K. Iwato, H. Asaoku, O. Tanabe, and A. Kuramoto. 1989. Interleukin-1 accelerates autocrine growth of myeloma cells through interleukin-6 in human myeloma. *Blood* **73**:2145–2148.

99. Kedes, D. H., and D. Ganem. 1997. Sensitivity of Kaposi's sarcoma-associated herpesvirus replication to antiviral drugs. Implications for potential therapy. *J. Clin. Invest.* **99**:2082–2086.

100. Kedes, D. H., D. Ganem, N. Ameli, P. Bacchetti, and R. Greenblatt. 1997. The prevalence of serum antibody to human herpesvirus 8 (Kaposi sarcoma-associated herpesvirus) among HIV-seropositive and high-risk HIV-seronegative women. *JAMA* **277**:478–481.

101. Kedes, D. H., E. Operskalski, M. Busch, R. Kohn, J. Flood, and D. Ganem. 1996. The seroepidemiology of human herpesvirus 8 (Kaposi's sarcoma-associated herpesvirus): distribution of infection in KS risk groups and evidence for sexual transmission. *Nat. Med.* **2**:918–924.

101a. Kledal, T. N., M. M. Rosenkilde, F. Coulin, G. Simmons, A. H. Johnson, S. Alouani, C. A. Power, H. R. Luttichau, J. Gerstoft, P. R. Clapham, I. Clark-Lewis, R. N. C. Wells, and T. W. Schwartz. 1997. A broad-spectrum chemokine antagonist encoded by Kaposi's sarcoma-associated herpesvirus. *Science* **277**:1656–1659.

102. Klein, B., and R. Bataille. 1992. Cytokine network in human multiple myeloma. *Hematol. Oncol. Clin. North Am.* **6**:273–284.

103. Klein, B., X. G. Zhang, M. Jourdan, J. Content, F. Houssiau, L. Aarden, M. Piechaczyk, and R. Bataille. 1989. Paracrine rather than autocrine regulation of myeloma-cell growth and differentiation by interleukin-6. *Blood* **73**:517–526.

104. Knowles, D. M., G. Inghirami, A. Ubriaco, and R. Dalla-Favera. 1989. Molecular genetic analysis of three AIDS-associated neoplasms of uncertain lineage demonstrates their B-cell derivation and the possible pathogenetic role of Epstein-Barr virus. *Blood* **73**:792–799.

105. Koelle, D. M., M.-L. Huang, B. Chandran, J. Vieira, M. Piepkorn, and L. Corey. 1997. Frequent detection of Kaposi's sarcoma-associated herpesvirus (human herpesvirus 8) DNA in saliva of human immunodeficiency virus-infected men: clinical and immunologic correlates. *J. Infect. Dis.* **176**:94–102.

106. Komanduri, K. V., J. A. Luce, M. S. McGrath, B. G. Herndier, and V. L. Ng. 1996. The natural history and molecular heterogeneity of HIV-associated primary malignant lymphomatous effusions. *J. AIDS Hum. Retrovir.* **13**:215–226.

107. Kowalik, T. F., J. DeGregori, J. K. Schwarz, and J. R. Nevins. 1995. E2F1 overexpression in quiescent fibroblasts leads to induction of cellular DNA synthesis and apoptosis. *J. Virol.* **69**:2491–2500.

107a. Lagunoff, M., and D. Ganem. 1997. The structure and coding organization of the genomic termini of Kaposi's sarcoma-associated herpesvirus (human herpesvirus 8). *Virology* **236**:147–154.

108. Lebbe, C., P. de Cremoux, M. Rybojad, C. Costa da Cunha, P. Morel, and F. Calvo. 1995. Kaposi's sarcoma and new herpesvirus. *Lancet* **345**:1180.

109. Lebbe, C., C. Pellet, R. Tatoud, F. Agbalika, P. Dosquet, J. P. Desgrez, P. Morel, and F. Calvo. 1997. Absence of human herpesvirus 8 sequences in prostate specimens. *AIDS* **11**:270.

110. Lefrere, J. J., M. C. Meyohas, M. Mariotti, J. L. Meynard, M. Thauvin, and J. Frottier. 1996. Detection of human herpesvirus 8 DNA sequences before the appearance of Kaposi's sarcoma in human immunodeficiency virus (HIV)-positive subjects with a known date of HIV seroconversion. *J. Infect. Dis.* **174**:283–287.

111. Leger-Ravet, M. B., M. Peuchmaur, O. Devergne, J. Audouin, M. Raphael, D. J. Van, P. Galanaud, J. Diebold, and D. Emilic. 1991. Interleukin-6 gene expression in Castleman's disease. *Blood* **78**:2923–2930.

112. Lennette, E. T., D. J. Blackbourne, and J. A. Levy. 1996. Antibodies to human herpesvirus type 8 in the general population and in Kaposi's sarcoma patients. *Lancet* **348**: 858–861.

113. Li, J. J., Y. Q. Huang, C. J. Cockerell, and K. A. Friedman. 1996. Localization of human herpes-like virus type 8 in vascular endothelial cells and perivascular spindle-shaped cells of Kaposi's sarcoma lesions by in situ hybridization. *Am. J. Pathol.* **148**: 1741–1748.

114. Li, M., H. Lee, D. W. Yoon, J. C. Albrecht, B. Fleckenstein, F. Neipel, and J. U. Jung. 1997. Kaposi's sarcoma-associated herpesvirus encodes a functional cyclin. *J. Virol.* **71**:1984–1991.

115. Lichtenstein, A., Y. Tu, C. Fady, R. Vescio, and J. Berenson. 1995. Interleukin-6 inhibits apoptosis of malignant plasma cells. *Cell. Immunol.* **162**:248–255.

116. Lin, J.-C., S.-C. Lin, E.-C. Mar, P. E. Pellett, F. R. Stamey, J. A. Stewart, and T. J. Spira. 1995. Is Kaposi's sarcoma-associated herpesvirus detectable in semen of HIV-infected homosexual men? *Lancet* **346**:1601–1602.

117. Lin, S. F., R. Sun, L. Heston, L. Gradoville, D. Shedd, K. Haglund, M. Rigsby, and G. Miller. 1997. Identification, expression, and immunogenicity of Kaposi's sarcoma-associated herpesvirus-encoded small viral capsid antigen. *J. Virol.* **71**:3069–3076.

118. Lunardi-Iskandar, Y., P. Gill, V. H. Lam, R. A. Zeman, F. Michaels, D. L. Mann, M. S. Reitz, M. Kaplan, Z. N. Berneman, D. Carter, J. L. Bryant, and R. C. Gallo. 1995. Isolation and characterization of an immortal neoplastic cell line (KS Y-1) from AIDS-associated Kaposi's sarcoma. *J. Natl. Cancer Inst.* **87**:974–981.

119. Luppi, M., P. Barozzi, A. Maiorana, T. Artusi, R. Trovato, R. Marasca, M. Savarino, N. L. Ceccherini, and G. Torelli. 1996. Human herpesvirus-8 DNA sequences in human immunodeficiency virus-negative angioimmunoblastic lymphadenopathy and benign lymphadenopathy with giant germinal center hyperplasia and increased vascularity. *Blood* **87**:3903–3909.

119a. Mackenzie, J., J. Sheldon, G. Morgan, G. Cook, T. F. Schulz, and R. F. Jarrett. 1997. HHV-8 and multiple myeloma in the U.K. *Lancet.* **350**:1144–1145. (Letter.)

120. **Mandler, R. N., D. P. Kerrigan, J. Smart, W. Kuis, P. Villiger, and M. Lotz.** 1992. Castleman's disease in POEMS syndrome with elevated interleukin-6. *Cancer* **69:** 2697–2703.

120a. **Marcelin, A. G., N. Dupin, D. Bouscary, P. Bossi, P. Cacoub, P. Ravaud, and V. Calvez.** 1997. HHV-8 and multiple myeloma in France. *Lancet* **350:**1144. (Letter.)

121. **Marchioli, C. C., J. L. Love, L. Z. Abbott, Y. Q. Huang, S. C. Remick, R. N. Surtento, R. E. Hutchison, D. Mildvan, K. A. Friedman, and B. J. Poiesz.** 1996. Prevalence of human herpesvirus 8 DNA sequences in several patient populations. *J. Clin. Microbiol.* **34:**2635–2638.

122. **Massung, R. F., L. I. Liu, J. Qi, J. C. Knight, T. E. Yuran, A. R. Kerlavage, J. M. Parsons, J. C. Venter, and J. J. Esposito.** 1994. Analysis of the complete genome of smallpox variola major virus strain Bangladesh-1975. *Virology* **201:**215–240.

123. **Matsushima, A., J. A. Strauchen, G. Lee, E. Scigilano, E. E. Hale, M. T. Weisse, D. Burstein, P. S. Moore, and Y. Chang.** Post-transplantation lymphoproliferative syndrome due to Kaposi's sarcoma-associated herpesvirus. Submitted for publication.

124. **McDonagh, D. P., J. Liu, M. J. Gaffey, L. J. Layfield, N. Azumi, and S. T. Traweek.** 1996. Detection of Kaposi's sarcoma-associated herpesvirus-like DNA sequence in angiosarcoma. *Am. J. Pathol.* **149:**1363–1368.

124a. **Medveczky, M. M., E. Horvath, T. Lund, and P. G. Medveczky.** 1997. In vitro sensitivity of the Kaposi's sarcoma-associated herpesvirus. *AIDS* **11:**1327–1332.

125. **Meinl, E., R. Hohlfeld, H. Wekerle, and B. Fleckenstein.** 1995. Immortalization of human T cells by Herpesvirus saimiri. *Immunol. Today.* **16:**55–58.

126. **Melbye, M., P. M. Cook, H. Hjalgrim, K. Begtrup, G. R. Simpson, R. J. Biggar, P. Ebbesen, and T. F. Schulz.** 1997. Transmission of human herpesvirus 8 (HHV 8) among homosexual men follows the pattern of a 'Kaposi's sarcoma agent.' *JAMA,* in press.

127. **Mesri, E. A., E. Cesarman, L. Arvanitakis, S. Rafii, M. A. S. Moore, D. N. Posnett, D. M. Knowles, and A. S. Asch.** 1996. Human herpesvirus-8/Kaposi's sarcoma-associated herpesvirus is a new transmissible virus that infects B cells. *J. Exp. Med.* **183:** 2385–2390.

128. **Miller, G., L. Heston, E. Grogan, L. Gradoville, M. Rigsby, R. Sun, D. Shedd, V. M. Kushnaryov, S. Grossberg, and Y. Chang.** 1997. Selective switch between latency and lytic replication of Kaposi's sarcoma herpesvirus and Epstein-Barr virus in dually infected body cavity lymphoma cells. *J. Virol.* **71:**314–324.

129. **Miller, G., M. Rigsby, L. Heston, E. Grogan, R. Sun, C. Metroka, S.-J. Gao, Y. Chang, and P. Moore.** 1996. Antibodies to butyrate inducible antigens of Kaposi's sarcoma-associated herpesvirus in HIV-1 infected patients. *N. Engl. J. Med.* **334:** 1292–1297.

130. **Minn, A. J., P. Velez, S. L. Schendel, H. Liang, S. W. Muchmore, S. W. Fesik, M. Fill, and C. B. Thompson.** 1997. Bcl-x(L) forms an ion channel in synthetic lipid membranes. *Nature* **385:**353–357.

131. **Miyamoto, M., T. Fujita, Y. Kimura, M. Maruyama, H. Harada, Y. Sudo, T. Miyata, and T. Taniguchi.** 1988. Regulated expression of a gene encoding a nuclear factor, IRF-1, that specifically binds to IFN-beta gene regulatory elements. *Cell* **54:**903–913.

132. **Mocroft, A., M. Youle, B. Gazzard, J. Morcinek, R. Halai, and A. N. Phillips.** 1996. Anti-herpesvirus treatment and risk of Kaposi's sarcoma in HIV infection. *AIDS* **10:** 1101–1105.

133. **Molden, J., C. Y., P. S. Moore, and M. A. Goldsmith.** 1997. A KSHV-encoded cytokine activates signaling through the shared gp130 receptor subunit. *J. Biol. Chem.* **272:** 19625–19631.

134. **Monini, P., L. de Lellis, and E. Cassai.** 1996. Absence of HHV-8 in prostate and semen. *N. Engl. J. Med.* **335:**1238–1239. (Letter.)

135. **Monini, P., L. de Lellis, M. Fabris, F. Rigolin, and E. Cassai.** 1996. Kaposi's sacoma-associated herpesvirus DNA sequences in prostate tissue and human semen. *N. Engl. J. Med.* **334**:1168–1172.

136. **Moore, P. S., C. Boshoff, R. A. Weiss, and Y. Chang.** 1996. Molecular mimicry of human cytokine and cytokine response pathway genes by KSHV. *Science* **274:** 1739–1744.

137. **Moore, P. S., and Y. Chang.** 1995. Detection of herpesvirus-like DNA sequences in Kaposi's sarcoma lesions from persons with and without HIV infection. *N. Engl. J. Med.* **332**:1181–1185.

138. **Moore, P. S., and Y. Chang.** KSHV-encoded oncogenes and oncogenesis. *J. Natl. Cancer Inst.*, in press.

138a. **Moore, P. S., and Y. Chang.** Antiviral activity of tumor suppressor pathways: clues from molecular piracy by KSHV. *Trends Genet.*, in press.

139. **Moore, P. S., S.-J. Gao, G. Dominguez, E. Cesarman, O. Lungu, D. M. Knowles, R. Garber, D. J. McGeoch, P. Pellett, and Y. Chang.** 1996. Primary characterization of a herpesvirus-like agent associated with Kaposi's sarcoma. *J. Virol.* **70**:549–558.

140. **Moore, P. S., L. Kingsley, S. D. Holmberg, T. Spira, L. J. Conley, D. Hoover, P. Gupta, H. Jaffe, and Y. Chang.** 1996. Kaposi's sarcoma-associated herpesvirus infection prior to onset of Kaposi's sarcoma. *AIDS* **10**:175–180.

141. **Morfeldt, L., and J. Torsander.** 1994. Long-term remission of Kaposi's sarcoma following foscarnet treatment in HIV-infected patients. *Scand. J. Infect. Dis.* **26**:749.

142. **Motokura, T., and A. Arnold.** 1993. PRAD1/Cyclin 1 proto-oncogenes: genomic organization, 5' DNA sequence, and sequence of a tumor-specific rearrangement breakpoint. *Gene Chromosome Cancer* **7**:89–95.

143. **Murphy, P. M.** 1997. AIDS—pirated genes in Kaposi's sarcoma. *Nature* **385**:296–297.

144. **Murray, P. G., L. J. Swinnen, C. M. Constandinou, J. M. Pyle, T. J. Carr, J. M. Hardwick, and R. F. Ambinder.** 1996. BCL-2 but not its Epstein-Barr virus-encoded homologue, BHRF1, is commonly expressed in posttransplantation lymphoproliferative disorders. *Blood* **87**:706–711.

145. **Nador, R. G., E. Cesarman, A. Chadburn, D. B. Dawson, M. Q. Ansari, J. Sald, and D. M. Knowles.** 1996. Primary effusion lymphoma: a distinct clinicopathologic entity associated with the Kaposi's sarcoma-associated herpes virus. *Blood* **88**:645–656.

146. **Nava, V. E., E. H.-Y. Cheng, M. Veliuona, S. Zou, R. J. Clem, M. L. Mayer, and J. M. Hardwick.** 1997. Herpesvirus saimiri encodes a functional homolog of the human *bcl-2* oncogene. *J. Virol.* **71**:4118–4122.

147. **Neil, J. C., E. R. Cameron, and E. W. Baxter.** 1997. p53 and tumour viruses: catching the guardian off-guard. *Trends Microbiol.* **5**:115–120.

148. **Neipel, F., J.-C. Albrecht, and B. Fleckenstein.** 1997. Cell-homologous genes in the Kaposi's sarcoma-associated rhadinovirus human herpesvirus 8: determinants of its pathogenicity? *J. Virol.* **71**:4187–4192.

149. **Neipel, F., J. C. Albrecht, A. Ensser, Y. Q. Huang, J. J. Li, K. A. Friedman, and B. Fleckenstein.** 1997. Human herpesvirus 8 encodes a homolog of interleukin-6. *J. Virol.* **71**:839–842.

149a. **Neyts, S., and E. Clercq.** 1997. Antiviral drug susceptibility of human herpesvirus 8. *Antimicrob. Agents Chemother.* **41**:2754–2756.

150. **Nicholas, J.** 1996. Determination and analysis of the complete nucleotide sequence of human herpesvirus. *J. Virol.* **70**:5975–5989.

151. **Nicholas, J., K. R. Cameron, and R. W. Honess.** 1992. Herpesvirus saimiri encodes homologues of G protein-coupled receptors and cyclins. *Nature* **355**:362–365.

152. **Nicholas, J., V. Ruvolo, J. Zong, D. Ciufo, H. G. Guo, M. S. Reitz, and G. S. Hayward.** 1997. A single 13-kilobase divergent locus in the Kaposi sarcoma-associated herpesvirus (human herpesvirus 8) genome contains nine open reading frames that are homologous to or related to cellular proteins. *J. Virol.* **71**:1963–1974.

153. **Nicholas, J., V. R. Ruvolo, W. H. Burns, G. Sandford, X. Wan, D. Ciufo, S. B. Hendrickson, H. G. Guo, G. S. Hayward, and M. S. Reitz.** 1997. Kaposi's sarcoma-associated human herpesvirus-8 encodes homologues of macrophage inflammatory protein-1 and interleukin-6. *Nat. Med.* **3:**287–292.

153a. **Nicholas, J., J. C. Zong, D. J. Alcendor, D. M. Cinto, L. J. Poole, R. J. Sarisky, J. C. Chiou, X. Zhang, X. Wan, H. G. Guo, M. S. Reitz, and G. S. Hayward.** Novel organizational features, captured cellular genes and strain variability within the genome of KSHV/HHV8. *J. Natl. Cancer Inst.,* in press.

154. **Noel, J. C., P. Hermans, J. Andre, I. Fayt, T. H. Simonart, A. Verhest, J. Haot, and A. Burny.** 1996. Herpesvirus-like DNA sequences and Kaposi's sarcoma: relationship with epidemiology, clinical spectrum, and histologic features. *Cancer* **77:**2132–2136.

155. **Oettle, A. G.** 1962. Geographic and racial differences in the frequency of Kaposi's sarcoma as evidence of environmental or genetic causes, p. 330–335. *In* L. V. Ackerman and J. F. Murray (ed.), *Symposium on Kaposi's Sarcoma,* vol. 18. Karger, Basel.

156. **Olsen, S. J., and P. S. Moore.** Kaposi's sarcoma-associated herpesvirus (KSHV/HHV8) and the etiology of KS. *In* H. Freidman, P. Medveczky, and M. Bendinelli (ed.), *Molecular Immunology of Herpesviruses,* in press. Plenum Publishing, New York.

157. **Olsen, S. J., W. Sherman, K. Tarte, B. Klein, A. Orazi, E. E. Hale, M. T. Weisse, and Y. Chang.** Examination of KSHV infection in multiple myeloma patients. Submitted for publication.

158. **Orenstein, J. M., S. Alkan, A. Blauvelt, K. T. Jeang, M. D. Weinstein, D. Ganem, and B. Herndier.** 1997. Visualization of human herpesvirus type 8 in Kaposi's sarcoma by light and transmission electron microscopy. *AIDS* **11:**F35–F45.

159. **Otsuki, T., S. Kumar, B. Ensoli, D. W. Kingma, T. Yano, S. M. Stetler, E. S. Jaffe, and M. Raffeld.** 1996. Detection of HHV-8/KSHV DNA sequences in AIDS-associated extranodal lymphoid malignancies. *Leukemia* **10:**1358–1362.

159a. **Parravicini, C., M. Corbellino, M. Paulli, U. Magrini, M. Lazzarino, P. S. Moore, and Y. Chang.** 1997. Expression of a virus-derived cytokine, KSHV vIL-6, in HIV-seronegative Castleman's disease. *Am. J. Pathol.* **151:**1517–1522.

160. **Pastore, C., A. Gloghini, G. Volpe, J. Nomdedeu, E. Leonardo, U. Mazza, G. Saglio, A. Carbone, and G. Gaidano.** 1995. Distribution of Kaposi's sarcoma herpesvirus sequences among lymphoid malignancies in Italy and Spain. *Br. J. Haematol.* **91:**918–920.

161. **Pawson, R., D. Catovsky, and T. F. Schulz.** 1996. Lack of evidence of HHV-8 in mature T-cell lymphoproliferative disorders. *Lancet* **348:**1450–1451. (Letter.)

162. **Penn, I.** 1979. Kaposi's sarcoma in organ transplant recipients: report of 20 cases. *Transplantation* **27:**8–11.

163. **Peterman, T. A., H. W. Jaffe, and V. Beral.** 1993. Epidemiologic clues to the etiology of Kaposi's sarcoma. *AIDS* **7:**605–611.

164. **Peterson, B. A., and G. Frizzera.** 1993. Multicentric Castleman's disease. *Semin. Oncol.* **20:**636–647.

165. **Popescu, N. C., D. B. Zimonjic, K. S. Leventon, J. L. Bryant, I. Y. Lunardi, and R. C. Gallo.** 1996. Deletion and translocation involving chromosome 3 (p14) in two tumorigenic Kaposi's sarcoma cell lines. *J. Natl. Cancer Inst.* **88:**450–455.

166. **Quelle, D. E., R. A. Ashmun, S. A. Shurtleff, J. Y. Kato, S. D. Bar, M. F. Roussel, and C. J. Sherr.** 1993. Overexpression of mouse D-type cyclins accelerates G1 phase in rodent fibroblasts. *Genes Dev.* **7:**1559–1571.

167. **Qunibi, W., M. Akhtar, K. Sheth, H. E. Ginn, F. O. Al, E. B. DeVol, and S. Taher.** 1988. Kaposi's sarcoma: the most common tumor after renal transplantation in Saudi Arabia. *Am. J. Med.* **84:**225–232.

168. **Raab-Traub, N., and K. Flynn.** 1986. The structure of the termini of the Epstein-Barr virus as a marker of clonal cellular proliferation. *Cell* **47:**883–889.

169. **Rabkin, C. S., G. Bedi, E. Musaba, R. Sunkutu, N. Mwansa, D. Sidransky, and R. J. Biggar.** 1995. AIDS-related Kaposi's sarcoma is a clonal neoplasm. *Clin. Cancer Res.* **1:**257–260.

170. **Rabkin, C. S., S. Janz, A. Lash, A. E. Coleman, E. Musaba, L. Liotta, R. J. Biggar, and Z. P. Zhuang.** 1997. Monoclonal origin of multicentric Kaposi's sarcoma lesions. *N. Engl. J. Med.* **336:**988–993.

171. **Rady, P. L., A. Yen, J. L. Rollefson, I., Orengo, S. Bruce, T. K. Hughes, and S. K. Tyring.** 1995. Herpesvirus-like DNA sequences in non-Kaposi's sarcoma skin lesions of transplant patients. *Lancet* **345:**1339–1340.

172. **Rainbow, L., G. M. Platt, G. R. Simpson, R. Sarid, S.-J. Gao, H. Stoiber, S. Herrington, P. S. Moore, and T. F. Schulz.** 1997. The 222- to 234-kilodalton latent nuclear protein (LNA) of Kaposi's sarcoma-associated herpesvirus (human herpesvirus 8) is encoded by orf73 and is a component of the latency-associated nuclear antigen. *J. Virol.* **71:**5915–5921.

173. **Renne, R., M. Lagunoff, W. Zhong, and D. Ganem.** 1996. The size and conformation of Kaposi's sarcoma-associated herpesvirus (human herpesvirus 8) DNA in infected cells and virions. *J. Virol.* **70:**8151–8154.

174. **Renne, R., W. Zhong, B. Herndier, M. McGrath, N. Abbey, D. Kedes, and D. Ganem.** 1996. Lytic growth of Kaposi's sarcoma-associated herpesvirus (human herpesvirus 8) in culture. *Nat. Med.* **2:**342–346.

175. **Rettig, M. B., H. J. Ma, R. A. Vescio, M. Pold, G. Schiller, D. Belson, A. Savage, C. Nishikubo, C. Wu, J. Fraser, J. W. Said, and J. R. Berenson.** 1997. Kaposi's sarcoma-associated herpesvirus infection of bone marrow dendritic cells from multiple myeloma patients. *Science* **276:**1851–1854.

176. **Rickinson, A. B.** 1996. Changing seroepidemiology of HHV-8. *Lancet* **348:**1110–1111.

177. **Robert, C., F. Agbalika, F. Blanc, and L. Dubertret.** 1996. HIV-negative patient with HHV-8 DNA follicular B-cell lymphoma associated with Kaposi's sarcoma. *Lancet* **347:**1042–1043. (Letter.)

178. **Roizman, B.** 1993. The family Herpesviridae, p. 1–9. *In* B. Roizman, R. J. Whitley, and C. Lopez (ed.), *The Human Herpesviruses.* Raven Press, Ltd., New York.

179. **Rose, T. M., K. B. Strand, E. R. Schultz, G. Schaefer, G. J. Rankin, M. E. Thouless, C. C. Tsai, and M. L. Bosch.** 1997. Identification of two homologs of the Kaposi's sarcoma-associated herpesvirus (human herpesvirus 8) in retroperitoneal fibromatosis of different macaque species. *J. Virol.* **71:**4138–4144.

180. **Rubin, M. A., J. P. Parry, and B. Singh.** 1998. Kaposi's sarcoma-associated herpesvirus DNA sequences: lack of detection in prostatic tissue of HIV-negative immunocompetent adults. *J. Urol.* **159:**146–148.

181. **Russo, J. J., R. A. Bohenzky, M. C. Chien, J. Chen, M. Yan, D. Maddalena, J. P. Parry, D. Peruzzi, I. S. Edelman, Y. Chang, and P. S. Moore.** 1996. Nucleotide sequence of the Kaposi sarcoma-associated herpesvirus (HHV8). *Proc. Natl. Acad. Sci. USA* **93:**14862–14867.

182. **Safai, B., V. Mike, G. Giraldo, E. Beth, and R. A. Good.** 1980. Association of Kaposi's sarcoma with second primary malignancies. *Cancer* **45:**1472–1479.

183. **Said, J. W., K. Chien, S. Takeuchi, T. Tasaka, H. Asou, S. K. Cho, S. de Vos, E. Cesarman, D. M. Knowles, and H. P. Koeffler.** 1996. Kaposi's sarcoma-associated herpesvirus (KSHV or HHV8) in primary effusion lymphoma: ultrastructural demonstration of herpesvirus in lymphoma cells. *Blood* **87:**4937–4943.

184. **Sanceau, J., J. Wijdenes, M. Revel, and J. Wietzerbin.** 1991. IL-6 and IL-6 receptor modulation by IFN-gamma and tumor necrosis factor-alpha in human monocytic cell line (THP-1). Priming effect of IFN-gamma. *J. Immunol.* **147:**2630–2637.

185. **Sander, C. A., M. Simon, U. Puchta, M. Raffeld, and P. Kind.** 1996. HHV-8 in lymphoproliferative lesions in skin. *Lancet* **348:**475–476. (Letter.)

186. **Sangfelt, O., S. Erickson, S. Einhorn, and D. Grander.** 1997. Induction of Cip/Kip

and Ink4 cyclin dependent kinase inhibitors by interferon-alpha in hematopoietic cell lines. *Oncogene* **14**:415–423.

187. **Sarid, R., O. Flore, R. A. Bohenzky, Y. Chang, and P. S. Moore.** 1998. Transcriptional mapping of the Kaposi's sarcoma-associated herpesvirus (KSHV/HHV8) genome in a body cavity-based lymphoma cell line (BC-1). *J. Virol.* **72**:1005–1012.

188. **Sarid, R., T. Sato, R. A. Bohenzky, J. J. Russo, and Y. Chang.** 1997. Kaposi's sarcoma-associated herpesvirus encodes a functional Bcl-2 homologue. *Nat. Med.* **3**:293–298.

189. **Sarid, R., J. S. Wiezorek, P. S. Moore, and Y. Chang.** Manuscript in preparation.

190. **Scala, G., I. Quinto, M. R. Ruocco, A. Arcucci, M. Mallardo, P. Caretto, G. Forni, and S. Ventura.** 1990. Expression of exogenous interleukin 6 gene in human Epstein-Barr virus B cells confers growth advantage and in vivo tumorigenicity. *J. Exp. Med.* **172**:61–68.

191. **Schalling, M., M. Ekman, E. E. Kaaya, A. Linde, and P. Biberfeld.** 1995. A role for a new herpesvirus (KSHV) in different forms of Kaposi's sarcoma. *Nat. Med.* **1**: 707–708.

192. **Schwarze, M. M., and R. G. Hawley.** 1995. Prevention of myeloma cell apoptosis by ectopic bcl-2 expression or interleukin 6-mediated up-regulation of bcl-xL. *Cancer Res* **55**:2262–2265.

193. **Senkevich, T. G., J. J. Bugert, J. R. Sisler, E. V. Koonin, G. Darai, and B. Moss.** 1996. Genome sequence of a human tumorigenic poxvirus: prediction of specific host response-evasion genes. *Science* **273**:813–816.

194. **Shan, B., and W. H. Lee.** 1994. Deregulated expression of E2F-1 induces S-phase entry and leads to apoptosis. *Mol. Cell. Biol.* **14**:8166–8173.

195. **Shchelkunov, S. N., R. F. Massung, and J. J. Esposito.** 1995. Comparison of the genome DNA sequences of Bangladesh-1975 and India-1967 variola viruses. *Virus Res.* **36**:107–118.

196. **Shen, Y., and T. E. Shenk.** 1995. Viruses and apoptosis. *Curr. Opin. Genet. Dev.* **5**: 105–111.

197. **Sherr, C. J.** 1995. D-type cyclins. *Trends Biochem. Sci.* **20**:187–190.

198. **Sherr, C. J.** 1996. Cancer cell cycles. *Science* **274**:1672–1677.

199. **Simpson, G. R., T. F. Schulz, D. Whitby, P. M. Cook, C. Boshoff, L. Rainbow, M. R. Howard, S.-J. Gao, R. A. Bohensky, P. Simmonds, C. Lee, A. de Ruiter, A. Hatziakis, R. S. Tedder, I. V. D. Weller, R. A. Weiss, and P. S. Moore.** 1996. Prevalence of Kaposi's sarcoma associated herpesvirus infection measured by antibodies to recombinant capsid protein and latent immunofluorescence antigen. *Lancet* **348**: 1133–1138.

200. **Sinclair, A. J., I. Palmero, A. Holder, G. Peters, and P. J. Farrell.** 1995. Expression of cyclin D2 in Epstein-Barr virus-positive Burkitt's lymphoma cell lines is related to methylation status of the gene. *J. Virol.* **69**:1292–1295.

201. **Sinclair, A. J., I. Palmero, G. Peters, and P. J. Farrell.** 1994. EBNA-2 and EBNA-LP cooperate to cause G0 to G1 transition during immortalization of resting human B lymphocytes by Epstein-Barr virus. *EMBO J.* **13**:3321–3328.

202. **Sirianni, M. C., S. Uccini, A. Angeloni, A. Faggioni, F. Cottoni, and B. Ensoli.** 1997. Circulating spindle cells: correlation with human herpesvirus-8 (HHV-8) infection and Kaposi's sarcoma. *Lancet* **349**:255. (Letter.)

203. **Smith, C. A.** 1995. A novel viral homologue of Bcl-2 and Ced-9. *Trends Cell Biol.* **5**: 344.

204. **Smith, M. S., C. Bloomer, R. Horvat, E. Goldstein, J. M. Casparian, and B. Chandran.** 1997. Detection of human herpesvirus 8 DNA in Kaposi's sarcoma lesions and peripheral blood of human immunodeficiency virus-positive patients and correlation with serologic measurements. *J. Infect. Dis.* **176**:84–93.

205. **Sofer, L. Y., and D. Resnitzky.** 1996. Apoptosis induced by ectopic expression of cyclin D1 but not cyclin E. *Oncogene* **13**:2431–2437.

206. **Soulier, J., L. Grollet, E. Oskenhendler, P. Cacoub, D. Cazals-Hatem, P. Babinet, M.-F. d'Agay, J.-P. Clauvel, M. Raphael, L. Degos, and F. Sigaux.** 1995. Kaposi's sarcoma-associated herpesvirus-like DNA sequences in multicentric Castleman's disease. *Blood* **86:**1276–1280.

207. **Staskus, K. A., W. Zhong, K. Gebhard, B. Herndier, H. Wang, R. Renne, J. Beneke, J. Pudney, D. J. Anderson, D. Ganem, and A. T. Haase.** 1997. Kaposi's sarcoma-associated herpesvirus gene expression in endothelial (spindle) tumor cells. *J. Virol.* **71:**715–719.

208. **Strathdee, S. A., P. J. Veugelers, and P. S. Moore.** 1996. The epidemiology of HIV-associated Kaposi's sarcoma: the unraveling mystery. *AIDS* **10:**S51–S57.

209. **Strauchen, J. A., A. D. Hauser, D. A. Burstein, R. Jiminez, P. S. Moore, and Y. Chang.** 1996. Body cavity-based malignant lymphoma containing Kaposi sarcoma-associated herpesvirus in an HIV-negative man with previous Kaposi sarcoma. *Ann. Intern. Med.* **125:**822–825.

210. **Sturzl, M., C. A. Balsig, Schreier, F. Neipel, C. Hohenadl, D. Cornali, G. Ascherl, S. Eaaner, N. H. Brockmeyer, M. Ekman, E. E. Kaaya, E. Tschachler, and P. Biberfeld.** 1997. Expression of HHV-8 latency-associated T0.7 RNA in spindle cells and endothelial cells of AIDS-associated, classical and African Kaposi's sarcoma (KS). *Int. J. Cancer* **72:**68–71.

211. **Su, I.-J., Y.-S. Hsu, Y.-C. Chang, and I.-W. Wang.** 1995. Herpesvirus-like DNA sequence in Kaposi's sarcoma from AIDS and non-AIDS patients in Taiwan. *Lancet* **345:**722–723.

212. **Sun, R., S.-F. Lin, L. Gradoville, and G. Miller.** 1996. Polyadenylated nuclear RNA encoded by Kaposi sarcoma-associated herpesvirus. *Proc. Natl. Acad. Sci. USA* **93:** 11883–11888.

212a. **Swanton, C., D. J. Mann, B. Fleckenstein, F. Neipal, G. Peters, and N. Jones.** 1997. Herpesviral cyclin/CDK6 complexes evade inhibition by CDK inhibitor proteins. *Nature* **390:**184–187.

213. **Taniguchi, T., H. Harada, and M. Lamphier.** 1995. Regulation of the interferon system and cell growth by the IRF transcription factors. *J. Cancer Res. Clin. Oncol.* **121:** 516–520.

213a. **Taniguchi, T., M. J. Lamphier, and N. Tanaka.** 1997. IRF-1: the transcription factor linking the interferon response and oncogenesis. *Biochim. Biophys. Acta* **1339:**M9-17.

214. **Tasaka, T., J. W. Said, R. Morosetti, D. Park, W. Verbeek, M. Nagai, J. Takahara, and H. P. Koeffler.** 1997. Is Kaposi's sarcoma-associated herpesvirus ubiquitous in urogenital and prostate tissues? *Blood* **89:**1686–1689.

215. **Telford, E. A., M. S. Watson, H. C. Aird, J. Perry, and A. J. Davison.** 1995. The DNA sequence of equine herpesvirus 2. *J. Mol. Biol.* **249:**520–528.

216. **Teodoro, J. G., and P. E. Branton.** 1997. Regulation of apoptosis by viral gene products. *J. Virol.* **71:**1739–1746.

217. **Thome, M., P. Schneider, K. Hofman, H. Fickenscher, E. Meinl, F. Neipel, C. Mattmann, K. Burns, J.-L. Bodmer, M. Schroter, C. Scaffidl, P. H. Krammer, M. E. Peter, and J. Tschopp.** 1997. Viral FLICE-inhibitory proteins (FLIPs) prevent apoptosis induced by death receptors. *Nature* **386:**517–521.

218. **Tomita, Y., N. Naka, K. Aozasa, E. Cesarman, and D. M. Knowles.** 1996. Absence of Kaposi's-sarcoma-associated herpesvirus-like DNA sequences (KSHV) in angiosarcomas developing in body-cavity and other sites. *Int. J. Cancer* **66:**141–142.

219. **Tommasino, M., and L. Crawford.** 1995. Human papillomavirus E6 and E7: proteins which deregulate the cell cycle. *Bioessays* **17:**509–518.

220. **Tosato, G., K. Jones, M. K. Breinig, H. P. McWilliams, and J. L. C. McKnight.** 1993. Interleukin-6 production in posttransplant lymphoproliferative disease. *J. Clin. Invest.* **91:**2806–2814.

221. **Tosato, G., G. Tanner, K. D. Jones, M. Revel, and S. E. Pike.** 1990. Identification of

interleukin 6 as an autocrine growth factor for Epstein-Barr virus immortalized B cells. *J. Virol.* **64:**3033–3041.

222. **Tsai, C. C., C.-C. Tsai, S. T. Roodman, and M.-D. Woon.** 1990. Mesenchymo-proliferative disorders (MPD) in simian AIDS associated with SRV-2 infection. *J. Med. Primatol.* **19:**189–202.

223. **Tsujimoto, Y., and C. M. Croce.** 1986. Analysis of the structure, transcripts, and protein products of bcl-2, the gene involved in human follicular lymphoma. *Proc. Natl. Acad. Sci. USA* **83:**5214–5218.

224. **Uthman, A., C. Brna, W. Weninger, and E. Tschachler.** 1996. No HHV8 in non-Kaposi's sarcoma mucocutaneous lesions from immunodeficient HIV-positive patients. *Lancet* **347:**1700–1701.

225. **Virgin, H. W., P. Latreille, P. Wamsley, K. Hallsworth, K. E. Weck, A. J. Dal Canto, and S. H. Speck.** 1997. Complete sequence and genomic analysis of murine gamma-herpesvirus 68. *J. Virol.* **71:**5894–5904.

226. **Viviano, E., F. Vitale, F. Ajello, A. M. Perna, M. R. Villafrate, F. Bonura, M. Arico, G. Mazzola, and N. Romano.** 1997. Human herpesvirus type 8 DNA sequences in biological samples of HIV-positive and negative individuals in Sicily. *AIDS* **11:**607–612.

227. **Vousden, K. H.** 1995. Regulation of the cell cycle by viral oncoproteins. *Semin. Cancer Biol.* **6:**109–116.

228. **Walter, P. R., E. Philippe, C. Nguemby-Mbina, and A. Chamlian.** 1984. Kaposi's sarcoma: presence of herpes-type particles in a tumor specimen. *Hum. Pathol.* **15:**1145–1146.

229. **Webster-Cyriaque, J., R. H. Edwards, E. B. Quinlivan, L. Patton, D. Wohl, and N. Raab-Traub.** 1997. Epstein-Barr virus and human herpesvirus 8 prevalence in human immunodeficiency virus-associated oral mucosa lesions. *J. Infect. Dis.* **175:**1324–1332.

230. **Weinberg, R. A.** 1997. The cat and mouse games that genes, viruses, and cells play. *Cell* **88:**573–575.

231. **Whitby, D., M. R. Howard, M. Tenant-Flowers, N. S. Brink, A. Copas, C. Boshoff, T. Hatziouannou, F. E. A. Suggett, D. M. Aldam, A. S. Denton, R. F. Miller, I. V. D. Weller, R. A. Weiss, R. S. Tedder, and T. F. Schulz.** 1995. Detection of Kaposi's sarcoma-associated herpesvirus (KSHV) in peripheral blood of HIV-infected individuals predicts progression to Kaposi's sarcoma. *Lancet* **364:**799–802.

231a. **Whitby, D., M. Luppi, P. Barrozzi, C. Boshoff, R. A. Weiss, and G. Forelli.** Human herpesvirus 8 seroprevalence in blood donors and patients with lymphoma from different regions of Italy. *J. Natl. Cancer Inst.*, in press.

232. **White, E.** 1994. p53, guardian of Rb. *Nature* **371:**21–22.

233. **White, E., P. Sabbatini, M. Debbas, W. S. Wold, D. I. Kusher, and L. R. Gooding.** 1992. The 19-kilodalton adenovirus E1B transforming protein inhibits programmed cell death and prevents cytolysis by tumor necrosis factor alpha. *Mol. Cell. Biol.* **12:**2570–2580.

234. **Wold, W. S. M., T. W. Hermiston, and A. E. Tollefson.** 1994. Adenovirus proteins that subvert host defenses. *Trends Microbiol.* **2:**437–443.

235. **Wu, X., and A. J. Levine.** 1994. p53 and E2F-1 cooperate to mediate apoptosis. *Proc. Natl. Acad. Sci. USA* **91:**3602–3606.

236. **Yee, C., A. Biondi, X. H. Wang, N. N. Iscove, J. de Souza, L. A. Aarden, G. G. Wong, S. C. Clark, H. A. Messner, and M. D. Minden.** 1989. A possible autocrine role for interleukin-6 in two lymphoma cell lines. *Blood* **74:**798–804.

237. **Yee, C., S. Sutcliffe, H. A. Messner, and M. D. Minden.** 1992. Interleukin-6 levels in the plasma of patients with lymphoma. *Leukemia Lymphoma* **7:**123–129.

238. **Yoshizaki, K., T. Matsuda, N. Nishimoto, T. Kuritani, L. Taeho, K. Aozasa, T. Nakahata, H. Kawai, H. Tagoh, T. Komori, S. Kishimoto, T. Hirano, and T. Kishimoto.** 1989. Pathogenic significance of interleukin-6 (IL-6/BSF-2) in Castleman's disease. *Blood* **74:**1360–1367.

239. **Zhong, W., and D. Ganem.** 1997. Characterization of ribonucleoprotein complexes containing an abundant polyadenylated nuclear RNA encoded by Kaposi's sarcoma-associated herpesvirus (human herpesvirus 8). *J. Virol.* **71:**1207–1212.

240. **Zhong, W., H. Wang, B. Herndier, and D. Ganem.** 1996. Restricted expression of Kaposi sarcoma-associated herpesvirus (human herpesvirus 8) genes in Kaposi sarcoma. *Proc. Natl. Acad. Sci. USA* **93:**6641–6646.

241. **Zhong, W. D., and D. Ganem.** 1997. Characterization of ribonucleoprotein complexes containing an abundant polyadenylated nuclear RNA encoded by Kaposi's sarcoma-associated herpesvirus (human herpesvirus 8). *J. Virol.* **71:**1207–1212.

242. **Ziegler, J. L., and E. Katongolc-Mbidde.** 1996. Kaposi's sarcoma in childhood: an analysis of 100 cases from Uganda and relationship to HIV infection. *Int. J. Cancer.* **65:**200–203.

242a. **Zimring, J. C., S. Goodbourn, and M. K. Offermann.** 1998. Human herpesvirus 8 encodes an interferon regulatory factor (IRF) homolog that represses IRF-1-mediated transcription. *J. Virol.* **72:**701–707.

243. **Zong, J. C., C. Metroka, M. S. Reitz, J. Nicholas, and G. S. Hayward.** 1997. Strain variability among Kaposi sarcoma-associated herpesvirus (human herpesvirus 8) genomes: evidence that a large cohort of United States AIDS patients may have been infected by a single common isolate. *J. Virol.* **71:**2505–2511.

Human Tumor Viruses
Edited by Dennis J. McCance
© 1998 American Society for Microbiology

4 | Molecular Biology of Epstein-Barr Virus
Richard Longnecker

Epstein-Barr virus (EBV), also designated human herpesvirus 4 (HHV-4), is one of eight human herpesviruses that establish latent infections in human hosts (232). The human herpesviruses are organized into three families (α, β, and γ) depending on biological characteristics and evolutionary relatedness. The three human alpha-herpesviruses, herpes simplex virus 1 (HSV-1), HSV-2, and varicella-zoster virus (VZV), are characterized by their rapid reproductive cycle and capacity to establish latent infections in sensory ganglia. The three human betaherpesviruses, human cytomegalovirus (CMV), human herpesvirus 6 (HHV-6), and HHV-7, typically have a long reproductive cycle and remain latent in a variety of tissues. The two human gamma-herpesviruses include EBV and the recently identified HHV-8 and are distinguished by their latent infection of lymphoblastoid cell lines (LCLs) either of T- or B-cell origin. At least two EBV types have been identified in human populations (1, 11, 85, 242, 252, 261, 267, 324, 328). These were formerly designated EBV types A and B, but they have recently been designated EBV-1 and EBV-2 to parallel the HSV-1 and HSV-2 nomenclature. However, unlike HSV-1 and HSV-2, there is extensive nucleotide homology and restriction endonuclease site conservation throughout most of the EBV-1 and EBV-2 genomes. Thus, the two types of EBV are considerably more closely related to each other than are HSV-1 and HSV-2.

The pathologies associated with EBV infection will be briefly discussed here, but refer to chapter 5 for more details and references. Considerable interest has focused on EBV since its discovery and its link with Burkitt's lymphoma (27, 28, 58). Along with HHV-8, EBV is the only herpesvirus with an etiological role in human malignancies, and EBV is a causative agent in endemic Burkitt's lymphoma and undifferentiated nasopharyngeal carcinoma (chapter 5 and reference 225). The virus has also been recognized as an important pathogen in individuals lacking cellular immunity due to either genetic defects, immune suppression for organ transplantation, or loss of immune function due to HIV infection (chapter 5 and reference 225). In immunosuppressed patients, EBV causes a variety of proliferative disorders including oral hairy leukoplakia in AIDS patients, immunoblastic lymphomas, and an unusual tumor of muscle origin in children who have AIDS or are under immune suppression after liver transplantation (chapter 5 and reference 225). EBV may also be a factor in a variety of other human malignancies including some T-cell lymphomas and Hodgkin's disease (chapter 5 and reference 225). In young boys with X-linked immunodeficiencies, EBV causes fulminant mononucleosis which results in death (chapter 5 and reference 225).

Infection with EBV usually occurs early in childhood and results in an asymptomatic infection. If primary infection occurs later, B-cell proliferation and the resulting severe immune response result in infectious mononucleosis. After primary infection, most individuals will harbor the virus for the remainder of their lives, and carriers develop cellular immunity against a variety of both lytic and latency-associated proteins (chapter 5 and reference 225). Periodically, virus is shed from latently infected individuals by the induction of lytic replication in B lymphocytes. The true site of latent infection has not been determined, but the virus likely resides in B lymphocytes. Recent studies have shown that EBV can be detected by PCR (both for viral DNA and viral mRNA) in circulating peripheral blood lymphocytes in carriers of EBV latent infections; virus has also been isolated by culturing peripheral lymphocytes (35, 54, 188, 218, 290, 320). It has not been determined if peripheral lymphocytes are the true site of latency. Other potential sites of EBV latency may include bone marrow, lymph nodes, or other lymphoid organs. Latency is not maintained by constant reinfection of circulating B lymphocytes, as evidenced in patients treated with acyclovir. Acyclovir, a nucleoside analog that can inhibit lytic replication in the oral epithelium, had no effect on the number of B cells in the peripheral blood population which harbors the virus (319). Further evidence of the hematopoetic site of EBV latency comes from engraftment of bone marrow cells, which can result in the loss of the resident virus or the appearance of a new virus strain from donor lymphocytes (85, 86). Lytic replication is presumed to occur when EBV-infected B lymphocytes traffic through and transmit infection to oral epithelium, providing a source for infection of other individuals. Latency in the ever-changing B-cell population requires the careful analysis of EBV gene function, which will be the topic of this chapter.

THE VIRAL GENOME

EBV has a linear double-stranded genome of approximately 172 kilobase pairs (kbp) (14). It was the first large DNA virus whose complete sequence was determined (14). The sequenced virus, B95-8, was from an EBV-infected marmoset cell line partially permissive for lytic replication. Upon further analysis, it was determined that this virus isolate contained a deletion of approximately 12 kb relative to other EBV strains (220). Sequence analysis of the Raji EBV strain revealed that the 12-kbp region deleted from B95-8 contained three potential open reading frames (207). The sequence analysis to date predicts around 85 to 95 open reading frames (14, 207). Subsequent to the sequencing of the EBV genome, the VZV, HSV, CMV, HHV-6, and Kaposi's sarcoma-linked herpesvirus genomes were cloned and sequenced (14, 34, 52, 82, 175, 176, 243).

Despite enormous differences in base composition, ranging from 43% guanine plus cytosine (VZV) to 68% guanine plus cytosine (HSV), it is apparent that the herpesvirus genomes have many open reading frames in common. Many products of EBV genes can be predicted on the basis of their amino acid homology with

HSV genes and genes from other herpesviruses such as HHV-8. The predictions and known functions are shown in Table 1. These similarities between proteins encoded by the different herpesviruses such as HSV-1 and HHV-8 underscore the relationships between the various herpesviruses. The homologous genes are primarily limited to those required for the cleavage and packaging of the viral genome and infection of susceptible host cells. Included among the conserved genes are virion structural proteins, enzymes involved in DNA replication, some regulators of gene expression, and glycoprotein genes. Lytic gene function has been described primarily in HSV because lytic replication is easily observed in tissue culture systems, allowing gene function studies. In contrast, EBV, which is largely latent in tissue culture systems, has been less amenable to studies of lytic gene function. The similarities at the DNA level are not sufficient to allow EBV DNA to hybridize with other known human herpesviruses, and despite conservation of gene function, antibody cross-reactivity is rare among gene products of the different herpesviruses.

The majority of EBV isolates in Western communities are type 1, while type 2 EBV isolates appear to be largely restricted to equatorial Africa and Papua New Guinea (324, 328). The differences between the two EBV types is primarily restricted to the genes that encode the nuclear proteins which are expressed in latent infection. EBNA2, EBNALP, EBNA3A, EBNA3B, and EBNA3C proteins have 56%, 78%, 84%, 80%, and 72% amino acid identity, respectively, between the two EBV types (1, 51, 252). Natural EBV isolates of the two types differ in their growth transformation phenotype in vitro. Type 1 transformed cells grow more rapidly, are less concentration dependent, and reach higher saturation densities than do type 2 isolates (226). These type-specific differences can primarily be attributed to the EBNA2 gene (46, 226). The proteins encoded by the latent membrane proteins (LMP1 and LMP2) vary to some degree between type 1 and type 2 isolates, but these differences do not appear to be significant (29, 110, 249). Interestingly, numerous reports describing the presence of a 30-bp deletion within the LMP1 coding domain have suggested this deletion may be important for the pathogenesis of more aggressive EBV-associated tumors (36, 38, 110, 135, 136, 205, 254) and reduced immunogenicity of the LMP1 protein (298), although closer examination has revealed that there appears to be no link between the presence of the LMP1 deletion and EBV-associated malignancies (120, 130, 268). Interestingly, there was a preferential association of the LMP1 deletion with type 2 EBV (130, 268).

INFECTION

EBV, in contrast to HSV, has a limited host range. This restriction is at least partially due to absence of the cellular receptor CD21 for EBV, which is also the receptor for the C3d component of complement (62, 66, 118, 197, 289). CD21 is a member of the immunoglobulin superfamily (189, 307). Other cellular factors are likely important as well since epithelial cells grown in vitro bind EBV but are not readily permissive for viral replication (2, 149). A related mRNA for CD21 has been found to be expressed in epithelial tissues, indicating that it might be the receptor for infection of epithelial cells by EBV (22).

Table 1 EBV genes associated with lytic infection[a]

EBV gene	HSV gene	KSHV gene	EBV protein name	E/D	Known or proposed function
BNRF1		ORF75			Virion protein
BCRF1					IL-10 homolog, host immune modulator
BCRF2					
BHRF1		ORF16			Bcl-2 homolog
BHLF1					
BFLF2	UL31	ORF69		E	Associates with nuclear matrix
BFLF1	UL32	ORF68		E	Virion protein, DNA cleavage/packaging
BFRF1	UL34	ORF67		E	Virion protein, capsid assembly
BFRF2		ORF66			
BFRF3	UL35	ORF65		E	Capsid protein
BPLF1	UL36	ORF64		E	Virion phosphoprotein, DNA release
BOLF1	UL37	ORF63		E	Cytoplasmic phosphoprotein
BORF1	UL38	ORF62		E	Capsid assembly, binds DNA
BORF2	UL39	ORF61		D	Ribonucleotide reductase (large subunit)
BaRF1	UL40	ORF60		D	Ribonucleotide reductase (small subunit)
BMRF1	UL42	ORF59		E	Polymerase-associated processivity factor
BMRF2	UL43	ORF58		D	
BSLF2/BMLF1	UL54	ORF57		E	Transactivator/repressor
BSLF1	UL52	ORF56		E	DNA replication, helicase/primase complex
BSRF1	UL51	ORF55		D	
BLRF1		ORF53			
BLRF2		ORF52			
BLLF1a			gp350		Virion binding to CR2 (CD21)
BLLF1b			gp220		Virion binding to CR2 (CD21)
BLLF3	UL50	ORF54		D	dUTPase
BZLF2			gp42		Complexes with gp25 and gp85, binds HLA class II
BZLF1					Z transactivator
RAZ					Z regulator
BRLF1		ORF50			R transactivator
BRRF1		ORF49			
BRRF2		ORF48			
BRKF2	UL1	ORF47	gp25	E	gL, complexes with gp42 and gp85
BKRF3	UL2	ORF46		D	Uracil-DNA glycosylase
BKRF4	UL3	ORF45		D	Nuclear phosphoprotein
BBLF4	UL5	ORF44		E	DNA replication, helicase/primase complex

Continued

Table 1 *Continued*

EBV gene	HSV gene	KSHV gene	EBV protein name	E/D	Known or proposed function
BBRF1	UL6	ORF43		E	Virion protein, DNA packaging
BBRF2	UL7	ORF42		E	
BBLF3	UL8	ORF41		E	DNA replication, helicase/primase complex
BBLF2	UL9	ORF40		E	DNA replication, helicase
BBRF3	UL10	ORF39		D	gM
BBLF1	UL11	ORF38		D	Myristylated virion protein
BGLF5	UL12	ORF37		D	Exonuclease
BGLF4	UL13	ORF36		D	Virion protein kinase
BGLF3	UL14	ORF34			
BGLF3.5		ORF35			
BGRF1	UL15	ORF29a		E	DNA packaging protein
BGLF2	UL16	ORF33		D	
BGLF1	UL17	ORF32		E	
BDLF4		ORF31			
BDRF1	UL15	ORF29b		E	DNA packaging protein
BDLF3		ORF28	gp150		
BDLF3.5		ORF30		E	
BDLF2		ORF27			
BDLF1	UL18	ORF26		E	Capsid protein
BcLF1	UL19	ORF25		E	Major capsid protein
BcRF1		ORF24			
BTRF1	UL21	ORF23		D	
BXLF2	UL22	ORF22	gp85	E	gH, complexes with gp25 and gp42
BXLF1	UL23	ORF21		D	Thymidine kinase
BXRF1	UL24	ORF20		D	Membrane protein, fusion
BVRF1	UL25	ORF19		E	Virion protein
BVRF2	UL26	ORF17		E	Serine protease
BILF2			gp78/55		
BILF1					Membrane protein?
BALF5	UL30	ORF9		E	DNA polymerase
BALF4	UL27	ORF8	gp110	E	gB, virus maturation
BALF3	UL28	ORF7		E	DNA cleavage/packaging
BALF2	UL29	ORF6		E	ssDNA binding protein
BALF1					
BARF0					
BARF1					
Raji LF3					
Raji LF2		ORF11			
Raji LF1					

a This table is based on the sequence data published in references 14, 175, 176, 207, and 243 and the table in reference 233. E/D refers to whether the gene is essential or dispensable for in vitro replication from experiments with EBV and other herpesviruses. In most cases this determination has not been made for EBV but rather for other herpesviruses such as HSV-1. ssDNA, single-stranded DNA.

Morphologically, the EBV virion is very similar to other herpesvirus virions, consisting of an envelope containing viral glycoproteins surrounding a nucleocapsid containing the linear double-stranded viral genome. The major EBV outer membrane glycoprotein, gp350/220, is the CD21 ligand (195, 285). Both proteins are encoded by the same mRNA, gp220 being spliced once while gp350 is unspliced (18). A peptide sequence within gp350/220 is homologous to a peptide sequence found within the C3d component of complement which binds CD21 (195, 285). Further evidence supporting an interaction of gp350/220 with CD21 of B lymphocytes is that peptides from gp350/220 block C3d binding to CD21 while peptides from C3d are able to block EBV infection of B lymphocytes (195, 196, 285, 286). In addition, deletion of two amino acids within the region of gp350/220 homologous with C3d abolishes binding of purified gp350 to CD21 (286). Thus, the high-affinity initial interaction of EBV with B lymphocytes is quite different from that in other herpesviruses in which virus binding has been studied. Studies with HSV have indicated that the initial stage of binding of the virion to the cell surface is accomplished by a rather nonspecific interaction of the virally encoded gC with cell surface proteoglycans (271).

After binding, EBV virions are endocytosed (285) and fusion of the viral membrane with cellular membranes occurs, resulting in the release of the virus nucleocapsid and tegument into the cytoplasm (31, 184, 194, 285). The EBV capsid is then transported to the nuclear pores by the cellular cytoskeleton, and the viral genome is released into the nucleus and is circularized (3, 116). Unknown factors determine whether the resulting infection is either latent or lytic.

In Vitro Lytic Replication

EBV lytic replication in latently infected lymphoblastoid cell lines (LCLs) can be induced by a variety of means. Induction of lytic replication in the EBV-infected cell line Raji by superinfection with EBV P3HR-1 and inhibition of cellular protein synthesis results in the accumulation of the immediate-early (IE) BZLF1 and BRLF1 mRNAs (20). Similarly, when lytic replication is induced in the EBV-infected cell line Akata by B-cell receptor (BCR) cross-linking in the presence of protein synthesis inhibitors, both the BRLF1 and BZLF1 messages are detected (283). Chemical treatment with phorbol esters also efficiently induces lytic replication (330). Phorbol esters mimic the action of the intracellular messenger diacylglycerol and activate the transcription factor AP-1. Cross-linking the BCR induces the activation of second messengers which can induce EBV lytic replication in some latently EBV-infected cell lines (283). Both phorbol ester and BCR cross-linking response elements have been identified upstream of these IE gene products (65, 141, 263), providing a mechanism for the induction of expression of these IE genes following activating treatments. Induction of lytic infection by superinfection with the P3HR-1 strain is a result of defective genomes present in P3HR-1, characterized by the fusion of the *Bam*HI Z fragment with the *Bam*HI W fragment of the viral genome (48, 49, 182). This rearrangement results in the constitutive expression of BZLF1 in latently infected LCLs following infection with the P3HR-1 virus.

Following induction of the lytic replication, BRLF1 and BSLF2/BMLF1, in addition to BZLF1, are expressed (40, 95, 153, 171, 284, 315). All of these viral proteins can activate other viral promoters (40, 95, 129, 151, 153, 284, 300). BZLF1 is a member of the basic leucine zipper family which binds to AP1-like elements in EBV early promoters (33, 61). Expression of BZLF1 is sufficient to trigger lytic replication in latently infected LCLs. Expression of BRLF1 or BSLF2/BMLF1 has no effect on switch from latent infection to lytic replication, although certain early viral promoters require both BZLF1 and BRLF1 for maximal transcriptional activation (40, 50, 219). In epithelial cells, the disruption of latency appears to require the expression of BRLF1 (325). A putative regulator, RAZ (R and Z), of BZLF1 has been identified which contains a portion of the BRLF1 gene spliced to the 3' end of the BZLF1 gene (65, 76). Expression of RAZ interferes with BZLF1 transactivation in a dimerization-independent fashion since substoichiometric levels of RAZ efficiently inhibit BZLF1 transactivation (262). The concerted action of the EBV IE genes culminates in the sequential activation of early gene expression followed by the lytic cascade of replication and late gene expression.

Once induced by IE gene expression, the EBV lytic phase is similar to that of other herpesviruses. The transcription of viral DNA occurs in the nucleus and viral proteins are synthesized in the cytoplasm. Viral protein synthesis is coordinately regulated in a cascade fashion. EBV early genes, like those of other herpesviruses, are defined by their expression in the presence of inhibitors of viral DNA synthesis. By these criteria, 30 EBV early messages and a similar number of late messages have been defined (14, 17, 20, 21, 79, 80, 98, 111–113, 115, 250). As with other herpesviruses, early and late genes are intermingled throughout the viral genome.

DNA replication

Like other herpesviruses, EBV encodes a large number of proteins involved in DNA synthesis. They are primarily the products of EBV early genes. Lytic replication is initiated within one of two copies of the lytic origin of replication, termed *oriLyt* (92, 259, 260). The 695-bp *oriLyt* consists of two core elements and one additional domain (92, 260). The first core element contains the promoter of the BHLF1 gene, which contains four BZLF1 binding sites (150, 152). The second essential element is defined by a central 225-bp region including two adenine- and thymidine-rich palindromes and an adjacent polypurine and polypyrimidine region (89, 244). The final domain can be functionally replaced in part by the CMV IE enhancer, suggesting that this domain is important only for transcriptional enhancement (55, 92).

Studies using a transient replication system have identified the EBV genes required for viral DNA synthesis (63, 64, 92). As expected, the identified genes were similar to those previously defined for HSV and CMV (32, 206, 233). The essential components include BALF2 (single-stranded DNA binding protein), BALF5 (DNA polymerase), BBLF2/3 (primase-associated factor), BBLF4 (DNA helicase), BMRF1 (polymerase accessory), and BSLF1 (primase). Interestingly, EBV does not encode a homolog of the origin binding protein found in HSV, but relies on BZLF1 as an initiator for DNA lytic replication (63, 64, 255, 259). In addition to the essential components, EBV DNA encodes two genes that augment viral DNA replication (63). BRLF1, similar to BZLF1, serves as a transactivator and acts in combination with BZLF1 to produce a synergistic response if binding sites are available for both activators (50, 219). BMLF1 serves a posttranscriptional function in stabilizing viral RNAs which encode proteins responsible for viral DNA synthesis and thus contributes to DNA replication in an indirect fashion (63).

EBV viral DNA is replicated to yield concatemers which are then cleaved into unit-length genomes and packaged to produce infectious virions. The sequences responsible for cleavage and packaging are found within the terminal repeat (TR) regions found at the ends of the linear viral genome (14). Similar sequences are contained within the termini of HSV. These sequences contained in the HSV genome have been shown to be the site of cleavage of viral DNA for the packaging of full-length genomes into virions (233).

Viral proteins

Membrane glycoproteins. Membrane glycoproteins encoded by the different herpesvirus genomes have been shown to play significant roles in a variety of important processes, including specific binding to the cell surface and fusion of viral and plasma membranes during entry, virion assembly, and egress (117, 233, 271). Among the alphaherpesviruses, four glycoproteins essential for plasma membrane binding or entry have been identified. Delivery of the viral capsid into the cell requires the actions of gB, gD, and the gH/gL complex, all of which are essential for infection (233, 271). Numerous host factors are also important for infection. Initial attachment of HSV is to cell surface glycosaminoglycans through interactions with gC and gB (117, 233, 271). Other cellular factors, such as herpesvirus entry mediator (HVEM), have recently been identified, demonstrating a complex series of interactions that must occur for HSV attachment and entry into susceptible cells (188a).

The nucleotide and amino acid sequences of the nine glycoproteins known to be encoded by EBV range from substantially conserved to virtually unrelated when compared with their respective homologs in the alphaherpesviruses (117). Reflecting its specialized cellular niche in B lymphocytes, EBV binding to the cell surface utilizes a specific interaction between the type 2 complement receptor (CR2) or CD21, and gp350/220, which has no known herpesvirus homolog (195, 285). Contained within gp350/220 is the amino acid sequence EDPGFFNVE, similar to amino acid sequence EDPGKQLNVE contained in the natural ligand for CD21, the C3dg fragment complement (195, 285). The first two short consensus repeats of CD21 serve as the binding region for both gp350/220 and C3dg (163, 173). Binding of gp350/220 to CD21 may stimulate B lymphocytes through a signal transduction cascade that facilitates early viral gene expression in resting B lymphocytes (227, 265, 266, 285).

The EBV gp85 protein is a member of the gH family of glycoproteins, which is common to all herpesviruses (90, 100, 117, 183). Studies of gH homologs, including gp85, have indicated a role for these glycoproteins in virus-directed fusion (90, 100, 117, 183). Monoclonal antibodies directed against gp85 prevent EBV from entering susceptible cells, but have no effect on virus binding (183, 272). Virosomes made from EBV proteins are able to bind and fuse with receptor-positive cells unless gp85 is absent (90). A second glycoprotein, gL, is required for the processing of HSV gH (271). The homolog of gL in EBV is gp25, and this associates with gp85, resulting in the formation of a stable complex and transport of gp85 to the cell surface (321). A third glycoprotein, gp42, which is unique to EBV, has also been found to lie in the gp85 complex (148). Interestingly, this third component of the gp85 complex appears to be essential for EBV infection of B lymphocytes, but not epithelial cells (148). A recent report indicates that gp42 binds HLA class II molecules, thus acting as an important cofactor for the infection of B lymphocytes (147). Thus, it would appear that gp25 is required as the chaperone of gp85 protein to the virion envelope, and that gp42 targets the complex to a second receptor on B lymphocytes that is not required for entry in epithelial cells (117, 147, 148).

Comparison of EBV gp110 with gB, its HSV homolog, indicates that gp110 shares colinear homology with gB and has similar structural motifs, but has different biochemical and biological properties (57, 83, 84, 209). Both gp110 and gB have been shown to be essential for the complete replicative cycle in their respective viral systems (106, 271), but are not able to complement each other (144). In lytically infected lymphocytes, gp110 localizes primarily to the endoplasmic reticulum (ER) and to the inner and outer nuclear membranes of B lymphocytes undergoing productive infection (57, 83, 84). The abundant stable 110-kDa form of gp110 has only endo H-sensitive high-mannose N-linked oligosaccharides, indicating that the gp110 product has not trafficked though the Golgi apparatus (83). This high-mannose form is analogous to the immature form of HSV-1 gB, which also accumulates in infected cells (210). The apparent lack of gp110 in the virion membrane (83, 84) (although some reports have indicated that a larger form of gp110 may be present in the virion [57, 83, 84]) and its intracellular localization suggest that the role of gp110 in EBV replication may be in assembly and budding of the virion from the nucleus of a lytically infected cell. Also indicative of the absence of a role in entry, antibodies isolated to date directed against gp110 do not neutralize virus infectivity (83). Recent experiments using gp110 deletion mutants have indicated that the cytoplasmic tail domain is very sensitive to mutation (144a). Deletion of only 16 amino acids resulted in loss of gp110 function (144a). This contrasts with HSV-1 gB, in which deletions of either 27 or 41 amino acids from the gB tail result in the production of viable virus (210). In addition, four basic arginine residues at amino acids 836 to 839 were identified which could function as a nuclear or ER retention signal (144a). Interestingly, a similar motif in HSV-1 gB in the same location, consisting of three arginines and one lysine, does not appear to be important for gB localization to the nuclear membrane and ER but rather a portion of the hydrophobic domain from amino acids 774 to 795 has been shown to be sufficient for nuclear membrane and ER localization (81). Electron microscopy of gp110 mutant-infected cells demonstrates the absence of nucleocapsids in the nucleus or enveloped virions in any cellular compartment, further suggesting that gp110 may provide signals that are responsible for the assembly and egress of EBV nucleocapsids from the nucleus of an EBV-infected cell (144a).

Many questions remain with respect to the role of EBV glycoproteins in lytic infection. The identification of multiple cellular proteins required for the entry of other herpesviruses suggests that entry of EBV into B lymphocytes and epithelial cells is likely more complicated than merely the binding of gp350/220 to CD21.

BCRF1. BCRF1 is expressed late in lytic replication and is 84% homologous at the amino acid level with human interleukin-10 (hIL-10) (301). BCRF1, also known as viral IL-10 (vIL-10), has many of the in vitro effects of IL-10 (109, 238). These include inhibition of the cytokines gamma interferon and IL-2 by activated $T_H 1$ lymphocytes and the ability to stimulate B lymphocytes to proliferate and differentiate (238). EBV mutants that do not express BCRF1 initiate and growth transform primary B lymphocytes in vitro (278). However, these BCRF1 mutant-infected LCLs lack the ability to block gamma interferon secretion in cultures of permissively infected LCLs incubated with autologous human peripheral blood lymphocytes (278), indicating that EBV may have acquired and retained hIL-10 activities necessary to modulate specific and nonspecific host responses to provide a survival advantage during viral lytic replication in the human host.

BHRF1. The BHFR1 gene of EBV encodes a protein which is homologous to Bcl2 and can functionally substitute for Bcl2 in promoting B-cell survival (13, 101, 288). BHRF1 is synthesized to high levels in EBV lytic replication (13, 101, 208). Expression of BHRF1 has also been detected in cultures undergoing EBV lytic replication and in latent EBV-infected cultures (13, 137, 208, 214). Despite detection of BHRF1 in cell lines grown in vitro, it has been demonstrated that deletion of BHRF1 does not affect the ability of EBV to initiate or maintain cell growth transformation (170). It is likely that the role of BHRF1 in EBV is to protect latently infected cells in vivo when they switch from latent infection to lytic infection when there is an absence of or very low Bcl2 expression, thus delaying apoptosis induced by lytic replication and ensuring productive lytic replication.

Latency

EBV-transformed and latently infected LCLs express nine proteins and two small RNAs when grown in tissue culture (Fig. 1). The proteins are expressed either in the nucleus (Epstein-Barr nuclear antigen [EBNA]) or in the plasma membrane (latent membrane protein [LMP]), and recent work has elucidated many of their functions (Table 2). These genes are not conserved in any other human herpesvirus, but homologs to several of these genes have been identified in the related herpesviruses which infect monkeys (67, 68, 157). Sequence analysis supports the notion that EBV genes expressed in latent infection may have been acquired from host cell DNA (125, 126).

The expression of different sets of viral genes in EBV-positive tumors and immortalized B-cell lines grown in vitro has indicated that EBV establishes several types of latent infections. The first definition of the different forms of latency resulted in the terminology of latency I, II, and III (239, 241). In latency I, defined by Burkitt's lymphoma, only the EBNA1 protein is expressed. In latency II, characterized by nasopharyngeal carcinoma, EBV-positive Hodgkin's, and T-cell lymphomas, EBNA1 plus the LMPs are expressed. In latency III, defined by LCLs grown in vitro, all the known latent genes are expressed. More recent studies analyzing expression of viral genes in immunocompetent individuals harboring EBV latent infections have shown that this nomenclature is inadequate (10, 35, 188, 218, 290), since only EBNA1 and LMP2A are expressed in latent infections of these individuals (35, 218, 290).

A. BamHI Map

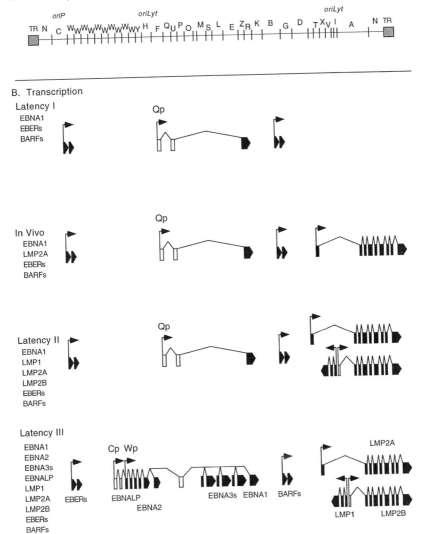

B. Transcription

Latency I
EBNA1
EBERs
BARFs

Qp

In Vivo
EBNA1
LMP2A
EBERs
BARFs

Qp

Latency II
EBNA1
LMP1
LMP2A
LMP2B
EBERs
BARFs

Qp

Latency III
EBNA1
EBNA2
EBNA3s
EBNALP
LMP1
LMP2A
LMP2B
EBERs
BARFs

EBERs

Cp Wp

EBNALP
EBNA2

EBNA3s EBNA1 BARFs

LMP2A

LMP1 LMP2B

Figure 1 Viral genes expressed in cells latently infected with EBV. The four forms of latent gene expression that have been demonstrated in EBV-infected cell lines, tumor biopsies, and in vivo in normal humans latently infected are shown. The four different programs are under the control of two transcriptional units which vary the expression of the six EBNAs or three LMPs. (A) *Bam*HI restriction enzyme map for the B95-8 sequence used to designate promoters and exons in each latent gene transcript. (B) The four different identified EBV latency programs. In latency I, EBNA 1 is expressed. In latency in immune-competent human hosts (In Vivo), the same EBNA1 transcript is expressed as well as LMP2A. In latency II, the previous proteins are expressed with the addition of LMP1 and LMP2B. Finally, in latency III all EBNAs and LMPs are expressed. The EBERs and BARFs (107, 256) are expressed in all types of EBV latencies. Solid boxes indicate exons, and arrows indicate sites of promoters. Letters indicate relevant *Bam*HI restriction fragment.

Table 2 EBV genes associated with in vitro latency and immortalization

Gene name	Known or proposed function	Alternate name	B95-8 sequence
EBNA1	Maintenance of viral episome		BKRF1
EBNA2	Regulator of viral gene transcription		BYRF1
EBNA3A	Regulator of viral gene transcription	EBNA3	BERF3-BERF4
EBNA3B	Regulator of viral gene transcription	EBNA4	BERF2a,b
EBNA3C	Regulator of viral gene transcription	EBNA6	BLRF3-BERF1
EBNALP	Inducer of cell cycle	EBNA5	
LMP1	Inducer of cell proliferation		
LMP2A	Regulator of viral latency	TP1	BNLF1a,b,c
LMP2B	Regulator of LMP2A function	TP2	
EBER1	Small RNA		
EBER2	Small RNA		

In latency III, in which 11 genes are expressed (Table 2), each of the six EBNA proteins is expressed from a multicistronic mRNA that is regulated by two adjacent promoters, Cp and Wp (25, 247, 248, 314) (Fig. 1). Early in infection of primary B cells, Wp is used exclusively to drive transcription of the EBNA genes, followed by a switch to Cp usage (312–314). Important for the regulation of Cp and Wp promoter activity is the dominant *cis* element *oriP*, which is upstream of Cp. This occurs by the binding of EBNA1 to the *oriP* element (217, 222, 274) along with EBNA2 binding to the EBNA2-dependent enhancer upstream of Cp (122, 275). Methylation of the EBNA2 response element may be important for down-modulation of Cp promoter activity in vivo and some EBV-related malignancies (7, 122, 230, 275). The Wp promoter appears to be active in many cell types, whereas the Cp promoter appears to be B-cell specific (47). Mutational analysis of the Wp and Cp promoters has suggested that transcription from Cp interferes with transcription initiation at the downstream Wp promoter (216). Interestingly, deletion analysis has demonstrated that the Cp promoter is not essential for in vitro transformation of primary B lymphocytes by EBV (277).

Expression of EBNA1 in latency I and II depends on the absence of expression of the other EBNA proteins. In latency I and II, Cp/Wp promoter activity is inactivated and EBNA1 is driven by an EBNA1-specific promoter termed Qp that is located more than 50 kbp from Cp and Wp (201, 245, 257). Qp is a TATA-less promoter which is negatively regulated by two EBNA1 binding sites downstream of the transcriptional initiation site (246, 247) and two upstream positive regulatory elements, QRE-1 and QRE-2 (199). QRE-2 is absolutely required for Qp activity (200). In addition, the transcription factor E2F may reverse EBNA1-mediated repression of the Qp by displacement of EBNA1 from Qp by binding to two E2F-like sites which overlap the EBNA1 binding sites (276). The Qp promoter is functional when introduced into a variety of cell types (199, 247), suggesting that Qp promoter activity may function like a housekeeping gene. EBNA1 is the only known negative regulator of Qp activity. In the absence of Cp- or Wp-mediated expression of EBNA1, the Qp promoter would become functional, thus insuring expression of EBNA1 to maintain the viral episome (200). Interestingly, a new cellular factor designated QRE-2 binding protein may function in initiating transcription at the Qp promoter (200).

Latent gene function

EBNA1. EBNA1 is a nuclear protein which contains phosphorylated serine residues on at least two distinct domains in the carboxy half of the molecule (99, 212, 215), the function of which is not known. Early investigations demonstrated that EBNA1 associates with chromosomes during mitosis (87, 203, 212), and it was shown that EBNA1 binds with high affinity to the specific DNA sequence GATAG-CATATGCTACCCAGA (9, 124, 132, 221). EBNA1 binding sequences are located at three sites in the EBV genome (124, 221, 273). The highest affinity site consists of 20 tandem direct 30-bp repeats. The second highest affinity site is 1 kbp to the right of the first site and consists of two cognate sequences in dyad symmetry and two in tandem. These two sites form the *oriP* domain (Fig. 2). A weaker third site is located in the Qp promoter element as discussed above (201, 257).

Interaction of EBNA1 with the specific sequence *oriP* within the viral genome enables the viral genome to exist as an episome in latently infected LCLs (8, 37, 42, 164, 322, 323). The EBV *oriP* contains multiple copies of an 18-bp EBNA1 recognition sequence, which are clustered in two noncontiguous functional elements, the family of repeats and the dyad symmetry (DS) element (223). The DS element contains four EBNA1 binding sites, a 65-bp region of dyad symmetry, and the site for initiation of DNA replication (77, 221, 223, 316). EBNA1 binds as a dimer, and both the DNA binding and dimerization domains map near the carboxy terminus of the protein (8, 23, 24, 37). The repeat unit of *oriP* acts as an enhancer in the presence of EBNA1, which is important for latent gene transcription from the Cp promoter (78, 222, 274, 316). In addition, as indicated above, lower-affinity EBNA1 binding sites near the Qp promoter are important for the expression of EBNA1 in the non-latency III programs (9, 124, 199, 245, 258, 269).

EBNA1 binds the 20-bp sequence GATAGCATATGCTACCCAGA as a homodimer. The structure of EBNA1 bound to DNA has been solved (23). Key elements in that structure determination are discussed below. The EBNA1 dimer is composed of multiple domains. The dimerization domain extends from amino acids 504 to 604 of each EBNA1 molecule (23, 24). Amino acids 470 to 503 include a

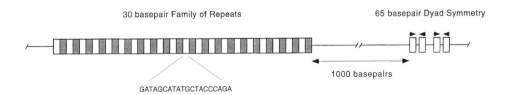

Figure 2 Structure of the *oriP* region from the EBV viral genome. Shown are the 30-bp family of repeats, the 65-bp dyad symmetry, and the B95-8 consensus binding site for EBNA1 within the 30-bp family of repeats. Open boxes indicate EBNA1 binding regions.

helix that projects into the major groove and an extended chain that moves along the minor groove and makes all the sequence-specific contacts with the DNA of the binding site. Amino acids 477 to 489 form the helix, which is oriented perpendicularly to the axis of the DNA with its amino-terminal amino acid (amino acid 477) contacting the DNA. Amino acids 490 to 504 form an extended loop that connects the helix to the first β-strand of the core β-barrel. Amino acids 461 to 469 form an extended chain that tunnels along the minor groove near the outside of the binding site, wraps around the DNA toward the center of the binding site, and exits between the second and third bases, away from the middle of the binding site. These amino acids appear to provide most of the sequence-specific DNA contacts. Amino acid 469 is connected to amino acid 477 by a loop that traverses along, and forms multiple hydrogen bonds to the DNA phosphate backbone.

EBNA1 binding to its cognate sequence appears to be cooperative (186, 221), suggesting either that EBNA1 molecules interact cooperatively in binding to DNA or that the DNA template is modified as a consequence of initial EBNA1 binding. Crystallography, electron microscopic, and biochemical data indicate that EBNA1 binding to the family of repeats and the DS element results in high-order structures that lead to bending and distortion of the DNA duplex (69–72, 177, 204, 273). At least one unknown cellular protein required for episome maintenance is not required for enhancer activity, because EBNA1 binding to the family of repeats enhances transcription in mouse or human cells, whereas EBNA1 promotes episome replication only in primate cells (138). This block can be overcome by using an origin sequence active in rodent cells with the EBNA1 enhancer function provided by the family of repeats (138). Cellular proteins which interact specifically with *oriP* have been identified, but the functional significance of these interactions is unknown at this date (202, 308, 309, 326). It has been suggested that these cellular binding proteins, which compete for EBNA1 binding at *oriP*, may modulate EBV DNA replication or EBV gene expression by displacing EBNA1 binding at *oriP* (202, 308, 309, 326). Functional cloning has allowed the identification of two partial cDNAs derived by differentially spliced transcripts of the same cellular gene (326). Analysis of the nucleotide sequence indicates that the encoded proteins have helix-loop-helix structures and have predicted phosphorylation sites for both protein kinase C and casein kinase (326). In latently infected cellular nuclei, EBNA1 is associated with the nucleoplasm and chromatin (87, 97, 203). This association may be important in segregation of the EBV episomes in dividing cells.

Interestingly, EBNA1 has been shown to cause tumors in transgenic mice when expression is directed to the B-lymphocyte compartment (310), suggesting that EBNA1 may have other functions in addition to its interaction with *oriP*. Analysis of the EBNA1-expressing, EBV-infected Burkitt cell line Akata, in which the EBV viral genome has been lost by tissue culture passage, indicates a loss of malignant phenotype, and animal models of in vivo EBV infection also suggest a role for EBNA1 in the malignant phenotype (264).

Recently it has been shown that a glycine-alanine repeat within the EBNA1 protein prevents presentation by major histocompatibility complex class I proteins of antigenic peptides derived from nonrepeating regions of the EBNA1 protein (146). This observation may explain the lack of cytotoxic responses to EBNA1 and the broad expression of EBNA1 in EBV latencies in both tumor tissues and the normal EBV-infected host.

EBNA2. EBNA2 is a multidomain protein and has been shown to be essential for in vitro transformation and immortalization of primary B lymphocytes by EBV (46, 93). It is a transactivator which has no DNA binding activity, but rather is brought to promoters that contain EBNA2 response elements by cellular DNA binding proteins. The transcription factors PU.1 and RBPJκ (also known as CBF1) are among the cellular sequence-specific binding proteins that are important for conferring EBNA2 responsiveness (88, 103, 123, 318). Linker insertion and deletion mutants have been used to define four regions of the EBNA2 protein that are essential for EBV transformation (45, 317). Analysis of the EBNA2 protein function has identified an acidic activation domain (region 4 [44, 45]) and a domain that interacts with RBP Jκ and PU.1 (region 2 [88, 103, 123, 318]). Within region 1, there are seven prolines that are essential for transformation (317). The importance of these prolines has yet to be determined. The arginine-glycine repeat region of region 3 contributes to transformation efficiency (297). Once bound by cellular proteins that are associated with specific sites in the viral or cellular genome, EBNA2 can then recruit basal transcription factors through its acidic activation domain. This activation domain binds to TAF40, TFIIB, TFIIH, and TFIIE via interaction with the novel coactivator p100 (294–296). EBNA2-responsive promoters containing RBP-Jκ and PU.1 binding sites include the LMP1 (59, 299), LMP2 (329), and Cp (275) viral promoters, as well as the cellular CD23 promoter (306). Analysis of these promoters has shown that EBNA2 activation cannot be limited to only RBPJκ binding sites because (i) the EBNA2 response elements bind uncharacterized factors (122, 123, 156, 329); (ii) RBPJκ is expressed in many cell lines and tissues (88), but the LMP1 EBNA2 response element functions only in B cells; and (iii) analysis of the LMP1 promoter has shown the binding of RBPJκ and PU.1, as well as six other B-cell proteins (123). Two of these have been shown not to be important for EBNA2 responsiveness by mutagenesis (123). These interactions are summarized in Fig. 3.

EBNA3A, EBNA3B, and EBNA3C. The EBNA3 gene family consists of three related genes that may have resulted by gene duplication (105, 211, 213, 252). As with the other EBNAs, the EBNA3 proteins are found within the nucleus in EBV-transformed cells (212). They are located in a tandem array in the EBV viral genome (Fig. 1) and all contain approximately 1,000 amino acids. Both EBNA3A and EBNA3C are essential for EBV transformation, while EBNA3B has been shown to be dispensable (292, 293). The EBNA3 proteins have glutamine-rich domains, and the EBNA3C domain can function as a transcriptor (172). In addition, the EBNA3 proteins contain regions homologous to the basic leucine zipper motif that is found in many mammalian transcription factors. EBNA3C transactivates some EBNA2-regulated genes (6, 305), can bind to RBPJκ in vitro and in yeast cells (172, 228), associates with RBPJκ in EBV-transformed LCLS (228), and blocks EBNA2 transactivation of both the LMP1 and LMP2 promoters in transiently transfected cells (143, 172, 228, 246). EBNA3A and EBNA3C can also block EBNA2 transactivation of the LMP2 promoter in transient transfection (143). All three EBNA3 proteins bind to RBPJκ and alter its ability to recognize its DNA binding

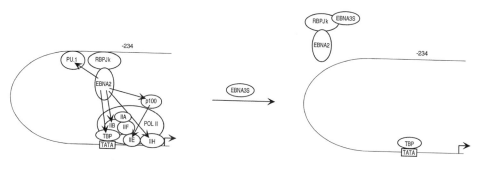

Figure 3 EBNA2 transactivation and EBNA3S modulation of RBPJκ (CBF1) in EBV infection. EBNA2, an acidic transcriptional transactivator, is directed to EBNA2 response elements by interaction with sequence-specific host DNA binding proteins such as RBPJκ. Once bound, EBNA2 stimulates transcription through interaction with components of TFIID. TFIID is a tightly associated protein complex of TBP (TATA binding protein) and eight or more TAFs (TBP-associated factors) (287). EBNA2 interacts with TAF40, TFIIB, the p62 and p80 subunits of TFIIH, and p100. p100 associates with the p56 and p34 subunits of TFIIE (294–296). The EBNA3S bind to RBPJκ to modulate the expression of EBNA2-responsive promoters by preventing the interaction of RBPJκ with DNA (229, 327).

sequence (Fig. 3) (229, 327). The domain important for binding RBPJκ is conserved among the EBNA3 proteins (229, 327).

EBNALP. EBV strains with mutations in EBNALP (EBNA leader protein) are partially defective with respect to immortalization of primary B lymphocytes (93, 168). Transformation efficiency is reduced approximately 10-fold and requires the presence of fibroblast feeder layers to augment transformation. EBNALP mutant virus-infected transformed B lymphocytes demonstrate delayed transit through the G1 phase of the cell cycle which can be complemented by introducing EBNALP by stable transfection (4). Transient transfection of EBNALP and EBNA2 into primary B lymphocytes, primed by CD21 crosslinking with gp350 envelope glycoprotein to mimic virus infection, progresses through the cell cycle as evidenced by the synthesis of cyclin D2 (266). EBNALP is phosphorylated on serine residues and in vitro is a substrate for both casein kinase II and the cyclin-dependent p34^{cdc42} kinase (133). This phosphorylation occurs in a cell cycle stage-specific manner with the protein being hyperphosphorylated in late G2 (133, 212). Like EBNA2, EBNALP is early upon infection of B lymphocytes (3, 5). EBNALP colocalizes with the retinoblastoma susceptibility gene (Rb) (121) and associates with tumor suppressor p53 in vitro experiments (282). Interaction of EBNALP with the 70-kDa family of heat shock proteins has also been demonstrated by immunoprecipitation and immunofluorescence studies (134, 169, 281). It is unlikely that EBNALP has such diverse functions. More recent studies may provide the best evidence for the true function of EBNALP in EBV transformation. In these studies, EBNALP was shown to stimulate EBNA2 transcriptional transactivation (94, 198). The stimulatory activity resided in the amino-terminal 66-amino-acid repeat domain (94, 198). Interestingly, the previously described EBNALP mutant virus contained this region (93, 168). It is therefore possible that EBNALP may be an essential gene for transformation because the EBNALP in the original mutant may have partial wild-type activity.

LMP1. LMP1 is an integral membrane protein with a predicted structure that consists of a short 10-amino-acid hydrophilic amino terminus, six hydrophobic transmembrane domains, and a 200-amino-acid carboxy terminus, rich in acidic residues. It lacks a signal peptide, but is inserted into membranes when translated in vitro (154, 155). Immunofluorescence microscopy, subcellular fractionation, and live cell protease cleavage experiments are consistent with LMP1 localization in the plasma membrane of infected cells with both the amino and carboxy terminus of the protein within the cytoplasm (104, 155, 166).

LMP1 forms patches in EBV-infected cells and non-EBV-infected cell lines when expressed by single gene transfer (104, 154, 155). A major portion of LMP1 is tightly associated with the cytoskeleton (154, 166, 190, 192), and this association with the cytoskeleton may be mediated by LMP1's association with vimentin (154). It is phosphorylated predominantly on serine residues, with some threonine phosphorylation (15, 167, 190, 192), and phosphorylated protein is bound to the cytoskeleton (154, 167, 190, 192). The half-life of unbound LMP1 is less than 2 h, whereas cytoskeletal-bound LMP1 has a half-life of greater than 3 h (190, 192). LMP1 has been expressed by gene transfection in a wide variety of cell types. In continuous rodent fibroblasts, LMP1 has transforming effects (16, 191, 302, 303) such as altering cell morphology, enabling cells to grow in medium with low serum, and causing loss of contact inhibition, loss of anchorage independence, and tumorigenic growth in nude mice (16, 191, 302, 303). In epithelial cells, LMP1 blocks differentiation (53, 60, 311) and protects B lymphocytes and epithelial cells from apoptosis by induction of A20 and *bcl-2* (73, 91, 102, 139, 174, 240).

In Burkitt's lymphoma cells, LMP1 expression induces most of the phenotypic differences observed with EBV infection. There is increased cell surface expression of CD23, CD39, CD40, and CD44 and of the cell adhesion molecules LFA-1, ICAM-1, and LFA-3; altered growth; and NF-κB activation (91, 139, 240, 302, 304). As a result of these changes, LMP1 expression mimics changes usually associated with B-cell activation, including cell clumping and increased villous projections. Two domains have been identified within the LMP1 carboxy-terminal domain that mediate activation of NF-κB (114, 187). The first domain (amino acids 187–231) is identical to the TRAF interaction domain (see below). The second domain (amino acids 352–386) is the stronger activation domain (114, 187), but it is the first domain shown to be sufficient for primary B-lymphocyte growth transformation (128).

Recent experiments, described below, suggest that LMP1 mimics constitutively activated tumor necrosis factor receptor (TNFR) family members which include the TNFR, the CD40 receptor, and the CD30 receptor (Fig. 4). The receptors all have in common the binding of signaling proteins called TNFR-associated factors (TRAFs) which activate NF-κB when engaged (236). In fact, the carboxy-terminal amino acids of LMP1 (amino acids 187 through 386) have been shown to interact with TRAF1, TRAF2, and TRAF3 (56, 193) through a domain lying between amino acids 187 and 231. A core TRAF binding motif, PXQXT/S, has been

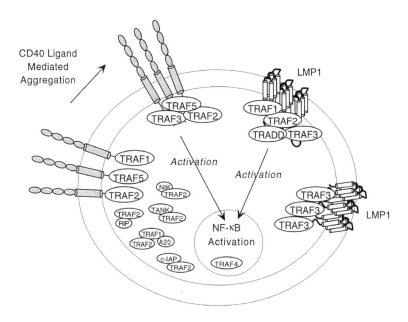

Figure 4 LMP1 activation of EBV-infected B cells. Signal transduction through TNF family receptors such as CD40 is initiated by clustering of the receptor by the binding of ligand. Receptor-associated proteins called TRAFs are then activated, leading to NK-κB activation (235). In EBV-immortalized B cells, LMP1 mimics a clustered TNF family receptor, thereby constitutively associating with the TRAFs and TRADD and activating the TNF signal transduction pathway, resulting in NF-κB activation (56, 120a, 127, 185, 193, 253). TRAF3 may interfere with LMP1 binding of TRAF1 and TRAF3, which may block LMP1 activation of NF-κB (56). The relevance of TRAF association with other cellular proteins such as A20, NIK, c-IAP, and RIP needs to be determined (39, 119, 165, 234–237, 270).

identified in the CD40 receptor, the CD30 receptor, and amino acids 205–209 of LMP1 (68), and mutagenesis studies in CD40 and LMP1 have confirmed the importance of this core TRAF binding domain (39, 56).

Experiments have indicated that both TRAF1 and TRAF3 bind most efficiently to the LMP1 TRAF binding site (56, 253). TRAF2 has been shown to be the most important for NF-κB activation, most likely by recruitment to LMP1, forming aggregates with TRAF1 (56). On the other hand, TRAF3 appears to be important for down-regulation of TRAF1 and TRAF2 activation by binding LMP1 (56). Finally, tumor necrosis factor receptor-associated death domain (TRADD) protein has been found to bind the carboxy-terminal portion of LMP1 involving tyrosine 384 and 385 (120a). This interaction is responsible for high-level NF-κB activation and efficient long-term LCL outgrowth (120a).

LMP2A and LMP2B. The LMP2 gene is simultaneously transcribed under the control of two promoters separated by 3 kb (140, 142, 251). The two resulting mRNAs are translated into two distinct LMP2 proteins, LMP2A and LMP2B (Fig.

5). Both proteins consist of 12 transmembrane domains of at least 16 amino acids, each of which traverses the B-cell plasma membrane, and a 27-amino-acid cytoplasmic carboxy-terminal tail. LMP2A contains an additional 119-amino-acid cytoplasmic amino-terminal domain not present in LMP2B. LMP2A protein localizes to numerous small patches in the plasma membrane of latently infected B lymphocytes (158, 159), and most of the antiphosphotyrosine reactivity within these B

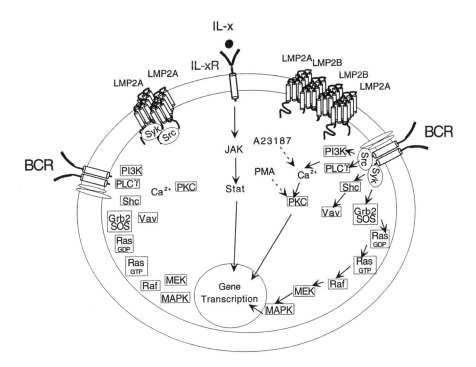

Figure 5 LMP2 effects on signal transduction through the B-cell antigen receptor complex (BCR). Ligation of the BCR induces the activation of the Src family and Syk PTKs, followed by activation of other transducing molecules (see reference 96 for a detailed review). In latently infected B cells, LMP2A is expressed and the multiple hydrophobic domains of LMP2A mediate aggregation in the plasma membrane, where the amino-terminal domains of LMP2A resemble cross-linked receptor tails and become tyrosine phosphorylated. The Src family PTKs and the Syk PTK bind. Other SH2-containing proteins may also bind. This complex then blocks signal transduction through the BCR, preventing activation of lytic replication following BCR ligation. The LMP2A complex does not block chemical inducers of gene transcription such as phorbol esters (PMA) or calcium ionophores (A23187). Activation to lytic replication may be mediated by an as yet unidentified pathway such as an interleukin cytokine pathway (IL-x and IL-xR) that is not blocked by LMP2A. LMP2B may aggregate in the plasma membrane with LMP2A. Lacking the amino-terminal domain of LMP2A, LMP2B may increase the spacing between LMP2A amino-terminal domains, resulting in release of the Src family and Syk PTKs from LMP2A and restoring normal signal transduction through the BCR.

cells is associated with these LMP2A patches (158). Within the 119-amino-acid N-terminal domain are eight tyrosine residues, two of which form an immuno-receptor tyrosine-based activation motif (ITAM) (30). This motif, first described by Reth (224), consists of paired tyrosine and leucine residues and is important in coupling intracellular protein tyrosine kinases (PTKs) with lymphocyte antigen receptors such as the BCR. The Src family PTKs and Syk PTK, which coimmunoprecipitate with the BCR complex, are also associated with LMP2A (26, 75, 178), and LMP2A induces the phosphorylation of Syk in vivo or in in vitro kinase reactions (178). The LMP2A residues that are phosphorylated and responsible for association with the PTKs are located in the amino-terminal domain of LMP2A. This was demonstrated by transfection of LMP2A cDNA mutant constructs into an EBV negative B-lymphoma cell line (26, 158).

EBV recombinant molecular genetic experiments have demonstrated that neither LMP2A or LMP2B is a mediator of the B-lymphocyte growth transformation that accompanies latent in vitro infection (131, 160–162). B lymphocytes transformed in vitro by the LMP2A and LMP2B mutants are indistinguishable in their growth in vitro from cells transformed by wild-type EBV, as measured by growth at low cell densities in low-serum media or when cells are grown in soft agar (160–162). In addition, LMP2A and LMP2B mutant-infected cell lines are not altered in tumor formation in SCID mice as compared with wild-type-infected cell lines (231). However, studies have demonstrated phenotypic differences between wild-type-infected LCLs and LMP2A mutant-infected LCLs. Wild-type infected LCLs are blocked in BCR-stimulated calcium mobilization, tyrosine phosphorylation, and lytic activation as compared with LMP2A mutant-infected LCLs (Fig. 5) (179, 181). The 119-amino-acid cytoplasmic amino-terminal domain, unique to LMP2A, is essential for this function (180). LMP2A functions as a negative regulator of the Src family PTK, Lyn, and the Syk PTK (179), while LMP2B may regulate LMP2A function by modulating spacing between individual LMP2A amino-terminal domains in the plasma membrane (Fig. 5). Constitutively clustered membrane patches of tyrosine-phosphorylated LMP2A might mimic activated receptor complexes, thereby competing for proteins with Src homology 2 domains (SH2). SH2 domains are noncatalytic domains conserved among cytoplasmic signaling proteins which bind tyrosine-phosphorylated proteins. Both deletion and point mutations of the LMP2A ITAM have determined that it is absolutely required for the ability of LMP2A to modulate B-cell signal transduction and the ability of LMP2A to bind the Syk PTK (74, 75).

EBERs. The EBERs (EBER1 and EBER2) are small nonpolyadenylated RNAs which are the most abundant EBV RNAs in latently infected cells. The EBERs have primary sequence similarity to adenovirus VA1, VA2, and U6 cell RNA (12, 108). They are expressed in almost all EBV-infected cell types and have been shown to be complexed with the La antigen, ribosomal protein L22, and the interferon-inducible protein kinase PKR (42, 43, 145, 291). Of the proteins that the EBERs have been shown to bind, the functional consequences have only been demonstrated for PKR. PKR has important roles in mediating antiviral and antiproliferative effects of interferons. VA1 RNA acts in the cytoplasm to inhibit activation of PKR, and the EBERS can substitute for VA1 and VA2 (19).

Molecular genetic studies have demonstrated that the EBERs can be deleted from the EBV viral genome without affecting in vitro transformation of primary B cells by EBV (280). Further studies using the EBER-deleted virus to analyze a potential EBER-mediated anti-interferon effect were unsuccessful (279). Therefore, the exact role of the EBV-encoded EBERs remains to be determined, although their consistent expression in EBV-infected cells has provided a convenient means to identify EBV-infected cells.

CONCLUSIONS

There has been dramatic progress in the understanding of the molecular biology of EBV. In particular, the functions of viral genes expressed in latent infection are being delineated so that there is a better understanding of both the structure and function of the viral proteins and the cellular proteins, viral proteins, and nucleic acids with which they interact. Further study will continue to benefit EBV research but also will benefit the understanding of normal cell growth mechanisms. Lytic gene function is a new and exciting area of EBV research. Despite similarities to lytic replication in other herpesviruses, it is apparent that EBV has evolved different strategies in regard to glycoprotein function, virus entry, and DNA replication. Investigation of these differences and a further elucidation of latent infection should be particularly informative in understanding the unique pathologies associated with EBV infections and may allow the development of novel therapeutics to treat such infections.

Acknowledgments
R.L. is supported by Public Health Service grants CA62234 and CA73507 from the National Cancer Institute. R.L. is a Scholar of the Leukemia Society of America. Sue Gerber and the members of the Longnecker Lab are gratefully acknowledged for critical review of this manuscript.

REFERENCES

1. **Adldinger, H. K., H. Dellus, U. K. Freese, J. Clarke, and G. W. Bornkamm.** 1985. A putative transforming gene of Jijoye virus differs from that of Epstein-Barr virus prototypes. *Virology* **141**:221–234.
2. **Ahearn, J. M., S. D. Hayward, J. C. Hickey, and D. T. Fearon.** 1988. Epstein-Barr virus (EBV) infection of murine L cells expressing recombinant human EBV/C3d receptor. *Proc. Natl. Acad. Sci. USA* **85**:9307–9311.
3. **Alfieri, C., M. Birkenbach, and E. Kieff.** 1991. Early events in Epstein-Barr virus infection of human B lymphocytes. *Virology* **181**:595–608. (Erratum, *Virology* **185**:946, 1991.)
4. **Allan, G. J., G. J. Inman, B. D. Parker, D. T. Rowe, and P. J. Farrell.** 1992. Cell growth effects of Epstein-Barr virus leader protein *J. Gen. Virol.* **73**:1547–1551.
5. **Allday, M. J., D. H. Crawford, and B. E. Griffin.** 1989. Epstein-Barr virus latent gene expression during the initiation of B cell immortalization. *J. Gen. Virol.* **70**:1755–1764.
6. **Allday, M. J., D. H. Crawford, and J. A. Thomas.** 1993. Epstein-Barr virus (EBV) nuclear antigen 6 induces expression of the EBV latent membrane protein and an activated phenotype in Raji cells. *J. Gen. Virol.* **74**:361–369.
7. **Allday, M. J., D. Kundu, S. Finerty, and B. E. Griffin.** 1990. CpG methylation of viral DNA in EBV-associated tumours. *Int. J. Cancer* **45**:1125–1130.

8. **Ambinder, R. F., M. A. Mullen, Y. N. Chang, G. S. Hayward, and S. D. Hayward.** 1991. Functional domains of Epstein-Barr virus nuclear antigen EBNA-1. *J. Virol.* **65:** 1466–1478.

9. **Ambinder, R. F., W. A. Shah, D. R. Rawlins, G. S. Hayward, and S. D. Hayward.** 1990. Definition of the sequence requirements for binding of the EBNA-1 protein to its palindromic target sites in Epstein-Barr virus DNA. *J. Virol.* **64:**2369–2379.

10. **Anagnostopoulos, I., M. Hummel, C. Kreschel, and H. Stein.** 1995. Morphology, immunophenotype, and distribution of latently and/or productively Epstein-Barr virus-infected cells in acute infectious mononucleosis: implications for the interindividual infection route of Epstein-Barr virus. *Blood* **85:**744–750.

11. **Apolloni, A., and T. B. Sculley.** 1994. Detection of A-type and B-type Epstein-Barr virus in throat washings and lymphocytes. *Virology* **202:**978–981.

12. **Arrand, J. R., L. S. Young, and J. D. Tugwood.** 1989. Two families of sequences in the small RNA-encoding region of Epstein-Barr virus (EBV) correlate with EBV types A and B. *J. Virol.* **63:**983–986.

13. **Austin, P. J., E. Flemington, C. N. Yandava, J. L. Strominger, and S. H. Speck.** 1988. Complex transcription of the Epstein-Barr virus BamHI fragment H rightward open reading frame 1 (BHRF1) in latently and lytically infected B lymphocytes. *Proc. Natl. Acad. Sci. USA* **85:**3678–3682.

14. **Baer, R., A. T. Bankier, M. D. Biggin, P. L. Deininger, P. J. Farrell, T. J. Gibson, G. Hatfull, G. S. Hudson, S. C. Satchwell, C. Seguin, P. S. Tuffnell, and B. G. Barrell.** 1984. DNA sequence and expression of the B95-8 Epstein-Barr virus genome. *Nature* **310:**207–211.

15. **Baichwal, V. R., and B. Sugden.** 1987. Posttranslational processing of an Epstein-Barr virus-encoded membrane protein expressed in cells transformed by Epstein-Barr virus. *J. Virol.* **61:**866–875.

16. **Baichwal, V. R., and B. Sugden.** 1988. Transformation of Balb 3T3 cells by the BNLF-1 gene of Epstein-Barr virus. *Oncogene* **2:**461–467.

17. **Bankier, A. T., P. L. Deininger, S. C. Satchwell, R. Baer, P. J. Farrell, and B. G. Barrell.** 1983. DNA sequence analysis of the EcoRI Dhet fragment of B95-8 Epstein-Barr virus containing the terminal repeat sequences. *Mol. Biol. Med.* **1:**425–445.

18. **Beisel, C., J. Tanner, T. Matsuo, D. Thorley Lawson, F. Kezdy, and E. Kieff.** 1985. Two major outer envelope glycoproteins of Epstein-Barr virus are encoded by the same gene. *J. Virol.* **54:**665–674.

19. **Bhat, R. A., and B. Thimmappaya.** 1983. Two small RNAs encoded by Epstein-Barr virus can functionally substitute for the virus-associated RNAs in the lytic growth of adenovirus 5. *Proc. Natl. Acad. Sci. USA* **80:**4789–4793.

20. **Biggin, M., M. Bodescot, M. Perricaudet, and P. Farrell.** 1987. Epstein-Barr virus gene expression in P3HR1-superinfected Raji cells. *J. Virol.* **61:**3120–3132.

21. **Biggin, M., P. J. Farrell, and B. G. Barrell.** 1984. Transcription and DNA sequence of the BamHI L fragment of B95-8 Epstein-Barr virus. *EMBO J.* **3:**1083–1090.

22. **Birkenbach, M., X. Tong, L. E. Bradbury, T. F. Tedder, and E. Kieff.** 1992. Characterization of an Epstein-Barr virus receptor on human epithelial cells. *J. Exp. Med.* **176:** 1405–1414.

23. **Bochkarev, A., J. A. Barwell, R. A. Pfuetzner, E. Bochkareva, L. Frappier, and A. M. Edwards.** 1996. Crystal structure of the DNA-binding domain of the Epstein-Barr virus origin-binding protein, EBNA1, bound to DNA. *Cell* **84:**791–800.

24. **Bochkarev, A., J. A. Barwell, R. A. Pfuetzner, W. Furey, Jr., A. M. Edwards, and L. Frappier.** 1995. Crystal structure of the DNA-binding domain of the Epstein-Barr virus origin-binding protein EBNA 1. *Cell* **83:**39–46.

25. **Bodescot, M., M. Perricaudet, and P. J. Farrell.** 1987. A promoter for the highly spliced EBNA family of RNAs of Epstein-Barr virus. *J. Virol.* **63:**3424–3430.

26. **Burkhardt, A. L., J. B. Bolen, E. Kieff, and R. Longnecker.** 1992. An Epstein-Barr virus transformation-associated membrane protein interacts with *src* family tyrosine kinases. *J. Virol.* **66:**5161–5167.

27. **Burkitt, D.** 1962. A children's cancer dependent on climatic factors. *Nature* **194:** 232–234.

28. **Burkitt, D. P.** 1983. The discovery of Burkitt's lymphoma. *Cancer* **51:**1777–1786.

29. **Busson, P., R. H. Edwards, T. Tursz, and N. Raab Traub.** 1995. Sequence polymorphism in the Epstein-Barr virus latent membrane protein (LMP)-2 gene. *J. Gen. Virol.* **76:**139–145.

30. **Cambler, J. C.** 1995. New nomenclature for the Reth motif (or ARH1/TAM/ARAM/YXXL). *Immunol. Today* **16:**110.

31. **Carel, J. C., B. L. Myones, B. Frazier, and V. M. Holers.** 1990. Structural requirements for C3d,g/Epstein-Barr virus receptor (CR2/CD21) ligand binding, internalization, and viral infection. *J. Biol. Chem.* **265:**12293–12299.

32. **Challberg, M. D.** 1986. A method for identifying the viral genes required for herpesvirus DNA replication. *Proc. Natl. Acad. Sci. USA* **83:**9094–9098.

33. **Chang, Y. N., D. L.-Y. Dong, G. S. Hayward, and S. D. Hayward.** 1990. The Epstein-Barr virus Zta transactivator: a member of the bZIP family with unique DNA-binding specificity and a dimerization domain that lacks the characteristic heptad leucine zipper motif. *J. Virol.* **64:**3358–3369.

34. **Chee, M. S., A. T. Bankier, S. Beck, R. Bohni, C. M. Brown, R. Cerney, T. Hornsnell, C. A. Hutchinson, T. Kouzarides, J. A. Martignetti, E. Preddie, S. C. Satchwell, P. Tomlinson, K. M. Weston, and B. G. Barrell.** 1990. Analysis of the protein coding content of the sequence of human cytomegalovirus strain AD169. *Curr. Top. Microbiol. Immunol.* **154:**125–169.

35. **Chen, F., J. Z. Zou, L. di Renzo, G. Winberg, L. F. Hu, E. Klein, G. Klein, and I. Ernberg.** 1995. A subpopulation of normal B cells latently infected with Epstein-Barr virus resembles Burkitt lymphoma cells in expressing EBNA-1 but not EBNA-2 or LMP1. *J. Virol.* **69:**3752–3758.

36. **Chen, M. L., C. N. Tsai, C. L. Liang, C. H. Shu, C. R. Huang, D. Sulitzeanu, S. T. Liu, and Y. S. Chang.** 1992. Cloning and characterization of the latent membrane protein (LMP) of a specific Epstein-Barr virus variant derived from the nasopharyngeal carcinoma in the Taiwanese population. *Oncogene* **7:**2131–2140.

37. **Chen, M. R., J. Zong, and S. D. Hayward.** 1994. Delineation of a 16 amino acid sequence that forms a core DNA recognition motif in the Epstein-Barr virus EBNA-1 protein. *Virology* **205:**486–495.

38. **Chen, W. G., Y. Y. Chen, M. M. Bacchi, C. E. Bacchi, M. Alvarenga, and L. M. Weiss.** 1996. Genotyping of Epstein-Barr virus in Brazilian Burkitt's lymphoma and reactive lymphoid tissue. Type A with a high prevalence of deletions within the latent membrane protein gene. *Am. J. Pathol.* **148:**17–23.

39. **Cheng, G., and D. Baltimore.** 1996. TANK, a co-inducer with TRAF2 of TNF- and CD40L-mediated NF-κB activation. *Genes Dev.* **10:**963–973.

40. **Chevallier Greco, A., E. Manet, P. Chavrier, C. Mosnier, J. Daillie, and A. Sergeant.** 1986. Both Epstein-Barr virus (EBV)-encoded trans-acting factors, EB1 and EB2, are required to activate transcription from an EBV early promoter. *EMBO J.* **5:**3243–3249.

41. **Chittenden, T., S. Lupton, and A. J. Levine.** 1989. Functional limits of *oriP*, the Epstein-Barr virus plasmid origin of replication. *J. Virol.* **63:**3016–3025.

42. **Clarke, P. A., M. Schwemmle, J. Schickinger, K. Hilse, and M. J. Clemens.** 1991.

Binding of Epstein-Barr virus small RNA EBER-1 to the double-stranded RNA-activated protein kinase DAI. *Nucleic Acids Res.* **19**:243–248.

43. **Clemens, M. J.** 1994. Functional significance of the Epstein-Barr virus-encoded small RNAs. *EBV Rep.* **5**:107–111.

44. **Cohen, J. I., and E. Kieff.** 1991. An Epstein-Barr virus nuclear protein 2 domain essential for transformation is a direct transcriptional activator. *J. Virol.* **65**:5880–5885.

45. **Cohen, J. I., F. Wang, and E. Kieff.** 1991. Epstein-Barr virus nuclear protein 2 mutations define essential domains for transformation and transactivation. *J. Virol.* **65**: 2545–2554.

46. **Cohen, J. I., F. Wang, J. Mannick, and E. Kieff.** 1989. Epstein-Barr virus nuclear protein 2 is a key determinant of lymphocyte transformation. *Proc. Natl. Acad. Sci. USA* **86**:9558–9562.

47. **Contreras-Brodin, B., A. Karisson, T. Nilsson, L. Rymo, and G. Klein.** 1996. B cell-specific activation of the Epstein-Barr virus encoded C promoter compared with the wide-range activation of the W promoter. *J. Gen. Virol.* **77**:1159–1162.

48. **Countryman, J., H. Jenson, R. Seibl, H. Wolf, and G. Miller.** 1987. Polymorphic proteins encoded within BZLF1 of defective and standard Epstein-Barr viruses disrupt latency. *J. Virol.* **61**:3672–3679.

49. **Countryman, J., and G. Miller.** 1985. Activation of expression of latent Epstein-Barr herpesvirus after gene transfer with a small cloned subfragment of heterogeneous viral DNA. *Proc. Natl. Acad. Sci. USA* **82**:4085–4089.

50. **Cox, M. A., J. Leahy, and J. M. Hardwick.** 1990. An enhancer within the divergent promoter of Epstein-Barr virus responds synergistically to the R and Z transactivators. *J. Virol.* **64**:313–321.

51. **Dambaugh, T., K. Hennessy, L. Chamnankit, and E. Kieff.** 1984. U2 region of Epstein-Barr virus DNA may encode Epstein-Barr nuclear antigen 2. *Proc. Natl. Acad. Sci. USA* **81**:7632–7636.

52. **Davison, A. J., and J. E. Scott.** 1986. The complete DNA sequence of varicella-zoster virus. *J. Gen. Virol.* **67**:1759–1816.

53. **Dawson, C. W., A. B. Rickinson, and L. S. Young.** 1990. Epstein-Barr virus latent membrane protein inhibits human epithelial cell differentiation. *Nature* **344**:777–780.

54. **Decker, L. L., L. D. Klaman, and D. A. Thorley Lawson.** 1996. Detection of the latent form of Epstein-Barr virus DNA in the peripheral blood of healthy individuals. *J. Virol.* **70**:3286–3289.

55. **DePamphilis, M. L.** 1988. Transcriptional elements as components of eucaryotic origin of DNA replication. *Cell* **52**:635–638.

56. **Devergne, O., E. Hatzivassiliou, K. M. Izumi, K. M. Kaye, M. F. Kleijnen, E. Kieff, and G. Mosialos.** 1996. Association of TRAF1, TRAF2, and TRAF3 with an Epstein-Barr virus LMP1 domain important for B-lymphocyte transformation: role in NF-κB activation. *Mol. Cell. Biol.* **16**:7098–7108.

57. **Emini, E. A., J. Luka, M. E. Armstrong, P. M. Keller, R. W. Ellis, and G. R. Pearson.** 1987. Identification of an Epstein-Barr virus glycoprotein which is antigenically homologous to the varicella-zoster virus glycoprotein II and the herpes simplex virus glycoprotein B. *Virology* **157**:552–555.

58. **Epstein, M. A., B. G. Achong, and Y. M. Barr.** 1964. Virus particles in cultured lymphoblasts from Burkitt's lymphoma. *Lancet* **1**:702–703.

59. **Fahraeus, R., A. Jansson, A. Sjoblom, T. Nilsson, G. Klein, and L. Rymo.** 1993. Cell phenotype-dependent control of Epstein-Barr virus latent membrane protein 1 gene regulatory sequences. *Virology* **195**:71–80.

60. **Fahraeus, R., L. Rymo, J. S. Rhim, and G. Klein.** 1990. Morphological transformation

of human keratinocytes expressing the LMP gene of Epstein-Barr virus. *Nature* **345:** 447–449.

61. **Farrell, P. J., D. T. Rowe, C. M. Rooney, and T. Kouzarides.** 1989. Epstein-Barr virus BZLF1 trans-activator specifically binds to a consensus AP-1 site and is related to c-fos. *EMBO J.* **8:**127–132.

62. **Fingeroth, J. D., J. J. Weis, T. F. Tedder, J. L. Strominger, P. A. Biro, and D. T. Fearon.** 1984. Epstein-Barr virus receptor of human B lymphocytes is the C3d receptor CR2. *Proc. Natl. Acad. Sci. USA* **81:**4510–4514.

63. **Fixman, E. D., G. S. Hayward, and S. D. Hayward.** 1992. *trans*-Acting requirements for replication of Epstein-Barr virus ori-Lyt. *J. Virol.* **66:**5030–5039.

64. **Fixman, E. D., G. S. Hayward, and S. D. Hayward.** 1995. Replication of Epstein-Barr virus oriLyt: lack of a dedicated virally encoded origin-binding protein and dependence on Zta in cotransfection assays. *J. Virol.* **69:**2998–3006.

65. **Flemington, E., and S. H. Speck.** 1990. Identification of phorbol ester response elements in the promoter of Epstein-Barr virus putative lytic switch gene BZLF1. *J. Virol.* **64:**1217–1226.

66. **Frade, R., M. Barel, B. Ehlin Henriksson, and G. Klein.** 1985. gp140, the C3d receptor of human B lymphocytes, is also the Epstein-Barr virus receptor. *Proc. Natl. Acad. Sci. USA* **82:**1490–1493.

67. **Franken, M., B. Annis, A. N. Ali, and F. Wang.** 1995. 5′ Coding and regulatory region sequence divergence with conserved function of the Epstein-Barr virus LMP2A homolog in herpesvirus papio. *J. Virol.* **69:**8011–8019.

68. **Franken, M., O. Devergne, M. Rosenzwelg, B. Annis, E. Kieff, and F. Wang.** 1996. Comparative analysis identifies conserved tumor necrosis factor receptor-associated factor 3 binding sites in the human and simian Epstein-Barr virus oncogene LMP1. *J. Virol.* **70:**7819–7826.

69. **Frappier, L., K. Goldsmith, and L. Bendell.** 1994. Stabilization of the EBNA1 protein on the Epstein-Barr virus latent origin of DNA replication by a DNA looping mechanism. *J. Biol. Chem.* **269:**1057–1062.

70. **Frappier, L., and M. O'Donnell.** 1991. Epstein-Barr nuclear antigen 1 mediates a DNA loop within the latent replication origin of Epstein-Barr virus. *Proc. Natl. Acad. Sci. USA* **88:**10875–10879.

71. **Frappier, L., and M. O'Donnell.** 1991. Overproduction, purification, and characterization of EBNA1, the origin binding protein of Epstein-Barr virus. *J. Biol. Chem.* **266:** 7819–7826.

72. **Frappier, L., and M. O'Donnell.** 1992. EBNA1 distorts *oriP*, the Epstein-Barr virus latent replication origin. *J. Virol.* **66:**1786–1790.

73. **Fries, K. L., W. E. Miller, and N. Raab-Traub.** 1996. Epstein-Barr virus latent membrane protein 1 blocks p53-mediated apoptosis through the induction of the A20 gene. *J. Virol.* **70:**8653–8659.

74. **Fruehling, S., S. K. Lee, R. Herrold, B. Frech, G. Laux, E. Kremmer, F. A. Grasser, and R. Longnecker.** 1996. Identification of latent membrane protein 2A (LMP2A) domains essential for the LMP2A dominant-negative effect on B-lymphocyte surface immunoglobulin signal transduction. *J. Virol.* **70:**6216–6226.

75. **Fruehling, S., and R. Longnecker.** 1997. The immunoreceptor tyrosine-based activation motif of Epstein-Barr virus LMP2A is essential for blocking BCR-mediated signal transduction. *Virology* **235:**241–251.

76. **Furnari, F. B., V. Zacny, E. B. Quinlivan, S. Kenney, and J. S. Pagano.** 1994. RAZ, an Epstein-Barr virus transdominant repressor that modulates the viral reactivation mechanism. *J. Virol.* **68:**1827–1836.

77. **Gahn, T. A., and C. L. Schildkraut.** 1989. The Epstein-Barr virus origin of plasmid

replication, oriP, contains both the initiation and termination sites of DNA replication. *Cell* **58:**527–535.

78. **Gahn, T. A., and B. Sugden.** 1995. An EBNA-1-dependent enhancer acts from a distance of 10 kilobase pairs to increase expression of the Epstein-Barr virus LMP gene. *J. Virol.* **69:**2633–2636.

79. **Gibson, T., P. Stockwell, M. Ginsburg, and B. Barrell.** 1984. Homology between two EBV early genes and HSV ribonucleotide reductase and 38K genes. *Nucleic Acids Res.* **12:**5087–5099.

80. **Gibson, T. J., B. G. Barrell, and P. J. Farrell.** 1986. Coding content and expression of the EBV B95-8 genome in the region from base 62,248 to base 82,920. *Virology* **152:** 136–148.

81. **Gilbert, R., K. Ghosh, L. Rasile, and H. P. Ghosh.** 1994. Membrane anchoring domain of herpes simplex virus glycoprotein gB is sufficient for nuclear envelope localization. *J. Virol.* **68:**2272–2285.

82. **Gompels, U. A., J. Nicholas, G. Lawrence, M. Jones, B. J. Thomson, M. E. Martin, S. Efstathiou, M. Craxton, and H. A. Macaulay.** 1995. The DNA sequence of human herpesvirus-6: structure, coding content, and genome evolution. *Virology* **209:**29–51.

83. **Gong, M., and E. Kieff.** 1990. Intracellular trafficking of two major Epstein-Barr virus glycoproteins, gp350/220 and gp110. *J. Virol.* **64:**1507–1516.

84. **Gong, M., T. Ooka, T. Matsuo, and E. Kieff.** 1987. Epstein-Barr virus glycoprotein homologous to herpes simplex virus gB. *J. Virol.* **61:**499–508.

85. **Gratama, J. W., E. T. Lennette, B. Lonnqvist, M. A. Oosterveer, G. Klein, O. Ringden, and I. Ernberg.** 1992. Detection of multiple Epstein-Barr viral strains in allogeneic bone marrow transplant recipients. *J. Med. Virol.* **37:**39–47.

86. **Gratama, J. W., M. A. Oosterveer, F. E. Zwaan, J. Lepoutre, G. Klein, and I. Emberg.** 1988. Eradication of Epstein-Barr virus by allogeneic bone marrow transplantation: implications for sites of viral latency. *Proc. Natl. Acad. Sci. USA* **85:**8693–8696.

87. **Grogan, E. A., W. P. Summers, S. Dowling, D. Shedd, L. Gradoville, and G. Miller.** 1983. Two Epstein-Barr viral nuclear neoantigens distinguished by gene transfer, serology, and chromosome binding. *Proc. Natl. Acad. Sci. USA* **80:**7650–7653.

88. **Grossman, S. R., E. Johannsen, X. Tong, R. Yalamanchill, and E. Kieff.** 1994. The Epstein-Barr virus nuclear antigen 2 transactivator is directed to response elements by the J kappa recombination signal binding protein. *Proc. Natl. Acad. Sci. USA* **91:** 7568–7572.

89. **Gruffat, H., O. Renner, D. Pich, and W. Hammerschmidt.** 1995. Cellular proteins bind to the downstream component of the lytic origin of DNA replication of Epstein-Barr virus. *J. Virol.* **69:**1878–1886.

90. **Haddad, R. S., and L. M. Hutt Fletcher.** 1989. Depletion of glycoprotein gp85 from virosomes made with Epstein-Barr virus proteins abolishes their ability to fuse with virus receptor-bearing cells. *J. Virol.* **63:**4998–5005.

91. **Hammarskjold, M. L., and M. C. Simurda.** 1992. Epstein-Barr virus latent membrane protein transactivates the human immunodeficiency virus type 1 long terminal repeat through induction of NF-κB activity. *J. Virol.* **66:**6496–6501.

92. **Hammerschmidt, W., and B. Sugden.** 1988. Identification and characterization of oriLyt, a lytic origin of DNA replication of Epstein-Barr virus. *Cell* **55:**427–433.

93. **Hammerschmidt, W., and B. Sugden.** 1989. Genetic analysis of immortalizing functions of Epstein-Barr virus in human B lymphocytes. *Nature* **340:**393–397.

94. **Harada, S., and E. Kieff.** 1997. Epstein-Barr virus nuclear protein LP stimulates EBNA-2 acidic domain-mediated transcriptional activation. *J. Virol.* **71:**6611–6618.

95. **Hardwick, J. M., P. M. Lieberman, and S. D. Hayward.** 1988. A new Epstein-Barr

virus transactivator, R, induces expression of a cytoplasmic early antigen. *J. Virol.* **62**:2274–2284.

96. **Harnett, M. M.** 1994. Antigen receptor signalling: from the membrane to the nucleus. *Immunol. Today* **15**:P1–P2.

97. **Harris, A., B. D. Young, and B. E. Griffin.** 1985. Random association of Epstein-Barr virus genomes with host cell metaphase chromosomes in Burkitt's lymphoma-derived cell lines. *J. Virol.* **56**:328–332.

98. **Hatfull, G., A. T. Bankier, B. G. Barrell, and P. J. Farrell.** 1988. Sequence analysis of Raji Epstein-Barr virus DNA. *Virology* **164**:334–340.

99. **Hearing, J. C., and A. J. Levine.** 1985. The Epstein-Barr virus nuclear antigen (BamHI K antigen) is a single-stranded DNA binding phosphoprotein. *Virology* **145**:105–116.

100. **Heineman, T., M. Gong, J. Sample, and E. Kleff.** 1988. Identification of the Epstein-Barr virus gp85 gene. *J. Virol.* **62**:1101–1107.

101. **Henderson, S., D. Huen, M. Rowe, C. Dawson, G. Johnson, and A. Rickinson.** 1993. Epstein-Barr virus-coded BHRF1 protein, a viral homologue of Bcl-2, protects human B cells from programmed cell death. *Proc. Natl. Acad. Sci. USA* **90**:8479–8483.

102. **Henderson, S., M. Rowe, C. Gregory, D. Croom Carter, F. Wang, R. Longnecker, E. Kieff, and A. Rickinson.** 1991. Induction of bcl-2 expression by Epstein-Barr virus latent membrane protein 1 protects infected B cells from programmed cell death. *Cell* **65**:1107–1115.

103. **Henkel, T., P. D. Ling, S. D. Hayward, and M. G. Peterson.** 1994. Mediation of Epstein-Barr virus EBNA2 transactivation by recombination signal-binding protein J kappa. *Science* **265**:92–95.

104. **Hennessy, K., S. Fennewald, M. Hummel, T. Cole, and E. Kieff.** 1984. A membrane protein encoded by Epstein-Barr virus in latent growth-transforming infection. *Proc. Natl. Acad. Sci. USA* **81**:7207–7211.

105. **Hennessy, K., F. Wang, E. W. Bushman, and E. Kieff.** 1986. Definitive identification of a member of the Epstein-Barr virus nuclear protein 3 family. *Proc. Natl. Acad. Sci. USA* **83**:5693–5697.

106. **Herrold, R. E., A. Marchini, S. Fruehling, and R. Longnecker.** 1996. Glycoprotein 110, the Epstein-Barr virus homolog of herpes simplex virus glycoprotein B, is essential for Epstein-Barr virus replication in vivo. *J. Virol.* **70**:2049–2054.

107. **Hitt, M. M., M. J. Allday, T. Hara, L. Karran, M. D. Jones, P. Busson, T. Tursz, I. Ernberg, and B. E. Griffin.** 1989. EBV gene expression in an NPC-related tumour. *EMBO J.* **8**:2639–2651.

108. **Howe, J. G., and M. D. Shu.** 1988. Isolation and characterization of the genes for two small RNAs of herpesvirus papio and their comparison with Epstein-Barr virus-encoded EBER RNAs. *J. Virol.* **62**:2790–2798.

109. **Hsu, D. H., R. de Waal Malefyt, D. F. Fiorentino, M. N. Dang, P. Vieira, J. de Vries, H. Spits, T. R. Mosmann, and K. W. Moore.** 1990. Expression of interleukin-10 activity by Epstein-Barr virus protein BCRF1. *Science* **250**:830–832.

110. **Hu, L. F., E. R. Zabarovsky, F. Chen, S. L. Cao, I. Ernberg, G. Klein, and G. Winberg.** 1991. Isolation and sequencing of the Epstein-Barr virus BNLF-1 gene (LMP1) from a Chinese nasopharyngeal carcinoma. *J. Gen. Virol.* **72**:2399–2409.

111. **Hudson, G. S., A. T. Bankier, S. C. Satchwell, and B. G. Barrell.** 1985. The short unique region of the B95-8 Epstein-Barr virus genome. *Virology* **147**:81–98.

112. **Hudson, G. S., P. J. Farrell, and B. G. Barrell.** 1985. Two related but differentially expressed potential membrane proteins encoded by the *Eco*RI Dhet region of Epstein-Barr virus B95-8. *J. Virol.* **53**:528–535.

113. **Hudson, G. S., T. J. Gibson, and B. G. Barrell.** 1985. The BamHI F region of the B95-8 Epstein-Barr virus genome. *Virology* **147**:99–109.

114. **Huen, D. S., S. A. Henderson, D. Croom Carter, and M. Rowe.** 1995. The Epstein-Barr virus latent membrane protein-1 (LMP1) mediates activation of NF-kappa B and cell surface phenotype via two effector regions in its carboxy-terminal cytoplasmic domain. *Oncogene* **10**:549–560.

115. **Hummel, M., D. Thorley Lawson, and E. Kieff.** 1984. An Epstein-Barr virus DNA fragment encodes messages for the two major envelope glycoproteins (gp350/300 and gp220/200). *J. Virol.* **49**:413–417.

116. **Hurley, E. A., and D. A. Thorley Lawson.** 1988. B cell activation and the establishment of Epstein-Barr virus latency. *J. Exp. Med.* **168**:2059–2075.

117. **Hutt-Fletcher, L. M.** 1995. Epstein-Barr virus glycoproteins—beyond gp350/220. *EBV Rep.* **2**:49–53.

118. **Hutt Fletcher, L. M., E. Fowler, J. D. Lambris, R. J. Feighny, J. G. Simmons, and G. D. Ross.** 1983. Studies of the Epstein Barr virus receptor found on Raji cells. II. A comparison of lymphocyte binding sites for Epstein Barr virus and C3d. *J. Immunol.* **130**:1309–1312.

119. **Ishida, T., T. Tojo, T. Aoki, N. Kobayashi, T. Ohishi, T. Watanabe, T. Yamamoto, and J. Inoue.** 1996. TRAF5, a novel tumor necrosis factor receptor-associated factor family protein, mediates CD40 signaling. *Proc. Natl. Acad. Sci. USA* **93**:9437–9442.

120. **Itakura, O., S. Yamada, M. Narita, and H. Kikuta.** 1996. High prevalance of a 30-base pair deletion and single-base mutations within the carboxy terminal end of the LMP-1 oncogene of Epstein-Barr virus in the Japanese population. *Oncogene* **13**:1549–1553.

120a. **Izumi, K., and E. Kieff.** 1997. The Epstein-Barr virus oncogene latent membrane protein 1 engages the tumor necrosis factor receptor-associated death domain protein to mediate B lymphocyte growth transformation and activate NF-κB. *Proc. Natl. Acad. Sci. USA* **94**:12592–12597.

121. **Jiang, W. Q., L. Szekely, V. Wendel Hansen, N. Ringertz, G. Klein, and A. Rosen.** 1991. Co-localization of the retinoblastoma protein and the Epstein-Barr virus-encoded nuclear antigen EBNA-5. *Exp. Cell. Res.* **197**:314–318.

122. **Jin, X. W., and S. H. Speck.** 1992. Identification of critical *cis* elements involved in mediating Epstein-Barr virus nuclear antigen 2-dependent activity of an enhancer located upstream of the viral *Bam*HI C promoter. *J. Virol.* **66**:2846–2852.

123. **Johannsen, E., E. Koh, G. Moslalos, X. Tong, E. Kieff, and S. R. Grossman.** 1995. Epstein-Barr virus nuclear protein 2 transactivation of the latent membrane protein 1 promoter is mediated by J kappa and PU.1. *J. Virol.* **69**:253–262.

124. **Jones, C. H., S. D. Hayward, and D. R. Rawlins.** 1989. Interaction of the lymphocyte-derived Epstein-Barr virus nuclear antigen EBNA-1 with its DNA-binding sites. *J. Virol.* **63**:101–110.

125. **Karlin, S., B. E. Blaisdell, and G. A. Schachtel.** 1990. Contrasts in codon usage of latent versus productive genes of Epstein-Barr virus: data and hypotheses. *J. Virol.* **64**:4264–4273.

126. **Karlin, S., E. S. Mocarski, and G. A. Schachtel.** 1994. Molecular evolution of herpesviruses: genomic and protein sequence comparisons. *J. Virol.* **68**:1886–1902.

127. **Kaye, K. M., O. Devergne, J. N. Harada, K. M. Izumi, R. Yalamzanchili, E. Kieff, and G. Mosialos.** 1996. Tumor necrosis factor receptor associated factor 2 is a mediator of NF-kappa B activation by latent infection membrane protein 1, the Epstein-Barr virus transforming protein. *Proc. Natl. Acad. Sci. USA* **93**:11085–11090.

128. **Kaye, K. M., K. M. Izumi, G. Mosialos, and E. Kieff.** 1995. The Epstein-Barr virus LMP1 cytoplasmic carboxy terminus is essential for B-lymphocyte transformation; fibroblast cocultivation complements a critical function within the terminal 155 residues. *J. Virol.* **69**:675–683.

129. **Kenney, S., J. Kamine, E. Holley Guthrie, J. C. Lin, E. C. Mar, and J. Pagano.** 1989. The Epstein-Barr virus (EBV) BZLF1 immediate-early gene product differentially affects latent versus productive EBV promoters. *J. Virol.* **63:**1729–1736.

130. **Khanim, F., Q. Y. Yao, G. Niedobltek, S. Sihota, A. B. Rickinson, and L. S. Young.** 1996. Analysis of Epstein-Barr virus polymorphisms in normal donors and in virus-associated tumors from different geographic locations. *Blood* **88:**3491–3501.

131. **Kim, O. J., and J. L. Yates.** 1993. Mutants of Epstein-Barr virus with a selective marker disrupting the TP gene transform B cells and replicate normally in culture. *J. Virol.* **67:**7634–7640.

132. **Kimball, A. S., G. Milman, and T. D. Tullius.** 1989. High-resolution footprints of the DNA-binding domain of Epstein-Barr virus nuclear antigen 1. *Mol. Cell. Biol.* **9:**2738–2742.

133. **Kitay, M. K., and D. T. Rowe.** 1996. Cell cycle stage-specific phosphorylation of the Epstein-Barr virus immortalization protein EBNA-LP. *J. Virol.* **70:**7885–7893.

134. **Kitay, M. K., and D. T. Rowe.** 1996. Protein-protein interactions between Epstein-Barr virus nuclear antigen-LP and cellular gene products: binding of 70-kilodalton heat shock proteins. *Virology* **220:**91–99.

135. **Knecht, H., E. Bachmann, P. Brousset, K. Sandvej, D. Nadal, F. Bachmann, B. F. Odermatt, G. Delsol, and G. Pallesen.** 1993. Deletions within the LMP1 oncogene of Epstein-Barr virus are clustered in Hodgkin's disease and identical to those observed in nasopharyngeal carcinoma. *Blood* **82:**2937–2942.

136. **Knecht, H., F. Martius, E. Bachmann, T. Hoffman, D. R. Zimmermann, S. Rothenberger, K. Sandvej, W. Wegmann, N. Hurwitz, B. F. Odermatt, H. Kummer, and G. Pallesen.** 1995. A deletion mutant of the LMP1 oncogene of Epstein-Barr virus is associated with evolution of angioimmunoblastic lymphadenopathy into B immunoblastic lymphoma. *Leukemia* **9:**458–465.

137. **Kocache, M. M., and G. R. Pearson.** 1990. Protein kinase activity associated with a cell cycle regulated, membrane-bound Epstein-Barr virus induced early antigen. *Intervirology* **31:**1–13.

138. **Krysan, P. J., and M. P. Calos.** 1993. Epstein-Barr virus-based vectors that replicate in rodent cells. *Gene* **136:**137–143.

139. **Laherty, C. D., H. M. Hu, A. W. Opipari, F. Wang, and V. M. Dixit.** 1992. The Epstein-Barr virus LMP1 gene product induces A20 zinc finger protein expression by activating nuclear factor kappa B. *J. Biol. Chem.* **267:**24157–24160.

140. **Laux, G., A. Economou, and P. J. Farrell.** 1989. The terminal protein gene 2 of Epstein-Barr virus is transcribed from a bidirectional latent promoter region. *J. Gen. Virol.* **70:**3079–3084.

141. **Laux, G., U. K. Freese, R. Fischer, A. Polack, E. Kofler, and G. W. Bornkamm.** 1988. TPA-inducible Epstein-Barr virus genes in Raji cells and their regulation. *Virology* **162:**503–507.

142. **Laux, G., M. Perricaudet, and P. J. Farrell.** 1988. A spliced Epstein-Barr virus gene expressed in immortalized lymphocytes is created by circularization of the linear viral genome. *EMBO J.* **7:**769–774.

143. **Le Roux, A., B. Kerdiles, D. Walls, J. F. Dedleu, and M. Perricaudet.** 1994. The Epstein-Barr virus determined nuclear antigens EBNA-3A, -3B, and -3C repress EBNA-2-mediated transactivation of the viral terminal protein 1 gene promoter. *Virology* **205:**596–602.

144. **Lee, S. K., T. Compton, and R. Longnecker.** 1997. Failure to complement infectivity of EBV and HSV-1 glycoprotein B (gB) deletion mutants with gBs from different human herpesvirus subfamilies. *Virology* **237:**170–181.

144a. **Lee, S. K., and R. Longnecker.** 1997. The Epstein-Barr virus glycoprotein 110 carboxyl-terminal tail domain is essential for lytic virus replication. *J. Virol.* **71:** 4092–4097.

145. **Lerner, M. R., N. C. Andrews, G. Miller, and J. A. Steitz.** 1981. Two small RNAs encoded by Epstein-Barr virus and complexed with protein are precipitated by antibodies from patients with lupus erythematosus. *Proc. Natl. Acad. Sci. USA* **78:**805–809.

146. **Levitskaya, J., M. Coram, V. Levitsky, S. Imreh, P. M. Steigerwald Mullen, G. Klein, M. G. Kurilla, and M. G. Masucci.** 1995. Inhibition of antigen processing by the internal repeat region of the Epstein-Barr virus nuclear antigen-1. *Nature* **375:** 685–688.

147. **Li, Q., M. K. Spriggs, S. Kovats, S. M. Turk, M. R. Comeau, B. Nepom, and L. M. Hutt-Fletcher.** 1997. Epstein-Barr virus uses HLA class II as a cofactor for infection of B lymphocytes. *J. Virol.* **71:**4657–4662.

148. **Li, Q., S. M. Turk, and L. M. Hutt Fletcher.** 1995. The Epstein-Barr virus (EBV) BZLF2 gene product associates with the gH and gL homologs of EBV and carries an epitope critical to infection of B cells but not of epithelial cells. *J. Virol.* **69:**3987–3994.

149. **Li, Q. X., L. S. Young, G. Niedobitek, C. W. Dawson, M. Birkenbach, F. Wang, and A. B. Rickinson.** 1992. Epstein-Barr virus infection and replication in a human epithelial cell system. *Nature* **356:**347–350.

150. **Lieberman, P. M., and A. J. Berk.** 1990. In vitro transcriptional activation, dimerization, and DNA-binding specificity of the Epstein-Barr virus Zta protein. *J. Virol.* **64:** 2560–2568.

151. **Lieberman, P. M., J. M. Hardwick, and S. D. Hayward.** 1989. Responsiveness of the Epstein-Barr virus *Not*I repeat promoter to the Z transactivator is mediated in a cell-type-specific manner by two independent signal regions. *J. Virol.* **63:**3040–3050.

152. **Lieberman, P. M., J. M. Hardwick, J. Sample, G. S. Hayward, and S. D. Hayward.** 1990. The Zta transactivator involved in induction of lytic cycle gene expression in Epstein-Barr virus-infected lymphocytes binds to both AP-1 and ZRE sites in target promoter and enhancer regions. *J. Virol.* **64:**1143–1155.

153. **Lieberman, P. M., P. O'Hara, G. S. Hayward, and S. D. Hayward.** 1986. Promiscuous *trans* activation of gene expression by an Epstein-Barr virus-encoded early nuclear protein. *J. Virol.* **60:**140–148.

154. **Liebowitz, D., R. Kopan, E. Fuchs, J. Sample, and E. Kieff.** 1987. An Epstein-Barr virus transforming protein associates with vimentin in lymphocytes. *Mol. Cell. Biol.* **7:**2299–2308.

155. **Liebowitz, D., D. Wang, and E. Kieff.** 1986. Orientation and patching of the latent infection membrane protein encoded by Epstein-Barr virus. *J. Virol.* **58:**233–237.

156. **Ling, P. D., D. R. Rawlins, and S. D. Hayward.** 1993. The Epstein-Barr virus immortalizing protein EBNA-2 is targeted to DNA by a cellular enhancer-binding protein. *Proc. Natl. Acad. Sci. USA* **90:**9237–9241.

157. **Ling, P. D., J. J. Ryon, and S. D. Hayward.** 1993. EBNA-2 of herpesvirus papio diverges significantly from the type A and type B EBNA-2 proteins of Epstein-Barr virus but retains an efficient transactivation domain with a conserved hydrophobic motif. *J. Virol.* **67:**2990–3003.

158. **Longnecker, R., B. Druker, T. M. Roberts, and E. Kieff.** 1991. An Epstein-Barr virus protein associated with cell growth transformation interacts with a tyrosine kinase. *J. Virol.* **65:**3681–3692.

159. **Longnecker, R., and E. Kieff.** 1990. A second Epstein-Barr virus membrane protein (LMP2) is expressed in latent infection and colocalizes with LMP1. *J. Virol.* **64:** 2319–2326.

160. **Longnecker, R., C. L. Miller, X. Q. Miao, A. Marchini, and E. Kieff.** 1992. The only domain which distinguishes Epstein-Barr virus latent membrane protein 2A

(LMP2A) from LMP2B is dispensable for lymphocyte infection and growth transformation in vitro; LMP2A is therefore nonessential. *J. Virol.* **66**:6461–6469.

161. **Longnecker, R., C. L. Miller, X. Q. Miao, B. Tomkinson, and E. Kieff.** 1993. The last seven transmembrane and carboxy-terminal cytoplasmic domains of Epstein-Barr virus latent membrane protein 2 (LMP2) are dispensable for lymphocyte infection and growth transformation in vitro. *J. Virol.* **67**:2006–2013.

162. **Longnecker, R., C. L. Miller, B. Tomkinson, X. Q. Miao, and E. Kieff.** 1993. Deletion of DNA encoding the first five transmembrane domains of Epstein-Barr virus latent membrane proteins 2A and 2B. *J. Virol.* **67**:5068–5074.

163. **Lowell, C. A., L. B. Klickstein, R. H. Carter, J. A. Mitchell, D. T. Fearon, and J. M. Ahearn.** 1989. Mapping of the Epstein-Barr virus and C3dg binding sites to a common domain on complement receptor type 2. *J. Exp. Med.* **170**:1931–1946.

164. **Lupton, S., and A. J. Levine.** 1985. Mapping genetic elements of Epstein-Barr virus that facilitate extrachromosomal persistence of Epstein-Barr virus-derived plasmids in human cells. *Mol. Cell. Biol.* **5**:2533–2542.

165. **Malinin, N. L., M. P. Boldin, A. V. Kovalenko, and D. Wallach.** 1997. MAP3K-related kinase involved in NF-kB induction by TNF, CD95 and IL-1. *Nature* **385**: 540–544.

166. **Mann, K. P., D. Staunton, and D. A. Thorley Lawson.** 1985. Epstein-Barr virus-encoded protein found in plasma membranes of transformed cells. *J. Virol.* **55**: 710–720.

167. **Mann, K. P., and D. Thorley Lawson.** 1987. Posttranslational processing of the Epstein-Barr virus-encoded p63/LMP protein. *J. Virol.* **61**:2100–2108.

168. **Mannick, J. B., J. I. Cohen, M. Birkenbach, A. Marchini, and E. Kieff.** 1991. The Epstein-Barr virus nuclear protein encoded by the leader of the EBNA RNAs is important in B-lymphocyte transformation. *J. Virol.* **65**:6826–6837.

169. **Mannick, J. B., X. Tong, A. Hemnes, and E. Kieff.** 1995. The Epstein-Barr virus nuclear antigen leader protein associates with hsp72/hsc73. *J. Virol.* **69**:8169–8172.

170. **Marchini, A., B. Tomkinson, J. I. Cohen, and E. Kieff.** 1991. BHRF1, the Epstein-Barr virus gene with homology to Bcl2, is dispensable for B-lymphocyte transformation and virus replication. *J. Virol.* **65**:5991–6000.

171. **Marschall, M., F. Schwarzmann, U. Leser, B. Oker, P. Alliger, H. Mairhofer, and H. Wolf.** 1991. The BL'LF4 trans-activator of Epstein-Barr virus is modulated by type and differentiation of the host cell. *Virology* **181**:172–179.

172. **Marshall, D., and C. Sample.** 1995. Epstein-Barr virus nuclear antigen 3C is a transcriptional regulator. *J. Virol.* **69**:3624–3630.

173. **Martin, D. R., A. Yuryev, K. R. Kalli, D. T. Fearon, and J. M. Ahearn.** 1991. Determination of the structural basis for selective binding of Epstein-Barr virus to human complement receptor type 2. *J. Exp. Med.* **174**:1299–1311.

174. **Martin, J. M., D. Veis, S. J. Korsmeyer, and B. Sugden.** 1993. Latent membrane protein of Epstein-Barr virus induces cellular phenotypes independently of expression of Bcl-2. *J. Virol.* **67**:5269–5278.

175. **McGeoch, D. J., M. A. Dalrymple, A. J. Davison, A. Dolan, M. C. Frame, D. McNab, L. J. Perry, J. E. Scott, and P. Taylor.** 1988. The complete DNA sequence of the long terminal unique region in the genome of herpes simplex virus 1. *J. Gen. Virol.* **69**: 1531–1574.

176. **McGeoch, D. J., A. Dolan, S. Donald, and F. J. Rixon.** 1985. Sequence determination and genetic content of the short unique region in the genome of herpes simplex type 1. *J. Mol. Biol.* **181**:1–13.

177. **Middleton, T., and B. Sugden.** 1992. EBNA1 can link the enhancer element to the

initiator element of the Epstein-Barr virus plasmid origin of DNA replication. *J. Virol.* **66**:489–495.

178. **Miller, C. L., A. L. Burkhardt, J. H. Lee, B. Stealey, R. Longnecker, J. B. Bolen, and E. Kieff.** 1995. Integral membrane protein 2 of Epstein-Barr virus regulates reactivation from latency through dominant negative effects on protein-tyrosine kinases. *Immunity* **2**:155–166.

179. **Miller, C. L., J. H. Lee, E. Kieff, A. L. Burkhardt, J. B. Bolen, and R. Longnecker.** 1994. Epstein-Barr virus protein LMP2A regulates reactivation from latency by negatively regulating tyrosine kinases involved in sLg-mediated signal transduction. *Infect. Agents Dis.* **3**:128–136.

180. **Miller, C. L., J. H. Lee, E. Kieff, and R. Longnecker.** 1994. An integral membrane protein (LMP2) blocks reactivation of Epstein-Barr virus from latency following surface immunoglobulin crosslinking. *Proc. Natl. Acad. Sci. USA* **91**:772–776.

181. **Miller, C. L., R. Longnecker, and E. Kieff.** 1993. Epstein-Barr virus latent membrane protein 2A blocks calcium mobilization in B lymphocytes. *J. Virol.* **67**:3087–3094.

182. **Miller, G., M. Rabson, and L. Heston.** 1984. Epstein-Barr virus with heterogeneous DNA disrupts latency. *J. Virol.* **50**:174–182.

183. **Miller, N., and L. M. Hutt Fletcher.** 1988. A monoclonal antibody to glycoprotein gp85 inhibits fusion but not attachment of Epstein-Barr virus. *J. Virol.* **62**:2366–2372.

184. **Miller, N., and L. M. Hutt Fletcher.** 1992. Epstein-Barr virus enters B cells and epithelial cells by different routes. *J. Virol.* **66**:3409–3414.

185. **Miller, W. E., G. Mosialos, E. Kieff, and N. Raab-Traub.** 1997. Epstein-Barr virus LMP1 induction of the epidermal growth factor receptor is mediated through a TRAF signaling pathway distinct from NF-kB activation. *J. Virol.* **71**:586–594.

186. **Milman, G., and E. S. Hwang.** 1987. Epstein-Barr virus nuclear antigen forms a complex that binds with high concentration dependence to a single DNA-binding site. *J. Virol.* **61**:465–471.

187. **Mitchell, T., and B. Sugden.** 1995. Stimulation of NF-kappa B-mediated transcription by mutant derivatives of the latent membrane protein of Epstein-Barr virus. *J. Virol.* **69**:2968–2976.

188. **Miyashita, E. M., B. Yang, K. M. Lam, D. H. Crawford, and D. A. Thorley Lawson.** 1995. A novel form of Epstein-Barr virus latency in normal B cells in vivo. *Cell* **80**: 593–601.

188a. **Montgomery, R. I., M. W. Warner, B. J. Lum, and P. G. Spear.** 1996. Herpes simplex virus 1 entry into cells mediated by a novel member of the TNF/NGF receptor family. *Cell* **87**:427–436.

189. **Moore, M. D., N. R. Cooper, B. F. Tack, and G. R. Nemerow.** 1987. Molecular cloning of the cDNA encoding the Epstein-Barr virus/C3d receptor (complement receptor type 2) of human B lymphocytes. *Proc. Natl. Acad. Sci. USA* **84**:9194–9198.

190. **Moorthy, R., and D. A. Thorley Lawson.** 1990. Processing of the Epstein-Barr virus-encoded latent membrane protein p63/LMP. *J. Virol.* **64**:829–837.

191. **Moorthy, R. K., and D. A. Thorley Lawson.** 1993. All three domains of the Epstein-Barr virus-encoded latent membrane protein LMP-1 are required for transformation of rat-1 fibroblasts. *J. Virol.* **67**:1638–1646.

192. **Moorthy, R. K., and D. A. Thorley Lawson.** 1993. Biochemical, genetic, and functional analyses of the phosphorylation sites on the Epstein-Barr virus-encoded oncogenic latent membrane protein LMP-1. *J. Virol.* **67**:2637–2645.

193. **Mosialos, G., M. Birkenbach, R. Yalamanchill, T. VanArsdale, C. Ware, and E. Kieff.** 1995. The Epstein-Barr virus transforming protein LMP1 engages signaling proteins for the tumor necrosis factor receptor family. *Cell* **80**:389–399.

194. **Nemerow, G. R., and N. R. Cooper.** 1984. Early events in the infection of human B lymphocytes by Epstein-Barr virus: the internalization process. *Virology* **132:**186–198.

195. **Nemerow, G. R., C. Mold, V. K. Schwend, V. Tollefson, and N. R. Cooper.** 1987. Identification of gp350 as the viral glycoprotein mediating attachment of Epstein-Barr virus (EBV) to the EBV/C3d receptor of B cells: sequence homology of gp350 and C3 complement fragment C3d. *J. Virol.* **61:**1416–1420.

196. **Nemerow, G. R., J. J. Mullen III, P. W. Dickson, and N. R. Cooper.** 1990. Soluble recombinant CR2 (CD21) inhibits Epstein-Barr virus infection. *J. Virol.* **64:**1348–1352.

197. **Nemerow, G. R., R. Wolfert, M. E. McNaughton, and N. R. Cooper.** 1985. Identification and characterization of the Epstein-Barr virus receptor on human B lymphocytes and its relationship to the C3d complement receptor (CR2). *J. Virol.* **55:**347–351.

198. **Nitsche, F., A. Bell, and A. Rickinson.** 1997. Epstein-Barr virus leader protein enhances EBNA-2-mediated transactivation of latent membrane protein 1 expression: a role for the W1W2 repeat domain. *J. Virol.* **71:**6619–6628.

199. **Nonkwelo, C., E. B. Henson, and J. Sample.** 1995. Characterization of the Epstein-Barr virus Fp promoter. *Virology* **206:**183–195.

200. **Nonkwelo, C., I. K. Ruf, and J. Sample.** 1997. The Epstein-Barr virus EBNA-1 promoter Qp requires an initiator-like element. *J. Virol.* **71:**354–361.

201. **Nonkwelo, C., J. Skinner, A. Bell, A. Rickinson, and J. Sample.** 1996. Transcription start sites downstream of the Epstein-Barr virus (EBV) Fp promoter in early-passage Burkitt lymphoma cells define a fourth promoter for expression of the EBV EBNA-1 protein. *J. Virol.* **70:**623–627.

202. **Oh, S. J., T. Chittenden, and A. J. Levine.** 1991. Identification of cellular factors that bind specifically to the Epstein-Barr virus origin of DNA replication. *J. Virol.* **65:**514–519.

203. **Ohno, S., J. Luka, T. Lindahl, and G. Klein.** 1977. Identification of a purified complement-fixing antigen as the Epstein-Barr virus nuclear antigen (EBNA) by its binding to metaphase chromosomes. *Proc. Natl. Acad. Sci. USA* **74:**1605–1609.

204. **Orlowski, R., and G. Miller.** 1991. Single-stranded structures are present within plasmids containing the Epstein-Barr virus latent origin of replication. *J. Virol.* **65:**677–686.

205. **Palefsky, J. M., J. Berline, M. E. Penaranda, E. T. Lennette, D. Greenspan, and J. S. Greenspan.** 1996. Sequence variation of latent membrane protein-1 of Epstein-Barr virus strains associated with hairy leukoplakia. *J. Infect. Dis.* **173:**710–714.

206. **Pari, G. S., and D. G. Anders.** 1993. Eleven loci encoding *trans*-acting factors are required for transient complementation of human cytomegalovirus *ori*Lyt-dependent DNA replication. *J. Virol.* **67:**6979–6988.

207. **Parker, B. D., A. Bankier, S. Satchwell, B. Barrell, and P. J. Farrell.** 1990. Sequence and transcription of Raji Epstein-Barr virus DNA spanning the B95-8 deletion region. *Virology* **179:**339–346.

208. **Pearson, G. R., J. Luka, L. Petti, J. Sample, M. Birkenbach, D. Braun, and E. Kieff.** 1987. Identification of an Epstein-Barr virus early gene encoding a second component of the restricted early antigen complex. *Virology* **160:**151–161.

209. **Pellett, P. E., M. D. Biggin, B. Barrell, and B. Roizman.** 1985. Epstein-Barr virus genome may encode a protein showing significant amino acid and predicted secondary structure homology with glycoprotein B of herpes simplex virus 1. *J. Virol.* **56:**807–813.

210. **Pereira, L.** 1994. Function of glycoprotein B homologues of the family herpesviridae. *Infect. Agents Dis.* **3:**9–28.

211. **Petti, L., and E. Kieff.** 1988. A sixth Epstein-Barr virus nuclear protein (EBNA3B) is

expressed in latently infected growth-transformed lymphocytes. *J. Virol.* **62:** 2173–2178.

212. **Petti, L., C. Sample, and E. Kieff.** 1990. Subnuclear localization and phosphorylation of Epstein-Barr virus latent infection nuclear proteins. *Virology* **176:**563–574.

213. **Petti, L., J. Sample, F. Wang, and E. Kieff.** 1988. A fifth Epstein-Barr virus nuclear protein (EBNA3C) is expressed in latently infected growth-transformed lymphocytes. *J. Virol.* **62:**1330–1338.

214. **Pfitzner, A. J., E. C. Tsai, J. L. Strominger, and S. H. Speck.** 1987. Isolation and characterization of cDNA clones corresponding to transcripts from the *Bam*HI H and F regions of the Epstein-Barr virus genome. *J. Virol.* **61:**2902–2909.

215. **Polvino Bodnar, M., J. Kiso, and P. A. Schaffer.** 1988. Mutational analysis of Epstein-Barr virus nuclear antigen 1 (EBNA 1). *Nucleic Acids Res.* **16:**3415–3435.

216. **Puglielli, M. T., N. Desai, and S. H. Speck.** 1997. Regulation of EBNA gene transcription in lymphoblastoid cell lines: characterization of sequences downstream of BCR2 (Cp). *J. Virol.* **71:**120–128.

217. **Puglielli, M. T., M. Woisetschlaeger, and S. H. Speck.** 1996. *oriP* is essential for EBNA gene promoter activity in Epstein-Barr virus-immortalized lymphoblastoid cell lines. *J. Virol.* **70:**5758–5768.

218. **Qu, L., and D. T. Rowe.** 1992. Epstein-Barr virus latent gene expression in uncultured peripheral blood lymphocytes. *J. Virol.* **66:**3715–3724.

219. **Quinlivan, E. B., E. A. Holley Guthrie, M. Norris, D. Gutsch, S. L. Bachenheimer, and S. C. Kenney.** 1993. Direct BRLF1 binding is required for cooperative BZLF1/BRLF1 activation of the Epstein-Barr virus early promoter, BMRF1. *Nucleic Acids Res.* **21:**1999–2007.

220. **Raab-Traub, N., T. Dambaugh, and E. Kieff.** 1980. DNA of Epstein-Barr virus. VIII. B95-8, the previous prototype, is an unusual deletion derivative. *Cell* **22:**257–267.

221. **Rawlins, D. R., G. Milman, S. D. Hayward, and G. S. Hayward.** 1985. Sequence-specific DNA binding of the Epstein-Barr virus nuclear antigen (EBNA-1) to clustered sites in the plasmid maintenance region. *Cell* **42:**859–868.

222. **Reisman, D., and B. Sugden.** 1986. *trans* activation of an Epstein-Barr viral transcriptional enhancer by the Epstein-Barr viral nuclear antigen 1. *Mol. Cell. Biol.* **6:** 3838–3846.

223. **Reisman, D., J. Yates, and B. Sugden.** 1985. A putative origin of replication of plasmids derived from Epstein-Barr virus is composed of two *cis*-acting components. *Mol. Cell. Biol.* **5:**1822–1832.

224. **Reth, M.** 1989. Antigen receptor clue. *Nature* **338:**383–384.

225. **Rickinson, A. B., and E. Kieff.** 1996. Epstein-Barr virus, p. 2397–2446. *In* B. N. Fields, D. M. Knipe, and P. M. Howley (ed.), *Fields Virology.* Lippincott-Raven Publishers, Philadelphia.

226. **Rickinson, A. B., L. S. Young, and M. Rowe.** 1987. Influence of the Epstein-Barr virus nuclear antigen EBNA 2 on the growth phenotype of virus-transformed B cells. *J. Virol.* **61:**1310–1317.

227. **Roberts, M. L., A. T. Luxembourg, and N. R. Cooper.** 1996. Epstein-Barr virus binding to CD21, the virus receptor, activates resting B cells via an intracellular pathway that is linked to B cell infection. *J. Gen. Virol.* **77:**3077–3085.

228. **Robertson, E. S., S. Grossman, E. Johannsen, C. Miller, J. Lin, B. Tomkinson, and E. Kieff.** 1995. Epstein-Barr virus nuclear protein 3C modulates transcription through interaction with the sequence-specific DNA-binding protein J kappa. *J. Virol.* **69:** 3108–3116.

229. **Robertson, E. S., J. Lin, and E. Kieff.** 1996. The amino-terminal domains of Epstein-Barr virus nuclear proteins 3A, 3B, and 3C interact with RBPJκ. *J. Virol.* **70:**3068–3074.

230. **Robertson, K. D., A. Manns, L. J. Swinnen, J. C. Zong, M. L. Gulley, and R. F. Ambinder.** 1996. CpG methylation of the major Epstein-virus latency promoter in Burkitt's lymphoma and Hodgkin's disease. *Blood* **88:**3129–3136.

231. **Rochford, R., C. L. Miller, M. J. Cannon, K. Izumi, E. Kieff, and R. Longnecker.** 1997. In vivo growth of Epstein-Barr virus transformed B cells with mutations in latent membrane protein 2 (LMP2). *Arch. Virol.* **142:**707–720.

232. **Roizman, B.** 1993. The family herpesviridae, p. 1–9. *In* B. Roizman, R. J. Whitely, and C. Lopez (ed.), *The Human Herpesviruses.* Raven Press, New York.

233. **Roizman, B., and A. E. Sears.** 1996. Herpes simplex viruses and their replication, p. 1043–1107. *In* B. N. Fields, D. M. Knipe, and P. M. Howley, *Fundamental Virology.* Lippincott-Raven, Philadelphia.

234. **Rothe, M., M.-G. Pan, W. J. Henzel, T. M. Ayres, and D. V. Goeddel.** 1995. The TNFR2-TRAF signaling complex contains two novel proteins related to baculoviral inhibitor of apoptosis proteins. *Cell* **83:**1243–1252.

235. **Rothe, M., V. Sarma, V. M. Dixit, and D. V. Goeddel.** 1995. TRAF2-mediated activation of NF-kB by TNF receptor 2 and CD40. *Science* **269:**1424–1427.

236. **Rothe, M., S. C. Wong, W. J. Henzel, and D. V. Goeddel.** 1994. A novel family of putative signal transducers associate with the cytoplasmic domain of the 75 kDa tumor necrosis factor receptor. *Cell* **78:**681–692.

237. **Rothe, M., J. Xiong, H.-B. Shu, K. Williamson, A. Goddard, and D. V. Goeddel.** 1996. I-TRAF is a novel TRAF-interacting protein that regulates TRAF-mediated signal transduction. *Proc. Natl. Acad. Sci. USA* **93:**8241–8246.

238. **Rousset, F., E. Garcia, T. Defrance, C. Peronne, N. Vezzio, D. H. Hsu, R. Kastelein, K. W. Moore, and J. Banchereau.** 1992. Interleukin 10 is a potent growth and differentiation factor for activated human B lymphocytes. *Proc. Natl. Acad. Sci. USA* **89:**1890–1893.

239. **Rowe, M., A. L. Lear, D. Croom Carter, A. H. Davies, and A. B. Rickinson.** 1992. Three pathways of Epstein-Barr virus gene activation from EBNA1-positive latency in B lymphocytes. *J. Virol.* **66:**122–131.

240. **Rowe, M., M. Peng-Pilon, D. S. Huen, R. Hardy, D. Croom-Carter, E. Lundgren, and A. B. Rickinson.** 1994. Upregulation of *bcl-2* by the Epstein-Barr virus latent membrane protein LMP1: a B-cell-specific response that is delayed relative to NF-κB activation and to induction of cell surface markers. *J. Virol.* **68:**5602–5612.

241. **Rowe, M., D. T. Rowe, C. D. Gregory, L. S. Young, P. J. Farrell, H. Rupani, and A. B. Rickinson.** 1987. Differences in B cell growth phenotype reflect novel patterns of Epstein-Barr virus latent gene expression in Burkitt's lymphoma cells. *EMBO J.* **6:**2743–2751.

242. **Rowe, M., L. S. Young, K. Cadwallader, L. Petti, E. Kieff, and A. B. Rickinson.** 1989. Distinction between Epstein-Barr virus type A (EBNA 2A) and type B (EBNA 2B) isolates extends to the EBNA 3 family of nuclear proteins. *J. Virol.* **63:**1031–1039.

243. **Russo, J. J., R. A. Bohenzky, M.-C. Chien, J. Chen, M. Yan, D. Maddalena, J. P. Parry, D. Peruzzi, I. S. Edelman, Y. Chang, and P. S. Moore.** 1996. Nucleotide sequence of the Kaposi sarcoma-associated herpesvirus (HHV8). *Proc. Natl. Acad. Sci. USA* **93:**14862–14867.

244. **Ryon, J. J., E. D. Fixman, C. Houchens, J. Zong, P. M. Lleberman, Y.-N. Chang, G. S. Hayward, and S. D. Hayward.** 1993. The lytic origin of herpesvirus papio is highly homologous to Epstein-Barr virus ori-Lyt: evolutionary conservation of transcriptional activation and replication signals. *J. Virol.* **67:**4006–4016.

245. **Sample, J., L. Brooks, C. Sample, L. Young, M. Rowe, C. Gregory, A. Rickinson, and E. Kieff.** 1991. Restricted Epstein-Barr virus protein expression in Burkitt lym-

phoma is due to a different Epstein-Barr nuclear antigen 1 transcriptional initiation site. *Proc. Natl. Acad. Sci. USA* **88**:6343–6347.

246. **Sample, J., E. B. Henson, and C. Sample.** 1992. The Epstein-Barr virus nuclear protein 1 promoter active in type I latency is autoregulated. *J. Virol.* **66**:4654–4661.

247. **Sample, J., M. Hummel, D. Braun, M. Birkenbach, and E. Kieff.** 1986. Nucleotide sequences of mRNAs encoding Epstein-Barr virus nuclear proteins: a probable transcriptional initiation site. *Proc. Natl. Acad. Sci. USA* **83**:5096–5100.

248. **Sample, J., and E. Kieff.** 1990. Transcription of the Epstein-Barr virus genome during latency in growth-transformed lymphocytes. *J. Virol.* **64**:1667–1674.

249. **Sample, J., E. F. Kieff, and E. D. Kieff.** 1994. Epstein-Barr virus types 1 and 2 have nearly identical LMP-1 transforming genes. *J. Gen. Virol.* **75**:2741–2746.

250. **Sample, J., G. Lancz, and M. Nonoyama.** 1986. Mapping of genes in BamHL fragment M of Epstein-Barr virus DNA that may determine the fate of viral infection. *J. Virol.* **57**:145–154.

251. **Sample, J., D. Liebowitz, and E. Kieff.** 1989. Two related Epstein-Barr virus membrane proteins are encoded by separate genes. *J. Virol.* **63**:933–937.

252. **Sample, J., L. Young, B. Martin, T. Chatman, E. Kieff, A. Rickinson, and E. Kieff.** 1990. Epstein-Barr virus types 1 and 2 differ in their EBNA-3A, EBNA-3B, and EBNA-3C genes. *J. Virol.* **64**:4084–4092.

253. **Sandberg, M., W. Hammerschmidt, and B. Sugden.** 1997. Characterization of LMP1's association with TRAF1, TRAF2, and TRAF3. *J. Virol.* **71**:4649–4656.

254. **Sandvej, K., S. C. Peh, B. S. Andresen, and G. Pallesen.** 1994. Identification of potential hot spots in the carboxy-terminal part of the Epstein-Barr virus (EBV) BNLF-1 gene in both malignant and benign EBV-associated diseases: high frequency of a 30-bp deletion in Malaysian and Danish peripheral T-cell lymphomas. *Blood* **84**:4053–4060.

255. **Sarisky, R. T., Z. Gao, P. M. Lieberman, E. D. Fixman, G. S. Hayward, and S. D. Hayward.** 1996. A replication function associated with the activation domain of the Epstein-Barr virus Zta transactivator. *J. Virol.* **70**:8340–8347.

256. **Sbih Lammali, F., D. Djennaoui, H. Belaoui, A. Bouguermouh, G. Decaussin, and T. Ooka.** 1996. Transcriptional expression of Epstein-Barr virus genes and proto-oncogenes in north African nasopharyngeal carcinoma. *J. Med. Virol.* **49**:7–14.

257. **Schaefer, B. C., J. L. Strominger, and S. H. Speck.** 1995. Redefining the Epstein-Barr virus-encoded nuclear antigen EBNA-1 gene promoter and transcription initiation site in group I Burkitt lymphoma cell lines. *Proc. Natl. Acad. Sci. USA* **92**:10565–10569.

258. **Schaefer, B. C., M. Woisetschlaeger, J. L. Strominger, and S. H. Speck.** 1991. Exclusive expression of Epstein-Barr virus nuclear antigen 1 in Burkitt lymphoma arises from a third promoter, distinct from the promoters used in latently infected lymphocytes. *Proc. Natl. Acad. Sci. USA* **88**:6550–6554.

259. **Schepers, A., D. Pich, and W. Hammerschmidt.** 1993. A transcription factor with homology to the AP-1 family links RNA transcription and DNA replication in the lytic cycle of Epstein-Barr virus. *EMBO J.* **12**:3921–3929.

260. **Schepers, A., D. Pich, J. Mankertz, and W. Hammerschmidt.** 1993. *cis*-Acting elements in the lytic origin of DNA replication of Epstein-Barr virus. *J. Virol.* **67**:4237–4245.

261. **Sculley, T. B., D. G. Sculley, J. H. Pope, G. W. Bornkamm, G. M. Lenoir, and A. B. Rickinson.** 1988. Epstein-Barr virus nuclear antigens 1 and 2 in Burkitt lymphoma cell lines containing either "A"- or "B"-type virus. *Intervirology* **29**:77–85.

262. **Segouffin, C., H. Gruffat, and A. Sergeant.** 1996. Repression by RAZ of Epstein-Barr virus bZIP transcription factor EB1 is dimerization independent. *J. Gen. Virol.* **77**:1529–1536.

263. **Shimizu, N., and K. Takada.** 1993. Analysis of the BZLF1 promoter of Epstein-Barr virus: identification of an anti-immunoglobulin response sequence. *J. Virol.* **67:** 3240–3245.

264. **Shimizu, N., A. Tanabe Tochikura, Y. Kuroiwa, and K. Takada.** 1994. Isolation of Epstein-Barr virus (EBV)-negative cell clones from the EBV-positive Burkitt's lymphoma (BL) line Akata: malignant phenotypes of BL cells are dependent on EBV. *J. Virol.* **68:**6069–6073.

265. **Sinclair, A. J., and P. J. Farrell.** 1995. Host cell requirements for efficient infection of quiescent primary B lymphocytes by Epstein-Barr virus. *J. Virol.* **69:**5461–5468.

266. **Sinclair, A. J., I. Palmero, G. Peters, and P. J. Farrell.** 1994. EBNA-2 and EBNA-LP cooperate to cause G0 to G1 transition during immortalization of resting human B lymphocytes by Epstein-Barr virus. *EMBO J.* **13:**3321–3328.

267. **Sixbey, J. W., P. Shirley, P. J. Chesney, D. M. Buntin, and L. Resnick.** 1989. Detection of a second widespread strain of Epstein-Barr virus. *Lancet* **2:**761–765.

268. **Smir, B. N., R. J. Hauke, P. J. Blerman, T. G. Gross, F. D'Amore, J. R. Anderson, and T. C. Greiner.** 1996. Molecular epidemiology of deletions and mutations of the latent membrane protein 1 of the Epstein-Barr virus in posttransplant lymphoproliferative disorders. *Lab. Invest.* **75:**575–588.

269. **Smith, P. R., and B. E. Griffin.** 1992. Transcription of the Epstein-Barr virus gene EBNA-1 from different promoters in nasopharyngeal carcinoma and B-lymphoblastoid cells. *J. Virol.* **66:**706–714.

270. **Song, H. Y., M. Rothe, and D. V. Goeddel.** 1996. The tumor necrosis factor-inducible zinc finger protein A20 interacts with TRAF1/TRAF2 and inhibits NK-kB activation. *Proc. Natl. Acad. Sci. USA* **93:**6721–6725.

271. **Spear, P. G.** 1993. Entry of alphaherpesviruses into cells. *Semin. Virol.* **4:**167–180.

272. **Strnad, B. C., M. R. Adams, and H. Rabin.** 1983. Glycosylation pathways of two major Epstein-Barr virus membrane antigens. *Virology* **127:**168–176.

273. **Su, W., T. Middleton, B. Sugden, and H. Echols.** 1991. DNA looping between the origin of replication of Epstein-Barr virus and its enhancer site: stabilization of an origin complex with Epstein-Barr nuclear antigen 1. *Proc. Natl. Acad. Sci.* **88:** 10870–10874.

274. **Sugden, B., and N. Warren.** 1989. A promoter of Epstein-Barr virus that can function during latent infection can be transactivated by EBNA-1, a viral protein required for viral DNA replication during latent infection. *J. Virol.* **63:**2644–2649.

275. **Sung, N. S., S. Kenney, D. Gutsch, and J. S. Pagano.** 1991. EBNA-2 transactivates a lymphoid-specific enhancer in the *Bam*HL C promoter of Epstein-Barr virus. *J. Virol.* **65:**2164–2169.

276. **Sung, N. S., J. Wilson, M. Davenport, N. D. Sista, and J. S. Pagano.** 1994. Reciprocal regulation of the Epstein-Barr virus *Bam*Hl-F promoter by EBNA-1 and an E2F transcription factor. *Mol. Cell Biol.* **14:**7144–7152.

277. **Swaminathan, S.** 1996. Characterization of Epstein-Barr virus recombinants with deletions of the BamHI C promoter. *Virology* **217:**532–541.

278. **Swaminathan, S., R. Hesselton, J. Sullivan, and E. Kieff.** 1993. Epstein-Barr virus recombinants with specifically mutated BCRF1 genes. *J. Virol.* **67:**7406–7413.

279. **Swaminathan, S., B. S. Huneycutt, C. S. Reiss, and E. Kieff.** 1992. Epstein-Barr virus-encoded small RNAs (EBERs) do not modulate interferon effects in infected lymphocytes. *J. Virol.* **66:**5133–5136.

280. **Swaminathan, S., B. Tomkinson, and E. Kieff.** 1991. Recombinant Epstein-Barr virus with small RNA (EBER) genes deleted transforms lymphocytes and replicates in vitro. *Proc. Natl. Acad. Sci. USA* **88:**1546–1550.

281. **Szekely, L., W. Q. Jiang, K. Pokrovskaja, K. G. Wiman, G. Klein, and N. Ringertz.**

1995. Reversible nucleolar translocation of Epstein-Barr virus-encoded EBNA-5 and hsp70 proteins after exposure to heat shock or cell density congestion. *J. Gen. Virol.* **76:**2423–2432.

282. **Szekely, L., G. Selivanova, K. P. Magnusson, G. Klein, and K. G. Wiman.** 1993. EBNA-5, an Epstein-Barr virus-encoded nuclear antigen, binds to the retinoblastoma and p53 proteins. *Proc. Natl. Acad. Sci. USA* **90:**5455–5459.

283. **Takada, K., and Y. Ono.** 1989. Synchronous and sequential activation of latently infected Epstein-Barr virus genomes. *J. Virol.* **63:**445–449.

284. **Takada, K., N. Shimizu, S. Sakuma, and Y. Ono.** 1986. *trans* Activation of the latent Epstein-Barr virus (EBV) genome after transfection of the EBV DNA fragment. *J. Virol.* **57:**1016–1022.

285. **Tanner, J., J. Weis, D. Fearon, Y. Whang, and E. Kieff.** 1987. Epstein-Barr virus gp350/220 binding to the B lymphocyte C3d receptor mediates adsorption, capping, and endocytosis. *Cell* **50:**203–213.

286. **Tanner, J., Y. Whang, J. Sample, A. Sears, and E. Kieff.** 1988. Soluble gp350/220 and deletion mutant glycoproteins block Epstein-Barr virus adsorption to lymphocytes. *J. Virol.* **62:**4452–4464.

287. **Tansey, W. P., and W. Herr.** 1997. TAFs: guilt by association? *Cell* **88:**729–732.

288. **Tarodi, B., T. Subramanian, and G. Chinnadural.** 1994. Epstein-Barr virus BHRF1 protein protects against cell death induced by DNA-damaging agents and heterologous viral infection. *Virology* **201:**404–407.

289. **Tedder, T. F., J. J. Weis, L. T. Clement, D. T. Fearon, and M. D. Cooper.** 1986. The role of receptors for complement in the induction of polyclonal B-cell proliferation and differentiation. *J. Clin. Immunol.* **6:**65–73.

290. **Tierney, R. J., N. Steven, L. S. Young, and A. B. Rickinson.** 1994. Epstein-Barr virus latency in blood mononuclear cells: analysis of viral gene transcription during primary infection and in the carrier state. *J. Virol.* **68:**7374–7385.

291. **Toczyski, D. P., A. G. Matera, D. C. Ward, and J. A. Steitz.** 1994. The Epstein-Barr virus (EBV) small RNA EBER1 binds and relocalizes ribosomal protein L22 in EBV-infected human B lymphocytes. *Proc. Natl. Acad. Sci. USA* **91:**3463–3467.

292. **Tomkinson, B., and E. Kieff.** 1992. Use of second-site homologous recombination to demonstrate that Epstein-Barr virus nuclear protein 3B is not important for lymphocyte infection or growth transformation in vitro. *J. Virol.* **66:**2893–2903.

293. **Tomkinson, B., E. Robertson, and E. Kieff.** 1993. Epstein-Barr virus nuclear proteins EBNA-3A and EBNA-3C are essential for B-lymphocyte growth transformation. *J. Virol.* **67:**2014–2025.

294. **Tong, X., R. Drapkin, D. Reinberg, and E. Kieff.** 1995. The 62- and 80-kDa subunits of transcription factor LLH mediate the interaction with Epstein-Barr virus nuclear protein 2. *Proc. Natl. Acad. Sci. USA* **92:**3259–3263.

295. **Tong, X., R. Drapkin, R. Yalamanchili, G. Mosialos, and E. Kieff.** 1995. The Epstein-Barr virus nuclear protein 2 acidic domain forms a complex with a novel cellular coactivator that can interact with TFIIE. *Mol. Cell. Biol.* **15:**4735–4744.

296. **Tong, X., F. Wang, C. J. Thut, and E. Kief.** 1995. The Epstein-Barr virus nuclear protein 2 acidic domain can interact with TFIIB, TAF40, and RPA70 but not with TATA-binding protein. *J. Virol.* **69:**585–588.

297. **Tong, X., R. Yalamanchili, S. Harada, and E. Kieff.** 1994. The EBNA-2 arginine-glycine domain is critical but not essential for B-lymphocyte growth transformation; the rest of region 3 lacks essential interactive domains. *J. Virol.* **68:**6188–6197.

298. **Trivedi, P., L. F. Hu, F. Chen, B. Christensson, M. G. Masucci, G. Klein, and G. Winberg.** 1994. Epstein-Barr virus (EBV)-encoded membrane protein LMP1 from a

nasopharyngeal carcinoma is non-immunogenic in a murine model system, in contrast to a B cell-derived homologue. *Eur. J. Cancer* **1**:84–88.

299. **Tsang, S. F., F. Wang, K. M. Izumi, and E. Kieff.** 1991. Delineation of the *cis*-acting element mediating EBNA-2 transactivation of latent infection membrane protein expression. *J. Virol.* **65**:6765–6771.

300. **Urier, G., M. Buisson, P. Chambard, and A. Sergeant.** 1989. The Epstein-Barr virus early protein EB1 activates transcription from different responsive elements including AP-1 binding sites. *EMBO J.* **8**:1447–1453.

301. **Vieira, P., R. de Waal Malefyt, M. N. Dang, K. E. Johnson, R. Kastelein, D. F. Fiorentino, J. E. deVrles, M. G. Roncarolo, T. R. Mosmann, and K. W. Moore.** 1991. Isolation and expression of human cytokine synthesis inhibitory factor cDNA clones: homology to Epstein-Barr virus open reading frame BCRFl. *Proc. Natl. Acad. Sci.* **88**: 1172–1176.

302. **Wang, D., D. Liebowitz, and E. Kieff.** 1985. An EBV membrane protein expressed in immortalized lymphocytes transforms established rodent cells. *Cell* **43**:831–840.

303. **Wang, D., D. Liebowitz, and E. Kieff.** 1988. The truncated form of the Epstein-Barr virus latent-infection membrane protein expressed in virus replication does not transform rodent fibroblasts. *J. Virol.* **62**:2337–2346.

304. **Wang, D., D. Liebowitz, F. Wang, C. Gregory, A. Rickinson, R. Larson, T. Springer, and E. Kieff.** 1988. Epstein-Barr virus latent infection membrane protein alters the human B-lymphocyte phenotype: deletion of the amino terminus abolishes activity. *J. Virol.* **62**:4173–4184.

305. **Wang, F., C. Gregory, C. Sample, M. Rowe, D. Liebowitz, R. Murray, A. Rickinson, and E. Kieff.** 1990. Epstein-Barr virus latent membrane protein (LMP1) and nuclear proteins 2 and 3C are effectors of phenotypic changes in B lymphocytes: EBNA-2 and LMP1 cooperatively induce CD23. *J. Virol.* **64**:2309–2318.

306. **Wang, F., H. Kikutani, S. F. Tsang, T. Kishimoto, and E. Kieff.** 1991. Epstein-Barr virus nuclear protein 2 transactivates a *cis*-acting CD23 DNA element. *J. Virol.* **65**: 4101–4106.

307. **Weis, J. J., L. E. Toothaker, J. A. Smith, J. H. Weis, and D. T. Fearon.** 1988. Structure of the human B lymphocyte receptor for C3d and the Epstein-Barr virus and relatedness to other members of the family of C3/C4 binding proteins. *J. Exp. Med.* **167**: 1047–1066. (Erratum, *J. Exp. Med.* **168**:1953–1954, 1988.)

308. **Wen, L. T., P. K. Lai, G. Bradley, A. Tanaka, and M. Nonoyama.** 1990. Interaction of Epstein-Barr viral (EBV) origin of replication (oriP) with EBNA-1 and cellular anti-EBNA-1 proteins. *Virology* **178**:293–296.

309. **Wen, L.-T., A. Tanaka, and M. Nonoyama.** 1989. Induction of anti-EBNA-1 protein by 12-O-tetradecanoylphorbol-13-acetate treatment of human lymphoblastoid cells. *J. Virol.* **63**:3315–3322.

310. **Wilson, J. B., J. L. Bell, and A. J. Levine.** 1996. Expression of Epstein-Barr virus nuclear antigen-1 induces B cell neoplasia in transgenic mice. *EMBO J.* **15**:3117–3126.

311. **Wilson, J. B., W. Weinberg, R. Johnson, S. Yuspa, and A. J. Levine.** 1990. Expression of the BNLF-1 oncogene of Epstein-Barr virus in the skin of transgenic mice induces hyperplasia and aberrant expression of keratin 6. *Cell* **61**:1315–1327.

312. **Woisetschlaeger, M., X. W. Jin, C. N. Yandava, L. A. Furmanski, J. L. Strominger, and S. H. Speck.** 1991. Role for the Epstein-Barr virus nuclear antigen 2 in viral promoter switching during initial stages of infection. *Proc. Natl. Acad. Sci. USA* **88**: 3942–3946.

313. **Woisetschlaeger, M., J. L. Strominger, and S. H. Speck.** 1989. Mutually exclusive use of viral promoters in Epstein-Barr virus latently infected lymphocytes. *Proc. Natl. Acad. Sci.* **86**:6498–6502.

314. **Woisetschlaeger, M., C. N. Yandava, L. A. Furmanski, J. L. Strominger, and S. H. Speck.** 1990. Promoter switching in Epstein-Barr virus during the initial stages of infection of B lymphocytes. *Proc. Natl. Acad. Sci.* **87:**1725–1729.

315. **Wong, K.-M., and A. J. Levine.** 1986. Identification and mapping of Epstein-Barr virus early antigens and demonstration of a viral gene activator that functions in *trans. J. Virol.* **60:**149–156.

316. **Wysokenski, D. A., and J. L. Yates.** 1989. Multiple EBNA1-binding sites are required to form an EBNA1-dependent enhancer and to activate a minimal replicative origin within *oriP* of Epstein-Barr virus. *J. Virol.* **63:**2657–2666.

317. **Yalamanchili, R., S. Harada, and E. Kieff.** 1996. The N-terminal half of EBNA2, except for seven prolines, is not essential for primary B-lymphocyte growth transformation. *J. Virol.* **70:**2468–2473.

318. **Yalamanchili, R., X. Tong, S. Grossman, E. Johannsen, G. Mosialos, and E. Kieff.** 1994. Genetic and biochemical evidence that EBNA 2 interaction with a 63-kDa cellular GTG-binding protein is essential for B lymphocyte growth transformation by EBV. *Virology* **204:**634–641.

319. **Yao, Q. Y., P. Ogan, M. Rowe, M. Wood, and A. B. Rickinson.** 1989. Epstein-Barr virus-infected B cells persist in the circulation of acyclovir-treated virus carriers. *Int. J. Cancer* **43:**67–71.

320. **Yao, Q. Y., A. B. Rickinson, and M. A. Epstein.** 1985. A re-examination of the Epstein-Barr virus carrier state in healthy seropositive individuals. *Int. J. Cancer* **35:**35–42.

321. **Yaswen, L. R., E. B. Stephens, L. C. Davenport, and L. M. Hutt Fletcher.** 1993. Epstein-Barr virus glycoprotein gp85 associates with the BKRF2 gene product and is incompletely processed as a recombinant protein. *Virology* **195:**387–396.

322. **Yates, J., N. Warren, D. Reisman, and B. Sugden.** 1984. A cis-acting element from the Epstein-Barr viral genome that permits stable replication of recombinant plasmids in latently infected cells. *Proc. Natl. Acad. Sci. USA* **81:**3806–3810.

323. **Yates, J. L., N. Warren, and B. Sugden.** 1985. Stable replication of plasmids derived from Epstein-Barr virus in various mammalian cells. *Nature* **313:**812–815.

324. **Young, L. S., Q. Y. Yao, C. M. Rooney, T. B. Sculley, D. J. Moss, H. Rupani, G. Laux, G. W. Bornkamm, and A. B. Rickinson.** 1987. New type B isolates of Epstein-Barr virus from Burkitt's lymphoma and from normal individuals in endemic areas. *J. Gen. Virol.* **68:**2853–2862.

325. **Zalani, S., E. Holley-Guthrie, and S. Kenney.** 1996. Epstein-Barr virus latency is disrupted by the immediate-early BRLF1 protein through a cell-specific mechanism. *Proc. Natl. Acad. Sci. USA* **93:**9194–9199.

326. **Zhang, S., and M. Nonoyama.** 1994. The cellular proteins that bind specifically to the Epstein-Barr virus origin of plasmid DNA replication belong to a gene family. *Proc. Natl. Acad. Sci.* **91:**2843–2847.

327. **Zhao, B., D. R. Marshall, and C. E. Sample.** 1996. A conserved domain of the Epstein-Barr virus nuclear antigens 3A and 3C binds to a discrete domain of Jκ. *J. Virol.* **70:** 4228–4236.

328. **Zimber, U., H. K. Adldinger, G. M. Lenoir, M. Vuillaume, M. V. Knebel Doeberitz, G. Laux, C. Desgranges, P. Wittmann, U. K. Freese, U. Schneider, and G. W. Bornkamm.** 1986. Geographical prevalence of two types of Epstein-Barr virus. *Virology* **154:**56–66.

329. **Zimber-Strobl, U., E. Kremmer, F. Grasser, G. Marschall, G. Laux, and G. W. Bornkamm.** 1993. The Epstein-Barr virus nuclear antigen 2 interacts with an EBNA2 responsive cis-element of the terminal protein 1 gene promoter. *EMBO J.* **12:**167–175.

330. **zur Hausen, H., F. O'Neil, and U. Freese.** 1978. Persisting oncogenic herpesviruses induced by the tumor promoter TPA. *Nature* **272:**373–375.

Human Tumor Viruses
Edited by Dennis J. McCance
© 1998 American Society for Microbiology

5

Pathogenesis of Epstein-Barr Virus
David Liebowitz

HISTORY AND BACKGROUND INFORMATION

Burkitt's Lymphoma and the Discovery of EBV

Denis Burkitt, an English surgeon, returned after World War II to equatorial Africa, where he developed an interest in an unusual lymphoma which was the most common childhood tumor that he encountered (8). Burkitt recognized the potential significance of the frequent occurrence of this tumor that was unknown outside of Africa. He visited medical missions and traveled throughout the region acquiring information about the clinical features and epidemiology of the tumor. He noted a unique linkage of the lymphoma to temperature and annual rainfall. Burkitt first published his experience in the late 1950s, initially to little attention (8). Anthony Epstein, a pathologist who had done earlier work on oncogenic avian retroviruses, attended a lecture given by Burkitt in 1961 and immediately started a collaboration to look for a virus in endemic Burkitt's lymphoma (BL). Shortly thereafter, a novel herpesvirus, which was serologically distinct from the other human herpesviruses and which would not grow in cells that supported the replication of other herpesviruses, was discovered to reactivate in the lymphoma cell cultures (20). The virus was named Epstein-Barr virus (EBV). Although viruses were already known to cause tumors in chickens and mice, this was the first evidence for a human tumor with an infectious cause.

EBV Transformation Biology

EBV infects and establishes a persistent infection in the host. Clinically, primary infection ranges from a mild self-limited illness in children to infectious mononucleosis in adolescents and adults. EBV has been linked to several human malignancies including endemic African BL, nasopharyngeal carcinoma, X-linked lymphoproliferative disease, Hodgkin's disease, posttransplantation lymphoproliferative disease, AIDS-related non-Hodgkin's lymphoma (NHL), certain peripheral T-cell NHLs, and smooth muscle tumors in immunosuppressed children.

The EBV genome is a double-stranded DNA molecule of more than 172 kbp in length which encodes more than 85 genes (58; see chapter 4). The viral genome consists of unique sequence DNA separated by internal repeat sequence DNA regions as well as terminal repeat (TR) sequence DNA at either end of the linear

175

viral genome. Upon infection of a target B lymphocyte, the linear genome contained within the virus circularizes via homologous recombination of the TR sequences. The exact site of recombination within the TR sequences is random, leading to a unique number of repeats in the newly fused TR region for each circularization event. Assessment of the length of the fused TR region by Southern blot analysis can identify the clonality of cells derived from a single progenitor containing a uniquely circularized EBV genome (102).

Nine EBV genes are expressed as proteins in EBV-transformed B-lymphoblastoid cell lines (LCLs) which divide continuously in vitro as a result of the expression of these viral genes. Six EBV nuclear antigens (EBNAs) and three membrane proteins (latent membrane proteins [LMPs]) have been identified in latently infected LCLs.

The LMP1 protein is responsible for the majority of the growth-altering effects of EBV on B lymphocytes (58; see chapter 4). In single gene transfection experiments, LMP1 has transforming effects in continuous rodent fibroblast cell lines (128, 129). In Rat-1 or NIH 3T3 cells, LMP1 alters cell morphology and enables cells to grow in medium supplemented with low serum. In Rat-1 cells, LMP1 also causes loss of contact inhibition, enabling cells to continue to divide after reaching confluence and loss of anchorage dependence, which results in a high soft-agar cloning efficiency (128). Rat-1 cells expressing LMP1, but not normal Rat-1 cells, form tumors very efficiently in nude mice (128). Similar observations were confirmed in other fibroblast cell lines (3).

LMP1 alters the growth of EBV-negative Burkitt tumor lymphoblasts when it is expressed at the appropriate level in such cells following gene transfer (64, 66, 130, 131). LMP1 induces cellular changes commonly associated with B-cell activation, including cell clumping, increased villous projections, and increased cell surface expression of CD23, CD30, CD39, CD40, CD44, and the cell adhesion molecules LFA-1, ICAM-1, and LFA-3. Recently, Kieff and coworkers demonstrated that LMP1 engages members of the tumor necrosis factor (TNF) receptor-associated factor (TRAF) family of signaling molecules via interaction with the C terminus of LMP1 (18, 55, 77). This shed new light on earlier observations that (i) LMP1 expression led to activation of the NF-κB transcription factor, (ii) the NF-κB activation domain(s) mapped to the C-terminal portion of the molecule, and (iii) the C-terminal portion of LMP1 was essential for B-lymphocyte transformation (52, 56, 66, 73). Thus, TRAF-mediated signal transduction, which leads to NF-κB activation, appears to be central to LMP1 effects on B-lymphocyte growth and activation. The tumor necrosis family of cytokines exerts pleiotropic effects on cell growth, determined in part by which members of the ligand family are present as well as by which members of the TNF receptor family are expressed on the cell surface. This may be due to specific ligand-receptor interactions resulting in different specific interactions with different subsets of TRAFs, ultimately leading to different intracellular growth signals being transduced. The variable response mediated through this complex system includes cell proliferation as well as differentiation and apoptosis. The role of LMP1 as a member of the TNF receptor family and as a TRAF-mediated signaling molecule leads to the hypothesis that LMP1 may alter the normal B lymphocyte's ability to terminally differentiate and undergo apoptosis, which is in part mediated by members of the TNF and TNF receptor molecule families. In addition to being a TNF receptor family member,

LMP1 induces cellular members of the TNF receptor family, including CD30 and CD40, that are involved in the pathogenesis of a variety of lymphoid malignancies. LMP1 has also been shown to protect B lymphocytes from apoptosis, which is mediated at least in part through the induction of Bcl-2 by LMP1 (33, 39).

Experimental Models for EBV Infection

The first major advance in the development of experimental models for EBV infection was the discovery that EBV infects primary human B lymphocytes in vitro and causes them to grow into immortalized LCLs (43, 96). The infection and cell growth transformation events evolve over a 1- to 2-day period, and a substantial fraction of the infected cells are transformed. The transformed cells grow in the peritoneum of profoundly immunodeficient severe-combined immunodeficient (SCID) mice (12, 78–80, 95, 108). Human B lymphocytes can be infected in vivo in SCID mice, and the infected cells will proliferate in the SCID mice. B lymphocytes from latently infected human peripheral blood will grow as tumors in SCID mice. The proliferating human cells grow free in the peritoneum, adhere to the peritoneum, grow as nodules in liver and spleen, and secrete human immunoglobulin (Ig) in the SCID animals. EBV gene expression in SCID mouse tumors exhibits a latency type III pattern (see chapter 4). Since other human hematopoietic cells will survive or grow in SCID mice, this system may be useful for studying EBV infection in bone marrow cells or the interaction between EBV-infected B lymphocytes and cells of the human immune system.

Soon after the discovery of EBV, considerable effort went into developing a primate model for EBV infection. No effect was evident when Old World primates were exposed to EBV by oral or parenteral routes. Old World primates were subsequently found to be uniformly infected with their own species-specific lymphocryptoviruses that are closely related to EBV. Due to antigen cross-reactivity, the animals were likely to have been immune to EBV infection. In contrast, New World primates are not naturally infected with EBV-related virus. In some New World primate species, such as cotton top tamarins, parenteral EBV infection results in an acute lymphoproliferative disease (113). Parenteral inoculation of 10^6 EB virions into cotton top tamarins results in death from lymphoma in most animals within 2 to 3 weeks (75, 76). The lymphoma cells are indistinguishable from cells transformed by EBV in vitro, and a latency type III expression pattern (see chapter 4) of EBNAs and LMPs is present in the lymphoma tissue (15, 136). Some animals recover or never develop disease, and subsequently have no persistent or latent EBV infection. Furthermore, attempts to infect the animals by oropharyngeal administration of virus have been without success. These last two points significantly limit the utility of this system as a model for human EBV pathogenesis.

Recently, a promising new model for EBV infection in Old World primates has been reported (74). An EBV-related rhesus monkey lymphocryptovirus was used to orally infect naive animals from a pathogen-free colony. This model reproduced key aspects of human EBV infection (see the next section below) including oral transmission, atypical lymphocytosis, lymphadenopathy, sustained serologic responses to lytic and latent EBV antigens, latent infection in peripheral blood B lymphocytes, and virus persistence in oropharyngeal secretions. Thus, this system may prove to be useful for studying the pathogenesis, prevention, and treatment of EBV infection and associated oncogenesis.

CLINICAL MANIFESTATIONS AND PATHOGENESIS OF EBV INFECTION

Primary and Latent EBV Infection in Normal Humans

In most of the world, primary EBV infection occurs asymptomatically during early childhood, before the age of 5 years. In contrast, primary infection in Western societies is often delayed by two or more decades and results in infectious mononucleosis (41, 42, 90). Our limited understanding of primary EBV infection comes from studies predominantly in patients with this disease.

The suspicion that infectious mononucleosis and therefore EBV, is transmitted orally (in the saliva) is apparent in its popular name, "kissing disease." The hallmarks of acute infectious mononucleosis are sore throat, fever, headache, malaise, lymphadenopathy, enlarged tonsils, and the finding of atypical lymphocytes in the peripheral blood. Other manifestations frequently include mild hepatitis, splenomegaly, and cerebritis. Less frequent are pneumonitis, cerebellitis, and autoimmune manifestations including skin rashes, blood dyscrasias, and peripheral neuropathy. The illness may last as long as several months, although illness of several weeks' duration is more typical. Persistence of illness beyond several months or recurrent illness is very rare, and recurrences beyond 2 years are not well documented.

In acute infectious mononucleosis, as many as 10% of the peripheral blood B lymphocytes are EBV infected. Most of the atypical lymphocytes in the peripheral blood, however, are NK or T cells (122, 123). Treatment of patients at this stage of illness with acyclovir results in inhibition of virus replication in the oropharynx but little if any effect on illness (21).

The development of CD8 T-cell-mediated cytotoxicity to latently infected lymphocytes correlates with the fall in the level of EBV-infected B lymphocytes from 1 in 10^1 to 1 in 10^6 (122–125). Latently infected B lymphocytes remain at the level of 1 in 10^6 to 1 in 10^7 thereafter. EBV-specific $CD8^+$ memory cytotoxic T lymphocytes (CTLs) persist in the peripheral blood indefinitely following acute EBV infection and prevent the outgrowth of latently infected B lymphocytes in vitro (82, 105). The precursor frequency of these EBV-specific CTLs is at least 1 in 10^3 to 10^4. The presence of these CTLs in nearly all healthy EBV-seropositive individuals, the high precursor frequency, and their ability to completely inhibit the outgrowth of autologous EBV-transformed LCLs in vitro are compatible with a role for CTL immunosurveillance in controlling persistent EBV infection in vivo (105).

Reactivated CTLs can be expanded in bulk populations or as clones in cultures containing irradiated autologous LCLs and interleukin 2-conditioned medium (for a review, see reference 81). Studies of these CTL clones have shown that in most individuals a substantial part of the EBV-specific CTL response is directed against epitopes in the EBNA-3A, EBNA-3B, and EBNA-3C family of proteins (57, 85). Some individuals had CTL responses against other latent EBV proteins including EBNA-2, EBNA-LP, LMP1, and LMP2. Notably, no CTL responses against EBNA-1 have been detected. A model for the virus-cell interactions that occur during EBV infection and persistence is shown in Fig. 1.

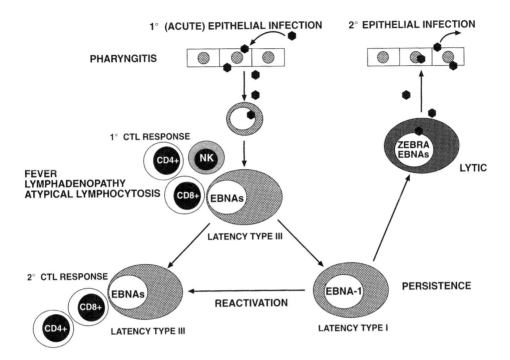

Figure 1   A model for the virus and cellular interactions that occur in primary EBV infection, persistence, and reactivation. Acute (primary [1°]) infection occurs in the oropharyngeal epithelium, and viral replication ensues. B lymphocytes are subsequently infected and display a latency type III viral gene expression pattern. At this point, the expanding population of EBV-infected B lymphoblasts are controlled by a primary CTL and NK cell response. EBV persistence is largely maintained in infected B cells that display a latency type I viral gene expression pattern. These cells may reactivate to a latency type II viral gene expression pattern, which is controlled by a secondary (memory) CTL response, or may be induced into lytic replication, where progeny virus may reinfect normal target tissues such as squamous epithelium and uninfected B lymphocytes. Details of the EBNA proteins and latent states are found in the text and in chapter 4.

EBV Proliferative Diseases

BL (Table 1)

In endemic areas of equatorial Africa, endemic BL is the most common childhood malignancy, accounting for approximately 80% of childhood cancers. The important clinicoepidemiological features of the disease, including the characteristic anatomical distribution to the jaw (72%) and the abdominal viscera (56%), the restricted geographic distribution, and the tight age distribution, were described by Burkitt shortly after his original account of this disease (9, 10).

Serologic data were collected in a prospective study in Uganda between 1970 and 1979 (26). More than 42,000 sera were collected from children 0 to 8 years of age from 1971 to 1972. During 7 years of follow-up, children who developed

Table 1 Viral and cellular features of well-characterized EBV-associated malignancies[a]

Malignancy	Subtype	% EBV positivity	Associated cytogenetic abnormalities	Cellular oncogene involvement	EBV gene expression pattern
Endemic BL	–	100	t(8;14), 85% t(2;8), 10% t(8;22), 5%	*myc*	Latency type I
NPC	Anaplastic	100	3p14	?	Latency type II
Hodgkin's disease	MC/LD	80–90	?	?	Latency type II
	NS	30–40	?	?	
PTLD	Polymorphic Monomorphic	>90	None	None	Latency type III
	DLC-IB	>90	+11 +9 3q27	? ? *bcl6*	
	SNC-BL	>90	t(8;14), t(2;8), t(8;22)	*myc*	
	DLC-CB	>90	?	?	
AIDS-associated	Primary CNS IB	>95	?	?	Latency type III
NHL	DLC/LC-IB	30–50	3q27	*bcl6*	Latency type III
	SNC	30–50	t(8;14), t(2;8), t(8;22)	*myc*	

[a] Abbreviations: MC, mixed cellularity; LD, lymphocyte depleted; NS, nodular sclerosing; DLC-IB, diffuse large cell, immunoblastic; SNC-BL, small noncleaved cell, Burkitt-like; DLC-CB, diffuse large cell, centroblastic; CNS IB, immunoblastic CNS lymphoma; LC, large cell.

endemic BL were infected with EBV months to years before the onset of the disease and had significantly higher IgG titers to the viral capsid antigen (IgG/VCA) than did controls who were matched for age, sex, location, and time of collection. The average IgG/VCA was two dilutions higher in children who developed endemic BL than in matched controls. IgG/VCA detection at this level indicated a 25-fold-increased relative risk for developing endemic BL.

Endemic BL has a narrow geographic distribution which extends across central Africa between 10 degrees north and 10 degrees south of the equator. Burkitt determined the chief climatic differences between the endemic and non-endemic regions to involve temperature and humidity (9, 10). Subsequent to the discovery of EBV in endemic BL cells, Burkitt noted that the endemic BL region coincided with areas of holoendemic falciparum malaria (54). Continuous reinfection with malaria causes polyclonal B-lymphocyte stimulation and, together with malnutrition, suppresses T-cell responses. Lower T-lymphocyte responses against EBV-infected cells may contribute to an expanded number of EBV-infected proliferating B lymphocytes, and this increase may allow for a higher probability of development of cytogenetic abnormalities. Analysis of the EBV TR sequences in endemic BL indicates that the tumors clonally arise from a single EBV-infected cell (7, 89, 102), suggesting that a "second hit" allows a single clone to become malignant.

Both endemic BL and sporadic BL are characterized by cytogenetic alterations involving translocations between chromosome 8 and either chromosome 14, 2, or 22 (17, 62, 70, 71, 119). The chromosome 8 site contains the *myc* proto-oncogene (8q24), while the other chromosomal sites contain Ig genes, with the Ig heavy-chain gene on chromosome 14 (14q32), the Ig kappa light-chain gene on chromosome 2 (2p11), and the Ig lambda light-chain gene on chromosome 22 (22q11). The most common translocation is the reciprocal t(8;14), which is present in approximately 80% of endemic BL cases, resulting in the translocation of *myc* coding sequences to the Ig heavy-chain constant region. The translocation breakpoints always leave the second and third *myc* coding exons intact. The Ig heavy-chain enhancer is on the reciprocally translocated fragment and thus does not affect *myc* expression. Rearrangements in the variants t(2;8) and t(8;22) usually result in translocation of the light chain genes to a position 3' to the *myc* coding sequences, often at distances greater than 50 kb. Variable effects of these translocations on *myc* expression have been noted. It is hypothesized that the resulting deregulation of *myc* expression affects cell proliferation.

In most endemic BL tumors, EBNA-1 is the only EBV protein expressed (35, 38, 107). As previously discussed, LMP1 is the major mediator of EBV-induced growth effects on B lymphocytes in vitro. The lack of expression of this protein in endemic BL tumors indicates that it is not required for maintenance of malignant cell growth. This suggests a role for EBV, and specifically LMP1, early in the development of malignancy and further suggests that oncogenesis is likely to be a multistep process.

One model based on the available data is as follows. Early in the evolution of endemic BL, after primary EBV infection, malnutrition and malaria result in polyclonal stimulation of B-cell and T-cell immunosuppression. This setting favors the proliferation of the EBV-infected B-lymphocyte population, resulting in longer survival, and a higher steady-state number, favoring the occurrence of the characteristic cytogenetic alterations involving the Ig and *myc* loci. As a result of changes in *myc* expression, the altered cell may have a growth advantage over the other cells in the expanded pool. Experimental evidence for the ability of EBV and altered *myc* expression to cooperate comes from studies in which EBV was used to transform human B lymphocytes in vitro followed by the introduction of a rearranged *myc* gene, cloned from an endemic BL cell line, into these cells (67). The EBV-transformed cells initially had very low cloning efficiencies in soft agar and did not form tumors in nude mice, but after gene transfer of a rearranged *myc* gene they grew more efficiently in soft agar and were tumorigenic in nude mice. Altered *myc* expression may replace EBV functions and allow cells to survive and proliferate while down-modulating EBNA and LMP expression. Since EBNAs and LMPs are targets of immune cytotoxic T lymphocytes, decreased expression of these viral proteins would be advantageous, enabling the cell to evade immunosurveillance. Further, LMP1 induces cell adhesion molecules, thereby enhancing HLA-restricted cyto-lytic T lymphocyte-mediated destruction of EBV-infected B cells (34, 86, 87). A cell whose growth is no longer dependent on external growth stimuli, and which has down-modulated EBNA and LMP1 expression, could escape immune detection and emerge as the dominant malignant clone.

PTLD (Table 1)

Posttransplant lymphoproliferative disease (PTLD) is a well-recognized risk of immunosuppressive therapy associated with solid organ transplantation or with T-cell-depleted allogeneic bone marrow transplantation. EBV is associated with the vast majority of cases, and the disease ranges histologically from polymorphic B-lymphocyte hyperplasia to malignant monoclonal lymphoma. Regardless of the histologic type or clonality, PTLD behaves in an aggressive fashion.

EBV-seronegative patients who acquire primary EBV infection in the post-transplant period have been shown to be at higher risk of developing PTLD than their seropositive counterparts, regardless of whether seropositive patients showed serologic evidence of reactivation of EBV infection (47, 48, 103). Active EBV infection is also associated with PTLD in seropositive transplant recipients, since nearly 100% of those who develop PTLD have serologic evidence of EBV reactivation compared with 33 to 48% in seropositive recipients as a whole (47, 48, 103). Thus, patients with active EBV infection are at risk for development of PTLD, and patients who acquire primary EBV infection in the posttransplant period are at especially high risk.

Studies on PTLD tumor tissue indicate that the EBV latent cycle genes coding for EBNA-1, -2, -3A, -3B, -3C, and -LP, LMP1, and LMP2 are all expressed. CD23 and cell adhesion molecules are also expressed in the tumor cells (28, 121, 134). As discussed previously, LMP1 is thought to be an important viral mediator of B-lymphocyte growth. Of note, LMP1 is also an important mediator of the immune response towards a cell since it is a major target for the cytolytic T-cell response against EBV-infected B lymphocytes and induces adhesion molecules which facilitate major histocompatibility complex (MHC)-restricted killing of EBV-infected B cells (34, 86, 87).

There is no consensus on a classification scheme for PTLD. A recently reported series of PTLD cases (22 patients) was separated into three categories based on morphologic and molecular criteria (60). The first category, termed plasmacytic hyperplasia, closely corresponds to the diffuse reactive plasma cell hyperplasia described in another classification scheme (88). This form of the disease is typically polyclonal, demonstrating multiple EBV infection events, and shows no evidence of oncogene or tumor suppressor gene alterations. The second category, termed polymorphic B-cell hyperplasia or polyclonal B-cell lymphoma, is usually monoclonal, with a single EBV infection event, lacking oncogene or tumor suppressor gene alterations. Finally, the third group of cases, called immunoblastic lymphoma or multiple myeloma, were always monoclonal with a single EBV infection event and featured frequent alterations in codon 61 of N-*ras* or mutations in exons 7 or 8 of *p53*. A single patient had a *myc* rearrangement on Southern blot.

We recently analyzed a series of 28 cases of PTLD (65). These data may shed new insight on the development and progression of this disease. In this series, 10 of the 28 patients were classified as polymorphic and the remaining 18 as monomorphic. All of the polymorphic cases were polyclonal or oligoclonal, and none of those analyzed had clonal cytogenetic abnormalities. Cytogenetic analysis of 12 of the 18 monomorphic cases revealed six distinct groups of abnormalities: 2 patients (17%) had non-clonal abnormalities or normal karyotypes; 3 patients (25%) had Burkitt's translocations involving the *myc* gene on chromosome 8; 3

patients (25%) had trisomy 9; 3 patients (25%) had trisomy 11; 2 patients (17%) had abnormalities involving the *bcl6* gene; and 1 patient (8%) had a rearrangement involving the Ig heavy chain locus. Further, groups of cytogenetic alterations were specifically associated with certain histologic subtypes of disease. Trisomy 11, trisomy 9, and *bcl6* alterations were found in large-cell, immunoblastic cases, while cases with small, noncleaved Burkitt-like histology had *myc* translocations. Tumors were EBV positive in 17 of 21 cases. LMP1 and EBNA-2 were expressed in 17 of 17 of the EBV-positive cases of PTLD examined, and LMP2A was expressed in 8 of 8 cases examined. Importantly, the pattern of viral antigen expression did not vary with the histology, clonality, or cytogenetic group of the tumor.

The results of these studies suggest a model for the pathogenesis of PTLD in which EBV drives the early proliferation of this disease. Later, as the tumors clonally evolve, specific cytogenetic alterations may occur, such as those noted above. Unlike endemic BL, in PTLD the persistent expression of certain EBV latent proteins, such as LMP1, regardless of tumor clonality or cytogenetic alterations suggests that these viral proteins not only contribute to the development of the disease, but also continue to provide signals which may be required for maintenance of the transformed state (Fig. 2).

The conventional therapeutic approach to PTLD in the transplant population has been to discontinue or reduce the dosage of immunosuppressive medication and to administer acyclovir. Reducing the immunosuppression may lead to increased immunosurveillance. A review of the literature reported that approximately two out of three transplant recipients treated by decreasing their immunosuppression survived, while the overall survival rate in transplant patients who developed PTLD was 31% (16). Acyclovir has generally not been effective in treating PTLD. In uncontrolled reports on the use of acyclovir in PTLD, the survival of patients treated with acyclovir (29%) did not differ from the overall survival rate (31%). The lack of efficacy of acyclovir in treating PTLD is not surprising since this agent inhibits viral replication, but has no effect on latently infected EBV-induced transformed B lymphocytes (114).

Anti-B-cell monoclonal antibodies have been used to treat PTLD (4, 23). Of 26 patients treated in an open, prospective, multicenter trial, 16 patients had complete responses and 2 had partial responses to anti-CD21 plus anti-CD24 antibodies (23). However, the treatment was only effective for patients with polyclonal B lymphoproliferations, and 7 of 7 patients who had monoclonal PTLD had no response to treatment. While encouraging, the study was not controlled and patients were not excluded from other treatment.

Recombinant alpha interferon has been reported to be effective in treating PTLD in bone marrow and organ transplant recipients as well as in patients with primary immunodeficiency. The only published results on the use of alpha interferon in PTLD report the treatment of five patients (111). Four of these patients had complete responses to treatment, and one patient had stable PTLD on interferon treatment when he died of cytomegalovirus pneumonia. Two of the five patients had monoclonal PTLD. The mechanism of action of alpha interferon on PTLD is not known. It may exert its effect through immune modulation (i.e., through up-regulation of MHC class II molecules) or through a direct antiproliferative effect on the EBV-infected B lymphocytes. Additionally, interferon specifically inhibits

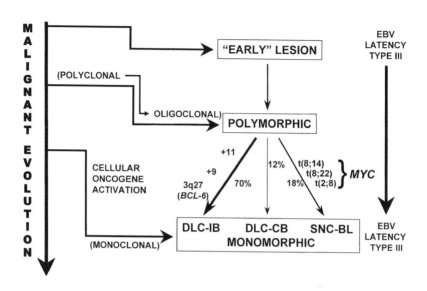

Figure 2 A model for the progression of typical EBV-positive B-cell PTLD from "early" proliferative lesions to polymorphic PTLD and then to monomorphic PTLD, based on viral gene expression and cytogenetic data. Polymorphic lesions are usually polyclonal or oligoclonal, display a latency type III viral gene expression pattern, and do not have any cytogenetic evidence for cellular oncogene activation. Monomorphic lesions are monoclonal, also have a latency type III viral gene expression pattern, and frequently have recurring cytogenetic abnormalities. Moreover, the histologic subtype of the monomorphic lesions correlates with specific recurring cytogenetic abnormalities: approximately 70% of the monomorphic B-cell lesions had diffuse large-cell, immunoblastic (DLC-IB) histology and frequently had +11, +9, or 3q27 (*bcl6*) recurring abnormalities; approximately 20% of the monomorphic B-cell lesions had small, non-cleaved, Burkitt-like (SNC-BL) histology and always had the t(8;14) or t(8;22) translocations involving *myc*; 10% of the monomorphic B-cell lesions had diffuse large-cell, centroblastic (DLC-CB) histology and did not have recurring abnormalities. The continued expression of EBV proteins, such as the LMP1 transforming gene in the monomorphic PTLD lesions, suggests that these proteins play an important role in the progression and maintenance of the transformed phenotype. The association of specific recurring abnormalities with different monomorphic histologic subtypes indicates that cellular genetic features may ultimately determine the phenotype of monomorphic lesions.

EBV-mediated B-lymphocyte transformation. Although alpha interferon does not inhibit the growth of EBV-transformed B lymphocytes in vitro, preincubation of mononuclear cells with pharmacologic levels of alpha interferon prior to infection with EBV blocks B-cell outgrowth (25).

Perhaps the most novel therapeutic approach for PTLD occurring in the post-allogeneic bone marrow transplant setting is the use of EBV-specific CTLs, generated from the original allogeneic donor, for adoptive immunotherapy (45, 106, 116, 117). The early studies suggests that adoptive immunotherapy with EBV-specific CTLs may be effective not only in treating established PTLD, but also as prophylaxis against PTLD in pediatric bone marrow transplant recipients who are at high risk for developing PTLD (45, 106, 117). It remains to be determined whether this approach can be adapted to treat EBV-related malignancies in other settings such as post-organ transplant PTLD or AIDS-related lymphoma. EBV-specific CTLs used in these settings would likely have to be derived from autologous T lymphocytes, which may be technically difficult due to the underlying immunosuppressed state.

OHL and AIDS-associated NHL (Table 1)

Oral hairy leukoplakia (OHL) is a wart-like lesion on the lateral aspects of the tongue in HIV-positive patients. This lesion is an epithelial-based focus of EBV replication (31, 32). Occasionally, OHL is seen in transplant recipients (30). Clinically the lesions respond well to acyclovir or desciclovir, with rapid restoration of the epithelial architecture, usually by 4 weeks (29, 31, 104). The lesions usually recur within several months upon discontinuation of treatment.

Viral DNA and lytic cycle antigens are present only in the outer, differentiating layers of the epithelium, with no viral DNA or latent or lytic antigens being present in the basal or superbasal layers (91, 110, 120). Latent antigens have been variably detected in the outer epithelial layers, with only EBNA-1 being consistently expressed. EBNA-1 is expressed during lytic cycle infection in B lymphocytes, and thus the same appears to be true in epithelial cells. Frequently, transformation-defective virus containing EBNA-2 deletions has been found in OHL by PCR and Southern analysis (115, 127). The lack of latent cycle (transformation-associated) EBV proteins and transformation-defective virus suggests that OHL is not a premalignant lesion. In fact, the data suggest that the structural alterations present in the epithelium, which include thickening of the epithelium due to a delay of differentiation, are related to viral replication. These changes may be related to the expression of the early lytic cycle BHRF1 protein, which is a homolog of the Bcl-2 cell survival protein (see chapter 4).

Non-Hodgkin's lymphoma (NHL) occurs with high frequency in HIV-infected patients and is one of the defining illnesses for AIDS (63, 139). AIDS-NHL is usually of B-lymphocyte origin and is a relatively late manifestation of HIV infection. Its frequency appears to be increasing as patients are living longer after their initial diagnosis with HIV infection. Similar to PTLD, AIDS-NHL ranges from a polyclonal lymphoproliferation to monoclonal lymphoma; however, unlike PTLD, AIDS-NHL is not uniformly EBV associated (37, 94, 112). Instead, only certain subsets of AIDS-NHL appear to be EBV related. In a report of 40 cases of AIDS-NHL, 35% were polyclonal and EBV negative, 27.5% were monoclonal and EBV negative, 7.5% were polyclonal and EBV positive, and 30% were monoclonal and EBV positive (112). Within the monoclonal subtypes, 64% of the EBV-negative lymphomas had *myc* gene rearrangements, while only 17% of the EBV-positive lymphomas had rearranged *myc* genes. In other studies, rearrangements at the *bcl6* locus have also been described in HIV-NHL, specifically in cases with diffuse large-cell histology (24).

Aside from being associated with certain subtypes of AIDS-NHL, EBV is associated with a high frequency of primary central nervous system (CNS) lym-

phomas in AIDS patients (68, 69). These data are most consistent with the hypothesis that multiple inciting events can lead to the development of AIDS-NHL. EBV
appears to be one etiologic factor that can contribute to the development of certain
subtypes of AIDS-NHL, with a strong association noted for primary CNS lymphoma (68, 69).

Recent observations may help to explain certain features of EBV pathogenesis
as it relates to AIDS-NHL. A recent study reported that in EBV-positive AIDS-
NHL cases, expression of the Bcl-6 protein was mutually exclusive with expression
of the EBV LMP1 protein (13). If this observation proves to be correct, it suggests
that there may be distinct pathways of molecular evolution that occur in the development of AIDS-NHL.

AIDS-NHL has an extremely poor prognosis, with median survivals on the
order of 5 to 6 months. Widely disseminated disease is usually present at the time
of diagnosis as is the presence of systemic B symptoms (i.e., fevers, chills, night
sweats, or weight loss). Approximately half of the patients die of progressive
lymphoma, while the other half die of opportunistic infection. The primary CNS
subtype of AIDS-NHL (the type most likely to be EBV associated) has the worst
prognosis of any group. In fact, it is often first diagnosed at autopsy.

Treatment of AIDS-NHL with multiple intensive combination chemotherapeutic regimens uniformly had low response and high relapse rates. Newer strategies employing less intensive regimens and early CNS prophylaxis appear to
achieve better responses which may prove to be long-term (for review, see references 109 and 118).

X-linked and sporadic lymphoproliferative disease

Rare deaths have been known to occur in the course of acute infectious mononucleosis in otherwise healthy people who have had no apparent difficulty with other
herpesvirus infections (99). Some of these occur in families with X-linked, recessive, highly penetrant genetic characteristics, while other cases are sporadic or
cluster in families, but are not X linked. The X-linked cases frequently occur in
children 2 to 5 years of age, while the non-X-linked cases tend to occur in early
adolescence. The common clinical course found in all these cases is an apparently
typical acute infectious mononucleosis that then follows a fulminant, progressive
course. These patients tend to have more severe cervical and systemic lymphadenopathy and a greater degree of splenomegaly. Severe hepatitis and hepatic failure
are often the dominant clinical problems. In other patients, lymphoproliferation
with widespread organ invasion or lymphocytic interstitial pneumonitis domi-
nates, and the clinical presentation is that of an acute lymphoma. Other patients
have illnesses marked by severe progressive infectious mononucleosis followed
by widespread necrosis and atrophy of bone marrow and lymphatic tissues. Even
among affected males of families with typical X-linked lymphoproliferative disease the course and severity of illness vary. The course is not invariably fatal, but
survivors often have permanent sequelae. Some have persistent agammaglobulin-
emia or specific aplasias, while others develop Burkitt-like lymphomas. An unusual predisposition to small intestinal lymphoma has also been noted.

Cases of X-linked lymphoproliferative disease have been collected in a worldwide registry to facilitate genetic and pathophysiologic investigation. A total of

240 males in 59 unrelated kindred were registered by 1991 (100, 101). Despite more than a decade of clinical studies directed toward defining an immune deficiency in affected males or in female carriers, no highly prevalent abnormality has been discovered. Some carriers have defects in the proliferative response to pokeweed mitogen. Others have selected deficiencies in antibody response to EBV infection and fail to develop EBNA antibodies. Diminished interferon production by EBV-immune T lymphocytes has been demonstrated among affected males. The currently favored hypothesis is an underlying defect in a cytokine or lymphokine. Substantial progress has been made in mapping the genetic defect and should facilitate immunologic and biochemical investigation. Tight linkage to locus Xq24-25 has been demonstrated, and detailed genetic analysis of the region is being performed (61, 97, 132). Related and matched, unrelated allogeneic bone marrow, stem cell, or cord blood transplants have corrected the disorder in affected males (19, 36, 98, 126, 133).

EBV-associated T-cell lymphomas

EBV, long considered to be strictly B lymphotropic, has occasionally been found to be present in T cells in rare, fulminant cases of infectious mononucleosis and in occasional cases of T-cell lymphoma. Recently, a more careful examination of T-cell lymphoproliferative disease has revealed a consistent association of at least three types of T-cell lymphomas with EBV. These are the virus-associated hemophagocytic syndrome (VAHS)-associated T-cell lymphocytosis/lymphoma, nasal T-cell lymphoma, and peripheral T-cell lymphomas of the angioimmunoblastic lymphadenopathy-like (AILD) appearance. EBV gene expression studies in the nasal T-cell lymphomas indicate a latency type II pattern (see chapter 4) in that EBNA-1, LMP1, and LMP2 are expressed and EBNA-2 is not (14, 72). More detailed studies are needed to define the pathogenic role that EBV plays in T-cell lymphoproliferative disease. It is presently unclear what host and viral factors contribute to the EBV infection of T cells, and whether this occurs at some level during the course of infection in all hosts or occurs only in certain persons who subsequently are at increased risk for developing an EBV-mediated T-cell lymphoproliferation.

HD (Table 1)

Many serologic and epidemiologic studies have suggested a link between Hodgkin's disease (HD) and EBV (53, 83, 84). However, the high EBV infection rate in the general population has made the interpretation of these studies problematic. Recently, EBV, DNA, RNA, and protein have been demonstrated in pathologic specimens from HD patients (1, 6, 44, 93). Several of these studies have demonstrated that EBV is confined to the binucleate Reed-Sternberg cell or its mononuclear variant Hodgkin cell. The mixed-cellularity and lymphocyte-depleted subtypes of HD are frequently EBV positive, the nodular sclerosing subtype is EBV positive in about one-third of cases, and the lymphocyte-predominant histologic subtypes are rarely EBV positive.

Immunohistochemical studies have demonstrated that EBV-positive HD has a latency type II (see chapter 4) gene expression pattern in that LMP1 is expressed in the Reed-Sternberg/Hodgkin cells of a large proportion of the mixed-cellularity and nodular sclerosing subtypes while EBNA2 expression could not be detected (44, 93). The potential interactions that exist between EBV and cellular genes are not clearly established in EBV-positive HD. Overall, it is clear that approximately 40 to 50% of HD cases in the Western world are etiologically linked to EBV.

NPC (Table 1)

Nasopharyngeal carcinoma (NPC) is rare among European and North American Caucasians, with age-adjusted incidence rates of less than 1 per 100,000. In contrast, NPC is one of the most common cancers in southern China, with age-adjusted incidence rates of 25.6 for men in Hong Kong and 54.7 for male Hong Kong "boat people" who are fishermen that live on their boats (46). Intermediate-incidence populations have been identified among North African Arabs. The etiologic factors identified for NPC include EBV, genetic susceptibility including an HLA-associated risk, potential tumor suppressor gene(s) located on chromosome 3, and environmental factors.

The association of EBV with NPC was first suggested based on serologic evidence (92). The association was subsequently confirmed by EBV nucleic acid hybridization studies on NPC biopsy material and by demonstrating that EBV DNA was present in the tumor cells (140). Furthermore, EBNA staining could be detected in the nuclei of NPC cells from fresh biopsies or from tumors that had been passaged in nude mice (50, 59). Moreover, Southern blot hybridization demonstrated that EBV DNA in NPC biopsies arises from a single EBV-infected cell by the criterion of homogeneity in TR sequence reiteration frequency (102). EBV is nearly always associated with poorly differentiated (anaplastic) NPC occurring in all races and also with a substantial proportion of moderately to well-differentiated NPC cases from southern China.

Studies on EBV latent protein expression in NPC tissue using immunohistochemistry and Western immunoblotting reveal that EBNA-1 is expressed in nearly all cases examined and that LMP1 is expressed in approximately two-thirds of cases of EBV-positive NPC (22, 135). LMP1 transcription may be in part under control of a different promoter in NPC (27). LMP2A and LMP2B mRNAs are also expressed, with LMP2A being detected in most cases examined (5, 11). EBNA-2, EBNA-LP, and EBNA-3s are not detected. EBNA-1 mRNA in NPC biopsies is transcribed from the Fp promoter the same as in latency type L BL cells (5). Thus, EBV gene expression in NPC has a latency type II pattern (see chapter 4).

Attention has recently focused on a new EBV transcript in cDNA libraries from RNA from NPC biopsies. This transcript is rightward in orientation. The site for initiation of transcription has not been established. The transcript is multiply spliced. An open reading frame (BARF0) encoded by an exon of the *Bam*A fragment appears to encode protein at some stage of EBV infection in vivo, since antibody to a fusion protein including part of the open reading frame is evident in EBV-immune humans. The BARF0 protein may be expressed in latent or lytic infection in epithelial cells or lymphocytes.

Cytogenetic studies on NPC xenografts identified abnormal markers involving chromosomes 1, 3, 11, 12, and 17 (49). A consistent deletion of the short arm of chromosome 3 (3p) led to more detailed analysis through molecular studies of this chromosome. Tumor and matched-blood leukocyte DNAs from 11 EBV-positive NPC cases were studied by restriction fragment length polymorphism (RFLP) analysis using six polymorphic chromosome 3-specific probes that map to known regions of the chromosome (51). Among evaluable cases (i.e., those with a poly-morphism in blood leukocyte DNA) there was a loss of heterozygosity (LOH), indicating a deletion of the genetic material from one of the chromosomes in the tumor DNA, in 10 of 10 cases for a probe mapping to 3p25 and in 9 of 9 cases for a probe mapping to 3p14. The LOH suggests that a tumor suppressor gene or recessive oncogene is present in the region that contributes to the malignant phe-notype when a normal copy of the gene is deleted from the other chromosome.

Although Chinese who have moved out of the endemic region maintain their high risk status, the geographic distribution of NPC to southeastern regions of China could be in part a consequence of environmental or cultural factors. Inges-tion of Cantonese-style salted fish, especially during childhood, correlates with increased risk for NPC (2, 46, 137). Among Malaysian Chinese, salted fish con-sumption carried a relative risk of 3.0, while daily consumption compared with nonconsumption carried a relative risk of 17.4 (2). Among Hong Kong Chinese, the relative risk for having salted fish as a weaning food was 7.5, and consuming salted fish at least once a week compared with less than once a month at 10 years of age had a relative risk of 37.7 (137). Other traditional Chinese salted foods were not associated with an increased risk for NPC. In order to put the strength of these epidemiologic data into perspective, the relative risk for cigarette smoking and developing lung cancer ranges from 3.95 to 17.5, depending on the number of cigarettes smoked and type of lung cancer. Up to 90% of cases of NPC in Hong Kong are attributable to childhood exposure to salted fish (137). Nasopharyngeal tumors can be induced in rats fed salted fish. Low levels of several carcinogenic volatile nitrosamines have been detected in Chinese salted fish, although their precise role in inducing NPC remains to be determined. Regardless of the com-pound(s) in salted fish which contributes to the development of NPC, the epide-miological evidence is compatible with the hypothesis that decreased salted fish ingestion might decrease the incidence of NPC in southern China.

NPC has a peak incidence in the 50- to 54-year age range, with the incidence starting to climb steadily to this point after 20 years of age. Males are affected more frequently, with a male:female ratio of 2.5–3.0:1. Because the nasopharynx is difficult to visualize, patients become aware of their tumor only after mass effects of the tumor are apparent, and they tend to present later than with other forms of head and neck cancer. Spread to regional lymph nodes is therefore com-mon and occurs in most patients with NPC. Patients often present with nasal stuffiness, discharge, nose bleeds, ear pain, or decreased hearing. Cranial nerve involvement is also common and may cause facial muscle signs, swallowing diffi-culty, facial pain, and aberrant sense of taste.

Population-based screening has validated the use of EBV serology for NPC detection in high-risk groups in southern China. This approach is based on obser-vation of a high incidence of IgA EBV EA or VCA antibody in NPC patients (41). Prospective screening was initially useful in detecting preclinical NPC (138). Over the next few years almost all cases of NPC occurred in patients with initially high IgA EA or VCA titers. Serology has also been useful in anticipating recurrences following therapy. The association between higher IgA EA and VCA titers and NPC could be due to lytic virus replication in a small fraction of cells in preclinical

NPC tumors or to an association between the level of EBV replication in the oropharynx, IgA response, and the subsequent evolution of NPC.

SUMMARY

EBV is a human DNA tumor virus with an extraordinarily diverse oncogenic potential. No other virus, human or otherwise, has shown the ability to contribute to malignant transformation in such a variety of cell backgrounds. Studies of EBV-related malignancies in immunodeficiency states (posttransplant and AIDS) have provided insight into the significance in vivo of EBV-specific T-cell immunity, which has been studied so elegantly in vitro, in controlling primary and persistent EBV infection. Recent studies on tumor tissue from these diseases have suggested that viral signaling, especially through the LMP1 protein, may directly contribute to the development and maintenance of the malignant phenotype. Novel strategies aimed at restoring antiviral T-cell immunity have shown promise in the post-bone marrow transplant setting in controlling EBV-related lymphoproliferations. Future effort into better characterizing the role of EBV in T-lymphocyte malignancies and in non-lymphoid malignancies (NPC and smooth muscle tumors) will likely identify new pathogenic mechanisms utilized by this versatile virus that have yet to be appreciated in the better-studied EBV-related B-lymphoproliferative diseases.

Acknowledgments

D.L. is supported by Public Health Service grant CA 73545 from the National Cancer Institute and is a Leukemia Society Translational Research Awardee (LSA #6240-98).

REFERENCES

1. **Anagnostopoulos, I., H. Herbst, G. Niedobitek, and H. Stein.** 1989. Demonstration of monoclonal EBV genomes in Hodgkin's disease and Ki-1-positive anaplastic large cell lymphoma by combined Southern blot and in situ hybridization [see comments]. *Blood* **74:**810–816.

2. **Armstrong, R. W., M. J. Armstrong, M. C. Yu, and B. E. Henderson.** 1983. Salted fish and inhalants as risk factors for nasopharyngeal carcinoma in Malaysian Chinese. *Cancer Res.* **43:**2967–2970.

3. **Baichwal, V. R., and B. Sugden.** 1988. Transformation of Balb 3T3 cells by the BNLF-1 gene of Epstein-Barr virus. *Oncogene* **2:**461–467.

4. **Blanche, S., F. Le Deist, F. Veber, G. Lenoir, A. M. Fischer, J. Brochier, C. Boucheix, M. Delaage, C. Grisceill, and A. Fischer.** 1988. Treatment of severe Epstein-Barr virus-induced polyclonal B-lymphocyte proliferation by anti-B-cell monoclonal antibodies. Two cases after HLA-mismatched bone marrow transplantation *Ann. Intern. Med.* **108:**199–203.

5. **Brooks, L., Q. Y. Yao, A. B. Rickinson, and L. S. Young.** 1992. Epstein-Barr virus latent gene transcription in nasopharyngeal carcinoma cells: coexpression of EBNA1, LMP1, and LMP2 transcripts. *J. Virol.* **66:**2689–2697.

6. **Brousset, P., S. Chittal, D. Schlaifer, J. Icart, C. Payen, F. Rigal Huguet, J. J. Voigt, and G. Deisol.** 1991. Detection of Epstein-Barr virus messenger RNA in Reed-Sternberg cells of Hodgkin's disease by in situ hybridization with biotinylated probes on specially processed modified acetone methyl benzoate xylene (ModAMeX) sections [see comments]. *Blood* **77:**1781–1786.

7. **Brown, N. A., C. R. Liu, Y. F. Wang, and C. R. Garcia.** 1988. B-cell lymphoproliferation and lymphomagenesis are associated with clonotypic intracellular terminal regions of the Epstein-Barr virus. *J. Virol.* **62:**962–969.

8. **Burkitt, D. P.** 1958. A sarcoma involving the jaws in African children. *Br. J. Surg.* **46:**218–223.

9. **Burkitt, D. P.** 1962. A children's cancer dependent on climactic factors. *Nature* **194:** 232–234.

10. **Burkitt, D. P.** 1962. Determining the climactic limitations of a children's cancer common in Africa. *Br. Med. J.* **2:**1019–1023.

11. **Busson, P., R. McCoy, R. Sadler, K. Gilligan, T. Tursz, and N. Raab Traub.** 1992. Consistent transcription of the Epstein-Barr virus LMP2 gene in nasopharyngeal carcinoma. *J. Virol.* **66:**3257–3262.

12. **Cannon, M. J., P. Pisa, R. I. Fox, and N. R. Cooper.** 1990. Epstein-Barr virus induces aggressive lymphoproliferative disorders of human B cell origin in SCID/hu chimeric mice. *J. Clin. Invest.* **85:**1333–1337.

13. **Carbone, A., G. Gaidano, A. Gloghini, C. Pastore, G. Saglio, U. Tirelli, R. Dalla Favera, and B. Falini.** 1997. BCL-6 protein expression in AIDS-related non-Hodgkin's lymphomas: inverse relationship with Epstein-Barr virus-encoded latent membrane protein-1 expression. *Am. J. Pathol.* **150:**155–165.

14. **Chen, C. L., R. H. Sadler, D. M. Walling, I. J. Su, H. C. Hsieh, and N. Raab Traub.** 1993. Epstein-Barr virus (EBV) gene expression in EBV-positive peripheral T-cell lymphomas. *J. Virol.* **67:**6303–6308.

15. **Cleary, M. L., M. A. Epstein, S. Finerty, R. F. Dorfman, G. W. Bornkamm, J. K. Kirkwood, A. J. Morgan, and J. Sklar.** 1985. Individual tumors of multifocal EB virus-induced malignant lymphomas in tamarins arise from different B-cell clones. *Science* **228:**722–724.

16. **Cohen, J. I.** 1991. Epstein-Barr virus lymphoproliferative disease associated with acquired immunodeficiency. *Medicine (Baltimore)* **70:**137–160.

17. **Dalla-Favera, R., M. Bregni, J. Erikson, D. Patterson, R. C. Gallo, and C. M. Croce.** 1982. Human c myc oncogene is located on the region of chromosome 8 that is translocated in Burkitt lymphoma cells. *Proc. Natl. Acad. Sci. USA* **79:**7824–7827.

18. **Devergne, O., E. Hatzivassiliou, K. M. Izumi, K. M. Kaye, M. F. Kleijnen, E. Kieff, and G. Mosallos.** 1996. Association of TRAF1, TRAF2, and TRAF3 with an Epstein-Barr virus LMP1 domain important for B-lymphocyte transformation: role in NF-κB activation. *Mol. Cell. Biol.* **16:**7098–7108.

19. **Ende, M., and F. I. Ende.** 1994. Correction of X-linked lymphoproliferative disease by stem-cell transplantation [letter; comment]. *N. Engl. J. Med.* **330:**1159.

20. **Epstein, M. A., B. Achong, and Y. Barr.** 1964. Virus particles in cultured lymphoblasts from Burkitt's lymphoma. *Lancet* **1:**702–703.

21. **Ernberg, I., and J. Andersson.** 1986. Acyclovir efficiently inhibits oropharyngeal excretion of Epstein-Barr virus in patients with acute infectious mononucleosis. *J. Gen. Virol.* **67:**2267–2272.

22. **Fahraeus, R., H. L. Fu, I. Ernberg, J. Finke, M. Rowe, G. Klein, K. Falk, E. Nilsson, M. Yadav, P. Busson, T. Tursz, and B. Kallin.** 1988. Expression of Epstein-Barr virus-encoded proteins in nasopharyngeal carcinoma. *Int. J. Cancer* **42:**329–338.

23. **Fischer, A., S. Blanche, J. Le Bidois, P. Bordigoni, J. L. Garnier, P. Nlaudet, F. Morinet, F. Le Deist, A. M. Fischer, C. Griscelli, and M. Hirn.** 1991. Anti-B-cell monoclonal antibodies in the treatment of severe B-cell lymphoproliferative syndrome following bone marrow and organ transplantation. *N. Engl. J. Med.* **324:** 1451–1456.

24. Gaidano, G., F. Lo Coco, B. H. Ye, D. Shibata, A. M. Levine, D. M. Knowles, and R. Dalla Favera. 1994. Rearrangements of the BCL-6 gene in acquired immunodeficiency syndrome-associated non-Hodgkin's lymphoma: association with diffuse large-cell subtype. *Blood* **84**:397–402.

25. Garner, J. G., M. S. Hirsch, and R. T. Schooley. 1984. Prevention of Epstein-Barr virus-induced B-cell outgrowth by interferon alpha. *Infect. Immun.* **43**:920–924.

26. Geser, A., G. M. Lenoir, M. Anvret, G. Bornkamm, G. Klein, E. H. Williams, D. H. Wright, and G. De The. 1983. Epstein-Barr virus markers in a series of Burkitt's lymphomas from the West Nile District, Uganda. *Eur. J. Cancer Clin. Oncol.* **19**: 1393–1404.

27. Gilligan, K., H. Sato, P. Rajadurai, P. Busson, L. Young, A. Rickinson, T. Tursz, and N. Raab Traub. 1990. Novel transcription from the Epstein-Barr virus terminal *Eco*RI fragment, DIJhet, in a nasopharyngeal carcinoma. *J. Virol.* **64**:4948–4956.

28. Gratama, J. W., M. M. Zutter, J. Minarovits, M. A. Oosterveer, E. D. Thomas, G. Klein, and I. Ernberg. 1991. Expression of Epstein-Barr virus-encoded growth-transformation-associated proteins in lymphoproliferations of bone-marrow transplant recipients. *Int. J. Cancer* **47**:188–192.

29. Greenspan, D., Y. G. De Souza, M. A. Conant, H. Hollander, S. K. Chapman, E. T. Lennette, V. Petersen, and J. S. Greenspan. 1990. Efficacy of desciclovir in the treatment of Epstein-Barr virus infection in oral hairy leukoplakia. *J. Acquired Immune Defic. Syndr.* **3**:571–578.

30. Greenspan, D., J. S. Greenspan, Y. de Souza, J. A. Levy, and A. M. Ungar. 1989. Oral hairy leukoplakia in an HIV-negative renal transplant recipient. *J. Oral Pathol. Med.* **18**:32–34.

31. Greenspan, J. S., and D. Greenspan. 1989. Oral hairy leukoplakia: diagnosis and management. *Oral Surg. Oral Med. Oral Pathol.* **67**:396–403.

32. Greenspan, J. S., D. Greenspan, E. T. Lennette, D. I. Abrams, M. A. Conant, V. Petersen, and U. K. Freese. 1985. Replication of Epstein-Barr virus within the epithelial cells of oral "hairy" leukoplakia, an AIDS-associated lesion. *N. Engl. J. Med.* **313**: 1564–1571.

33. Gregory, C. D., C. Dive, S. Henderson, C. A. Smith, G. T. Williams, J. Gordon, and A. B. Rickinson. 1991. Activation of Epstein-Barr virus latent genes protects human B cells from death by apoptosis. *Nature* **349**:612–614.

34. Gregory, C. D., R. J. Murray, C. F. Edwards, and A. B. Rickinson. 1988. Downregulation of cell adhesion molecules LFA-3 and ICAM-1 in Epstein-Barr virus-positive Burkitt's lymphoma underlies tumor cell escape from virus-specific T cell surveillance. *J. Exp. Med.* **167**:1811–1824.

35. Gregory, C. D., M. Rowe, and A. B. Rickinson. 1990. Different Epstein-Barr virus-B cell interactions in phenotypically distinct clones of a Burkitt's lymphoma cell line. *J. Gen. Virol.* **71**:1481–1495.

36. Gross, T. G., A. H. Filipovich, M. E. Conley, E. Pracher, K. Schmiegelow, J. D. Verdirame, M. Vowels, L. L. Williams, and T. A. Seemayer. 1996. Cure of X-linked lymphoproliferative disease (XLP) with allogeneic hematopoietic stem cell transplantation (HSCT): report from the XLP registry. *Bone Marrow Transplant.* **17**:741–744.

37. Hamilton Dutoit, S. J., G. Pallesen, M. B. Franzmann, J. Karkov, F. Black, P. Skinhoj, and C. Pedersen. 1991. AIDS-related lymphoma. Histopathology, immunophenotype, and association with Epstein-Barr virus as demonstrated by in situ nucleic acid hybridization. *Am. J. Pathol.* **138**:149–163.

38. Hatzubai, A., M. Anafi, M. G. Masucci, J. Dillner, R. A. Lerner, G. Klein, and D. Sulitzeanu. 1987. Down-regulation of the EBV-encoded membrane protein (LMP) in Burkitt lymphomas. *Int. J. Cancer* **40**:358–364.

39. **Henderson, S., M. Rowe, C. Gregory, D. Croom Carter, F. Wang, R. Longnecker, E. Kieff, and A. Rickinson.** 1991. Induction of bcl-2 expression by Epstein-Barr virus latent membrane protein 1 protects infected B cells from programmed cell death. *Cell* **65:**1107–1115.

40. **Henle, G., and W. Henle.** 1970. Observations on childhood infections with the Epstein-Barr virus. *J. Infect. Dis.* **121:**303–310.

41. **Henle, G., and W. Henle.** 1976. Epstein-Barr virus-specific IgA serum antibodies as an outstanding feature of nasopharyngeal carcinoma. *Int. J. Cancer* **17:**1–7.

42. **Henle, G., W. Henle, and V. Diehl.** 1968. Relation of Burkitt's tumour associated herpes-type virus to infectious mononucleosis. *Proc. Natl. Acad. Sci. USA* **59:**94–101.

43. **Henle, W., V. Diehl, G. Kohn, H. zur Hausen, and G. Henle.** 1967. Herpes-type virus and chromosome marker in normal leukocytes after growth with irradiated Burkitt cells. *Science* **157:**1064–1065.

44. **Herbst, H., F. Dallenbach, M. Hummel, G. Niedobitek, S. Pilerl, N. Muller Lantzsch, and H. Stein.** 1991. Epstein-Barr virus latent membrane protein expression in Hodgkin and Reed-Sternberg cells. *Proc. Natl. Acad. Sci. USA* **88:**4766–4770.

45. **Heslop, H. E., C. Y. Ng, C. Li, C. A. Smith, S. K. Loftin, R. A. Krance, M. K. Brenner, and C. M. Rooney.** 1996. Long-term restoration of immunity against Epstein-Barr virus infection by adoptive transfer of gene-modified virus-specific T lymphocytes. *Nat. Med.* **2:**551–555.

46. **Ho, J.** 1978. An epidemiologic and clinical study of nasopharyngeal carcinoma. *Int. J. Radiat. Oncol. Biol. Phys.* **4:**183–197.

47. **Ho, M., R. Jaffe, G. Miller, M. K. Breinig, J. S. Dummer, L. Makowka, R. W. Atchison, F. Karrer, M. A. Nalesnik, and T. E. Starzl.** 1988. The frequency of Epstein-Barr virus infection and associated lymphoproliferative syndrome after transplantation and its manifestations in children. *Transplantation* **45:**719–727.

48. **Ho, M., G. Miller, R. W. Atchison, M. K. Breinig, J. S. Dummer, W. Andiman, T. E. Starzl, R. Eastman, B. P. Griffith, R. L. Hardesty, H. T. Bahnson, T. R. Hakala, and J. T. Rosenthal.** 1985. Epstein-Barr virus infections and DNA hybridization studies in posttransplantation lymphoma and lymphoproliferative lesions: the role of primary infection. *J. Infect. Dis.* **152:**876–886.

49. **Huang, D. P., J. H. C. Ho, W. K. Chan, W. H. Lau, and M. Lui.** 1989. Cytogenetics of undifferentiated nasopharyngeal carcinoma xenografts from southern Chinese. *Int. J. Cancer* **43:**936–839.

50. **Huang, D. P., J. H. C. Ho, W. Henle, and G. Henle.** 1974. Demonstration of Epstein-Barr virus associated nuclear antigen in nasopharyngeal carcinoma cells from fresh biopsies. *Int. J. Cancer* **14:**580–588.

51. **Huang, D. P., K. W. Lo, P. H. Choi, A. Y. Ng, S. Y. Tsao, G. K. Yiu, and J. C. Lee.** 1991. Loss of heterozygosity on the short arm of chromosome 3 in nasopharyngeal carcinoma. *Cancer Genet. Cytogenet.* **54:**91–99.

52. **Huen, D., S. Henderson, D. Croom-Carter, and M. Rowe.** 1995. The Epstein-Barr virus latent membrane protein-1 (LMP1) mediates activation of NF-kappa B and cell surface phenotype via two effector regions in its carboxy-terminal cytoplasmic domain. *Oncogene* **10:**549–560.

53. **Johansson, B., G. Klein, W. Henle, and G. Henle.** 1970. Epstein-Barr virus (EBV)-associated antibody patterns in lymphoma and leukemia. *Int. J. Cancer* **6:**450–62.

54. **Kafuko, G. W., and D. P. Burkitt.** 1970. Burkitt's lymphoma and malaria. *Int. J. Cancer* **6:**1–9.

55. **Kaye, K. M., O. Devergne, J. N. Harada, K. M. Izumi, R. Yalamanchili, E. Kieff, and G. Mosalios.** 1996. Tumor necrosis family receptor associated factor 2 is a mediator

of NF-kB activation by latent infection membrane protein 1, the Epstein-Barr virus transforming protein. *Proc. Natl. Acad. Sci. USA* **93:**11085–11090.

56. **Kaye, K. M., K. M. Izumi, and E. Kieff.** 1993. Epstein-Barr virus latent membrane protein 1 is essential for B-lymphocyte growth transformation. *Proc. Natl. Acad. Sci. USA* **90:**9150–9154.

57. **Khanna, R., S. R. Burrows, M. G. Kurilla, C. A. Jacob, I. S. Misko, T. B. Sculley, E. Kieff, and D. J. Moss.** 1992. Localization of Epstein-Barr virus cytotoxic T cell epitopes using recombinant vaccinia: implications for vaccine development. *J. Exp. Med.* **176:**169–176.

58. **Kieff, E.** 1996. Epstein-Barr virus and its replication, p. 2343–2396. *In* B. N. Fields, D. M. Knipe, and P. M. Howley (ed.), *Fields Virology*, 3rd ed., vol. 2. Lippincott-Raven, Philadelphia.

59. **Klein, G., B. C. Giovanella, T. Lindahl, P. J. Fialkow, S. Singh, and J. S. Steihlin.** 1974. Direct evidence for the presence of Epstein-Barr virus DNA and nuclear antigen in malignant epithelial cells from patients with poorly differentiated carcinoma of the nasopharynx. *Proc. Natl. Acad. Sci. USA* **71:**4737–4741.

60. **Knowles, D. M., E. Cesarman, A. Chadburn, G. Frizzera, J. Chen, E. A. Rose, and R. E. Michler.** 1995. Correlative morphologic and molecular genetic analysis demonstrates three distinct categories of posttransplantation lymphoproliferative disorders. *Blood* **85:**552–565.

61. **Lanyi, A., B. Li, S. Li, C. B. Talmadge, B. Brichacek, J. R. Davis, B. A. Kozel, B. Trask, G. van den Engh, E. Uzvolgyi, E. J. Stanbridge, D. L. Nelson, C. Chinault, H. Heslop, T. G. Gross, T. A. Seemayer, G. Klein, D. T. Purtllo, and J. Sumegi.** 1997. A yeast artificial chromosome (YAC) contig encompassing the critical region of the X-linked lymphoproliferative disease (XLP) locus. *Genomics* **39:**55–65.

62. **Leder, P., J. Battey, G. Lenoir, C. Moulding, W. Murphy, H. Potter, T. Stewart, and R. Taub.** 1983. Translocations among antibody genes in human cancer. *Science* **222:** 765–771.

63. **Levine, A. M.** 1987. Non-Hodgkin's lymphomas and other malignancies in the acquired immune deficiency syndrome. *Semin. Oncol.* **14:**34–39.

64. **Liebowitz, D., and E. Kieff.** 1989. Epstein-Barr virus latent membrane protein: induction of B-cell activation antigens and membrane patch formation does not require vimentin. *J. Virol.* **63:**4051–4054.

65. **Liebowitz, D., M. M. LeBeau, and O. I. Olopade.** 1997. Evidence for signal transduction through the TNF receptor pathway in post-transplant lymphoproliferative disorders. *J. Acquired Immune Defic. Syndr. Hum. Retrovirol.* **14:**A51.

66. **Liebowitz, D., J. Mannick, K. Takada, and E. Kieff.** 1992. Phenotypes of Epstein-Barr virus LMP1 deletion mutants indicate transmembrane and amino-terminal cytoplasmic domains necessary for effects in B-lymphoma cells. *J. Virol.* **66:**4612–4616.

67. **Lombardi, L., E. W. Newcomb, and R. Dalla Favera.** 1987. Pathogenesis of Burkitt lymphoma: expression of an activated c-myc oncogene causes the tumorigenic conversion of EBV-infected human B lymphoblasts. *Cell* **49:**161–170.

68. **MacMahon, E. M., J. D. Glass, S. D. Hayward, R. B. Mann, P. S. Becker, P. Charache, J. C. McArthur, and R. F. Ambinder.** 1991. Epstein-Barr virus in AIDS-related primary central nervous system lymphoma. *Lancet* **338:**969–973.

69. **MacMahon, E. M., J. D. Glass, S. D. Hayward, R. B. Mann, P. Charache, J. C. McArthur, and R. F. Ambinder.** 1992. Association of Epstein-Barr virus with primary central nervous system lymphoma in AIDS. *AIDS Res. Hum. Retroviruses* **8:**740–742.

70. **Manilov, G., and Y. Manilova.** 1972. Marker band in one chromosome 14 from Burkitt lymphomas. *Nature* **237:**33–34.

71. **Manilov, G., Y. Manilova, G. Klein, G. Lenoir, and A. Levan.** 1986. Alternative

involvement of two cytogenetically distinguishable breakpoints on chromosome 8 in Burkitt's lymphoma associated translocations. *Cancer Genet. Cytogenet.* **20**:95–99.

72. **Minarovits, J., L. F. Hu, S. Imai, Y. Harabuchi, A. Kataura, S. Minarovits Kormuta, T. Osato, and G. Klein.** 1994. Clonality, expression and methylation patterns of the Epstein-Barr virus genomes in lethal midline granulomas classified as peripheral angiocentric T cell lymphomas. *J. Gen. Virol.* **75**:77–84.

73. **Mitchell, T., and B. Sugden.** 1995. Stimulation of NF-kappa B-mediated transcription by mutant derivatives of the latent membrane protein of Epstein-Barr virus. *J. Virol.* **69**:2968–2976.

74. **Moghaddam, A., M. Rosenzweig, D. Lee Parritz, B. Annis, R. P. Johnson, and F. Wang.** 1997. An animal model for acute and persistent Epstein-Barr virus infection. *Science* **276**:2030–2033.

75. **Morgan, A. J., S. Finerty, K. Lovgren, F. T. Scullion, and B. Morein.** 1988. Prevention of Epstein-Barr (EB) virus-induced lymphoma in cottontop tamarins by vaccination with the EB virus envelope glycoprotein gp340 incorporated into immune-stimulating complexes. *J. Gen. Virol.* **69**:2093–2096.

76. **Morgan, A. J., M. Mackett, S. Finerty, J. R. Arrand, F. T. Scullion, and M. A. Epstein.** 1988. Recombinant vaccinia virus expressing Epstein-Barr virus glycoprotein gp340 protects cottontop tamarins against EB virus-induced malignant lymphomas. *J. Med. Virol.* **25**:189–195.

77. **Moslalos, G., M. Birkenbach, R. Yatamanchili, T. VanArsdale, C. Ware, and E. Kieff.** 1995. The Epstein-Barr virus transforming protein LMP1 engages signaling proteins for the tumor necrosis factor receptor family. *Cell* **80**:389–399.

78. **Mosier, D. E., R. J. Gulizia, S. M. Baird, and D. B. Wilson.** 1988. Transfer of a functional human immune system to mice with severe combined immunodeficiency. *Nature* **335**:256–259.

79. **Mosier, D. E., G. R. Picchio, S. M. Baird, R. Kobayashi, and T. J. Kipps.** 1992. Epstein-Barr virus-induced human B-cell lymphomas in SCID mice reconstituted with human peripheral blood leukocytes. *Cancer Res.* **52**:5552s–5553s.

80. **Mosier, D. E., G. R. Picchio, M. B. Kirven, J. L. Garnier, B. E. Torbett, S. M. Baird, R. Kobayashi, and T. J. Kipps.** 1992. EBV-induced human B cell lymphomas in hu-PBL-SCID mice. *AIDS Res. Hum. Retroviruses* **8**:735–740.

81. **Moss, D. J., S. R. Burrows, R. Khanna, I. S. Misko, and T. B. Sculley.** 1992. Immune surveillance against Epstein-Barr virus. *Semin. Immunol.* **4**:97–104.

82. **Moss, D. J., A. B. Rickinson, and J. H. Pope.** 1978. Long-term T-cell-mediated immunity to Epstein-Barr virus in man. I. Complete regression of virus-induced transformation in cultures of seropositive donor leukocytes. *Int. J. Cancer* **22**:662–668.

83. **Mueller, N.** 1987. Epidemiologic studies assessing the role of the Epstein-Barr virus in Hodgkin's disease. *Yale J. Biol. Med.* **60**:321–332.

84. **Mueller, N., A. Evans, N. L. Harris, G. W. Comstock, E. Jellum, K. Magnus, N. Orentreich, B. F. Polk, and J. Vogelman.** 1989. Hodgkin's disease and Epstein-Barr virus. Altered antibody pattern before diagnosis. *N. Engl. J. Med.* **320**:689–695.

85. **Murray, R. J., M. G. Kurilla, J. M. Brooks, W. A. Thomas, M. Rowe, E. Kieff, and A. B. Rickinson.** 1992. Identification of target antigens for the human cytotoxic T cell response to Epstein-Barr virus (EBV): implications for the immune control of EBV-positive malignancies. *J. Exp. Med.* **176**:157–168.

86. **Murray, R. J., D. Wang, L. S. Young, F. Wang, M. Rowe, E. Kieff, and A. B. Rickinson.** 1988. Epstein-Barr virus-specific cytotoxic T-cell recognition of transfectants expressing the virus-coded latent membrane protein LMP. *J. Virol.* **62**:3747–3755.

87. **Murray, R. J., L. S. Young, A. Calender, C. D. Gregory, M. Rowe, G. M. Lenoir, and A. B. Rickinson.** 1988. Different patterns of Epstein-Barr virus gene expression

and of cytotoxic T-cell recognition in B-cell lines infected with transforming (B95.8) or nontransforming (P3HR1) virus strains. *J. Virol.* **62**:894–901.

88. **Nalesnik, M. A., R. Jaffe, T. E. Starzl, A. J. Demetris, K. Porter, J. A. Burnham, L. Makowka, M. Ho, and J. Locker.** 1988. The pathology of posttransplant lymphoproliferative disorders occurring in the setting of cyclosporine A-prednisone immunosuppression. *Am. J. Pathol.* **133**:173–192.

89. **Nerl, A., F. Barriga, G. Inghirami, D. M. Knowles, J. Neequaye, I. T. Magrath, and R. Dalla Favera.** 1991. Epstein-Barr virus infection precedes clonal expansion in Burkitt's and acquired immunodeficiency syndrome-associated lymphoma [see comments]. *Blood* **77**:1092–1095.

90. **Niederman, J. C., R. W. McCollum, G. Henle, and W. Henle.** 1968. Infectious mononucleosis. *JAMA* **203**:139–143.

91. **Niedobitek, G., L. S. Young, R. Lau, L. Brooks, D. Greenspan, J. S. Greenspan, and A. B. Rickinson.** 1991. Epstein-Barr virus infection in oral hairy leukoplakia: virus replication in the absence of a detectable latent phase. *J. Gen. Virol.* **72**:3035–3046.

92. **Old, L. J., A. E. Boyse, H. F. Oettgen, E. de Harven, G. Geering, E. Willamson, and P. Clifford.** 1966. Precipitating antibody in human serum to an antigen present in cultured Burkitt's lymphoma cells. *Proc. Natl. Acad. Sci. USA* **56**:1699–1704.

93. **Pallesen, G., S. J. Hamilton Dutoit, M. Rowe, and L. S. Young.** 1991. Expression of Epstein-Barr virus latent gene products in tumour cells of Hodgkin's disease [see comments]. *Lancet* **337**:320–322.

94. **Pedersen, C., J. Gerstoft, J. D. Lundgren, P. Skinhoj, C. Bottzauw, C. Geisler, S. J. Hamilton Dutoit, S. Thorsen, I. Lisse, E. Ralfkiaer, et al.** 1991. HIV-associated lymphoma: histopathology and association with Epstein-Barr virus genome related to clinical, immunological and prognostic features. *Eur. J. Cancer* **27**:1416–1423.

95. **Pisa, P., M. J. Cannon, E. K. Pisa, N. R. Cooper, and R. I. Fox.** 1992. Epstein-Barr virus induced lymphoproliferative tumors in severe combined immunodeficient mice are oligoclonal. *Blood* **79**:173–179.

96. **Pope, J. H., M. K. Horne, and W. Scott.** 1968. Transformation of fetal human leucocytes in vitro by filtrates of a human leukemic cell line containing herpes-like virus. *Int. J. Cancer* **3**:857–866.

97. **Porta, G., S. MacMillan, R. Nagaraja, S. Mumm, I. Zucchi, G. Pilla, S. Maio, T. Featherstone, and D. Schlessinger.** 1997. 4.5-Mb YAC STS contig at 50-kb resolution, spanning Xq25 deletions in two patients with lymphoproliferative syndrome. *Genome Res.* **7**:27–36.

98. **Pracher, E., E. R. Panzer Grumayer, A. Zoubek, C. Peters, and H. Gadner.** 1994. Successful bone marrow transplantation in a boy with X-linked lymphoproliferative syndrome and acute severe infectious mononucleosis. *Bone Marrow Transplant.* **13**:655–658.

99. **Purtilo, D. T.** 1985. Epstein-Barr virus-induced diseases in the X-linked lymphoproliferative syndrome and related disorders. *Biomed. Pharmacother.* **39**:52–58.

100. **Purtilo, D. T., and H. L. Grierson.** 1991. Methods of detection of new families with X-linked lymphoproliferative disease. *Cancer Genet. Cytogenet.* **51**:143–153.

101. **Purtilo, D. T., H. L. Grierson, J. R. Davis, and M. Okano.** 1991. The X-linked lymphoproliferative disease: from autopsy toward cloning the gene 1975–1990. *Pediatr. Pathol.* **11**:685–710.

102. **Raab Traub, N., and K. Flynn.** 1986. The structure of the termini of the Epstein-Barr virus as a marker of clonal cellular proliferation. *Cell* **47**:883–889.

103. **Randhawa, P. S., R. S. Markin, T. E. Starzl, and A. J. Demetris.** 1990. Epstein-Barr virus-associated syndromes in immunosuppressed liver transplant recipients.

Clinical profile and recognition on routine allograft biopsy. *Am. J. Surg. Pathol.* **14:** 538–547.

104. **Resnick, L., J. S. Herbst, D. V. Ablashi, S. Atherton, B. Frank, L. Rosen, and S. N. Horwitz.** 1988. Regression of oral hairy leukoplakia after orally administered acyclovir therapy. *JAMA* **259:**384–388.

105. **Rickinson, A. B., M. Rowe, I. J. Hart, Q. Y. Yao, L. E. Henderson, H. Rabin, and M. A. Epstein.** 1984. T-cell-mediated regression of "spontaneous" and of Epstein-Barr virus-induced B-cell transformation in vitro: studies with cyclosporin A. *Cell Immunol.* **87:**646–658.

106. **Rooney, C. M., C. A. Smith, C. Y. Ng, S. Loftin, C. Li, R. A. Krance, M. K. Brenner, and H. E. Heslop.** 1995. Use of gene-modified virus-specific T lymphocytes to control Epstein-Barr-virus-related lymphoproliferation. *Lancet* **345:**9–13.

107. **Rowe, M., D. T. Rowe, C. D. Gregory, L. S. Young, P. J. Farrell, H. Rupani, and A. B. Rickinson.** 1987. Differences in B cell growth phenotype reflect novel patterns of Epstein-Barr virus latent gene expression in Burkitt's lymphoma cells. *EMBO J.* **6:**2743–2751.

108. **Rowe, M., L. S. Young, J. Crocker, H. Stokes, S. Henderson, and A. B. Rickinson.** 1991. Epstein-Barr virus (EBV)-associated lymphoproliferative disease in the SCID mouse model: implications for the pathogenesis of EBV-positive lymphomas in man. *J. Exp. Med.* **173:**147–158.

109. **Sandler, A. S., and L. Kaplan.** 1996. AIDS lymphoma. *Curr. Opin. Oncol.* **8:**377–385.

110. **Sandvej, K., L. Krenacs, S. J. Hamilton Dutoit, J. L. Rindum, J. J. Pindborg, and G. Pallesen.** 1992. Epstein-Barr virus latent and replicative gene expression in oral hairy leukoplakia. *Histopathology* **20:**387–395.

111. **Shapiro, R. S., A. Chauvenet, W. McGuire, A. Pearson, A. W. Craft, P. McGlave, and A. Filipovich.** 1988. Treatment of B-cell lymphoproliferative disorders with interferon alfa and intravenous gamma globulin [letter]. *N. Engl. J. Med.* **318:**1334.

112. **Shiramizu, B., B. Herndier, T. Meeker, L. Kaplan, and M. McGrath.** 1992. Molecular and immunophenotypic characterization of AIDS-associated, Epstein-Barr virus-negative, polyclonal lymphoma [see comments]. *J. Clin. Oncol.* **10:**383–389.

113. **Shope, T., D. Dechairo, and G. Miller.** 1973. Malignant lymphomas in cotton top marmosets following inoculation of Epstein-Barr virus. *Proc. Natl. Acad. Sci. USA* **70:** 2487–2491.

114. **Sixbey, J. W., and J. S. Pagano.** 1985. Epstein-Barr virus transformation of human B lymphocytes despite inhibition of viral polymerase. *J. Virol.* **53:**299–301.

115. **Sixbey, J. W., P. Shirley, M. Sloas, N. Raab Traub, and V. Israele.** 1991. A transformation-incompetent, nuclear antigen 2-deleted Epstein-Barr virus associated with replicative infection. *J. Infect. Dis.* **163:**1008–1015.

116. **Smith, C. A., C. Y. Ng, H. E. Heslop, M. S. Holladay, S. Richardson, E. V. Turner, S. K. Loftin, C. Li, M. K. Brenner, and C. M. Rooney.** 1995. Production of genetically modified Epstein-Barr virus-specific cytotoxic T cells for adoptive transfer to patients at high risk of EBV-associated lymphoproliferative disease. *J. Hematother.* **4:**73–79.

117. **Smith, C. A., C. Y. Ng, S. K. Loftin, C. Li, H. E. Heslop, M. K. Brenner, and C. M. Rooney.** 1996. Adoptive immunotherapy for Epstein-Barr virus-related lymphoma. *Leuk. Lymphoma* **23:**213–220.

118. **Straus, D. J.** 1997. Human immunodeficiency virus-associated lymphomas. *Med. Clin. North. Am.* **81:**495–510.

119. **Taub, R., I. Kirsch, C. Morton, G. Lenoir, D. Swan, S. Tronick, S. Aaronson, and P. Leder.** 1982. Translocation of the c-myc gene into the immunoglobulin heavy chain locus in human Burkitt's lymphoma and murine plasmacytoma cells. *Proc. Natl. Acad. Sci. USA* **79:**7837–7841.

120. **Thomas, J. A., D. H. Felix, D. Wray, J. C. Southam, H. A. Cubie, and D. H. Crawford.** 1991. Epstein-Barr virus gene expression and epithelial cell differentiation in oral hairy leukoplakia. *Am. J. Pathol.* **139:**1369–1380.

121. **Thomas, J. A., N. A. Hotchin, M. J. Allday, P. Amlot, M. Rose, M. Yacoub, and D. H. Crawford.** 1990. Immunohistology of Epstein-Barr virus-associated antigens in B cell disorders from immunocompromised individuals. *Transplantation* **49:**944–953.

122. **Tomkinson, B. E., R. Maziarz, and J. L. Sullivan.** 1989. Characterization of the T cell-mediated cellular cytotoxicity during acute infectious mononucleosis. *J. Immunol.* **143:**660–670.

123. **Tomkinson, B. E., D. K. Wagner, D. L. Nelson, and J. L. Sullivan.** 1987. Activated lymphocytes during acute Epstein-Barr virus infection. *J. Immunol.* **139:**3802–3807.

124. **Tosato, G.** 1987. The Epstein-Barr virus and the immune system. *Adv. Cancer Res.* **49:**75–125.

125. **Tosato, G., and R. M. Blaese.** 1985. Epstein-Barr virus infection and immunoregulation in man. *Adv. Immunol.* **37:**99–149.

126. **Vowels, M. R., R. L. Tang, V. Berdoukas, D. Ford, D. Thierry, D. Purtilo, and E. Gluckman.** 1993. Brief report: correction of X-linked lymphoproliferative disease by transplantation of cord-blood stem cells [see comments]. *N. Engl. J. Med.* **329:**1623–1625.

127. **Walling, D. M., S. N. Edmiston, J. W. Sixbey, M. Abdel Hamid, L. Resnick, and N. Raab Traub.** 1992. Coinfection with multiple strains of the Epstein-Barr virus in human immunodeficiency virus-associated hairy leukoplakia. *Proc. Natl. Acad. Sci. USA* **89:**6560–6564.

128. **Wang, D., D. Liebowitz, and E. Kleff.** 1985. An EBV membrane protein expressed in immortalized lymphocytes transforms established rodent cells. *Cell* **43:**831–840.

129. **Wang, D., D. Liebowitz, and E. Kleff.** 1988. The truncated form of the Epstein-Barr virus latent-infection membrane protein expressed in virus replication does not transform rodent fibroblasts. *J. Virol.* **62:**2337–2346.

130. **Wang, D., D. Liebowitz, F. Wang, C. Gregory, A. Rickinson, R. Larson, T. Springer, and E. Kleff.** 1988. Epstein-Barr virus latent infection membrane protein alters the human B-lymphocyte phenotype: deletion of the amino terminus abolishes activity. *J. Virol.* **62:**4173–4184.

131. **Wang, F., C. Gregory, C. Sample, M. Rowe, D. Liebowitz, R. Murray, A. Rickinson, and E. Kleff.** 1990. Epstein-Barr virus latent membrane protein (LMP1) and nuclear proteins 2 and 3C are affectors of phenotypic changes in B lymphocytes: EBNA-2 and LMP1 cooperatively induce CD23. *J. Virol.* **64:**2309–2318.

132. **Wang, Q., Y. Ishikawa Brush, A. P. Monaco, D. L. Nelson, C. T. Caskey, S. P. Pauly, G. M. Lenoir, and B. S. Sylla.** 1993. Physical mapping of Xq24-25 around loci closely linked to the X-linked lymphoproliferative syndrome locus: an overlapping YAC map and linkage between DXS12, DXS42, and DXS37. *Eur. J. Hum. Genet.* **1:**64–71.

133. **Williams, L. L., C. M. Rooney, M. E. Conley, M. K. Brenner, R. A. Krance, and H. E. Heslop.** 1993. Correction of Duncan's syndrome by allogeneic bone marrow transplantation [see comments]. *Lancet* **342:**587–588.

134. **Young, L., C. Alfieri, K. Hennessy, H. Evans, O. H. C, K. C. Anderson, J. Ritz, R. S. Shapiro, A. Rickinson, E. Kieff, and J. Cohen.** 1989. Expression of Epstein-Barr virus transformation-associated genes in tissues of patients with EBV lymphoproliferative disease. *N. Engl. J. Med.* **321:**1080–1085.

135. **Young, L. S., C. W. Dawson, D. Clark, H. Rupani, P. Busson, T. Tursz, A. Johnson, and A. B. Rickinson.** 1988. Epstein-Barr virus gene expression in nasopharyngeal carcinoma. *J. Gen. Virol.* **69:**1051–1065.

136. **Young, L. S., S. Finerty, L. Brooks, F. Scullion, A. B. Rickinson, and A. J. Morgan.**

1989. Epstein-Barr virus gene expression in malignant lymphomas induced by experimental virus infection of cottontop tamarins. *J. Virol.* **63:**1967–1974.

137. **Yu, M. C., J. H. Ho, S. H. Lai, and B. E. Henderson.** 1986. Cantonese-style salted fish as a cause of nasopharyngeal carcinoma: report of a case-control study in Hong Kong. *Cancer Res.* **46:**956–961.

138. **Zeng, Y., L. G. Zhang, Y. C. Wu, Y. S. Huang, N. Q. Huang, J. Y. Li, Y. B. Wang, M. K. Jiang, Z. Fang, and N. N. Meng.** 1985. Prospective studies on nasopharyngeal carcinoma in Epstein-Barr virus IgA/VCA antibody-positive persons in Wuzhou City, China. *Int. J. Cancer* **36:**545–547.

139. **Ziegler, J. L., J. A. Beckstead, P. A. Volberding, D. I. Abrams, A. M. Levine, R. J. Lukes, P. S. Gill, R. L. Burkes, P. R. Meyer, C. E. Metroka, J. Mouradian, A. Moore, S. A. Riggs, J. J. Butler, F. C. Cabanillas, E. Hersh, G. R. Newell, L. J. Laubenstein, D. Knowles, C. Odajnyk, B. Raphael, B. Koziner, C. Urmacher, and B. D. Clarkson.** 1984. Non-Hodgkin's lymphoma in 90 homosexual men. Relation to generalized lymphadenopathy and the acquired immunodeficiency syndrome. *N. Engl. J. Med.* **311:**565–570.

140. **zur Hausen, H., H. Schulte-Holthausen, G. Klein, W. Henle, G. Henle, P. Clifford, and L. Santesson.** 1970. EB-virus DNA in biopsies of Burkitt tumors and anaplastic carcinomas of the nasopharynx. *Nature* **228:**1056–1057.

Human Tumor Viruses
Edited by Dennis J. McCance
© 1998 American Society for Microbiology

6 | Regulation of Transcription and Replication by Human Papillomaviruses
Laimonis A. Laimins

Papillomaviruses are small double-stranded DNA viruses which target epithelial tissues for infection. These viruses exhibit a high degree of host specificity and only rarely cross species (55). Over 70 different human papillomavirus (HPV) types have been identified, and all exhibit a tropism for epithelia at specific body locations (30). For instance, HPV-1 is the causative agent of plantar warts and is found primarily in infections of epithelia on the soles of feet (50). Similarly, HPV-2, -4, and -7 induce hyperproliferative lesions (warts) in cutaneous epithelia on hands (50). One-third of all HPV types specifically infect the genital tract, and a subset of these are associated with over 90% of all cervical malignancies (105). This latter group includes HPV-16, -18, -31, -33, and -45 and is referred to as high-risk HPV types (55). In contrast, HPV-6 and -11, which also infect genital epithelia, are not associated with the development of anogenital malignancies and are termed low-risk types. In this review, I will concentrate on the mechanisms by which HPVs regulate viral transcription and replication during the productive life cycle. Chapter 7 will describe the viral oncoproteins and how they contribute to the development of anogenital malignancies.

Despite the differences in target tissues and the kinds of hyperproliferative diseases that HPVs produce, the genomic organization of all types of HPV is quite similar. The genes expressed prior to vegetative viral replication are termed early genes and consist of six to eight open reading frames (ORFs) (Fig. 1). The E6 and E7 genes encode proteins which are primarily localized to the nucleus and bind the cell cycle regulators p53 and Rb (26, 35, 51, 100). Proteins from the high-risk HPV types are able to immortalize many primary human cells, including keratinocytes. The E1 and E2 proteins function in viral replication and may also modulate viral expression (87, 116, 118). Following HPV productive replication, late gene expression is activated. The gene products encoded by late genes consist of the E4 ORF, which is synthesized as a fusion protein with the N terminus of the E1 ORF, and the L1 and L2 capsid proteins. The E1^E4 fusions are cytoplasmic proteins which are expressed in differentiated suprabasal cells and are believed to induce collapse of keratin filaments to facilitate egress of progeny virions from keratinized cells; however, they may have additional uncharacterized functions (33, 97). The L1 protein is the major component of the viral capsid while L2 is a minor constituent.

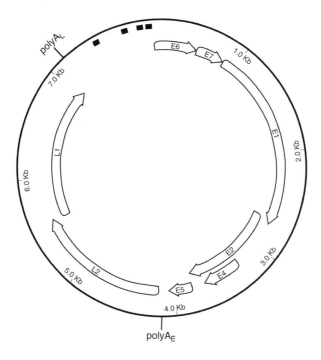

Figure 1 Genomic map of high-risk HPV-31. ORFs designated E indicate ORFs expressed early during infection as well as in basal cells. ORFs designated L indicate late capsid genes. Small black boxes indicate E2 binding sites in the URR. This genomic organization is similar to those of other genital papillomaviruses.

Transcription of early genes occurs from a single strand of HPV DNA and is controlled by sequences located in regions of approximately 1 kb in length which do not code for any viral proteins. These regions are alternatively called NCR (noncoding region), LCR (long control region), or URR (upstream regulatory region). For simplicity, I will refer to these sequences as the URR. In addition to controlling early transcription, the URR also contains the viral origin of replication as well as a series of E2 binding sites (55). Most of the HPVs contain four E2 binding sites in the URR which are present at conserved locations. While the E2 protein provides the major transcriptional activation function in bovine papillomavirus type 1 (BPV-1) (55), no early HPV promoter which is positively regulated by E2 has been identified. E2 also functions in replication, and this may be its primary role in the human viruses (20, 28, 40).

Papillomaviruses are among the small group of viruses that link their productive life cycle to the differentiation program of the target cell (Fig. 2). Papillomaviruses infect the stem cell population of stratified epithelia, the basal cells, and establish viral genomes as low-copy episomes (69). Infection by HPVs occurs through microlesions of the epithelia which expose basal cells to viral entry. Following entry, genomes are established as low-copy episomes which are replicated coordinately with cellular chromosomes. A low level of transcription of viral genes occurs in infected basal cells, and significant transcription occurs in differentiated daughter cells. Vegetative genome replication, late gene expression, and virion assembly are restricted to the nuclei of highly differentiated epithelial cells present in the uppermost epithelial layers. In this manner, a reservoir of viral genomes is stably maintained in the basal layer and allows for the continuous generation of infected daughter cells which go on to synthesize progeny virions. Papillomaviruses are dependent on polymerases and other cellular proteins for viral replication. In normal epithelia, cells exit the cell cycle after migrating from the basal layer, and this process is blocked in HPV infections. When HPV-infected cells migrate from the basal layer, they become locked in late G1/S and are induced to reenter the S phase in the most differentiated layers to allow for viral DNA replication (69).

Low-grade HPV infections of the cervix are referred to as condyloma, or CIN I (cervical intraepithelial neoplasia grade I), and actively produce HPV virions (64). In these lesions, basal-like cells are restricted to the lower third of the epithelia, and suprabasal cells contain dense nuclear chromatin which is surrounded by large vacuoles (64). These cells are referred to as koilocytes and are the sites of synthesis of progeny virions. Progression from low- to high-grade lesions such as CIN III, or carcinoma, is accompanied by a loss of epithelial differentiation with the appearance of mitotically active cells in suprabasal layers (105). In these high-grade lesions, the viral DNA is often found integrated into host chromosomes, but many lesions also contain episomal copies. It has been hypothesized that integration of HPV genomes was an important step in the progression to malignancy, but the detection of episomal copies in many high-grade lesions indicates that this may not be an essential step (105). Since high-grade lesions exhibit a lack of differentiation, no virions are produced. The induction of carcinomas

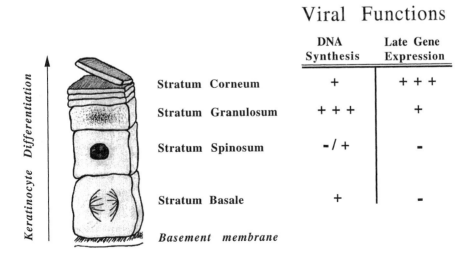

	Viral Functions	
	DNA Synthesis	**Late Gene Expression**
Stratum Corneum	+	+ + +
Stratum Granulosum	+ + +	+
Stratum Spinosum	- / +	-
Stratum Basale	+	-
Basement membrane		

Figure 2 Cartoon of differentiating epithelia and a table of viral functions which are induced upon epithelial differentiation. The various epithelial layers are indicated.

following papillomavirus infection does not provide any growth advantage to the virus and is likely a consequence of the manner in which these viruses target cell cycle regulators.

The ability to propagate HPVs in animals as well as in tissue culture has allowed for an analysis of the viral life cycle in the laboratory. The xenograft system allows for the growth of HPVs in nude mice and was pioneered by John Kreider and his colleagues at Penn State Medical School (65). In this system, human foreskin explants are incubated with genital wart material and then implanted under the renal capsule of nude mice. Following growth for several months, the explants are removed and progeny virions are isolated and used to infect naive explant cultures. The xenograft system is able to produce large amounts of HPV-11 but appears to be restricted to explant material. Also using nude mice, Sterling et al. were able to transfer cells from an HPV-16-positive cell line into skin flaps and detected the synthesis of HPV-16 virions (108). While these methods are able to synthesize HPV virions, both are technically challenging and are restricted in their ability to study viral functions.

Success has recently been achieved in growing papillomaviruses in tissue culture through the use of organotypic raft cultures. Raft cultures duplicate epithelial differentiation in vitro and allow for virion production. The raft system was initially developed by Asselineau et al. and has gained considerable use in the papillomavirus field (4). In this method, cells are first grown as submerged monolayer cultures and then transferred onto a collagen plug which contains fibroblast feeders. The collagen-cell matrix is then placed onto a wire grid which is maintained at an air-liquid interface. The cells are thus exposed to air, and subsequent feeding occurs from underneath by diffusion through the collagen. Cultivation of epithelial cells by this method results in differentiation that occurs over a period of 2 weeks. When keratinocytes derived from biopsy specimens were grown in this system, they exhibited histological changes similar to those observed in HPV lesions in vivo (85). The ability to duplicate HPV-induced morphologies in raft cultures stimulated attempts to reproduce the viral life cycle. Using a combination of xenograft and raft cultures, Dollard et al. were able to excise xenografts which had been grown for several months in nude mice, transfer them to raft cultures, and observe the continued synthesis of HPV-11 (32). Meyers et al. examined a biopsy-derived cell line which contained episomal copies of HPV-31 and found that it induced virion synthesis in raft cultures following treatment with phorbol esters (86). Treatment with phorbol esters induced a more complete program of epithelial differentiation, leading to the activation of late gene expression. Recently, these methods have been extended to synthesize HPV-31 and -18 virions from transfected DNA templates (38). Use of these techniques should permit a detailed genetic analysis of HPV function during the productive life cycle.

In this chapter I will describe our current knowledge of the mechanisms regulating HPV gene expression and replication. The first studies which examined papillomaviruses concentrated on the bovine papillomaviruses (BPVs) with the hope that BPV-1 would serve as a useful model for the high-risk human viruses. While many similarities exist between the two viral types, a significant number of differences are also present. For this reason, I will concentrate the discussion on the HPV types and refer to BPVs only when no information is available in the human system.

REGULATION OF VIRAL EXPRESSION

Early and Late Viral Transcripts

In order to understand how viral gene expression is regulated during a productive infection, it is first important to describe the transcripts which are synthesized. Information on papillomavirus expression was initially obtained by in situ analysis of biopsy specimens and R-looping studies performed on purified RNAs by using electron microscopy. Crum et al. demonstrated that transcripts from the E6 region were fairly evenly distributed throughout the lower two-thirds of infected epithelia while expression of L1/L2 and E4/E5 was significantly increased in the uppermost layers (27). Using R-loop analysis, Chow and her colleagues demonstrated that transcripts from the E4/E5 and E6/E7 regions are the predominate species (23). Additional evidence suggested that transcripts from the E7 region increased upon stratification, and this is likely the result of activation of the late promoter which is located in the middle of the E7 ORF (57, 109). Viral expression appears to be low in the basal layer and is increased upon differentiation. Such a mechanism would minimize the exposure of viral proteins to the immune system which is active in the dermis and the lower part of the epithelium. In carcinomas in situ, cells remain undifferentiated throughout all layers, and viral transcription is often restricted to the E6 and E7 genes as a result of integration into host chromosomes (101).

Factors regulating transcription of early viral genes

During the productive life cycle, most viral transcripts are expressed as polycistronic messages (5, 23, 57, 104). In the genital HPVs, two major start sites for transcription have been identified: the first lies upstream of the E6 ORF and regulates early expression while the second is in the middle of the E7 ORF and controls the synthesis of late transcripts (5, 23, 48, 53, 57, 101). Additional minor transcripts initiating upstream of E1 and in the URR may also exist (91). An example of the major transcripts expressed during the productive life cycle of HPV-31 is shown in Fig. 3, and other studies have identified similar transcripts in HPV-11 and -16. The majority of early viral transcripts (designated in Fig. 3 as class I to III transcripts) initiate upstream of the E6 ORF and encode multiple early proteins. The major early transcripts (class I) contain either a full-length E6 protein or a short alternatively spliced E6 protein called E6*. It is not clear if the E6* protein is actually translated or exists only to increase the spacing upstream of the initiator AUG for E7, the second ORF on these transcripts. The third ORF on the major early transcripts is a fusion between the five N-terminal amino acids from E1 and the E4 ORF (89). The E4 ORF lacks an initiating AUG and relies on the N terminus of E1 for this function. There is no evidence that E1^E4 proteins are actually translated from this early transcript. The final ORF on these transcripts is E5 and, likewise, no information is available as to whether it is synthesized from this transcript. A collection of less abundant early transcripts encoding E2, E5, and a novel E6^E4 fusion product have also been identified and are shown in Fig. 3 (designated class II). One major HPV early transcript which encodes E1 is initiated upstream of E6 and encodes all early ORFs (class III) (58). This transcript also serves as the precursor for spliced early mRNAs. Additional low-abundance transcripts may also initiate from a promoter upstream of E1 (91).

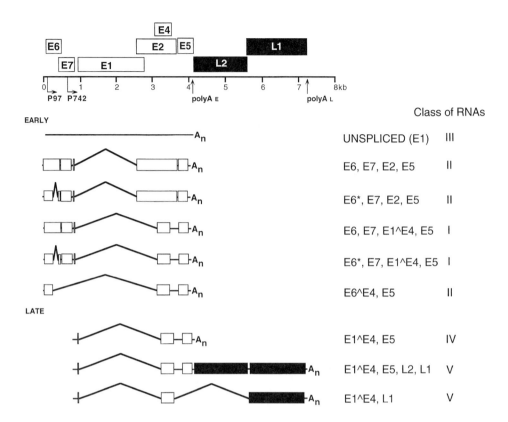

Figure 3 Transcript map of HPV-31. Transcripts expressed constitutively throughout the epithelia initiate at the early p97 promoter and terminate at the early polyadenylation signal located at the end of the E5 ORF. Late transcripts initiate at the late promoter, p742, and terminate at either the early or the late polyadenylation signals.

Late transcripts

The late promoter of the genital HPVs has been mapped to the middle of the E7 gene (nucleotide 742 in HPV-31), and late transcripts appear to be of two types (48, 57, 91). One class (designated class IV [Fig. 3]) encodes E1^E4/E5 and terminates at the early polyadenylation signal. The second class also initiates at p742 and contains E1^E4 and L1 or E5/L2/L1 (class V). These transcripts terminate at the late polyadenylation signal downstream of L1. It has been suggested that the late promoter may also function as an immediate early promoter to direct expression of E1 and E2 during the initial phase of infection (19). However, in transient replication assays, which mimic the early phase, expression from p97 appears to predominate (110).

Early promoter and tissue specificity

The early HPV promoter which directs initiation of transcription at sites upstream of the E6 ORF has been extensively studied in the genital papillomaviruses. This promoter is referred to as p97 in HPV-16 and -31 or p105 in HPV-18. The transcription factors which regulate the early promoters bind to sequences located within the URRs (18, 107, 120). An analysis of URRs of different HPV types has revealed a number of binding sites which are common to all types as well as some which are unique. It is possible that this variation in the spectrum of factors which regulate transcription contributes to the tissue tropism of papillomaviruses. Factor binding sites common among all types include those for TFIID binding to canonical TATA boxes located approximately 30 bp upstream of the early start sites. Adjacent to this sequence and present in most HPV types are Sp1 and AP-1 binding sites. All URR sequences contain a series of AP-1 sites although the number and exact locations vary according to type (16, 17, 67, 68, 80, 115). Other factor binding sites include those for TEF-1, TEF-2, Oct-1, NF-1, AP-2, YY1, and glucocorticoid responsive elements (GRE) (2, 7, 16, 18, 46, 61, 90). An additional factor, called KRF, was identified by gel shift analyses and found to bind the upstream regulatory regions of HPV-18. KRF was initially believed to be expressed primarily in keratinocytes and was a possible determining factor for regulating the tropism of HPVs (80). Subsequent studies have shown that it is present in many cell types (68). Another factor, termed CEII, has been implicated in HPV-11 cell-type-specific regulation (22).

The search for determinants of the epithelial tropism of HPVs remains an area of active study. In transient expression assays, URR reporter constructs are most active in cells of squamous epithelial origin and significantly less active in lines derived from other tissues (25, 45, 46). As a result, it has been postulated that keratinocyte-specific enhancers are present in the URR and that these contribute to tissue tropism (46). Recent evidence suggests that a complex of YY1 and C/EBP acts to determine cell type specificity (8). While numerous putative YY1 binding sites can be found in HPVs, the one in the so-called switch region is specifically thought to be involved in determining tissue tropism. In another model, Chang et al. have postulated that it is the combination of ubiquitous factors working in concert that determines the cell-type-specific expression (18). Additional studies have implicated the ratio of Sp1/Sp3 factors as contributing to HPV-16 epithelial specific expression (3).

In addition to positive elements located in the promoter-proximal region of HPV, a negative or silencer element has been identified at the distal end of the HPV-6 URR. This element binds the transcriptional repressor, CCAAT displacement protein (CDP), and has been postulated to play a role in controlling differentiation-dependent viral expression (93). Upon epithelial differentiation, the repressive activity of CDP is reduced, contributing to the activation of late expression. Other negative regulators, including those for C/EBPβ, have been shown to regulate HPV-11 expression (121). Furthermore, Kyo et al. demonstrated that the composition of the Fos and Jun family members varies as a function of epithelial differentiation, which may contribute to regulating HPV early gene expression (67).

Much less information is available concerning the factors which regulate late gene expression. The tight linkage of viral late gene expression and epithelial differentiation suggests that differentiation-specific cellular factors control this process. Recently, advantage has been taken of the ability to synthesize HPV genomes from transfected DNA templates to demonstrate that the state of the viral DNA is also important in activating late expression (38). When HPV-31 genomes containing a mutation in the E1 ORF were transfected into normal human keratinocytes, the viral DNA was found to integrate as concatemers into host chromosomes with their late regions intact. When these cell lines were examined in raft cultures, the cells differentiated fully but failed to induce significant levels of late gene expression. This indicated that viral genomes must be maintained as episomes for the induction of late gene expression and suggests that viral replication may be required. In simian virus 40 (SV40), it is believed that late gene expression may be activated following vegetative replication by the titration of factors which repress late gene expression (122). Such repressors could also be present on HPV DNA and act to actively suppress late gene expression in basal cells. Following vegetative HPV replication in suprabasal cells, such repressors could be titrated away from viral DNA, resulting in the activation of late gene expression. With the availability of methods for genetic analysis of HPV functions during the viral life cycle, a detailed examination of these processes is now feasible.

Posttranscriptional regulation

The regulation of late gene expression is a complex process involving control of both initiation of transcription and RNA stability. Studies from Carl Baker's laboratory first identified inhibitory sequences located in the 3' UTR (untranslated region) of the BPV-1 L1 gene located between the end of L1 and the late AAUAAA sequence which had the ability to decrease expression of heterologous reporter constructs in transient assays (42). Similar studies have identified comparable elements in the 3' UTRs of HPV-1 and -16 late genes (62, 114). Demonstration that posttranscriptional mechanisms are functional in vivo came from studies using the CIN 612 cell line which maintains stable episomal copies of HPV-31 (58). When this cell line was grown in raft cultures, the activation of the late promoter was observed but transcripts were limited to those encoding E1^E4/E5. No significant expression of the L1/L2 capsid genes was detected. However, when phorbol esters were added to growth media, a more complete program of epithelial differentiation was induced, resulting in the appearance of late transcripts. The level of transcription initiation at the late promoter was not changed by the addition of phorbol esters, suggesting that the appearance of late transcripts was due to stabilization of late mRNAs. Such a process could be mediated by the differentiation-dependent loss of instability factors, which in basal cells are bound to sequences in the 3' UTR. Studies by Dietrich et al. identified a 65-kDa protein which specifically bound to late inhibitory sequences of HPV-16 and whose binding was lost upon addition of phorbol esters (31). In BPV-1, the inhibitory sequence can function as a splice donor (43), but no such activities have been shown for the human virus counterparts. In addition to the presence of instability sequences, a differentiation-dependent switch in usage of polyadenylation sequences occurs. Late transcripts initiate in the early region and are transcribed through the early polyadenylation signal to terminate at the late signal. This suggests that mechanisms exist that suppress the early polyadenylation signal and allow passage of polymerases into the late region. Alternatively, it is possible that the early signal is inefficient in termination and transcripts consistently pass through the early signal only to be rapidly turned over as a result of L1 3′ UTR sequences. A similar differentiation-dependent switch in usage of tandem polyadenylation sites is seen during B-cell differentiation, leading to a change from membrane-bound to secreted forms of immunoglobulin (36).

One of the most remarkable features of HPV transcription is that all early and late transcripts are polycistronic (5, 23, 57). Early transcripts initiate and termi-nate at common sites, and various viral genes are expressed from alternatively spliced messages. It is likely that viral gene expression is regulated through alter-native splicing. In papillomaviruses, alternative splicing provides the opportunity to shuffle exons. Examples of this are the E1ˆE4 and E6ˆE4 fusion proteins. The major splice donor in HPVs is located at the 3′ end of the E1 ORF, and most viral transcripts utilize this sequence. It is therefore unlikely that significant regulation occurs through utilization of this site. Another donor site is located downstream of the E4 ORF and is used only to generate E1ˆE4/L1 transcripts. This donor/acceptor is a good candidate for regulation. The regulating presence of multiple ORFs on viral messages suggests that mechanisms exist for initiation of translation of internal ORFs. One possibility is that internal ribosome entry se-quences mediate this process, though none have been identified. It is therefore likely that additional control is exerted at the level of translation since many more cells are positive for late transcripts than synthesize capsid proteins (32). Taken together, these observations indicate that multiple mechanisms exist for regulating HPV gene expression during the viral life cycle.

Role of E1 and E2 in regulation

The E2 protein from BPV-1 was first characterized as a potent transcriptional activator which regulates viral transcription (106). The E2 proteins from HPVs are very similar in structure and function as activators in transient expression assays. All full-length E2 proteins (E2-TA) form dimers through amino acids located in the C terminus, while the N terminus functions as the activation domain (52, 83, 84). All E2 proteins bind to palindromic sequences consisting of ACCGN4CGGT motifs (1). Seventeen E2 binding sites are present in the BPV-1 genome (74) while the genital HPVs contain only four sites. In HPV-1 and -8, an intermediate number of E2 sites are present. In BPV-1, E2 is a potent activator of early gene expression (55), but no early promoter positively activated by E2 in the genital papillomavirus types has been identified. However, in HPV-8, E2 has been implicated in the activation of the late promoter (111). A similar role for E2 in the genital papillo-maviruses remains to be demonstrated.

In BPV-1, truncated forms of E2 proteins have been identified. These proteins, referred to as E2 repressors, consist of either a fusion between a small ORF desig-nated E8 to the C-terminal DNA binding domain (E8-E2) or the C-terminal portion of E2 alone (E2-TR) (56, 70). These repressors are expressed either from spliced transcripts for E8-E2 or from a promoter within the E2 gene for E2-TR. The repres-sors can inhibit the transactivation function of the full-length E2 protein either by competition for binding to DNA (70) or through the formation of heterodimers with the full-length proteins (6). It is believed that repressors provide a way to tightly regulate the activity of E2.

In contrast to the major role for BPV-1 E2 proteins as transcriptional activators, the HPV full-length E2 proteins have been shown to act only as repressors of URR transcription (11, 29). When expression vectors for E2 were cotransfected with HPV URR reporter plasmids, repression of the early viral promoter has been observed (25, 59, 113). This repression resulted from binding of E2 to one of the four binding sites located adjacent to the TATA box of the early viral promoter. Binding of E2 molecules to this site prevents binding of TFIID to the TATA sequence, which results in reduced early gene expression (113). This repression by E2 has been postulated to play a role in down regulating transcription of the transforming genes E6 and E7, and that integration of viral genomes which often disrupts E2 expression leads to their increased expression and progression to malignancy. It is also possible that repression acts to modulate expression of the viral replication genes as part of the copy number control mechanism (110a). HPV E2 is also a replication factor which forms a complex with the origin recognition protein E1 to facilitate viral replication. Studies with the BPVs have demonstrated that E1 can also modulate E2 activity through complex formation (72). At low concentrations E1 activates E2 function while in high amounts it represses activity. Since a transcriptional role for HPV E2 has yet to be defined, it is not clear if a similar role for E1 exists in the human viruses. A more detailed discussion of the role of these proteins in replication is provided in the next section.

Information on the structure of the C-terminal DNA binding domain of E2 bound to its recognition site has been obtained from crystallographic studies. The E2 DNA binding domain forms a dimeric antiparallel β-barrel structure (52). The binding site DNA is bent smoothly around the E2 dimer, with alpha-helical domains binding to the major groove. Each E2 monomer forms one-half of the barrel structure. A similar structure has recently been identified in the Epstein-Barr virus replication protein, EBNA-1, though the DNA is bound through different amino acids adjacent to the β-barrel. Efforts to determine the structure of the N terminus of E2 have been largely unsuccessful due to the tendency of the activation domain to form aggregates. The processes of transcription and replication are tightly coupled in papillomaviruses, as evidenced by the dual roles of E2. In order to further understand the interplay of these two viral functions, I will address the control of viral replication in the next section.

VIRAL REPLICATION

The replication of papillomaviruses is a complex process which can be divided into three distinct parts. The first is the establishment phase in which a single or small number of viral particles enter a cell and the genome quickly replicates up to 50 to 100 copies per cell. During the second phase, which is referred to as maintenance, viral episomes replicate in synchrony with the cellular chromosomes (44) of dividing basal cells and copy number is maintained constant at 50 to 100 per cell (34). Finally, as infected cells leave the basal layer and differentiate, productive viral replication or amplification occurs in the uppermost layers of the epithelia, resulting in the synthesis of thousands of copies per cell. Transition between these three replication phases is believed to be highly regulated, though little is known about the mechanisms involved. The most detailed information is available on the first phase of the viral replication cycle, and I will begin by describing this process.

Transient Replication and the Establishment Phase

The first accurate characterization of the *cis* and *trans* requirements for papillomavirus replication was performed by Ustav and Stenlund using transient transfection assays, which most closely mimic the establishment phase of the viral life cycle (118). Such assays are now commonly used in the field and consist of transfecting cells with vectors expressing viral replication proteins together with origin fragments. After 2 to 3 days, low-molecular-weight DNA is isolated and Southern analysis is performed for replicated viral DNA. In this manner, both E1 and E2 were shown to be required for replication (118). In addition, the minimal BPV-1 origin was localized to sequences within the URR that are proximal to the translation initiation codon of the E6 ORF (Fig. 1) (119). The minimal origin sequences consist of an A/T-rich sequence flanked by E2 binding sites, and similar elements are found in all papillomaviruses though the exact arrangement varies (21, 40, 76, 112). For instance, in the human genital viruses, the A/T-rich region is flanked by two E2BS on the downstream side and a third site upstream. One study has suggested that additional sequences in the late region enhance transient replication of the entire BPV-1 genome, but only a limited characterization of this region has been described (94). Holt and Wilson demonstrated that the A/T-rich sequences contain a palindromic binding site for E1 (54). In BPV-1 this site is 18 bp in length while in HPV-31 and ~11 it is approximately 30 bp in size (40, 76). The E2 binding sites which flank the E1 recognition sequence vary in affinity and contain the sequence ACCGN4CGGT. In transient replication assays, the E1 binding site together with a single E2 site or two E2 sites by themselves are sufficient for transient replication (21, 76, 112). This suggests that an intact E1 site is not absolutely required for transient replication though it increases efficiency. The relative distance between the E1 and E2 sites can vary depending on the relative binding affinity. Interestingly, while E1 and E2 are required for replication of most HPV types, E1 alone is sufficient for transient replication of HPV-1 (47, 112).

Most assays examining transient replication have used expression vectors for E1 and E2 together with minimal origin fragments (118). Such experiments may not accurately reflect effects in the intact viral genome (28). Recent studies by Stubenrauch et al. have examined the role of the E2 binding sites in the context of the entire HPV-31 genome and demonstrated a significant amount of redundancy in E2 binding sites (110a). While mutation of the second E2 binding site (Fig. 4) had a minimal effect on replication, elimination of either site 1 or site 3 reduced replication efficiency to 30 to 50% of the wild type (Fig. 4). Interestingly,

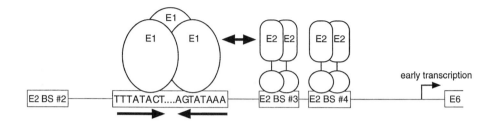

Figure 4 Origin of replication of HPV-31, indicating the E1 binding site and the two flanking E2 sites. E1 proteins bind to E1 sites as multimeric complexes which can form larger complexes with adjacent E2 dimers.

mutation of the site which is closest to the start of HPV early transcription (site 4) resulted in a fourfold increase in replication. Transcript analysis and cotransfection with expression vectors revealed an apparent negative feedback control of E1 expression mediated by E2 through its binding of the E6-proximal E2 site (site 4). Binding to this site decreases expression of the major early transcripts encoding E1 which initiate at the early p97 promoter. This is consistent with studies described above which identified E2 as a repressor of HPV early expression (60, 116). It appears that repression by E2 may function to negatively regulate E1 expression as part of a regulatory loop to control copy number. Modulation of copy number by increasing the levels of E1 and E2 proteins will be discussed below.

In Vitro Replication

In addition to transient replication assays, in vitro replication studies have also provided significant insights. In vitro replication has been achieved for both BPV-1 and HPV-11 by using permissive cell nuclear extracts which were supplemented with either purified bacterial or baculovirus-expressed E1 and E2 proteins (66, 123). In vitro studies indicated that BPV-1 E1 by itself is sufficient for replication and that E2 is not required (103, 124). This is consistent with the HPV-1 in vivo studies, which demonstrated that expression of E1 alone is sufficient to mediate replication (47). In contrast, in HPV-11 a strong dependence on both E1 and E2 has been demonstrated (66). In vitro studies also suggest that E2 binding sites are not required for BPV-1 origin function, though origin-specific replication is achieved only with high concentrations of E1 (103, 124). These in vitro studies suggest that E1 functions as the primary papillomavirus replication factor and that E2 provides auxiliary functions. For HPV-11 E2 is absolutely required (66) while for BPV-1 E2 appears to provide specificity as E2 suppressed nonspecific replication (12). This may reflect a fundamental difference between the two viral types or variations between the experimental protocols.

E1 and E2 Replication Proteins

The E1 and E2 replication proteins individually bind to their respective binding sites and are also able to form a heteromeric complex with each other. On its own, E1 is a weak DNA binding protein and complex formation with E2 greatly increases its binding affinity to the origin (39, 78, 87, 103). The E1/E2 complexes can form in solution as well as assemble on DNA through cooperative binding to the origin. The presence of either E1 or E2 bound to DNA greatly increases the binding affinity of the other protein to the cognate sequences. If complexes pre-form in solution prior to binding to DNA, then either E1 and E2 protein could first bind their respective sites and form a tether or the E1/E2 complex could form in solution and bind the E2 site alone (39). It is not yet clear which arrangements are functional in vivo and it is possible that both types of interactions occur.

The E1 protein functions both as an origin recognition protein and as a heli-case (124). The C-terminal portion of SV40 T antigen mediates similar replication functions and exhibits homology to the E1 proteins (24). These regions of homol-ogy include those for ATP binding and helicase functions. T antigen also binds to Rb and p53, functions which are apparently not shared by E1. Like T antigen, E1 can recruit cellular replication enzymes such as DNA polymerase α to papillo-mavirus origins (92). It has been suggested that E1 exists in several multimeric forms (79, 102, 123), of which the hexameric is most similar to that seen in T-antigen binding to the origin (81). E2 may also act to recruit cellular enzymes as it has been shown to bind the cellular replication protein RPA (73). If E2 functions primarily as an auxiliary protein, then it could act to alter chromatin structure around the origin or provide a function similar to NF-1 in adenovirus replication (88). In adenovirus replication, the cellular factor NF-1 interacts with the terminal protein (pTP) complex to increase its binding affinity to viral DNA, and it is possible that E2 functions in a similar manner.

An extensive mutational analysis of the E2 proteins has been performed to identify sequences important for replication. Approximately one-third of the amino acids in E2 have been altered and a functional map has been developed (13, 49, 98). Sequences in the N terminus of E2 mediate E1 association, while amino acids in the C terminus are necessary for binding to DNA. Interestingly, mutations of E2 which retain replication activity in transient assays but fail to activate tran-scription have been identified (13, 98). These mutants demonstrate that transacti-vation is dispensable for transient replication. Using the ability to synthesize re-combinant HPV-31 from transfected DNA templates, Stubenrauch and Laimins further demonstrated that transactivation is also not required for stable mainte-nance of episomes or induction of late viral functions (110).

The amino acids in E1 which are important for replication function are less well defined. The portion of BPV-1 E1 which binds DNA has been localized to amino acids 180 to 270, with substitutions at amino acids 241, 247, and 180 being deficient in this activity (117). Less is known about the E2 interaction domain of E1 in BPV-1, as contradictory information has been published. Thorner et al. found C-terminal sequences (amino acids 458 to 605) to be important at low tempera-tures, while more N-terminal sequences were important at higher temperatures (117). In contrast, Bensen and Howley (10) defined amino acids 1 to 250 in HPV-16 E1 and similar observations have been made with HPV-31 E1 and E2 interaction domains (40). Clearly, a finer mapping analysis is required and it is also possible that two interaction domains exist. Little is known about which specific sequences define the helicase domain in E1, and while some homology exists with SV40 large T antigen (96), the helicase domain remains poorly defined. In contrast, an ATP binding domain is well defined in large T antigen (93) and a homologous region exists in many HPV E1 proteins. It is anticipated, given the role of E1 as a helicase, that the ATP binding activity will be found to be essential for replication.

Maintenance of Viral Episomes

Unlike the early establishment phase of replication, our knowledge of the mechanisms controlling stable maintenance is only beginning to evolve. Two systems have been used to study this process. Transfection of C127 cells with BPV-1 genomes results in focus formation, and transformed cells maintain genomes as episomes (34, 71). Similarly, lipofection of normal human keratinocytes with cloned HPV-18 or -31 sequences results in immortalization of cells which stably maintain viral episomes (38, 41). Replication of stably maintained episomes is restricted to the S phase and occurs in synchrony with chromosomal replication. It was originally suggested that each of the 50 to 100 genome copies replicate once per cell cycle; however, subsequent studies have demonstrated that this is a random process similar to bacterial plasmid replication (44).

One of the most interesting questions regarding maintenance is how copy number is controlled in infected basal cells in vivo. Studies by Frattini and Laimins (40) demonstrated that increasing the levels of the replication proteins E1 and E2 resulted in an increase in stable copy number of HPV-31 epiosmes. This indicated that the levels of active E1 and E2 proteins could regulate copy number. This is consistent with observations in BPV-1 where phosphorylation mutations in E2 result in an increased number of stably maintained genomes (82). If copy number is regulated by modulating the activities of the replication proteins, a feedback loop regulating expression of replication proteins would be useful. For instance, if during mitosis one cell received less than the optimal number of genomes, then low levels of E2 would lead to increased expression of E1 and a corresponding increase in replication in the next S phase. In addition, the levels of E2 repressors could be used to control E2 activities. Such a model clearly requires more rigorous testing.

Information on the requirements for stable maintenance has also been obtained from BPV-1 transformation of mouse C127 cells. It has been demonstrated that plasmids containing only 69% of the BPV-1 genome can be stably maintained in transfected cells (75). The 69% fragment contains the entire URR and early region but lacks late sequences. In addition, some heterologous genes, such as rat insulin, could be cloned downstream of this 69% fragment and recombinant genomes could be maintained extrachromosomally (99). In contrast, other genes were not compatible with episomal maintenance, and the source of this incompatibility has been suggested to lie in the presence of cellular enhancer elements (77). Unfortunately, the reason why such a transcriptional element is required is not clear.

A more systematic approach to studying the requirements for maintenance was undertaken by Piirsoo et al. using Chinese hamster ovary cell lines in which the expression of E1 and E2 was directed from heterologous expression vectors (95). In these cells, episomal maintenance could be achieved when six BPV-1 E2 binding sites were placed upstream of the E1 binding site. Since the genital HPVs contain only four E2 binding sites, a different spectrum of *cis* sequences is required for maintenance of human viral genomes. Recent studies by Stubenrauch and Laimins have demonstrated that three of the four E2 binding sites in the HPV-31 URR are essential for stable maintenance of episomes (110) while the E2 binding site located adjacent to the E1 site gene was found to be dispensable (Fig. 4). This suggests that in the context of the entire viral genome, it is the particular arrangement of HPV E2 sites that is important and not merely their number.

An equally important part of maintenance is the proper segregation of plasmids between daughter cells. Following replication of infected basal cells, viral DNA is distributed to daughter cells during mitosis. Equal partition of episomes could occur if there are distinct attachment sites for viral episomes on specific condensed chromosomes. Alternatively, if segregation is a random process, then any disparity in copy number could be restored during the next S phase if mechanisms exist for increasing or decreasing episomal replication. CIN 612 cells which stably maintain HPV-31 episomes have been grown continuously for over 1 year and have maintained a constant copy number (9). This indicates that mechanisms to control copy number exist. Several studies indicate that the requirements for stable maintenance of episomes are more stringent than those for replication in transient assays, and this may be a reflection of the need for proper segregation (63, 110).

Differentiation-Dependent Genome Amplification

The mechanisms which regulate differentiation-dependent vegetative viral DNA replication in suprabasal cells remain largely points of speculation. As infected cells migrate from the basal layer, a subset are induced to reenter the S phase, resulting in amplification of viral genomes to thousands of copies per cell. This process is not restricted to viral DNA as cellular chromosomes are also replicated, as indicated by thymidine incorporation (19). Coordinate with vegetative replication is the induction of late gene expression and capsid synthesis.

Studies by Burnett and colleagues provided the first insight into possible mechanisms controlling amplification. When mouse C127 cells which contained episomal copies of BPV-1 were incubated at confluence for extended periods of time, a number of "jackpot" cells appeared (14). These jackpot cells contained thousands of copies of viral DNA as well as high levels of E2 and E5 proteins (15). While this model system is not physiological, the observations suggest that increased expression of viral proteins could be involved in the amplification process. Expanding on this idea, Frattini et al. constructed recombinant adenoviruses expressing the HPV-31 E1 and E2 genes and then used these viruses to infect CIN 612 cells which stably maintain HPV episomes (37). Following infection, an increase in template number of 10-fold was observed. Interestingly, infection by the E2 viruses induced an S-phase arrest leading to re-replication of cellular DNA.

This suggests that viral proteins may induce an aberrant S phase in terminally differentiated epithelial cells. This interesting model awaits the demonstration that E2 expression or function is increased upon differentiation.

SUMMARY

Our knowledge of the mechanisms which regulate papillomavirus expression and replication are still evolving, and many important insights have already been gained. Future work will involve a closer examination of these regulatory activities during the normal life cycle of papillomaviruses in differentiating epithelia. Of particular importance will be an examination of the mechanisms by which late gene expression is regulated both at the level of initiation of transcription and posttranscriptionally. With respect to replication, it will be important to determine what processes control amplification in suprabasal cells. Given the tight linkage of late gene expression and amplification to epithelial differentiation, it is likely that cellular factors play important roles. These future studies should expand our understanding of how infection by HPVs can in some cases result in the development of malignancy.

Acknowledgments
This work was supported by grants from NCI and NIAID to L.A.L. I thank Jennifer Thomas, Mark Frattini, and Scott Terhune for assistance with the illustrations.

REFERENCES

1. **Androphy, E., D. R. Lowy, and J. T. Schiller.** 1987. Bovine papillomavirus E2 trans-activating gene binds to specific sites in papillomavirus DNA product. *Nature* **325:** 70–73.
2. **Apt, D., T. Chong, Y. Liu, and H.-U. Bernard.** 1993. Nuclear factor I and epithelial cell-specific transcription of human papillomavirus type 16. *J. Virol.* **67:**4455–4463.
3. **Apt, D., R. Watts, G. Suske, and H. -U. Bernard.** 1996. High Sp1/Sp3 levels in epithelial cells during epithelial differentiation and cellular transformation correlate with the activation of the HPV16 promoter. *Virology* **224:**281–291.
4. **Asselineau, D., B. A. Bernard, C. Bailly, M. Darmon, and M. Prunieras.** 1986. Human epidermis reconstructed by culture: is it normal? *J. Invest. Dermatol.* **86:** 181–186.
5. **Baker, C. C., W. C. Phelps, V. Lindgren, M. J. Braun, M. A. Gonda, and P. M. Howley.** 1987. Structural and transcriptional analyses of human papillomavirus type 16 sequences in cervical carcinoma cell lines. *J. Virol.* **61:**962–971.
6. **Barsoum, J., S. S. Prakash, P. Han, and E. J. Androphy.** 1992. Mechanism of action of the papillomavirus E2 repressor: repression in the absence of DNA binding. *J. Virol.* **66:**3941–3945.
7. **Bauknecht, T., P. Angel, H.-D. Royer, and H. zur Hausen.** 1992. Identification of a negative regulatory domain in the human papillomavirus type 18 promoter: interaction with the transcriptional repressor YY1. *EMBO J.* **11:**4607–4617.
8. **Bauknecht, T., R. H. See, and Y. Shi.** 1996. A novel C/EBP β-YY1 complex controls the cell-type-specific activity of the human papillomavirus type 18 upstream regulatory region. *J. Virol.* **70:**7695–7705.
9. **Bedell, M. A., J. B. Hudson, T. R. Golub, M. E. Turyk, M. Hosken, G. D. Wilbanks, and L. A. Laimins.** 1991. Amplification of human papillomavirus genomes in vitro is dependent on epithelial differentiation. *J. Virol.* **65:**2254–2260.

10. **Bensen, J. D., and P. M. Howley.** 1995. Amino-terminal domains of the bovine papillomavirus type 1 E1 and E2 protein participate in complex formation. *J. Virol.* **69:**4364–4372.

11. **Bernard, B. A., C. Bailly, M. -C. Lenoir, M. Darmon, F. Thierry, and M. Yaniv.** 1989. The human papillomavirus type 18 (HPV18) E2 gene product is a repressor of the HPV18 regulatory region in human keratinocytes. *J. Virol.* **63:**4317–4324.

12. **Bonne-Andrea, C., F. Tillier, G. McShan, V. Wilson, and P. Clertant.** 1997. Bovine papillomavirus type 1 DNA replication: the transcriptional activator E2 acts in vitro as a specificity factor. *J. Virol.* **71:**6805–6815.

13. **Brokaw, J. L., M. Blanco, and A. A. McBride.** 1996. Amino acids critical for the functions of the bovine papillomavirus type 1 E2 transactivator. *J. Virol.* **70:**23–29.

14. **Burnett, S., U. Kiessling, and U. Petterson.** 1989. Loss of bovine papillomavirus DNA replication control in growth-arrested transformed cells. *J. Virol.* **63:**2215–2225.

15. **Burnett, S., A. C. Strom, N. Jaerbourg, A. Alderborn, J. Dillner, L. Moreno, U. Petterson, and U. Kiessling.** 1990. Induction of bovine papillomavirus E2 gene expression and early region transcription by cell growth arrest: correlation with viral DNA amplification and evidence for differential promoter induction. *J. Virol.* **64:** 5529–5541.

16. **Butz, K., and F. Hoppe-Seyler.** 1993. Transcriptional control of human papillomavirus (HPV) oncogene expression: composition of the HPV type 18 upstream regulatory region. *J. Virol.* **67:**6476–6486.

17. **Chan, W. K., T. Chong, H.-U. Bernard, and G. Klock.** 1990. Transcription of the transforming genes of the oncogenic human papillomavirus type 16 is stimulated by tumor promoters through AP1 binding. *Nucleic Acids Res.* **18:**763–769.

18. **Chang, T., W. Chan, and U. Bernard.** 1989. Transcriptional activation of HPV 16 by the nuclear factor 1, AP-1, steroid receptors and possibly a novel transcription factor. *Nucleic Acids Res.* **18:**465–470.

19. **Cheng, S., S.-D. Schimdt-Grimminger, T. Murant, T. R. Broker, and L. T. Chow.** 1995. Differentiation-dependent up-regulation of the HPV E7 gene reactivates cellular DNA replication in suprabasal differentiated keratinocytes. *Genes Dev.* **9:**2335–2349.

20. **Chiang, C.-M., G. Dong, T. R. Broker, and L. T. Chow.** 1992. Control of human papillomavirus type 11 origin of replication by the E2 family of transcription regulatory factors. *J. Virol.* **66:**5224–5231.

21. **Chiang, C.-M., M. Usatv, A. Stenlund, T. F. Ho, T. R. Broker, and L. T. Chow.** 1992. Viral E1 and E2 proteins support replication of homologous and heterologous papillomaviral origins. *Proc. Natl. Acad. Sci. USA* **89:**5799–5803.

22. **Chin, M. T., T. R. Broker, and L. T. Chow.** 1989. Identification of a novel constitutive enhancer element and an associated binding protein: implications for human papillomavirus type 11 enhancer regulation. *J. Virol.* **63:**2967–2976.

23. **Chow, L. T., M. Nasseri, S. M. Wolinsky, and T. R. Broker.** 1987. Human papillomavirus types 6 and 11 mRNAs from genital condylomata acuminata. *J. Virol.* **61:** 2581–2588.

24. **Clertant, P., and I. Seif.** 1984. A common function for polyoma virus large T-antigen and papillomavirus E1 proteins? *Nature* **311:**276–279.

25. **Cripe, T. P., T. H. Haugen, J. P. Turk, F. Tatabai, P. G. Schmid, M. Durst, L. Gissman, A. Roman, and L. Turek.** 1987. Transcriptional regulation of the HPV 16 E6/E7 promoter by a keratinocyte-dependent enhancer and by E2 transactivator and repressor gene products: implications for cervical cancer. *EMBO J.* **6:**3745–3753.

26. **Crook, T., J. A. Tidy, and K. H. Vousden.** 1991. Degradation of p53 can be targeted by HPV E6 sequences distinct from those required for p53 binding and transactivation. *Cell* **67:**547–556.

27. **Crum, C. P., G. Nuovo, D. Friedman, and S. J. Silverstein.** 1988. Accumulation RNA homologous to human papillomavirus type 16 open reading frames in genital precancers. *J. Virol.* **62:**84–90.

28. **Del Vecchio, A. M., H. Romanczuk, P. M. Howley, and C. C. Baker.** 1992. Transient replication of human papillomavirus DNAs. *J. Virol* **66:**5949–5958.

29. **Desaintes, C., C. Demeret, S. Goyat, M. Yaniv, and F. Thierry.** 1997. Expression of papillomavirus E2 protein in Hela cells leads to apoptosis. *EMBO J.* **16:**504–514.

30. **de Villiers, E.-M.** 1989. Heterogeneity of the human papillomavirus group. *J. Virol.* **63:**4898–4903.

31. **Dietrich, G., I. Kennedy, B. Levins, M. Stanley, and J. Clements.** 1997. A cellular 65 kDa protein recognizes the negative element of human papillomavirus late mRNAs. *Proc. Natl. Acad. Sci. USA* **94:**163–168.

32. **Dollard, S. C., J. L. Wilson, L. M. Demeter, W. Bonnez, R. C. Reichman, T. R. Broker, and L. T. Chow.** 1992. Production of human papillomavirus and modulation of the infectious program in epithelial raft cultures. *Genes Dev.* **6:**1131–1142.

33. **Doorbar, J., S. Ely, J. Sterling, C. McLean, and L. Crawford.** 1991. Specific interaction between HPV-16 E1-E4 and cytokeratins results in collapse of the epithelial cell intermediate filament network. *Nature* **352:**824–827.

34. **Dvorestsky, I., R. Shober, S. Chattopadhyay, and D. Lowy.** 1980. A quantitative in vitro focus assay for bovine papillomavirus. *Virology* **103:**369–375.

35. **Dyson, N., P. M. Howley, K. Munger, and E. Harlow.** 1989. The human papillomavirus 16 E7 oncoprotein is able to bind to the retinoblastoma gene product. *Science* **243:**934–937.

36. **Edwalds-Gilbert, G., and C. Milcarek.** 1995. Regulation of poly(A) site use during mouse B-cell development involves a change in the binding of a general polyadenylation factor in a B-cell stage-specific manner. *Mol. Cell. Biol.* **15:**6420–6429.

37. **Frattini, M., S. D. Hurst, H. Lim, S. Swaminathan, and L. A. Laimins.** 1997. Abrogation of a mitotic checkpoint by E2 proteins from oncogenic human papillomaviruses correlates with increased turnover of the p53 tumor suppressor protein. *EMBO J.* **16:**318–331.

38. **Frattini, M., H. Lim, and L. A. Laimins.** 1996. In vitro synthesis of oncogenic human papillomaviruses requires episomal templates for differentiation-dependent late expression. *Proc. Natl. Acad. Sci. USA* **93:**3062–3067.

39. **Frattini, M. G., and L. A. Laimins.** 1994. Binding of the human papillomavirus E1 origin-recognition protein is regulated through complex formation with the E2 enhancer-binding protein. *Proc. Natl. Acad. Sci. USA* **91:**12398–12402.

40. **Frattini, M. G., and L. A. Laimins.** 1994. The role of the E1 and E2 proteins in the replication of human papillomavirus type 31b. *Virology* **204:**799–804.

41. **Frattini, M. G., H. B. Lim, J. Doorbar, and L. A. Laimins.** 1997. Induction of human papillomavirus type 18 late gene expression and genomic amplification in organotypic cultures from transfected DNA templates. *J. Virol.* **71:**7068–7072.

42. **Furth, P. A., and C. C. Baker.** 1991. An element in the bovine papillomavirus late 3′ untranslated region reduces polyadenylated cytoplasmic RNA levels. *J. Virol.* **65:**5806–5812.

43. **Furth, P. A., W.-T. Choe, J. H. Rex, J. C. Byrne, and C. C. Baker.** 1994. Sequences homologous to 5′ splice sites are required for the inhibitory activity of papillomavirus late 3′ untranslated regions. *Mol. Cell. Biol.* **14:**5278–5289.

44. **Gilbert, D., and S. Cohen.** 1987. Bovine papillomavirus plasmids replicate randomly in mouse fibroblasts throughout S phase of the cell cycle. *Cell* **50:**59–68.

45. **Gius, D., S. Grossman, M. A. Bedell, and L. A. Laimins.** 1988. Inducible and constitu-

tive enhancers in the noncoding region of human papillomavirus type 18. *J. Virol.* **62**:665–672.

46. **Gloss, B., H. U. Bernard, K. Seedorf, and G. Klock.** 1987. The upstream regulatory region of the human papillomavirus-16 contains an E2 protein-independent enhancer which is specific for cervical carcinoma cells and is regulated by glucocorticoid hormones. *EMBO J.* **6**:3735–3743.

47. **Gopalakrishan, V., and S. A. Kahn.** 1994. E1 protein of human papillomavirus type 1a is sufficient for viral DNA replication. *Proc. Natl. Acad. Sci. USA* **91**:9597–9601.

48. **Grassmann, K., B. Rapp, H. Maschek, K. U. Petry, and T. Iftner.** 1996. Identification of a differentiation-inducible promoter in the E7 open reading frame of human papillomavirus type 16 (HPV-16) in raft cultures of a new cell line containing high copy numbers of episomal HPV-16 DNA. *J. Virol.* **70**:2339–2349.

49. **Grossel, M., J. Barsoum, S. Prakash, and E. A. Androphy.** 1996. The BPV-1 E2 DNA-contact helix is required for transcriptional activation but not replication in mammalian cells. *Virology* **217**:301–310.

50. **Grussendorf-Conen, E.-I.** 1987. Papillomavirus-induced tumors of the skin: cutaneous warts and epidermodysplasia verruciformis, p. 158–181. *In* K. Syrjanen, L. Gissmann, and L. Koss (ed.), *Papillomaviruses and Human Disease.* Springer-Verlag, Berlin.

51. **Heck, D. V., C. L. Yee, P. M. Howley, and K. M. Munger.** 1992. Efficiency of binding the retinoblastoma protein correlates with the transforming capacity of the E7 oncoproteins of human papillomaviruses. *Proc. Natl. Acad. Sci. USA* **89**:4442–4446.

52. **Hegde, R. S., S. Grossman, L. A. Laimins, and P. Sigler.** 1992. Crystal structure at 1.7A of the BPV-1 E2 DNA-binding domain bound to its target DNA. *Nature* **359**:505–512.

53. **Higgins, G. D., D. M. Uzelin, G. E. Phillips, P. McEvoy, R. Marin, and C. J. Burrell.** 1992. Transcription patterns of human papillomavirus type 16 in genital intraepithelial neoplasia: evidence for promoter usage within the E7 open reading frame during epithelial differentiation. *J. Gen. Virol.* **73**:2047–2057.

54. **Holt, S. E., and V. G. Wilson.** 1995. Mutational analysis of the 18-base-pair inverted repeat element at the bovine papillomavirus origin of replication: identification of critical sequences for E1 for binding and in vivo replication. *J. Virol.* **69**:6525–6532.

55. **Howley, P. M.** 1996. Papillomavirinae: the viruses and their replication, p. 947–978. *In* B. N. Fields, D. M. Knipe, P. M. Howley, and R. M. Chanock (ed.), *Fundamental Virology.* Raven Press, New York.

56. **Hubbert, N. L., J. T. Schiller, D. R. Lowy, and E. J. Androphy.** 1988. BPV-transformed cells contain multiple E2 proteins. *Proc. Natl. Acad. Sci. USA* **85**:5864–5868.

57. **Hummel, M., J. B. Hudson, and L. A. Laimins.** 1992. Differentiation-induced and constitutive transcription of human papillomavirus type 31b in cell lines containing viral episomes. *J. Virol.* **66**:6070–6080.

58. **Hummel, M., H. B. Lim, and L. A. Laimins.** 1995. Human papillomavirus 31b late gene expression is regulated through protein kinase C-mediated changes in RNA processing. *J. Virol.* **69**:3381–3388.

59. **Hwang, E.-S., L. Naeger, and D. DiMaio.** 1996. Activation of the endogenous p53 growth inhibitory pathway in Hela cervical carcinoma cells by expression of BPV-1 E2. *Oncogene* **12**:795–803.

60. **Hwang, E.-S., D. J. Riese II, J. Settleman, L. A. Nilson, J. Honig, S. Flynn, and D. DiMaio.** 1993. Inhibition of cervical carcinoma cell line proliferation by the introduction of a bovine papillomavirus regulatory gene. *J. Virol.* **67**:3720–3729.

61. **Ishiji, T., M. Lace, S. Parkinnen, R. Anderson, T. Haugen, T. Cripe, J. Xiao, I. Davidson, P. Chambon, and L. Turek.** 1992. Transcriptional enhancer factor (TEF-

1) and its cell-specific co-activator activate HPV-16 E6 and E7 oncogene transcription in keratinocytes and cervical carcinoma cells. *EMBO J.* **11**:2271–2281.

62. **Kennedy, I. M., J. K. Haddow, and J. B. Clements.** 1991. A negative regulatory element in the human papillomavirus type 16 genome acts at the level of late mRNA stability. *J. Virol.* **65**:2093–2097.

63. **Klumpp, D. J. F. Stubenrauch, and L. A. Laimins.** 1997. Differential effects of the splice acceptor at nucleotide 3295 of human papillomavirus type 31 on stable and transient viral replication. *J. Virol.* **71**:8186–8194.

64. **Koss, L. G.** 1987. Cytologic and histologic manifestations of HPV infection of the female genital tract and their clinical significance. *Cancer* **60**:1942–1950.

65. **Kreider, J. W., M. K. Howett, A. E. Leure-Dupree, R. J. Zaino, and J. A. Weber.** 1987. Laboratory production in vivo of infectious human papillomavirus type 11. *J. Virol.* **61**:590–593.

66. **Kuo, S. -R., J. -S. Liu, T. R. Broker, and L. T. Chow.** 1994. Cell-free replication of human papillomavirus DNA with homologous viral E1 and E2 proteins and human cell extracts. *J. Biol. Chem.* **269**:24058–24065.

67. **Kyo, S., D. Klumpp, M. Inoue, T. Kanaya, and L. A. Laimins.** 1997. Expression of Ap-1 during cellular differentiation determines HPV E6/E7 expression in stratified epithelial cells. *J. Gen. Virol.* **78**:401–411.

68. **Kyo, S., A. Tam, and L. A. Laimins.** 1995. Transcriptional activity of HPV 31b enhancer is regulated through synergistic interactions of AP1 with two novel cellular factors. *Virology* **211**:184–197.

69. **Laimins, L. A.** 1993. The biology of human papillomaviruses: from warts to cancer. *Infect. Agents Dis.* **2**:74–86.

70. **Lambert, P., B. Spalholtz, and P. M. Howley.** 1987. A transcriptional repressor encoded by BPV-1 shares a common carboxyl terminal domain with the E2 transactivator. *Cell* **50**:69–78.

71. **Law, M. -F., D. R. Lowy, I. Dvoretsky, and P. M. Howley.** 1981. Mouse cells transformed by bovine papillomavirus contain only extrachromosomal viral DNA sequences. *Proc. Natl. Acad. Sci. USA* **78**:2727–2731.

72. **Le Moal, M. A., M. Yaniv, and F. Thierry.** 1994. The bovine-papillomavirus type 1 (BPV1) replication protein E1 modulates transcriptional activation by interacting with BPV1 E2. *J. Virol.* **68**:1085–1093.

73. **Li, R., and M. Botchan.** 1993. The acidic transcriptional activation domains of VP16 and p53 bind the replication protein A and stimulate in vitro BPV-1 replication. *Cell* **73**:1207–1221.

74. **Li, R., J. Knight, G. Bream, A. Stenlund, and M. Botchan.** 1989. Specific recognition nucleotides and their DNA context determine the affinity of E2 protein for 17 binding sites in the BPV-1 genome. *Genes Dev.* **3**:510–526.

75. **Lowy, D. R., L. Dvoretsky, and R. Shober.** 1980. In vitro tumorigenic transformation by a defined sub-genomic fragment of bovine papilloma virus DNA. *Nature* **287**:72–74.

76. **Lu, J. Z.-J., Y.-N. Sun, R. C. Rose, W. Bonnez, and D. J. McCance.** 1993. Two E2 binding sites (E2BS) alone or one E2BS plus an A/T-rich region are minimal requirements for the replication of the human papillomavirus type II origin. *J. Virol.* **67**:7131–7139.

77. **Lusky, M., and M. Botchan.** 1986. Transient replication of bovine papillomavirus type 1: cis and trans requirements. *Proc. Natl. Acad. Sci. USA* **83**:3609–3613.

78. **Lusky, M., and E. Fontaine.** 1991. Formation of the complex of BPV-1 E1 and E2 proteins is modulated by E2 phosphorylation and depends upon sequences. *Proc. Natl. Acad. Sci. USA* **88**:6363–6367.

79. **Lusky, M., J. Hurwitz, and Y.-S. Seo.** 1994. The bovine papillomavirus E2 protein modulates the assembly but is not stably maintained in a replication-competent multimetric E1-replication origin complex. *Proc. Natl. Acad. Sci. USA* **91:**8895–8899.

80. **Mack, D., and L. A. Laimins.** 1991. Keratinocyte-specific transcription factor, KRF-1, interacts with AP-1 to activate HPV 18 expression in squamous epithelial cells. *Proc. Natl. Acad. Sci. USA* **88:**9102–9106.

81. **Mastrangelo, I., P. Hough, J. Wall, M. Dodson, F. Dean, and J. Hurwitz.** 1989. ATP-dependent assembly of double hexamers of SV40 T antigen at the viral origin of replication. *Nature* **338:**658–662.

82. **McBride, A. A., J. B. Bolen, and P. M. Howley.** 1989. Phosphorylation sites of the E2 transcriptional regulatory proteins of bovine papillomavirus type 1. *J. Virol.* **63:** 5076–5085.

83. **McBride, A. A., J. C. Byrne, and P. M. Howley.** 1989. E2 polypeptides encoded by BPV-1 form dimers through the common carboxyl-terminal domain: transactivation is mediated by the conserved amino-terminal domain. *Proc. Natl. Acad. Sci. USA* **86:** 510–514.

84. **McBride, A. A., R. Schlegel, and P. M. Howley.** 1988. The carboxyl-terminal domain shared by the BPV-1 E2 transactivator and repressor proteins contains a specific DNA binding activity. *EMBO J.* **7:**533–539.

85. **McCance, D., R. Kopan, E. Fuchs, and L. A. Laimins.** 1988. Human papillomavirus type 16 alters human epithelial cell differentiation in vitro. *Proc. Natl. Acad. Sci. USA* **85:**7169–7173.

86. **Meyers, C., M. G. Frattini, J. B. Hudson, and L. A. Laimins.** 1992. Biosynthesis of human papillomavirus from a continuous cell line upon epithelial differentiation. *Science* **257:**971–973.

87. **Mohr, I. J., R. Clark, S. Sun, E. Androphy, P. MacPherson, and M. Bothchan.** 1990. Targeting the E1 replication factor to the papillomavirus origin of replication by complex formation with the E2 transactivator. *Science* **250:**1694–1699.

88. **Mul, Y., and P. van der Vliet.** 1992. Nuclear factor I enhances adenovirus DNA replication by increasing the stability of a preinitiation complex. *EMBO J.* **11:**751–760.

89. **Nasseri, M., R. Hirochika, T. R. Broker, and L. T. Chow.** 1987. A human papillomavirus type II transcript encoding an E1-E4 protein. *Virology* **159:**433–439.

90. **O'Conner, M., and H. U. Bernard.** 1995. Oct-1 activates the epithelial-specific enhancer of HPV 16 via synergistic interaction with NF1 at a conserved composite regulatory element. *Virology* **207:**77–88.

91. **Ozbun, M. A., and C. Meyers.** 1997. Characterization of late transcripts expressed during vegetative replication of human papillomavirus type 31b. *J. Virol.* **71:** 5161–5172.

92. **Park, P., W. Copeland, L. Yang, T. Wang, M. Botchan, and I. Mohr.** 1994. The cellular DNA polymerase alpha-primase is required for papillomavirus replication and associates with the viral E1 helicase. *Proc. Natl. Acad. Sci. USA* **91:**8700–8704.

93. **Pattison, S., D. G. Skalnik, and A. Roman.** 1997. CCAAT displacement protein, a regulator of differentiation-specific gene expression, binds a negative regulatory element within the 5′ end of the human papillomavirus type 6 long control region. *J. Virol.* **71:**2013–2022.

94. **Pierrefite, V., and F. Cuzin.** 1995. Replication efficiency of bovine papillomavirus type 1 DNA depends on *cis*-acting sequences distinct from the replication origin. *J. Virol.* **69:**7682–7687.

95. **Piirsoo, M., E. Ustav, T. Mandel, A. Stenlund, and M. Ustay.** 1996. Cis and trans requirements for stable episomal maintenance of the BPV-1 replicator. *EMBO J* **15:** 1–11.

96. **Pipas, J. M.** 1992. Common and unique features of T antigens encoded by the poly-omavirus group. *J. Virol.* **66:**3979–3985.

97. **Roberts, S., I. Ashmole, G. Johnson, J. Kreider, and P. Gallimore.** 1993. Cutaneous and mucosal HPV E4 proteins form intermediate filament-like structures in epithelial cells. *Virology* **197:**176–187.

98. **Sakai, H., T. Yasugi, J. D. Benson, J. J. Dowhanick, and P. M. Howley.** 1996. Targeted mutagenesis of the human papillomavirus type 16 E2 transactivation domain reveals separable transcriptional activation and DNA replication functions. *J. Virol.* **70:** 1602–1611.

99. **Sarver, N., P. Gruss, M. F. Law, G. Khoury and P. M. Howley.** 1981. Bovine papillo-mavirus deoxyribonucleic acid: a novel eukaryotic cloning vector. *Mol. Cell. Biol.* **1:** 486–496.

100. **Scheffner, M., B. A. Werness, J. M. Huibregste, A. J. Levine, and P. M. Howley.** 1990. The E6 oncoprotein encoded by HPV 16 and 18 promotes the degradation of p53. *Cell* **63:**1129–1136.

101. **Schwarz, E., U. Freese, L. Gissman, W. Mayer, A. Roggenbauch, A. Stremlau, and H. zur Hausen.** 1985. Structure and transcription of HPV sequences in cervical carci-noma cells. *Nature* **314:**111–119.

102. **Sedman, J., and A. Stenlund.** 1995. Co-operative interaction between the E1 initiator and the transcriptional activator E2 is required for replicator specific DNA replication in vivo and in vitro. *EMBO J.* **14:**6218–6228.

103. **Seo, Y. -S., F. Muller, M. Lusky, E. Gibbs, H. -Y. Kim, B. Phillips, and J. Hurwitz.** 1993. BPV-1 encoded E2 proteins enhance binding of E1 to the BPV replication origin. *Proc. Natl. Acad. Sci. USA* **90:**2865–2869.

104. **Sherman, L., N. Alloul, I. Golan, M. Durst, and A. Baram.** 1992. Expression and splice patterns of HPV 16 mRNAs in pre-cancerous lesions and carcinomas of the cervix in keratinocytes immortalized by HPV 16 and in lines established from cervical cancer. *Int. J. Cancer* **50:**356–364.

105. **Singer, A., L. Ho, G. Terry, and T. S. Kwie.** 1995. Association of human papillomavi-rus with cervical cancer and precancer, p. 105–129. *In* A. Mindel (ed.), *Genital Warts: Human Papillomavirus Infection.* Edward Arnold, London.

106. **Spalholz, B., Y. C. Yang, and P. M. Howley.** 1985. Transactivation of a bovine papillo-mavirus transcriptional regulatory element by the E2 gene product. *Cell* **42:**183–191.

107. **Steinberg, B. M., K. J. Auborn, J. L. Brandsma, and L. B. Taichman.** 1989. Tissue site-specific enhancer function of the upstream regulatory region of human papillo-mavirus type 11 in cultured keratinocytes. *J. Virol.* **63:**957–960.

108. **Sterling, J., M. Stanley, G. Gatward, and T. Minson.** 1990. Production of human papillomavirus type 16 virions in a keratinocyte cell line. *J. Virol.* **64:**6305–6307.

109. **Stoler, M. H., S. M. Wolinsky, A. Whitbeck, T. R. Broker, and L. T. Chow.** 1989. Differentiation-linked human papillomavirus types 6 and 11 transcription in genital condylomata revealed by in situ hybridization with message-specific RNA probes. *Virology* **172:**331–340.

110. **Stubenrauch, F., and L. A. Laimins.** Unpublished data.

110a. **Stubenrauch, F., H. B. Lim, and L. A. Laimins.** 1998. Differential requirements for conserved E2 binding sites in the life cycle of oncogenic human papillomavirus type 31. *J. Virol.* **72:**1071–1077.

111. **Stubenrauch, F., J. Malejczyk, P. G. Fuchs, and H. Pfister.** 1992. Late promoter of human papillomavirus type 8 and its regulation. *J. Virol.* **66:**3485–3493.

112. **Sverdrup, F., and S. A. Khan.** 1995. Two E2 binding sites alone are sufficient to function as the minimal origin of replication of human papillomavirus type 18 DNA. *J. Virol.* **69:**1319–1323.

113. **Tan, S.-H., E. C. Leong, P. A. Walker, and H.-U. Bernard.** 1994. The human papillomavirus type 16 E2 transcription factor binds with low cooperativity to two flanking sites and represses the E6 promoter through displacement of Sp1 and TFIID. *J. Virol.* **68:**6411–6420.

114. **Tan, W., and S. Schwartz.** 1995. The Rev protein of human immunodeficiency virus type 1 counteracts the effect of an AU-rich negative element in the human papillomavirus type 1 late 3' untranslated region. *J. Virol.* **69:**2932–2945.

115. **Thierry, F., G. Spyrou, M. Yaniv, and P. Howley.** 1992. Two AP-1 sites binding JunB are essential for human papillomavirus type 18 transcription in keratinocytes. *J. Virol.* **66:**3740–3748.

116. **Thierry, F., and M. Yaniv.** 1987. The BPV-1 E2 trans-acting protein can be either an activator or a repressor of the HPV18 regulatory region. *EMBO J.* **6:**3391–3397.

117. **Thorner, L. K., D. A. Lim, and M. R. Botchan.** 1993. DNA-binding domain of bovine papillomavirus type 1 E1 helicase: structural and functional aspects. *J. Virol.* **67:** 6000–6014.

118. **Ustav, M., and A. Stenlund.** 1991. Transient replication of BPV-1 requires two polypeptides encoded by the E1 and E2 open reading frames. *EMBO J.* **10:**449–457.

119. **Ustav, M., E. Ustav, P. Szymanski, and A. Stenlund.** 1991. Identification of the origin of replication of BPV-1 and characterization of the viral origin recognition factor E1. *EMBO J.* **10:**4321–4329.

120. **Vande Pol, S. B., and P. M. Howley.** 1990. A bovine papillomavirus constitutive enhancer is negatively regulated by the E2 repressor through competitive binding for a cellular factor. *J. Virol.* **64:**5420–5429.

121. **Wang, H., K. Liu, F. Yuan, L. Berdichevsky, L. B. Taichman and K. Auborn.** 1996. C/EBPβ is a negative regulator of human papillomavirus type 11 in keratinocytes. *J. Virol.* **70:**4839–4844.

122. **Wiley, S. R. R. Kraus, F., Zuo, E. Murray, K. Loritz, and J. E. Mertz.** 1993. SV40 early to late switch involves titration of cellular transcriptional repressors. *Genes Dev.* **7:**2206–2219.

123. **Yang, L., R. Li, I. Mohr, R. Clark, and M. Botchan.** 1991. Activation of BPV-1 replication in vitro by the transcription factor E2. *Nature* **353:**628–632.

124. **Yang, L., I. Mohr, R. Li, S. Nottoli, S. Sun, and M. Botchan.** 1993. The E1 protein of BPV-1 is an ATP-dependent DNA helicase. *Proc. Natl. Acad. Sci. USA* **90:**5086–5090.

Human Tumor Viruses
Edited by Dennis J. McCance
© 1998 American Society for Microbiology

7 | Activities of the Transforming Proteins of Human Papillomaviruses
M. A. Nead and D. J. McCance

Human papillomaviruses (HPVs) infect both mucosal and cutaneous surfaces, resulting in benign or malignant disease depending on the type of infecting virus. In this chapter we will deal exclusively with viruses which infect the genital tract, although viruses that infect in this region can also infect the oral cavity (see chapter 1). HPVs are the etiological agents of various lower genital tract cancers, which are important globally. A clear causal association now exists between infection with the oncogenic HPV types and the development of cervical dysplasia, which may progress to malignant disease. Table 1 in chapter 1 lists the HPVs infecting the genital tract and associated lesions. The most common benign viruses are HPV types 6 and 11 (HPV-6 and -11), and the viruses mostly commonly associated with malignant disease are HPV types 16 and 18 (HPV-16 and -18). The premalignant and malignant lesions of the cervix arise from the transformation zone, which is an area of immature metaplasia between the mature epithelium of the exocervix and the columnar epithelium of the endocervical canal. The biology and natural history of the transformation zone are controlled by female hormones, and the zone appears at puberty and disappears or is greatly reduced at menopause. Little is known of the biology of the transformation zone, and it cannot be produced in tissue culture in the way that mature epithelium can be constructed using the raft system (see chapter 6 and reference 79). While the virus infects all areas of the male and female genital tract, it appears that infection of the transformation zone has more serious consequences, since cervical cancer is more common than other lower genital tract cancers which arise from mature epithelium such as the vulva, vaginal wall, and penis. The reasons for this are unclear, but it may be that the immature cells making up the transformation zone are more sensitive to external insults such as a virus infection than are cells of the mature epithelium. What is clear is that there are a large number of men and women between the ages of 18 and 30 years of age who are infected, but who develop transitory disease which will spontaneously regress, possibly due to an activated immune response.

The virus is thought to infect basal epithelial cells through micro-lesions and replicates at a low level in these cells. As the cells move up through the stratified epithelium and differentiate, amplified viral DNA replication is initiated and the full viral cycle is completed in the area where terminal differentiation normally takes place. However, the virus needs all of the replicative machinery of the cell, which would presumably be absent or in short supply in terminally differentiating keratinocytes. Therefore, the virus needs to stimulate cells into the synthesis phase (S phase) of the cell cycle for successful completion of viral replication resulting in a productive infection. To achieve this S-phase stimulation, the virus codes for at least three proteins, E5, E6, and E7, which, as will be discussed below, have properties consistent with a role in stimulating passage through G1 into S phase. It may be that the result of this stimulation varies between HPV types. For example, S-phase induction in tissues containing benign viruses such as HPV-6 or -11 results in a hyperproliferative lesion, with cells retaining the capacity to differentiate, while malignant HPV types such as HPV-16 or -18 produce S-phase cells but eventually inhibit differentiation, resulting in a dysplastic lesion and reduction in viral DNA replication and infectious virus. In both cases the virus has induced S phase and so can propagate using the cell's replicative machinery. This chapter will describe the properties of HPV-16 E5, E6, and E7 and will suggest ways in which they might collectively induce S phase in epithelial cells that are being programmed for terminal differentiation. Reference will be made to other HPVs where appropriate, and some comparisons with bovine papillomaviruses (BPVs) will be made, especially when discussing the properties of E5. The first section will deal with the E6 protein, its interaction with cellular proteins, and the possible consequences of these interactions with regard to HPV life cycle.

STRUCTURE AND FUNCTION OF E6 PROTEINS

E6, a multifunctional HPV protein, has been shown to activate transcription from some promoters and inhibit from others (42, 70, 112) and binds to at least two cellular proteins, E6-associated protein (E6AP) (60, 61) and E6 binding protein (E6BP) (21). (E6BP is identical to the cellular ERC-55 protein and is potentially a Ca^{2+} binding protein.) The E6 protein from the oncogenic viruses can immortalize primary human mammary epithelial cells (7) and can act with E7 to immortalize primary human keratinocytes (53). In vivo research on E6 has been difficult because the protein is produced in small amounts in cells and is difficult to detect. Purification for structural studies such as crystallization has been hampered because of the insolubility of the protein. Therefore, apart from the immortalization and transformation studies, the work describing the interaction of E6 with cellular proteins has been carried out in vitro using glutathione-*S*-transferase (GST) fusion proteins and in vitro translated proteins.

Structure

HPV-16 E6 is 151 amino acids (aa) in length and contains two zinc fingers which bind Zn^{2+} in vitro and are apparently essential for most of the properties exhibited by the protein. There is an upstream tandem ATG site which, if utilized, would produce a protein of 158 aa, although it is not clear if both forms are produced in vivo. The presence of two zinc fingers in E6 and one in E7 suggests that the former may be a duplication of the latter. The protein is approximately 18 kDa in size and can be found in both the nuclear and cytoplasmic compartments of cells (1). The actual sequence is not well conserved between types, although all contain the two zinc fingers. As described in chapter 6, there are different spliced mRNAs encoding truncated proteins (31, 110, 114), but at present no convincing detection of these proteins has been observed in transfected cells or infected lesions. While the functions of E6 are essential for certain HPV properties, it is clear that not all papillomaviruses require this protein. For instance, three bovine papillomaviruses (BPVs), types 3, 4, and 6, which are epitheliotropic viruses, do not appear to code for E6 (93). (It should be noted that BPV types 1, 2, and 5 infect fibroblasts as well as epithelial cells in the natural host, producing fibropapillomas. This is the only example of papillomaviruses infecting fibroblasts during natural infection.) Further investigations on the physical properties of E6 through crystallography will be very difficult as the protein is insoluble as a purified protein, although one possibility might be to copurify E6 with one of the cellular interactive proteins, which may increase solubility.

Interaction of E6 with Cellular Proteins

HPV E6 has been shown to interact directly with two cellular proteins, E6AP (61) and E6BP (21). In addition, E6 can form a tertiary complex with E6AP and p53 and even a ternary one composed of E6, E6AP, E6BP, and p53 (21). While the exact consequences of the interactions remain unclear, the nature of the known properties of some of the cellular proteins, especially p53, indicates that E6 acts in the G1 phase of the cell cycle and may inhibit normal differentiation and encourage cells to move into S phase.

E6, p53, and E6AP interactions

When it was discovered that simian virus 40 large T protein bound to both the retinoblastoma protein (Rb) and p53, other viral oncoproteins were investigated for similar interactions. HPV E6 was found to bind to p53 (126) and E7 to Rb (40) (see next section). The interaction of E6 from the high-risk oncogenic types like HPV-16 and -18 causes the degradation of p53 through the ubiquitin proteolysis pathway (106), which is a major degradative mechanism for protein breakdown and amino acid recycling. The efficient binding and subsequent degradation of p53 is dependent on the interaction of E6 with a cellular protein, initially called the E6-associated protein (E6AP), which turns out to be part of the ubiquitin proteolytic cascade and is a ubiquitin-protein ligase (61). The sequence of events appears to be as follows (105). E6 binds to the E6AP protein, which then recruits p53. Since the E6AP has a ubiquitin-ligase activity, it in turn recruits the E2 protein (part of the ubiquitin complex), which ubiquitinates lysines on p53 and leads to the degradation of the protein through the rest of the pathway. While the low-risk HPV types such as HPV-6 and -11 do not cause degradation of p53, there is some dispute as to whether they bind to p53 (32, 44). One group observed that HPV-6 E6 bound to p53 at 40% of the level of binding of HPV-16 E6 (32), while another group failed to find any detectable binding to HPV-6 E6 (44). HPV-11 E6 was found to bind weakly to p53 (71). While the controversy has not been resolved, there is consensus that in the case of HPV-6 E6, there is little or no degradation of p53 even when binding is observed.

The domains of HPV-16 E6 important for p53 interaction and degradation have been mapped by at least two groups (32, 44), but again there are inconsisten-

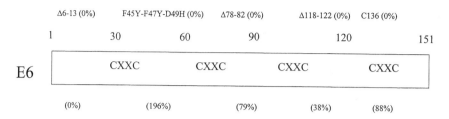

Figure 1 Diagram of HPV-16 E6 and some of the mutations used in two studies to determine the domains of E6 important for binding to p53. The mutations are shown on the top line, and the percentage binding is given in parentheses for each study as indicated. The data as indicated are from references 43 (above the box) and 32 (below the box). A more complete description of the E6 mutations and p53 binding can be found in reference 82.

cies in the data. All the results are from in vitro experiments using in vitro rabbit reticulocyte lysate translated proteins, which contain the essential E6AP protein, and the translated proteins are pulled down by GST fusion proteins. Initial results suggested that the N-terminal domain of HPV-16 E6 was responsible for the binding to p53 (32), but others have found that mutations throughout the length of the protein have profound effects on the ability of E6 to bind to E6AP (44). Figure 1 shows some of the mutations of E6 used and the percentage of binding for each of the mutations. For instance, F45Y-F47Y-D49H eliminated binding in one study (44), but retained complete binding in another study (32), while mutations R8S-P9A-R10T produced a protein which failed to bind p53 in either study. While there are discrepancies in binding studies, all mutations in both studies abrogate the degradation of p53. The consequence of the ubiquitination of p53 is not clearly understood, but it is consistent with the idea that reducing the effect of p53 in G1 may allow cells to progress through to S phase and so be in a position to support viral DNA replication. The effects of E6 on G1 will also be complemented by the activities of E5 and E7 (see later sections).

Since p53 can act as a transcriptional activator or repressor, the effects of E6 on these functions were studied. Investigations using wild-type E6 and E6 mutations and reporter plasmids containing either p53 consensus binding sites or promoters repressed by p53 showed that wild-type E6 could derepress promoters down-regulated by p53, or inhibit activation of promoters up-regulated by p53. The mutations indicated that the activities were dependent on binding to p53 (72, 98).

Interaction of E6 and E6BP
Using the yeast two-hybrid system, Chen et al. (21) showed that HPV-16 E6 bound to a cellular protein of 210 aa and molecular size of approximately 55 kDa. The protein was found to contain four potential calcium binding motifs, EF-hand domains, and an endoplasmic reticulum retention signal of HDEL (one-letter amino acid code). The EF-hand structural motif, first described for parvalbumin, consists of two perpendicularly placed α-helices (E and F) with an interhelical loop (for a

review of EF-hand proteins, see reference 64). EF-hand-containing proteins consist of single or multiple pairs of helix-loop-helix motifs, which change conformation on the binding of Ca^{2+}. Calcium regulates a wide variety of cellular processes including cell cycle control and differentiation. For instance, differentiation of both mouse and human keratinocytes can be induced by increasing the levels of Ca^{2+} in the medium. The sequence of E6BP is identical to that of ERC-55, which has recently been cloned, but its function remains unknown. The fact that E6BP binds Ca^{2+}, which is known to stimulate cellular differentiation, has led to speculation that E6 may disrupt its activity and contribute towards the inhibition of differentiation observed in HPV-16-infected tissues.

Immortalization and Transformation of Cells by E6

Unlike HPV-16 E7, E6 on its own does not induce immortalization of primary human foreskin keratinocytes. However, immortalization by E7 is very inefficient, and E6 will cooperate to increase the level of immortalization (11). HPV-16 E6 will on its own immortalize human mammary epithelial cells. The significance of this observation is unclear, since there is no convincing evidence that HPVs are involved in the development of mammary cancer, and it indicates that (i) these cells may be sensitive to the effects of E6 on p53 and (ii) it is important to use the correct target cell when investigating the biological consequences of HPV oncoproteins.

Cervical Cancer and Mutations in p53 and Rb

Early observations (34, 35) suggested that cancer cells from patients with cervical malignancies with evidence of HPV infection contained no mutations in either p53 or the retinoblastoma protein Rb, while cancer cells without evidence of infection had mutations in either or both cellular proteins. The theory was that in HPV-infected cancers the cells did not acquire mutations in these cellular proteins because the activities of the two viral proteins, E6 and E7, abrogated the activity of both p53 and Rb. Later studies indicated that only the primary tumors had wild-type p53 and Rb, but that metastases contained mutations in p53, suggesting that this may have contributed to secondary disease (33). However, further studies have not supported the initial findings, in that primary cancers exhibiting HPV infection did contain mutations in p53 (19, 26). Therefore, the idea that in infected cells E6 and E7 abrogate the activities of p53 and Rb is not so simple as first suggested. There appear to be a number of different pathways that lead to malignant conversion, but perhaps all or most are activated by HPV infection.

STRUCTURE AND FUNCTION OF E7

The HPV E7 protein plays a critical role in altering the cellular environment for the benefit of viral replication. It has a number of described functions which are consistent with its having an important role during G1 phase of the cell cycle and in stimulating cells through the G1/S boundary. Such properties as the ability to bind and abrogate the functions of Rb, binding and inhibiting the function of the cyclin-dependent kinase inhibitor p21, and interaction with the AP-1 family of transcription factors are consistent with a major role in G1. E7 appears to be the major oncoprotein of HPVs. It can transform rodent fibroblasts and is essential for the immortalization of primary human keratinocytes. However, immortalization of keratinocytes is more efficient in the presence of E6.

Structure

HPV-16 E7 consists of 98 aa with a predicted molecular size of 11 kDa, although the apparent molecular sizes of E7 from HPV-6 and HPV-16 are 14 kDa and 17 kDa, respectively. This discrepancy is due to variation in the number of charged residues present in the N-terminal 10 aa (4, 5) in HPV-16 compared to HPV-6. The E7 protein shares homology with adenovirus E1a and simian virus 40 T protein (83), leading to a common nomenclature of the protein's domains (see chapter 2). The first 20 aa of E7 comprise conserved region 1 (CR1), while CR2 consists of aa 20 to 39 (Fig. 2). CR3 (40–98 aa) is less well defined, but contains two CXXC motifs capable of binding zinc (102) and is important for dimerization of E7 (Fig. 2). E7 has been shown to dimerize through the metal binding domain in CR3 (28, 80, 131). Dimerization and proper folding of the E7 protein depend on the presence of zinc and intact CXXC motifs (28, 80, 131). These motifs are also important for transformation (25, 41, 65, 80, 95, 118), immortalization (65), c-Jun binding (3), and E2F release (58, 91, 127), but not for pRb binding (65). Distinguishing between general protein folding versus dimerization requirements for interactions and activities that require intact CXXC motifs poses a particular challenge.

This section will focus on the binding partners of E7 and discuss the regions involved and potential implications of the interaction. For a more complete understanding of E7, additional reviews are also recommended (54, 84).

Interaction of E7 with Cellular Proteins

E7 has been shown to bind to a number of cellular proteins, and even though the protein is only 98 aa, all of the interactions may be important for the life cycle of

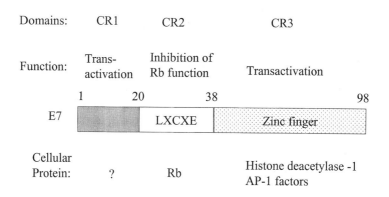

Figure 2 Diagrammatic representation of the HPV-16 E7 protein and the three domains CR1, CR2, and CR3. The known functions of each domain are indicated, and the cellular proteins which bind to each domain are noted below the diagram.

the virus. It is probable that E7 binds to different cellular proteins at different and important stages in the cell cycle, especially from G0 through to S phase, and therefore can alter a number of different cell pathways.

RB binding

The best-documented E7 binding interaction involves the retinoblastoma tumor suppressor protein (Rb). Rb regulates the transition from G0 through G1 into S phase, becoming increasingly phosphorylated as cells approach S phase. The hyperphosphorylated form of Rb is thought to be inactive, as many of its repressive activities are abrogated. For instance, upon phosphorylation, Rb releases bound transcription factors such as E2F, which then activate promoter elements in genes required for DNA synthesis. The E7 protein from papillomaviruses is able to bind Rb in a hypophosphorylated state, causing the premature release of E2F and thereby promoting cell cycling.

The primary region of E7 involved in binding to pRb falls within conserved region 2 (CR2). For adenoviral E1a, pRb binding occurs in CR1 and CR2, while for HPV E7, CR1 is not involved. The domain within CR2 that associates with pRb has the amino acid sequence motif LXCXE, where the Xs represent a variety of amino acids, the identity of which depends on the specific viral type. This motif is found in a number of Rb-associated proteins, including cyclins D1, D2, and D3 and several cellular transcription factors (see references 122 and 125 for reviews), suggesting a competitive role for E7 in infected cells. However, the major difference in the strength of binding of HPV-6 E7, which binds weakly, and HPV-16 E7, which binds more strongly, is an aspartic acid at position 21 in the latter, while there is a glycine at the equivalent position (aa 22) of HPV-6 E7.

Recent studies have shown that the carboxy-terminal region of E7 (CR3) also plays a role in efficient binding to Rb, as well as promoting the release of E2F (91). To effectively displace E2F from Rb and allow activation of transcription, E7 would have to either bind Rb in a similar domain as E2F or else produce a conformational change in Rb such that E2F could no longer bind. CR2 of E7 has been shown to bind to the pocket region of Rb, which consists of two separate regions, aa 394–571 and 649–772 (57, 59, 67). As mentioned above, CR3 has also been shown to bind to Rb, and the region for this interaction is between 803 and 841 aa of Rb (91). Peptides consisting solely of the amino terminus of E7 fail to displace E2F (58, 127), while peptides containing at least a portion of CR3 are able to displace E2F from Rb and allow transcriptional activation (58, 91). The requirement for a region of E7 that interacts with the carboxy terminal of Rb to displace E2F is consistent with the carboxy terminus of Rb playing a role in E2F binding (100).

Another protein, however, histone deacetylase 1 (HDAC-1), has recently been shown to bind to both Rb and the CR3 region of HPV-16, -31, and -33 E7, and this interaction may explain the necessity of the CR3 domain (14). The histone deacetylase proteins are thought to be important for the repression of transcription from various promoters and to act by altering chromatin structure. Therefore the repressive activity of Rb on certain transcription factors may be mediated by its ability to bind HDAC-1. It may be that E7 competes with Rb for HDAC-1, resulting in the derepression of the inhibitory activity of Rb on E2F-containing promoters

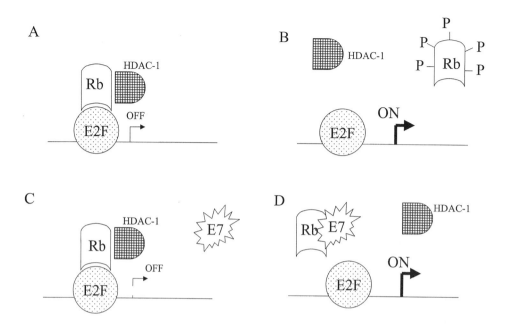

Figure 3   The retinoblastoma protein (Rb) represses the transactivation of E2F (panels A and C), and this repression is removed during normal cell cycling at the latter part of the G1 phase of the cell cycle when Rb is phosphorylated (B). Rb binds to the histone deacetylase 1 protein (HDAC-1), which may help in the repression of transcription from an E2F promoter. E7 is known to depress Rb, and it is thought to disrupt the Rb/E2F complex (D). In addition, E7 can bind HDAC-1 and so may derepress by competing away from Rb the inhibitory activity of HDAC-1 (D).

(Fig. 3). Therefore, E7 may need two domains to activate transcription factors which are repressed by Rb: one domain to bind Rb and the other to bind and remove the HDAC-1 activity. While the full consequences of the interaction of both Rb and E7 with HDAC-1 remain to be determined, the targeting of the deacetylase enzymes may be a common pathway used by viral proteins to increase transcription of cellular and viral genes without directly binding DNA.

Initial observations suggested that the Rb binding ability of the E7 protein determined the potential of the viral strain to cause cancer or, more measurably, to immortalize human keratinocytes and transform rodent fibroblast cells in tissue culture. The theory was that E7 proteins that bound Rb with greater affinity would also displace E2F more efficiently and lead to cell cycle disruption. Consistent with this is the fact that E7 proteins from the high-risk viruses were shown to bind Rb with greater affinity than those from the low-risk viruses (9, 47, 85). The Rb binding ability correlated not only with oncogenic potential, but also with cell transformation ability (9, 11, 121). Domain swapping and mutational studies showed the amino-terminal region of E7 to be responsible for the differing transforming ability between high- and low-risk viruses (86, 90, 121). Switching the

CR2 domain of HPV-16 E7 with that of HPV-6 E7 leads to an almost complete inhibition of transformation activity (55). Point mutations in HPV-6 E7 that create an HPV-16 E7-like CR2 domain increased the viral transforming ability (55, 104). In contrast, mutations in CR1 of HPV-16 E7 that retain Rb binding show a decreased transforming activity (8, 15, 95), while mutations that disrupt the ability to bind Rb still permit keratinocyte immortalization (65). Taken together, these data suggest the requirement of increased pRb binding for transformation but not for immortalization. However, it should be remembered that HPVs infect epithelial cells only, so the important property is the ability to immortalize keratinocytes rather than transform rodent fibroblasts.

Further evidence suggesting multiple roles for E7 comes from the cottontail rabbit papillomavirus (CRPV) system, which serves as a useful model of papilloma pathogenesis. Mutations in CRPV are easily screened for the ability to produce warts upon injection into rabbits. The E7 protein from CRPV binds to pRb (37, 109), though with variable reported binding efficiency. Disruption of Rb binding through CR2 mutations resulted in diminished, though it did not abolish, wart production (37), while an in-frame insertion mutation in CR1 at aa 9 completely abrogated wart formation while retaining pRb binding ability (13). Together, the mutations and insertions suggest that Rb binding is not necessary for benign disease, but it is not clear if Rb is required for the progression of these benign warts to squamous cell carcinomas. In parallel with the HPV data, this suggests that E7 binding to pRb is not sufficient to explain the virus's molecular pathogenesis.

In addition, recent evidence suggests that the function of E7 is not entirely coded for by Rb binding and CR2-dependent functions. First, naturally occurring mutations in E7 from the low-risk virus HPV-6, isolated from malignant tumors, show changes in aa 38, 52, or 88, but not in CR2 (50). Second, the low-risk cutaneous virus HPV1 encodes an E7 that binds strongly to Rb, yet lacks transforming activity in primary rodent cells (27) and does not transform rodent cell lines (109). Finally, the high-risk cutaneous viruses HPV-8 and HPV-47 are associated with carcinomas in epidermodysplasia verruciformis patients, yet bind Rb poorly and have negligible in vitro transforming ability (56, 63, 109).

p107 and p130
E7 has also been shown to interact with the two other members of the pRb family, namely, p107 and p130 (40). The pRb family members share considerable homology in the A and B regions of the pocket domain, although they differ significantly in their N-terminal sequences. The CR2 domain of E7 appears to mediate binding to p107 and p130; peptides containing aa 2–32 bound to both family members, while a peptide spanning aa 2–20 bound neither member (36, 40).

AP-1 binding
E7 with a mutation that abolishes Rb binding can still immortalize human keratinocytes in the context of full-length virus (65). This suggested that other interactions with cellular proteins were important. Cells were screened with a GST-16 E7 fusion protein to determine which protein interactions might be responsible for the immortalization ability of HPV-16 E7. A number of protein bands were identified as binding to GST-16 E7 and not to GST (3), and a 39-kDa band was

identified as c-Jun. A member of the AP-1 family of transcription factors, c-Jun is able to homodimerize or heterodimerize with other AP-1 family members, namely JunB, JunD, Fra-1, Fra-2, c-Fos, FosB, and FosB2. The variety of dimers that are formed can bind with different efficiencies to the AP-1 consensus site (TGAG/CTCA) or can recognize slightly different DNA sequences, producing different levels of transcriptional activation. AP-1 factors appear to be involved both in proliferative responses and in differentiation of keratinocytes and myeloid cells. There are a number of genes whose products are involved in DNA synthesis or the regulation of that process, such as proliferating cell nuclear factor (PCNA), topoisomerase I, thymidylate transferase, and thymidine kinase, that are regulated by AP-1 factors. E7 binds to c-Jun and enhances c-Jun-induced transcription three- to fourfold over levels reached by c-Jun alone (3). Mutations that abolish pRb binding produce the same level of activation as wild-type E7. Interestingly, the ability of E7 mutants to enhance the transactivation level of c-Jun paralleled their immortalization ability (3, 65). Initial data obtained using GST fusion proteins suggested that CR3 is the region of E7 that interacts with c-Jun, since disruption of the zinc finger abrogated binding. E7 from HPV-16 binds well to c-Jun, while the E7 protein from HPV-6 does not (unpublished data).

HPV-16 E7 was shown, with the use of c-Jun deletion mutants, to bind to c-Jun between aa 224 and 286 (3). This region was further narrowed, using the yeast two-hybrid system, to aa 224–249, just to the amino-terminal side of the DNA binding dimerization domain of c-Jun, and includes several phosphorylation sites which alter DNA binding. HPV-16 E7 is able to bind to an in vitro-produced c-Jun that has all of these phosphorylation sites mutated (unpublished data), suggesting that phosphorylation at these sites is not necessary for E7 to bind. E7 could thus potentially be binding to the form of c-Jun that binds DNA, altering either the binding ability of c-Jun or the partner to which c-Jun dimerizes, and possibly affecting the interaction of the AP-1 complexes with other proteins. The significance of AP-1 binding on the biology of HPVs will be discussed in a later section.

Cell cycle proteins
Other cell cycle-regulating proteins that E7 has been shown to bind to or alter the expression of include the cyclins, cyclin-dependent kinases (cdk), and their inhibitors. In a normal cell cycle, progression through G1 requires the D-type cyclins to interact with the cyclin kinases, cdk4 and cdk6. The complexes and activity of these cyclins and kinases accumulate into mid-G1 and are responsible for phosphorylating Rb. In cells containing E7, cyclin D1 mRNA and protein levels are reduced (74, 99), and the cdk inhibitor $p16^{INK}$ is elevated, possibly as a result of the elevated levels of E2F activating the promoter of $p16^{INK}$ (69, 103). E7 could be down-regulating cyclin D1 production by inhibiting the normal activity of Rb, which has been postulated to play a role in cyclin D1 promoter regulation (74, 99). Furthermore, HPV-16 E6 and E7 together were shown to disrupt cyclin D-cdk4 complexes and promote $cdk4-p16^{INK}$ complex formation, while E7 from a low-risk virus was unable to do either (128).

E7 has been shown to bind to cyclin A as well as cdk2 (40, 123). Cyclin A is induced at G1/S phase and regulates S-phase transition and passage through G2 (48, 89). During S phase, cyclin A and cdk2 exist in a complex containing p107 and E2F (20, 38, 111). HPV-16 E7 is able to bind to this complex via CR2 and does so with greater affinity than E7 from low-risk HPV-6, though neither protein dissociates the complex (6, 88, 128). The cyclin that controls passage through G1, cyclin E, is also targeted by E7, which binds a complex containing p107, cyclin E, and cdk2 (81, 129). E7 up-regulates some of these proteins, including cyclin A (92, 129) and cyclin E (45, 129), and in addition another cyclin kinase, cdc-2 (92), which activates cyclin A. The two cyclins can be up-regulated even in the absence of growth factors, suggesting that E7 could be up-regulating these proteins to push the cell through the cell cycle in the absence of external growth stimuli. This observation led to the hypothesis that while it down-regulates cyclin D1, E7 could replace some cyclin D1 functions, as cyclin A and E were both up-regulated in the presence of low levels of cyclin D1 (129). CR2 is required for up-regulation of cyclins A and E, though CR1 is also required for up-regulation of cyclin A (129). CR3 could conceivably play a role in up-regulating cyclin A levels as well, since the cdk inhibitor p27^{KIPI} has been shown to inhibit cyclin A gene expression (113), while E7 has been shown to relieve this inhibition by binding p27^{KIPI} through CR3 (130).

Recent investigations have shown that HPV-16 E7 binds to the kinase inhibitor p21^{CIP-1} (46, 66). This kinase inhibitor binds to and inhibits the activity of cdk2 and is active during keratinocyte differentiation. cdk2 is involved in phosphorylation of a number of substrates when bound to cyclins D or E, and phosphorylation of these substrates allows cells to transverse the G1/S-phase boundary. Therefore induction of p21^{CIP-1}, possibly through p53 activation, will result in inhibition of cyclin-associated kinase activity and cell cycle arrest. E6 and E7 could abrogate this cell arrest in combination by inhibiting p53 activation and binding and inhibiting p21^{CIP-1} function. During keratinocyte differentiation in the presence of E7 the levels of p21^{CIP-1} actually increase. This increase would seem to be contrary to a role for E7 in pushing cells through G1 into S phase. However, the same studies showed that the ability of the p21^{CIP-1} to inhibit kinase activity was greatly reduced, even in the presence of increased amounts of p21^{CIP-1}. The mechanism of inhibition is unclear as E7 does not disrupt cyclin-cdk2-p21^{CIP-1} complexes.

Overall, investigations point to the fact that E7 can act during the G1 phase of the cell cycle by up-regulating the levels of cyclins and inhibiting the activity of kinase inhibitors, with the result that substrates may be inappropriately phosphorylated late in G1 and cells may be stimulated to enter prematurely into S phase. These activities complement other functions of E7, discussed above, which result in the specific stimulation of the expression of genes whose products also encourage G1/S-phase transition. E7 from the high-risk HPV types also affects the S phase and early G2 phase of the cell cycle by up-regulating cyclin B and p34^{cdc2} through apparent translational and posttranslational control mechanisms (92, 116), but does not disrupt cyclin B-cdc2 complexes (128). Since degradation of cyclin B is required for cells to exit mitosis, elevated levels of this cyclin could prolong mitosis and the period during which HPV can use cellular machinery to replicate its DNA. It has to be remembered that the virus is attempting to replicate in a differentiating epithelium, which is a very dynamic region as cells are moving up through the epithelium and being shed off into the environment. If the virus is able to stimulate S phase, but inhibit mitosis, the virus can propagate and the small amount of infectious virus produced per cell will not be diluted out by cell division.

TBP and TAFs

As mentioned, in the course of the cell cycle, pRb becomes hyperphosphorylated, releasing the transcription factor E2F and activating transcription of certain genes. E7 produces the same activation effect by causing the release of E2F from pRb in a phosphorylation-independent manner. Another method by which E7 has been postulated to enhance transcriptional activation of certain genes is through its interaction with members of the transcription factor machinery. In particular, E7 has been shown to bind to the TATA-binding protein (TBP) (77, 78, 96) and to TBP-associated factor-110 (TAF-110) (78). Efficient interaction with TBP requires three domains, the pRb binding region, the case in kinase II domain (CKII), and the carboxy-terminal zinc finger region (96). Either phosphorylation of the CKII domain or replacement of the two serines in the CKII domain (positions 31 and 32) with acidic amino acids enhances the TBP/E7 interaction (77). While it is not important for pRb binding, conflicting reports suggest that the CKII domain may be important for transformation (9, 43, 55, 95). The E7 from HPV-16 is a more efficient CKII substrate than E7 from HPV-6 (9), perhaps suggesting that the high-risk viral E7 proteins may interact with TBP more efficiently. It is worth noting that the interaction of E7 and TBP need not be stimulatory to transcription. Indeed, it has been suggested that this interaction is responsible for the ability of E7 to repress p53 transcriptional activity (76).

E7 and pRB/c-Jun Interactions

Since E7 is able to bind to both AP-1 family members and Rb, and Rb and c-Jun can bind to each other, it is possible that E7 could alter the interaction of these proteins. HPVs infect differentiating keratinocytes. Lacking the machinery necessary for their own replication, the viruses are dependent on the cell, and since a differentiating host is unlikely to provide the milieu the virus requires, HPV must disrupt the differentiation process and stimulate cells into S phase. Rb has previously been implicated in the differentiation of multiple cell types (22, 51), and there is evidence for a positive role for Rb in keratinocyte differentiation (87). In addition, the AP-1 family of transcription factors has also been implicated in the differentiation of keratinocytes and myeloid cells (2). Rb is able to bind to c-Jun and increases both DNA binding and the transactivation level of c-Jun. The hypophosphorylated form of Rb binds the leucine zipper of c-Jun through two separate domains, the B pocket and the C terminus. HPV-16 E7 can disrupt the Rb–c-Jun complex. Recall that E7 has two domains that interact with pRb, one in CR2 and one in the zinc finger region, and that this interaction occurs through the small pocket and carboxy-terminal regions of pRb. The overlapping binding sites for c-Jun and E7 in Rb help explain the ability of E7 to disrupt Rb–c-Jun complexes.

In a keratinocyte differentiation system, the raft system (see chapter 6), the presence of HPV-16 E7 reduces c-Jun levels, normally seen in the granular layer, to very low levels. Keratinocytes containing the HPV-16 genome with an E7 protein mutated in the Rb binding domain stain at levels near wild type. The mutated E7 clone contains only a single amino acid change which disrupts Rb binding but leaves c-Jun binding intact. Since there is a dramatic difference in Jun staining in raft cultures containing wild-type E7, it can be hypothesized that E7 is disrupting some Rb-mediated regulation of c-Jun expression in differentiating keratinocytes. Therefore, E7 may alter the differentiation process by disrupting the Rb–c-Jun interaction and may also up-regulate AP-1 factor-responsive genes, independent of the effect on Rb, which are important for G1/S-phase transition.

Immortalization and Transformation Properties of E7

The ability of HPVs to immortalize primary human foreskin keratinocytes is a reflection of their oncogenic potential in humans. For instance, the oncogenic types 16, 18, and 31 can all immortalize keratinocytes, but the benign viruses such as HPV-6 and -11 cannot. Full-length genomes of HPV-16 immortalize keratinocytes very efficiently, whereas E6 and E7 together, while they can immortalize cells, do so at low levels (10% of full length [10]). E7 appears to be the major component of immortalization since it alone has been shown to immortalize keratinocytes at very low levels when expressed from an exogenous promoter. In addition, in the context of the full genome, the presence of E7 is essential for immortalization (65). So far, there have been no reports of E6 alone being able to immortalize foreskin keratinocytes, although this protein has been shown to immortalize mammary epithelial cells (7). As mentioned above, the significance of this is unclear as there is no evidence that HPV is involved in any breast cancers.

HPVs can transform rodent fibroblasts, and again the ability to carry this out is reflected in the level of oncogenicity observed in vivo. The major transforming protein is E7, and this alone can transform immortalized rodent fibroblasts such as NIH 3T3, C127, A31, and RaT1 cells. Primary rat embryo fibroblasts can only be transformed by E7 in the presence of an activated *ras* protein such as Harvey *ras* (24). The low-risk types do not exhibit this transforming potential, and the difference appears to be in the N-terminal region of the protein, both in the Rb binding (CR2) and CR1 domains (8). A deletion of aa 6–10 in the CR1 domain inhibited transformation (15), while a substitution of the histidine at position 2 with a proline, also in CR1, caused a significant reduction in transforming potential (8). Mutations which abrogate Rb binding (25, 41), or a change of the aspartic acid at aa 21, resulted in inhibition of transformation (55, 104). Chimeras between HPV-6 and HPV-16 E7 proteins showed that when amino acids of HPV-16 E7 between aa 16 and 30 were replaced with equivalent HPV-6 E7 amino acids there was almost complete loss of transformation (55). However, mutations which disrupt the zinc finger in the CR3 domain are also inactive (25, 41, 80, 95), although these latter mutations, as mentioned previously, probably have an effect on the overall structure of E7.

Therefore, there is a difference in the requirements for immortalization of primary human keratinocytes and transformation of rodent fibroblasts in that the former activity is independent of binding Rb whereas the latter is Rb binding dependent. The requirements for immortalization remain to be determined.

Summary

The list of proteins that E7 binds or interacts with lengthens continually. One important endeavor in HPV research is to determine the sequential progression of proteins bound by E7 as an infection evolves and the contribution of each interaction to the life cycle of the virus and to the potential transformation of host cells. In addition, most E7 biology has been reported in cell types other than keratinocytes, due both to the difficulties associated with working with these primary cells and to the lack of an in vivo model of HPV infection. Advances in the understanding of keratinocyte biology and differentiation will certainly further our understanding of the role of this 98-aa protein in disrupting normal cellular function.

STRUCTURE AND FUNCTION OF E5 PROTEINS

The E5 gene of most HPVs is located just downstream of the E2 open reading frame. The gene is not well conserved at the DNA level between HPVs or animal viruses, although there is a conservation of the physicochemical properties in that the proteins are all highly hydrophobic and membrane bound (18, 29). The proteins from both animal and human sources can transform mammalian cells with varying degrees of efficiency. Most research has been carried out with the BPV-1 E5 for two very good reasons. First, this was the first papillomavirus to be completely sequenced and have its E5 gene cloned, and second, its physicochemical properties (it has a hydrophilic tail [18]) make it easier to work with and it is produced in detectable quantities in transfected cells. This section will concentrate on the E5 from HPVs but, where appropriate, comparisons to the BPV-1 E5 will be made.

Structure of E5

E5 is a small protein, of 84 aa in the case of HPV-16 (16) and only 44 aa for BPV-1 E5, making the latter the smallest known oncoprotein (17, 39, 101, 107, 108). BPV-1 E5 runs as a 7-kDa protein in reducing gels, but since it can dimerize through cysteine residues, in nonreducing conditions it runs at 14 kDa. The HPV-16 E5 has been shown, in one study, to dimerize even in reducing conditions regardless of its cysteine residues (68). In the case of the benign virus HPV-6, there are two open reading frames in the region of E5 designated E5a and E5b; however, only work on E5a has been reported (23, 124). As mentioned above, the E5 protein is hydrophobic, with three very hydrophobic peaks and less hydrophobic troughs (16). The proteins from many of the HPV types show little or no hydrophilic regions in hydrophobicity plots, while BPV-1 E5 has a hydrophilic tail of approximately 14 aa. The nature of the protein means that it is membrane bound and distributed predominantly in the endoplasmic reticulum and Golgi apparatus, but also in the cytoplasmic membrane (18, 29, 115).

The hydrophobic nature of E5 also means that purification is difficult because the protein is insoluble and good antibody reagents directed against E5 are limited. Antipeptide antibodies directed against N and C termini have been used successfully (62), and peptides have also been added as tags to the N terminus of the protein (29), taking advantage of the commercially available antibodies to the peptide tags.

Interactions of E5 with Growth Factor Receptors and Vacuolar ATPase

Unlike the E6 and E7 proteins, which abrogate the effects of negative cellular regulators, the major biological function of the E5 proteins appears to be through cellular growth factor receptors, whose activity it up-regulates. The original E5/receptor interaction was described for BPV-1 E5, where the viral protein was found to stabilize the epidermal growth factor receptor (EGFR) in the presence of the ligand epidermal growth factor (EGF), with the result that the complex was not broken down and a hyperstimulating signal was observed (75). Subsequent to this study, BPV-1 E5 was found in a complex with the platelet-derived growth factor (PDGF) receptor (PDGFR) (94). It is important to be aware of the kinetics of the interaction of both receptors and their respective ligands. When the ligand binds to the receptors it causes dimerization, and the receptors autophosphorylate each other through a kinase domain in the cytoplasmic tail and are rapidly endocytosed (Fig. 4A). For instance, in the case of the EGFR, 90% of receptors are internalized within 10 min of the addition of EGF. The phosphorylated receptors bind to adaptor proteins, which in turn activate signaling pathways through Ras, phosphatidylinositol 3-kinase, or phospholipase C-γ. The endosomes containing the receptor/ligand complex are then acidified and become part of the lysosomal compartment of the cell, and within 90 min all the receptors are degraded, resulting in a short-lived stimulatory signal. It was also found that BPV-1 E5 could activate the PDGFR in the absence of the ligand PDGF (94). How this is achieved is not clear, but it may be caused by the ability of the E5 to dimerize (18) with itself and to bind to PDGFR, and so in effect to cause dimerization of the receptor and its subsequent autophosphorylation (49). However, the PDGFR is not found on human keratinocytes where the major growth factor receptor is EGFR, so most of the work on HPV-16 E5 has been carried out with EGFR.

Several studies have shown a cooperative interaction between E5 and EGF in the transformation of rodent fibroblasts (23, 73, 97, 120) for both the benign (HPV-6 E5a) and oncogenic (HPV-16 E5) types. There is conflicting evidence on whether E5 binds directly to the EGFR. In one study, HPV-6 E5 was shown to bind to the EGFR when both were expressed in COS-1 cells, but HPV-16 E5 was not observed to bind (30), whereas in another study there was evidence of binding to EGFR using the same cell type (62). In the former study, both E5 proteins were epitope tagged and anti-tagged antibodies were used to coimmunoprecipitate the proteins, while in the latter, E5 was not tagged and anti-E5 peptide antibodies were used. The reason for this discrepancy in the binding results remains to be determined, but the fact that one set of E5 proteins was tagged with a hydrophilic peptide may have resulted in the protein being inefficiently immunoprecipitated due to masking of the peptide by the hydrophobic protein.

The E5 proteins from both HPV-16 and BPV-1 also bind to the 16-kDa subunit of the vacuolar ATPase (29, 124). The ATPase is a large protein complex made up of a number of subunits of which the 16-kDa protein forms the pore through which H$^+$ ions pass to acidify the contents of endosomes. The effect on the acidification of endosomes has been tested only with HPV-16 E5 in human keratinocytes (Fig. 4B) (119). To maximize the effect, the cells were starved of EGF, and then the ligand was added and the pH of endosomes was measured at times thereafter.

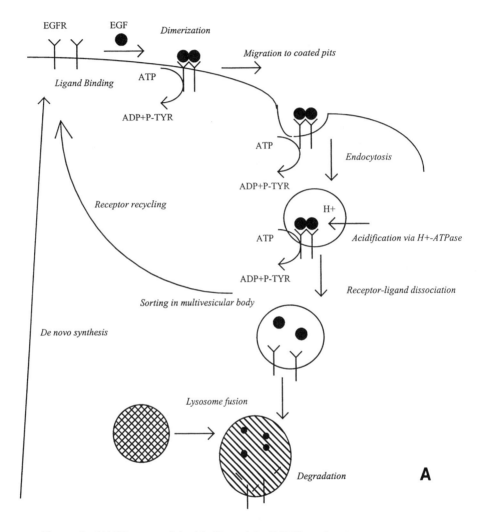

Figure 4 (A) Diagram of the binding of the EGF ligand to its receptor, EGFR, and the subsequent pathway of the down-regulation and destruction of receptor and ligand through the lysosomal pathway. In human keratinocytes in culture, about 5% of receptors recycle to the surface of the cell.

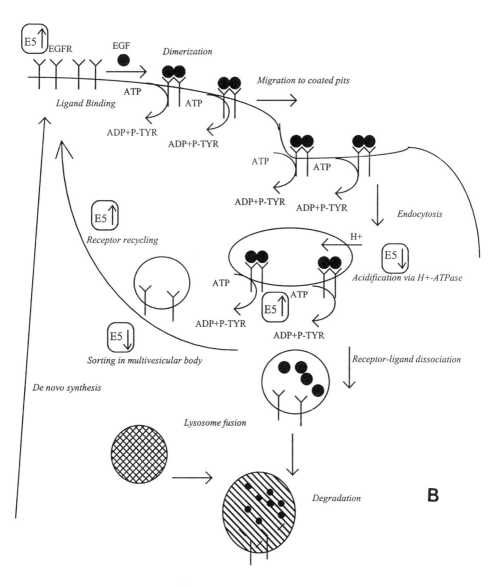

Figure 4 (B) The same pathway but this time in the presence of HPV-16 E5. The effects of E5 are indicated by the direction of the arrows inside the E5 boxes. Thus, for example, acidification of the endosomal compartment is inhibited in the present of E5 as indicated by the arrow, while recycling of the receptor is increased to 40% in the presence of E5.

In normal keratinocytes, the pH of the endosome decreased to below 5 in 20 min, but in E5-expressing keratinocytes the drop in pH was delayed for up to 3 h and in many cases did not go below pH 6. The inhibition of pH change results in the retention of the receptor in the endosomal compartment and recycling of 40% of the receptors back to the surface in the absence of the ligand (Fig. 4B). This probably accounts for the increased numbers of EGFR observed on E5-expressing keratinocytes. The fact that the ligand is removed from the receptor is a consequence of the dissociation pH being 6.5, a level reached even in the presence of E5. The inhibition of acidification is probably not dependent on the EGFR, although this has not been tested. Instead it is likely that the acidification of any endosome will be effected by the presence of HPV-16 E5. The direct testing of the binding of E5 to the 16-kDa subunit has not been shown in human keratinocytes due to the low levels of E5 protein and its very hydrophobic nature. However, coexpression of both E5 and the yeast vacuolar ATPase subunit in COS-1 cells has shown a complex containing both E5 and the 16-kDa protein (29). It is assumed, although not proven, that this is a direct interaction and there is no third protein involved.

There are downstream effects of E5 expression in rodent cells. For instance, there is an increase in c-Fos and c-Jun expression in cells, especially in the presence of EGF (12, 73). The upstream regulatory region (URR) of most HPV types contain an AP-1 binding site, and in the presence of E5 there is an increase in transcription from this promoter in E5-containing NIH 3T3 cells (12). This is a potentially important observation because one of the earliest mRNAs produced by HPV-16 DNA in infected tissues potentially codes for E5, so this protein may be important for the production of early messages coding for E6 and E7. There is also an increase in the mitogen-activated protein (MAP) kinase kinase and MAP kinase activity (52) in cells stimulated by EGF, and this increase is enhanced and prolonged in the presence of E5. Therefore it appears that the various pathways activated by the interaction of EGF with the EGF receptor are enhanced and prolonged in the presence of HPV-16 E5. It is, however, not clear at the moment whether all of this can be accounted for by the inhibition of acidification of the endosomal contents, leading to prolonged signaling by the EGF/EGFR complex, or whether E5 can act on the pathways independent of the effects on endosomal pH.

Since EGF is a growth requirement for human keratinocytes and is necessary through part of the G1 phase of the cell cycle, the activity of E5 suggests that it acts to stimulate cells through this phase of the cell cycle and into S phase.

Transformation and Immortalization Properties of E5

As was mentioned above, HPV-16 E5 can transform immortalized rodent fibroblasts, and the frequency of transformation is increased in the presence of EGF, but not PDGF. The transformation assays have all been carried out using soft agar to measure growth in suspension. HPV-16 E5 is unable to form transformed foci on cell monolayers and therefore is considered to have only moderate transforming activity, unlike the BPV-1 E5 gene, which can form transformed foci on rodent monolayer cultures. It should be remembered that the E5 protein of BPV-1 is the major transforming protein, unlike the situation with HPVs, where E7 has the major transforming role.

HPV-16 E5 is unable to immortalize primary human keratinocytes (119, 120), but there are reports of an extended lifespan of cells expressing E5 from an exogenous promoter (119). In addition, the full-length HPV-16 genome with a premature stop codon in the E5 gene is only able to immortalize keratinocytes at 10% of the frequency of the wild-type genome, suggesting that E5 plays a complementary role in the immortalization process (117). Therefore, in the life cycle of the virus E5 plays an important role along with E6 and E7 proteins.

SUMMARY OF THE ACTION OF E6, E7, AND E5 ON KERATINOCYTE DIFFERENTIATION AND CELL CYCLE PROGRESSION

For viral replication, the keratinocytes must be stimulated to move into S phase, which may be accompanied by inhibition of terminal differentiation, at least temporarily. This scenario is necessary because the virus requires the cell's replicative machinery for replication of its own genome. The shift into S phase need not cause the cell to go through a complete cycle, and in fact this is probably not desirable for the virus as it would dilute out newly produced genomes and the capsid proteins needed to produce new viral particles. Therefore, to achieve the cell cycle progression, the virus must uncouple G1/S-phase progression from the differentiation process and stimulate passage through G1 into S phase. However, there are still some differentiation-specific signals that are required by the virus to switch from the major early to the major late promoter for transcription of capsid protein mRNA (see chapter 6). Therefore, in HPV-16-infected tissue there is an alteration in differentiation, and if the lesion persists there is a chance of complete inhibition of differentiation and the development of progressive disease with the production of aneuploid cells and invasive cancer. Both the E6 and E7 proteins are candidate proteins for the G1-to-S-phase progression. E6 can bind to E6BP, which is a Ca^{2+} binding protein, and is possibly involved in differentiation (i.e., E6 may disrupt the normal activity). E7 binds to Rb and to members of the AP-1 family of transcription factors, both of which have been implicated in cell differentiation. Rb has been shown to be induced by the muscle differentiation transcription factor MyoD, while c-Jun, an AP-1 member, is up-regulated on the differentiation of human keratinocytes.

The signals for keratinocytes to enter a differentiation pathway must appear some time in G1 phase, most probably around the G0/G1 point. Thus, even if cells are prevented from differentiating, they will still need stimulation to move through G1. A candidate for inducing progression through G1 is the E5 protein, which can stimulate DNA synthesis in keratinocytes in the absence of EGF in serum-starved cells; furthermore, addition of EGF leads to additional activity. As mentioned above, E5 and EGF can act cooperatively in transformation of rodent cells and in induction of the MAP kinase pathway. E6 has been shown to decrease the half-life of p53, which is an important protein for monitoring the cell as it moves through the G1 phase. Degrading this protein prematurely may take away an important negative signaling pathway, thus allowing the cell to move into S phase. Finally, E7 can interact with the AP-1 family of transcription factors and can up-regulate transcriptional activity. This could be important since a number of important genes required for DNA synthesis are regulated by AP-1 factors. The up-regulation of transcription factors may be mediated by the ability of E7 to bind histone deacetylases and remove from the promoter this inhibitory activity. Therefore, cooperation between the early proteins of HPV will alter the differentiation process and at the same time stimulate progression through G1 into S phase.

REFERENCES

1. **Androphy, E., N. L. Hubbert, J. T. Schiller, and D. R. Lowy.** 1987. Identification of the HPV-16 E6 protein from transformed mouse cells and human cervical carcinoma cell lines. *EMBO J.* **6:**989–992.

2. **Angel, P., and M. Karin.** 1991. The role of Jun, Fos and the AP-1 complex in cell-proliferation and transformation. *Biochim. Biophys. Acta* **1072:**129–157.

3. **Antinore, M. J., M. J. Birrer, D. Patel, L. Nader, and D. J. McCance.** 1996. The human papillomavirus type 16 E7 gene product interacts with and trans-activates the AP1 family of transcription factors. *EMBO J.* **15:**1950–1960.

4. **Armstrong, D. J., and A. Roman.** 1992. Mutagenesis of human papillomavirus types 6 and 16 E7 open reading frames alters the electrophoretic mobility of the expressed proteins. *J. Gen. Virol.* **73:**1275–1279.

5. **Armstrong, D. J., and A. Roman.** 1993. The anomalous electrophoretic behavior of the human papillomavirus type 16 E7 protein is due to the high content of acidic amino acid residues. *Biochem. Biophys. Res. Commun.* **192:**1380–1387.

6. **Arroyo, M., S. Bagchi, and P. Raychaudhuri.** 1993. Association of the human papillomavirus type 16 E7 protein with the S-phase-specific E2F-cyclin A complex. *Mol. Cell. Biol.* **13:**6537–6546.

7. **Band, V., D. Zajchowki, V. Kulesa, and R. Sager.** 1990. Human papillomavirus DNAs immortalize normal epithelial cells and reduce their growth factor requirements. *Proc. Natl. Acad. Sci. USA* **87:**463–467.

8. **Banks, L., C. Edmonds, and K. H. Vousden.** 1990. Ability of the HPV16 E7 protein to bind RB and induce DNA synthesis is not sufficient for efficient transforming activity in NIH3T3 cells. *Oncogene* **5:**1383–1389.

9. **Barbosa, M. S., C. Edmonds, C. Fisher, J. T. Schiller, D. R. Lowy, and K. H. Vousden.** 1990. The region of the HPV E7 oncoprotein homologous to adenovirus E1a and Sv40 large T antigen contains separate domains for Rb binding and casein kinase II phosphorylation. *EMBO J.* **9:**153–160.

10. **Barbosa, M. S., and R. Schlegel.** 1989. The E6 and E7 genes of HPV-18 are sufficient for inducing two-stage in vitro transformation of human keratinocytes. *Oncogene* **4:**1529–1532.

11. **Barbosa, M. S., W. C. Vass, D. R. Lowy, and J. T. Schiller.** 1991. In vitro biological activities of the E6 and E7 genes vary among human papillomaviruses of different oncogenic potential. *J. Virol.* **65:**292–298.

12. **Bouvard, V., G. Matlashewski, Z. M. Gu, A. Storey, and L. Banks.** 1994. The human papillomavirus type 16 E5 gene cooperates with the E7 gene to stimulate proliferation of primary cells and increases viral gene expression. *Virology* **203:**73–80.

13. **Brandsma, J. L., Z. H. Yang, S. W. Barthold, and E. A. Johnson.** 1991. Use of a rapid, efficient inoculation method to induce papillomas by cottontail rabbit papillomavirus DNA shows that the E7 gene is required. *Proc. Natl. Acad. Sci. USA* **88:**4816–4820.

14. **Brehm, A., E. A. Miskka, D. J. McCance, J. L. Reid, A. J. Bannister, and T. Kouzarides.** Retinoblastoma recruits histone deacetylase activity to repress transcription. *Nature,* in press.

15. **Brokaw, J. L., C. L. Yee, and K. Munger.** 1994. A mutational analysis of the amino

terminal domain of the human papillomavirus type 16 E7 oncoprotein. *Virology* **205:** 603–607.

16. **Bubb, V., D. J. McCance, and R. Schlegel.** 1988. DNA sequence of the HPV-16 E5 ORF and the structural conservation of its encoded protein. *Virology* **163:**243–246.

17. **Burkhardt, A., D. DiMaio, and R. Schlegel.** 1987. Genetic and biochemical definition of the bovine papillomavirus E5 transforming protein. *EMBO J.* **6:**2381–2385.

18. **Burkhardt, A., M. Willingham, C. Gay, K. T. Jeang, and R. Schlegel.** 1989. The E5 oncoprotein of bovine papillomavirus is oriented asymmetrically in Golgi and plasma membranes. *Virology* **170:**334–339.

19. **Busby-Earle, R. M., C. M. Steel, A. R. Williams, B. Cohen, and C. C. Bird.** 1994. p53 mutations in cervical carcinogenesis—low frequency and lack of correlation with human papillomavirus status. *Br. J. Cancer* **69:**732–737.

20. **Cao, L., B. Faha, M. Dembski, L. H. Tsai, E. Harlow, and N. Dyson.** 1992. Independent binding of the retinoblastoma protein and p107 to the transcription factor E2F. *Nature* **355:**176–179.

21. **Chen, J. J., C. E. Reid, V. Band, and E. J. Androphy.** 1995. Interaction of papillomavirus E6 oncoproteins with a putative calcium-binding protein. *Science* **269:**529–531.

22. **Chen, P. L., D. J. Riley, Y. Chen, and W. H. Lee.** 1996. Retinoblastoma protein positively regulates terminal adipocyte differentiation through direct interaction with C/EBPs. *Genes Dev.* **10:**2794–2804.

23. **Chen, S. L., and P. Mounts.** 1990. Transforming activity of E5a protein of human papillomavirus type 6 in NIH 3T3 and C127 cells. *J. Virol.* **64:**3226–3233.

24. **Chesters, P. M., and D. J. McCance.** 1989. Human papillomavirus types 6 and 16 in cooperation with Ha-ras transform secondary rat embryo fibroblasts. *J. Gen. Virol.* **70:**353–365.

25. **Chesters, P. M., K. H. Vousden, C. Edmonds, and D. J. McCance.** 1990. Analysis of human papillomavirus type 16 open reading frame E7 immortalizing function in rat embryo fibroblast cells. *J. Gen. Virol.* **71:**449–453.

26. **Chiba, I., M. Shindoh, M. Yasuda, Y. Yamazaki, A. Amemiya, Y. Sato, K. Fujinaga, K. Notani, and H. Fukuda.** 1996. Mutations in the p53 gene and human papillomavirus infection as significant prognostic factors in squamous cell carcinomas of the oral cavity. *Oncogene* **12:**1663–1668.

27. **Ciccolini, F., G. Di Pasquale, F. Carlotti, L. Crawford, and M. Tommasino.** 1994. Functional studies of E7 proteins from different HPV types. *Oncogene* **9:**2633–2638.

28. **Clemens, K. E., R. Brent, J. Gyuris, and K. Munger.** 1995. Dimerization of the human papillomavirus E7 oncoprotein in vivo. *Virology* **214:**289–293.

29. **Conrad, M., V. J. Bubb, and R. Schlegel.** 1993. The human papillomavirus type 6 and 16 E5 proteins are membrane-associated proteins which associate with the 16-kilodalton pore-forming protein. *J. Virol.* **67:**6170–6178.

30. **Conrad, M., D. Goldstein, T. Andresson, and R. Schlegel.** 1994. The E5 protein of HPV-6, but not HPV-16, associates efficiently with cellular growth factor receptors. *Virology* **200:**796–800.

31. **Cornelissen, M., H. L. Smits, M. A. Briet, J. G. van den Tweel, A. P. Struyk, J. van der Noorda, and J. Schegget.** 1990. Uniformity of the splicing pattern of the E6/E7 transcripts in human papillomavirus type 16-transformed human fibroblasts, human cervical premalignant lesions and carcinomas. *J. Gen. Virol.* **71:**1243–1246.

32. **Crook, T., J. A. Tidy, and K. H. Vousden.** 1991. Degradation of p53 can be targeted by IIPV E6 sequences distinct from those required for p53 binding and trans-activation. *Cell* **67:**547–556.

33. **Crook, T., and K. H. Vousden.** 1992. Properties of p53 mutations detected in primary

and secondary cervical cancers suggest mechanisms of metastasis and involvement of environmental carcinogens. *EMBO J.* **11**:3935–3940.

34. **Crook, T., D. Wrede, J. A. Tidy, W. P. Mason, D. J. Evans, and K. H. Vousden.** 1992. Clonal p53 mutation in primary cervical cancer: association with human-papillomavirus-negative tumours. *Lancet* **339**:1070–1073.

35. **Crook, T., D. Wrede, and K. H. Vousden.** 1991. p53 point mutation in HPV negative human cervical carcinoma cell lines. *Oncogene* **6**:873–875.

36. **Davies, R., R. Hicks, T. Crook, J. Morris, and K. Vousden.** 1993. Human papillomavirus type 16 E7 associates with a histone H1 kinase and with p107 through sequences necessary for transformation. *J. Virol.* **67**:2521–2528.

37. **Defeo-Jones, D., G. A. Vuocolo, K. M. Haskell, M. G. Hanobik, D. M. Kiefer, E. M. McAvoy, M. Ivey-Hoyle, J. L. Brandsma, A. Oliff, and R. E. Jones.** 1993. Papillomavirus E7 protein binding to the retinoblastoma protein is not required for viral induction of warts. *J. Virol.* **67**:716–725.

38. **Devoto, S. H., M. Mudryj, J. Pines, T. Hunter, and J. R. Nevins.** 1992. A cyclin A-protein kinase complex possesses sequence-specific DNA binding activity: p33cdk2 is a component of the E2F-cyclin A complex. *Cell* **68**:167–176.

39. **DiMaio, D., D. Guralski, and J. T. Schiller.** 1986. Translation of open reading frame E5 of bovine papillomavirus is required for its transforming activity. *Proc. Natl. Acad. Sci. USA* **83**:1797–1801.

40. **Dyson, N., P. Guida, K. Munger, and E. Harlow.** 1992. Homologous sequences in adenovirus E1A and human papillomavirus E7 proteins mediate interaction with the same set of cellular proteins. *J. Virol.* **66**:6893–6902.

41. **Edmonds, C., and K. H. Vousden.** 1989. A point mutational analysis of human papillomavirus type 16 E7 protein. *J. Virol.* **63**:2650–2656.

42. **Etscheid, B. G., S. A. Foster, and D. A. Galloway.** 1994. The E6 protein of human papillomavirus type 16 functions as a transcriptional repressor in a mechanism independent of the tumor suppressor protein, p53. *Virology* **205**:583–585.

43. **Firzlaff, J. M., B. Luscher, and R. N. Eisenman.** 1991. Negative charge at the casein kinase II phosphorylation site is important for transformation but not for Rb protein binding by the E7 protein of human papillomavirus type 16. *Proc. Natl. Acad. Sci. USA* **88**:5187–5191.

44. **Foster, S. A., G. W. Demers, B. G. Etscheid, and D. A. Galloway.** 1994. The ability of human papillomavirus E6 proteins to target p53 for degradation in vivo correlates with their ability to abrogate actinomycin D-induced growth arrest. *J. Virol.* **68**:5698–5705.

45. **Foster, S. A., and D. A. Galloway.** 1996. Human papillomavirus type 16 E7 alleviates a proliferation block in early passage human mammary epithelial cells. *Oncogene* **12**:1773–1779.

46. **Funk, J. O., S. Waga, J. B. Harry, E. Espling, B. Stillman, and D. A. Galloway.** 1997. Inhibition of CDK activity and PCNA-dependent DNA replication by p21 is blocked by interaction with the HPV-16 E7 oncoprotein. *Genes Dev.* **11**:2090–2100.

47. **Gage, J. R., C. Meyers, and F. O. Wettstein.** 1990. The E7 proteins of the nononcogenic human papillomavirus type 6b (HPV-6b) and of the oncogenic HPV-16 differ in retinoblastoma protein binding and other properties. *J. Virol.* **64**:723–730.

48. **Girard, F., U. Strausfeld, A. Fernandez, and N. J. Lamb.** 1991. Cyclin A is required for the onset of DNA replication in mammalian fibroblasts. *Cell* **67**:1169–1179.

49. **Goldstein, D. J., T. Andresson, J. J. Sparkowski, and R. Schlegel.** 1992. The BPV-1 E5 protein, the 16 kDa membrane pore-forming protein and the PDGF receptor exist in a complex that is dependent on hydrophobic transmembrane interactions. *EMBO J.* **11**:4851–4859.

50. **Grassman, K., S. P. Wilezynski, N. Cook, B. Rapp, and T. Iftner.** 1996. HPV6 variants from malignant tumors with sequence alterations in the regulatory region do not reveal differences in the activities of the oncogene promoters but do contain amino acid exchanges in the E6 and E7 proteins. *Virology* **223:**185–197.

51. **Gu, W., J. W. Schneider, G. Condorelli, S. Kaushal, V. Mahdavi, and B. Nadal-Ginard.** 1993. Interaction of myogenic factors and the retinoblastoma protein mediates muscle cell commitment and differentiation. *Cell* **72:**309–324.

52. **Gu, Z., and G. Matlashewski.** 1995. Effect of human papillomavirus type 16 oncogenes on MAP kinase activity. *J. Virol.* **69:**8051–8056.

53. **Halbert, C., G. W. Demers, and D. A. Galloway.** 1992. The E6 and E7 genes of human papillomavirus type 6 have weak immortalizing activity in human epithelial cells. *J. Virol.* **66:**2125–2134.

54. **Halpern, A., and M. Munger.** 1995. *HPV-16 E7: Primary Structure and Biological Properties*, vol. 2. Theoretical Biology and Biophysics Group, Los Alamos, N. Mex.

55. **Heck, D. V., C. L. Yee, P. M. Howley, and K. Munger.** 1992. Efficiency of binding the retinoblastoma protein correlates with the transforming capacity of the E7 oncoproteins of the human papillomaviruses. *Proc. Natl. Acad. Sci. USA* **89:**4442–4446.

56. **Hiraiwa, A., T. Kiyono, K. Segawa, K. R. Utsumi, M. Ohashi, and M. Ishibashi.** 1993. Comparative study on E6 and E7 genes of some cutaneous and genital papillomaviruses of human origin for their ability to transform 3Y1 cells. *Virology* **192:**102–111.

57. **Hu, Q. J., N. Dyson, and E. Harlow.** 1990. The regions of the retinoblastoma protein needed for binding to adenovirus E1A or SV40 large T antigen are common sites for mutations. *EMBO J.* **9:**1147–1155.

58. **Huang, P. S., D. R. Patrick, G. Edwards, P. J. Goodhart, H. E. Huber, L. Miles, V. M. Garsky, A. Oliff, and D. C. Heimbrook.** 1993. Protein domains governing interactions between E2F, the retinoblastoma gene product, and human papillomavirus type 16 E7 protein. *Mol. Cell. Biol.* **13:**953–960.

59. **Huang, S., N. P. Wang, B. Y. Tseng, W. H. Lee, and E. H. Lee.** 1990. Two distinct and frequently mutated regions of retinoblastoma protein are required for binding to SV40 T antigen. *EMBO J.* **9:**1815–1822.

60. **Huibregtse, J., M. Schneffer, and P. M. Howley.** 1991. A cellular protein mediates association of p53 with the E6 oncoprotein of human papillomavirus types 16 or 18. *EMBO J.* **10:**4129–4135.

61. **Huibregtse, J., M. Schneffner, and P. M. Howley.** 1993. Localization of the E6-AP regions that direct human papillomavirus E6 binding, association with p53, and ubiquitination of associated proteins. *Mol. Cell. Biol.* **13:**4918–4927.

62. **Hwang, E. S., T. Nottoli, and D. Dimaio.** 1995. The HPV16 E5 protein: expression, detection, and stable complex formation with transmembrane proteins in COS cells. *Virology* **211:**227–233.

63. **Iftner, T., S. Bierfelder, Z. Csapo, and H. Pfister.** 1988. Involvement of human papillomavirus type 8 genes E6 and E7 in transformation and replication. *J. Virol.* **62:**3655–3661.

64. **Ikura, M.** 1996. Calcium binding and conformational response in EF-hand proteins. *Trends Biochem. Sci.* **21:**14–17.

65. **Jewers, R. J., P. Hildebrandt, J. W. Ludlow, B. Kell, and D. J. McCance.** 1992. Regions of human papillomavirus type 16 E7 oncoprotein required for immortalization of human keratinocytes. *J. Virol.* **66:**1329–1335.

66. **Jones, D. L., R. M. Alani, and K. Munger.** 1997. The human papillomavirus E7 oncoprotein can uncouple cellular differentiation and proliferation in human kera-

tinocytes by abrogating p21Cip1-mediated inhibition of cdk2. *Genes Dev.* **11:** 2101–2111.

67. **Kaelin, W. G., Jr., M. E. Ewen, and D. M. Livingston.** 1990. Definition of the minimal simian virus 40 large T antigen- and adenovirus E1A-binding domain in the retinoblastoma gene product. *Mol. Cell. Biol.* **10:**3761–3769.

68. **Kell, B., J. R. Jewers, J. Cason, F. Pkarian, J. N. Kaye, and J. M. Best.** 1994. Detection of E5 oncoprotein in human papillomavirus type 16-positive cervical scrapes using antibodies raised to synthetic peptides. *J. Gen. Virol.* **75:**2451–2456.

69. **Khleif, S. N., J. DeGregori, C. L. Yee, G. A. Otterson, F. J. Kaye, J. R. Nevins, and P. M. Howley.** 1996. Inhibition of cyclin D-CDK4/CDK6 activity is associated with an E2F-mediated induction of cyclin kinase inhibitor activity. *Proc. Natl. Acad. Sci. USA* **93:**4350–4354.

70. **Lamberti, C., L. C. Morrissey, S. R. Grossman, and E. J. Androphy.** 1990. Transcriptional activation by the papillomavirus E6 zinc finger oncoprotein. *EMBO J.* **9:** 1907–1913.

71. **Lechner, M. S., and L. A. Laimins.** 1994. Inhibition of p53 DNA binding by human papillomavirus E6 proteins. *J. Virol.* **68:**4262–4273.

72. **Lechner, M. S., D. H. Mack, A. B. Finicle, T. Crook, K. H. Vousden, and L. A. Laimins.** 1992. Human papillomavirus E6 proteins bind p53 in vivo and abrogate p53-mediated repression of transcription. *EMBO J.* **11:**3045–3052. (Erratum, *EMBO J.* **11:**4248, 1992.)

73. **Leechanachi, P., L. Banks, F. Moreau, and G. Matlashewski.** 1992. The E5 gene from human papillomavirus type 16 is an oncogene which enhances growth factor-mediated signal transduction to the nucleus. *Oncogene* **7:**19–25.

74. **Lukas, J., H. Muller, J. Bartkova, D. Spitkovsky, A. A. Kjerulff, P. Jansen-Durr, M. Strauss, and J. Bartek.** 1994. DNA tumor virus oncoproteins and retinoblastoma gene mutations share the ability to relieve the cell's requirement for cyclin D1 function in G1. *J. Cell Biol.* **125:**625–638.

75. **Martin, P., W. C. Vass, J. T. Schiller, D. R. Lowy, and T. J. Velu.** 1989. The bovine papillomavirus E5 transforming protein can stimulate the transforming activity of EGF and CSF-1 receptors. *Cell* **59:**21–32.

76. **Massimi, P., and L. Banks.** 1997. Repression of p53 transcriptional activity by the HPV E7 proteins. *Virology* **227:**255–259.

77. **Massimi, P., D. Pim, A. Storey, and L. Banks.** 1996. HPV-16 E7 and adenovirus E1a complex formation with TATA box binding protein is enhanced by casein kinase II phosphorylation. *Oncogene* **12:**2325–2330.

78. **Mazzarelli, J. M., G. B. Atkins, J. V. Geisberg, and R. P. Ricciardi.** 1995. The viral oncoproteins Ad5 E1A, HPV16 E7 and SV40 TAg bind a common region of the TBP-associated factor-110. *Oncogene* **11:**1859–1864.

79. **McCance, D. J., R. Kopan, E. Fuchs, and L. A. Laimins.** 1988. Human papillomavirus type 16 alters human epithelial cell differentiation in vitro. *Proc. Natl. Acad. Sci. USA* **85:**7169–7173.

80. **McIntyre, M. C., M. G. Frattini, S. R. Grossman, and L. A. Laimins.** 1993. Human papillomavirus type 18 E7 protein requires intact Cys-X-X-Cys motifs for zinc binding, dimerization, and transformation but not for Rb binding. *J. Virol.* **67:**3142–3150.

81. **McIntyre, M. C., M. N. Ruesch, and L. A. Laimins.** 1996. Human papillomavirus E7 oncoproteins bind a single form of cyclin E in a complex with cdk2 and p107. *Virology* **215:**73–82.

82. **Meyers, G., and E. Androphy.** 1995. *The E6 Protein*, vol. 2. Theoretical Biology and Biophysics Group, Los Alamos, N. Mex.

83. **Moran, E., and M. B. Mathews.** 1987. Multiple functional domains in the adenovirus E1A gene. *Cell* **48:**177–178.

84. **Munger, K., and W. C. Phelps.** 1993. The human papillomavirus E7 protein as a transforming and transactivating factor. *Biochim. Biophys. Acta* **1155:**111–123.

85. **Munger, K., B. A. Werness, N. Dyson, W. C. Phelps, E. Harlow, and P. M. Howley.** 1989. Complex formation of human papillomavirus E7 proteins with the retinoblastoma tumor suppressor gene product. *EMBO J.* **8:**4099–4105.

86. **Munger, K., C. L. Yee, W. C. Phelps, J. A. Pietenpol, H. L. Moses, and P. M. Howley.** 1991. Biochemical and biological differences between E7 oncoproteins of the high- and low-risk human papillomavirus types are determined by amino-terminal sequences. *J. Virol.* **65:**3943–3948.

87. **Nead, N., L. Baglia, M. J. Antinore, J. W. Ludlow, and D. J. McCance.** Rb binds c-Jun and activates transcription: a possible role in keratinocyte differentiation. Submitted for publication.

88. **Pagano, M., M. Durst, S. Joswig, G. Draetta, and P. Jansen-Durr.** 1992. Binding of the human E2F transcription factor to the retinoblastoma protein but not to cyclin A is abolished in HPV-16-immortalized cells. *Oncogene* **7:**1681–1686.

89. **Pagano, M., R. Pepperkok, F. Verde, W. Ansorge, and G. Draetta.** 1992. Cyclin A is required at two points in the human cell cycle. *EMBO J.* **11:**961–971.

90. **Pater, M. M., H. Nakshatri, C. Kisaka, and A. Pater.** 1992. The first 124 nucleotides of the E7 coding sequences of HPV16 can render the HPV11 genome transformation competent. *Virology* **186:**348–351.

91. **Patrick, D. R., A. Oliff, and D. C. Heimbrook.** 1994. Identification of a novel retinoblastoma gene product binding site on human papillomavirus type 16 E7 protein. *J. Biol. Chem.* **269:**6842–6850.

92. **Pei, X. F.** 1996. The human papillomavirus E6/E7 genes induce discordant changes in the expression of cell growth regulatory proteins. *Carcinogenesis* **17:**1395–1401.

93. **Pennie, W., G. J. Grindlay, M. Cairney, and M. S. Campo.** 1993. Analysis of the transforming functions of bovine papillomavirus type 4. *Virology* **193:**614–620.

94. **Petti, L., L. A. Nilson, and D. DiMaio.** 1991. Activation of the platelet-derived growth factor receptor by the bovine papillomavirus E5 transforming protein. *EMBO J.* **10:**845–855.

95. **Phelps, W. C., K. Munger, C. L. Yee, J. A. Barnes, and P. M. Howley.** 1992. Structure-function analysis of the human papillomavirus type 16 E7 oncoprotein. *J. Virol.* **66:**2418–2427.

96. **Phillips, A. C., and K. H. Vousden.** 1997. Analysis of the interaction between human papillomavirus type 16 E7 and the TATA-binding protein, TBP. *J. Gen. Virol.* **78:**905–909.

97. **Pim, D., M. Collins, and L. Banks.** 1992. Human papillomavirus type 16 E5 gene stimulates the transforming activity of the epidermal growth factor receptor. *Oncogene* **7:**27–32.

98. **Pim, D., A. Storey, M. Thomas, P. Massimi, and L. Banks.** 1994. Mutational analysis of HPV-18 E6 identifies domains required for p53 degradation in vitro, abolition of p53 transactivation in vivo and immortalization of primary BMK cells. *Oncogene* **9:**1869–1876.

99. **Pusch, O., T. Soucek, E. Wawra, E. Hengstschlager-Ottnad, G. Bernaschek, and M. Hengstschlager.** 1996. Specific transformation abolishes cyclin D1 fluctuation throughout the cell cycle. *FEBS Lett.* **385:**143–148.

100. **Qin, X. Q., T. Chittenden, D. M. Livingston, and W. G. Kaelin, Jr.** 1992. Identification of a growth suppression domain within the retinoblastoma gene product. *Genes Dev.* **6:**953–964.

101. **Rabson, M. S., C. Yee, Y. C. Yang, and P. M. Howley.** 1986. Bovine papillomavirus type 1 3′ early region transformation and plasmid maintenance functions. *J. Virol.* **60:**626–634.

102. **Rawls, J. A., R. Pusztai, and M. Green.** 1990. Chemical synthesis of human papillomavirus type 16 E7 oncoprotein: autonomous protein domains for induction of cellular DNA synthesis and for *trans* activation. *J. Virol.* **64:**6121–6129.

103. **Reznikoff, C. A., T. R. Yeager, C. D. Belair, E. Savelieva, J. A. Puthenveettil, and W. M. Stadler.** 1996. Elevated p16 at senescence and loss of p16 at immortalization in human papillomavirus 16 E6, but not E7, transformed human uroepithelial cells. *Cancer Res.* **56:**2886–2890.

104. **Sang, B. C., and M. S. Barbosa.** 1992. Single amino acid substitutions in "low-risk" human papillomavirus (HPV) type 6 E7 protein enhance features characteristic of the "high-risk" HPV E7 oncoproteins. *Proc. Natl. Acad. Sci. USA* **89:**8063–8067.

105. **Scheffner, M., J. M. Huibregtse, R. D. Vierstra, and P. M. Howley.** 1993. The HPV-16 E6 and E6-AP complex functions as a ubiquitin-protein ligase in the ubiquitination of p53. *Cell* **75:**495–505.

106. **Scheffner, M., B. A. Werness, J. M. Huibregtse, A. J. Levine, and P. M. Howley.** 1990. The E6 oncoprotein encoded by human papillomavirus types 16 and 18 promotes the degradation of p53. *Cell* **63:**1129–1136.

107. **Schiller, J. T., W. C. Vass, K. H. Vousden, and D. R. Lowy.** 1986. E5 open reading frame of bovine papillomavirus type 1 encodes a transforming gene. *J. Virol.* **57:**1–6.

108. **Schlegel, R., M. Wade-Glass, M. S. Rabson, and Y. C. Yang.** 1986. The E5 transforming gene of bovine papillomavirus encodes a small, hydrophobic polypeptide. *Science* **233:**464–467.

109. **Schmitt, A., J. B. Harry, B. Rapp, F. O. Wettstein, and T. Iftner.** 1994. Comparison of the properties of the E6 and E7 genes of low- and high-risk cutaneous papillomaviruses reveals strongly transforming and high Rb-binding activity for the E7 protein of the low-risk human papillomavirus type 1. *J. Virol.* **68:**7051–7059.

110. **Schneider-Gadicke, A., and E. Schwarz.** 1986. Different human cervical carcinoma cell lines show similar transcription patterns of human papillomavirus type 18 genes. *EMBO J.* **5:**2285–2292.

111. **Schwarz, J. K., S. H. Devoto, E. J. Smith, S. P. Chellappan, L. Jakoi, and J. R. Nevins.** 1993. Interactions of the p107 and Rb proteins with E2F during the cell proliferation response. *EMBO J.* **12:**1013–1020.

112. **Sedman, S., M. S. Barbosa, W. C. Vass, N. I. Hubbert, J. A. Haas, D. R. Lowy, and J. T. Schiller.** 1991. The full length E6 protein of human papillomavirus type 16 has transforming and transactivating activities and cooperates with E7 to immortalize keratinocytes in culture. *J. Virol.* **65:**4860–4866.

113. **Shulze, A., K. Zerfass-Thome, J. Berges, S. Middendorp, P. Jansen-Durr, and B. Henglein.** 1996. Anchorage-dependent transcription of the cyclin A gene. *Mol. Cell. Biol.* **16:**4632–4638.

114. **Snijders, P., A. J. C. van den Brule, H. F. J. Schrijnemakers, P. M. C. Raaphorst, C. J. L. M. Meijer, and J. M. M. Walboomers.** 1992. Human papillomavirus type 33 in a tonsillar carcinoma generates its putative E7 mRNA via two E6* transcript species which are terminated at different early region poly(A) sites. *J. Virol.* **66:**3172–3178.

115. **Sparkowski, J., J. Anders, and R. Schlegel.** 1995. E5 oncoprotein retained in the endoplasmic reticulum/cis Golgi still induces PDGF receptor autophosphorylation but does not transform cells. *EMBO J.* **14:**3055–3063.

116. **Steinmann, K. E., X. F. Pei, H. Stoppler, R. Schlegel, and R. Schlegel.** 1994. Elevated expression and activity of mitotic regulatory proteins in human papillomavirus-immortalized keratinocytes. *Oncogene* **9:**387–394.

117. **Stoppler, M. C., S. W. Straight, G. Tsao, R. Schlegel, and D. J. McCance.** 1996. The E5 gene of HPV-16 enhances keratinocyte immortalization by full-length DNA. *Virology* **223**:251–254.

118. **Storey, A., N. Almond, K. Osborn, and L. Crawford.** 1990. Mutations of the human papillomavirus type 16 E7 gene that affect transformation, transactivation and phosphorylation by the E7 protein. *J. Gen. Virol.* **71**:965–970.

119. **Straight, S. W., B. Herman, and D. J. McCance.** 1995. The E5 oncoprotein of human papillomavirus type 16 inhibits the acidification of endosomes in human keratinocytes. *J. Virol.* **69**:3185–3192.

120. **Straight, S. W., P. M. Hinkle, R. J. Jewers, and D. J. McCance.** 1993. The E5 oncoprotein of human papillomavirus type 16 transforms fibroblasts and effects the downregulation of the epidermal growth factor receptor in keratinocytes. *J. Virol.* **67**:4521–4532.

121. **Takami, Y., T. Sasagawa, T. M. Sudiro, M. Yutsudo, and A. Hakura.** 1992. Determination of the functional difference between human papillomavirus type 6 and 16 E7 proteins by their 30 N-terminal amino acid residues. *Virology* **186**:489–495.

122. **Taya, Y.** 1997. RB kinases and RB-binding proteins: new points of view. *Trends Biochem. Sci.* **22**:14–17.

123. **Tommasino, M., J. P. Adamczewski, F. Carlotti, C. F. Barth, R. Manetti, M. Contorni, F. Cavalieri, T. Hunt, and L. Crawford.** 1993. HPV16 E7 protein associates with the protein kinase p33CDK2 and cyclin A. *Oncogene* **8**:195–202.

124. **Valle, G., and L. Banks.** 1995. The human papillomavirus HPV-6 and HPV-16 E5 proteins co-operate with HPV-16 E7 in the transformation of primary rodent cells. *J. Gen. Virol.* **76**:1239–1245.

125. **Wang, J. Y.** 1997. Retinoblastoma protein in growth suppression and death protection. *Curr. Opin. Genet. Dev.* **7**:39–45.

126. **Werness, B. A., A. J. Levine, and P. M. Howley.** 1990. Association of human papillomavirus types 16 and 18 E6 proteins with p53. *Science* **248**:76–79.

127. **Wu, E. W., K. E. Clemens, D. V. Heck, and K. Munger.** 1993. The human papillomavirus E7 oncoprotein and the cellular transcription factor E2F bind to separate sites on the retinoblastoma tumor suppressor protein. *J. Virol.* **67**:2402–2407.

128. **Xiong, Y., D. Kuppuswamy, Y. Li, E. M. Livanos, M. Hixon, A. White, D. Beach, and T. D. Tlsty.** 1996. Alteration of cell cycle kinase complexes in human papillomavirus E6- and E7-expressing fibroblasts precedes neoplastic transformation. *J. Virol.* **70**:999–1008.

129. **Zerfass, K., A. Schulze, D. Spitkovsky, V. Friedman, B. Henglein, and P. Jansen-Durr.** 1995. Sequential activation of cyclin E and cyclin A gene expression by human papillomavirus type 16 E7 through sequences necessary for transformation. *J. Virol.* **69**:6389–6399.

130. **Zerfass-Thome, K., A. Schulze, W. Zwerschke, B. Vogt, K. Helin, J. Bartek, B. Henglein, and P. Jansen-Durr.** 1997. p27kip1 blocks cyclin E-dependent transactivation of cyclin A gene expression. *Mol. Cell. Biol.* **17**:407–415.

131. **Zwerschke, W., S. Joswig, and P. Jansen-Durr.** 1996. Identification of domains required for transcriptional activation and protein dimerization in the human papillomavirus type-16 E7 protein. *Oncogene* **12**:213–220.

Human Tumor Viruses
Edited by Dennis J. McCance
© 1998 American Society for Microbiology

8 | Hepatitis B Virus Replication, Liver Disease, and Hepatocellular Carcinoma
William S. Mason, Alison A. Evans, and W. Thomas London

All members of the hepatitis B virus (HBV) (hepadnavirus) family infect the liver, their major site of reproduction. Other family characteristics include a novel genome structure, a unique mode of replication, the ability to induce transient (3 to 12 months) as well as chronic (generally lifelong) infections, and, in chronically infected mammalian hosts, a high risk that chronic infection will lead to liver cancer. The ongoing interest in the hepadnaviruses stems from the fact that HBV is a major public health problem. Chronic carrier rates range from 0.1% to 25% in different parts of the world, and as many as 1 million people per year die as a consequence of chronic infection, most often of hepatocellular carcinoma (HCC). Though a vaccine has been available for over a decade, this is still not effectively utilized in many high-risk populations. In addition, treatments for chronically infected individuals are of limited efficacy.

Attempts to understand how these viruses replicate have been highly successful. However, much about the biology of infection has been difficult to study, at least in part because liver disease takes so many years to evolve. Human liver tumors generally arise after several decades of chronic infection. In addition, host and genetic factors probably contribute to the development of liver cancer, which only arises in about 10 to 25% of long-term carriers. The nature of these other risk factors is still largely unknown and, again, difficult to study. Antiviral therapies have been delivered to selected patients for many years, but are limited in efficacy and are likely to remain so until we have a clearer understanding of how transient infections are cleared and why chronic infections sometimes resolve. Indeed, with the resolution of many fundamental questions about virus replication in tissue culture systems, most of the issues that confront the researcher over the coming years will require a broad knowledge of liver biology and immunology.

One factor that has been of great importance to the study of HBV replication, and which will become of increasing importance in characterizing the biology of infection, is the existence of hepadnaviruses in other species. Members of this virus family have been found not just in humans (7), but in great apes (95), ground squirrels (83, 133, 134), woodchucks (129), ducks (86), herons (124), and geese (J. Newbold, personal communication; GenBank accession no. M95589). The viruses infecting mammals are closely related to each other in sequence and genome organization, as are the viruses of birds. These two families of viruses are accordingly subdivided into *Orthohepadnaviridae* and *Avihepadnaviridae*, respectively. The viruses are all highly species specific, and attempts at cross-species transmission are usually unsuccessful. However, it has been possible to transmit to closely related species, for instance, human HBV from humans to great apes (5), duck hepatitis B virus (DHBV) to geese (82), ground squirrel hepatitis virus (GSHV) to woodchucks (114) and chipmunks (140), and woodchuck hepatitis virus (WHV) to Chinese marmots (164).

The basic replication mechanisms of all of these viruses are similar. However, DHBV has proven the most useful for studies of hepadnavirus replication. The reasons for this are several. First, about 10% of the ducks in many commercial flocks are chronically infected, providing easy accessibility to investigators. This accessibility was especially important in early studies, as cell culture systems for virus expression were not available until about 10 years ago. Second, DHBV is maintained in nature primarily by vertical transmission into the developing egg, with infection of the developing liver by 4 to 5 days of embryogenesis (98, 131, 146). This route of infection appears to induce immune tolerance to the virus. Thus, DHBV-specific liver disease does not develop and antiviral antibodies have not, so far, been reported. In the absence of antibodies, it was much easier to isolate viral replicative intermediates from infected duck liver than from any of the mammalian models, where antiviral immunoglobulins are almost always found. Third, this is the only model from which primary liver cell cultures can be readily produced (143). In experimental systems, primary liver cell cultures are the only cells that have virus receptors and are therefore able to support a complete cycle of virus replication. Fourth, very large amounts of DHBV are produced when recombinant DHBV DNA is transfected into LMH cells (a liver cell line of chicken origin [19, 62]), making genetic studies of this particular hepadnavirus comparatively easy. In the following sections, replication is discussed in some detail. However, it should be kept in mind that some of the fine details have been confirmed only for the duck virus.

HEPADNAVIRUS PROTEINS

To appreciate the hepadnavirus replication scheme, it is important, first, to appreciate that hepadnaviruses have one of the smallest of all animal virus genomes (ca. 3 to 3.3 kbp). Thus, if there were not overlapping reading frames, these viruses would only encode about 100 kDa of protein. Instead, through the use of multiple overlapping genes, the coding capacity is increased about 50% (for review, see reference 36). This is illustrated in Fig. 1, using the genome of HBV as a typical mammalian hepadnavirus.

The genome itself has a relaxed circular conformation which is held together by a short cohesive overlap between the 5' ends of the two DNA strands. One strand, of minus polarity, has a protein covalently attached to its 5' end through a phosphodiester linkage to a tyrosine. This protein is the virus-encoded reverse transcriptase (RT) (see below). The other strand, of plus polarity, has a capped RNA, 17 to 18 bases long, linked to its 5' end and is always incomplete within

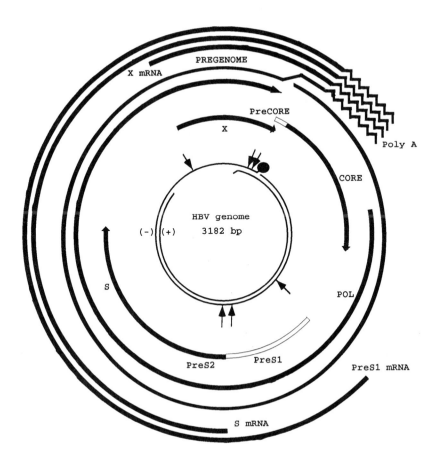

Figure 1   The HBV genome. The placement of the various open reading frames and of the viral mRNAs relative to the relaxed circular virion DNA is shown. The RNAs are represented by solid lines on the outside of the diagram, and the coding regions are represented by lines with arrowheads. The short solid arrows indicate the start sites of the various viral RNAs. All end at a common polyadenylation site within the core gene.

virus particles. The genomes of mammalian hepadnaviruses encode at least four groups of mRNA, each from its own promoter (for review, see reference 112). None of the RNAs illustrated in Fig. 1 is spliced. However, recent data suggest that an additional spliced transcript might be involved in expression of the viral envelope proteins (97). The group of RNAs with the greatest length, as shown in Fig. 1, include the pregenome and the slightly larger precore mRNA. Both are terminally redundant. The precore mRNA starts just upstream of the first AUG of the precore/core open reading frame, whereas the pregenome starts just downstream of this AUG. The pregenome thus serves as the mRNA for the core protein, which begins from a downstream AUG. The core protein is the single subunit of the viral nucleocapsid. The pregenome also serves as an mRNA for the viral RT, the single product of the viral *pol* gene, albeit at low efficiency. Functionally, the

RT can be divided into an amino-terminal domain encoding a protein primer of reverse transcription, followed by a spacer domain and the carboxy-terminal RT and RNase H domains (for review, see reference 116).

The precore mRNA encodes the protein product of the entire precore/core open reading frame, which gives rise to the viral e antigen (for HBV, this antigen is designated HBeAg) (101). The polypeptide coded for by the precore mRNA, which has a signal peptide directing it to the endoplasmic reticulum, is cleaved in that location. Cleavage of the amino-terminal signal peptide and a carboxy-terminal region precedes secretion of the e antigen. The role of e antigen in the virus life cycle is unclear, although it has been suggested that this antigen may be able to induce tolerance to immunodominant epitopes of the core protein. Thus, transplacental passage of e antigen might facilitate vertical transmission of the virus, which is thought to occur later, at birth (90). However, no direct test of this possibility has been available.

The next largest RNA is the preS1 mRNA, which begins just upstream of a single large open reading frame encoding the viral envelope proteins. This mRNA directs synthesis of an envelope protein (termed L; 401 amino acids) which spans the entire open reading frame. A smaller mRNA, the S mRNA, actually has distinct start sites spanning an internal AUG (112) in the open reading frame. Thus, two different proteins are made. One, initiating from this AUG and extending to the end of the reading frame, is termed the M protein (281 amino acids). S mRNAs starting downstream of this AUG direct expression of the smaller S protein (226 amino acids), which also extends to the end of the reading frame and is the most abundantly produced of the three envelope proteins. All three are found in viral particles. In addition, as discussed below, these proteins are produced in greater abundance than required for virion production. Traditionally, the envelope gene and the corresponding regions of the envelope proteins are divided into preS1, preS2, and S domains (Fig. 1).

The final protein product of the mammalian hepadnaviruses is X. (The hepadnaviruses of ducks and other fowl have not been found to contain this open reading frame.) A variety of functions have been ascribed to this protein, but it is still unclear which of these are active in virus infection and which are laboratory artifacts (for review, see reference 30). What seems clear is that X expression has a modest activating effect on viral RNA synthesis. There is also evidence that X is required for virus expression in vivo (12, 165).

MECHANISM OF REPLICATION

The hepadnaviruses all replicate via reverse transcription. However, despite some similarities to retrovirus replication, an important difference is that hepadnavirus replication does not involve formation of an integrated provirus. The hepadnavirus replication strategy, with an emphasis on the fate of viral nucleic acids, is diagrammed in Fig. 2 (see figure legend for details). As noted earlier, within virus particles, the plus strand is always incomplete and about 90% of the viral DNA is found in a relaxed circular conformation with a short cohesive overlap between the 5′ ends of the two DNA strands (Fig. 1). However, the remainder of the viral genomes have a linear conformation. These two conformations are alternative products of the viral DNA synthesis pathway and, as recently discovered, both are infectious, though only the former consistently gives rise to infectious progeny viruses (162, 163).

Upon infection, viral DNA is transported to the nucleus and processed into covalently closed circular (CCC) DNA. Conversion of relaxed circular to CCC DNA involves removal of the covalently attached protein from the minus strand and RNA from the plus strand and repair of the single-stranded region of the plus strand. However, as suggested by Yang and Summers (162, 163), it is also possible that, rather than a simple repair of the single-stranded gap, synthesis can extend through the cohesive overlap region to produce a linear, double-stranded DNA with a large terminal redundancy, which is then circularized via homologous recombination. For the linear virion DNA, CCC DNA formation occurs via illegitimate recombination, producing mutations that may destroy the ability to produce new, or at least infectious, viral DNA (162, 163). Assuming that the CCC DNA that is formed has a wild-type or functionally wild-type sequence, the steps diagrammed in Fig. 2 will then occur. These steps lead to production of new copies of viral DNA, which are synthesized within viral nucleocapsids localized to the cytoplasm. Nonfunctional mutants created from the linear genomes may also be replicated by using proteins provided through coinfection with wild-type virus, provided that *cis*-acting regulatory sequences are preserved (e.g., epsilon).

Following completion of the DNA minus strand and partial completion of the plus strand, the nucleocapsids undergo changes, presumably in their surface structure, which allow interaction with viral envelope proteins and virion formation via budding into the endoplasmic reticulum (10, 69, 107, 156). Early in infection, envelope protein concentration is apparently insufficient for virion assembly, and mature nucleocapsids migrate to the nucleus to form additional CCC DNA (127, 128, 143), which generally reaches copy numbers of 5 to 50 per cell. (CCC DNA is not replicated semiconservatively [143].) Ultimately, viral envelope proteins are produced in vast excess over the amount needed to assemble virions, apparently to ensure the shutdown of CCC DNA formation. The surplus envelope proteins assemble into surface antigen particles (termed HBsAg for HBV) which are secreted from the cell in excess (>100-fold) over virions. S protein (Fig. 1) alone is sufficient for HBsAg production, though the other envelope proteins are normal components of these particles (50). Additional accumulation of CCC DNA probably occurs during established infections, if stress on the cell results in an imbalance between viral DNA synthesis and envelope protein production (143). Summers and colleagues have shown, however, that accumulation of CCC DNA may be cytocidal if allowed to proceed unchecked (68, 128).

An unusual aspect of viral DNA synthesis, as noted above, is the formation of virus with linear rather than circular genomes. This occurs when the primer of plus-strand DNA synthesis fails to translocate to DR2 (see Fig. 2). The importance of this linear species in the biology of an infection is still unknown. Recent evidence suggests that it may be an important precursor to the integrated forms of viral DNA which appear during the course of a chronic infection (39, 40). Liver tumors arising in carriers are clonal, as evidenced by the presence of viral DNA, which integrated before clonal expansion, and most tumors lack replicating virus. The almost universal presence of the integrated DNA suggests that integration

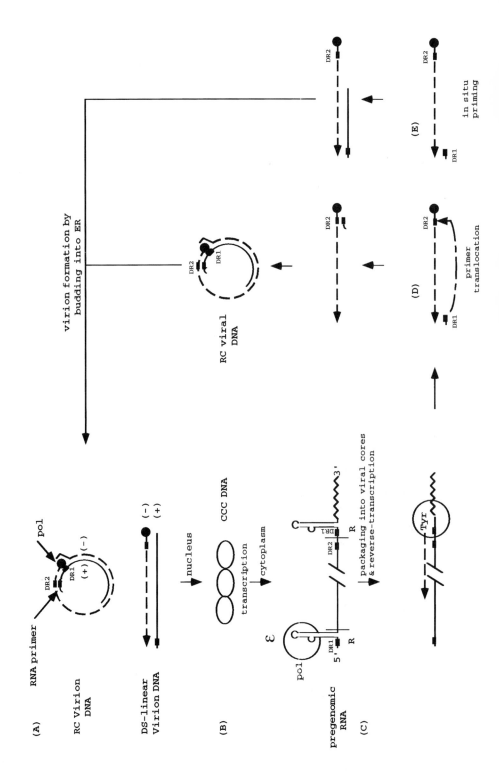

may have a role in the outgrowth of at least some tumors. This has been confirmed in some cases (see below).

VIRUS REPLICATION AND CELL PROLIFERATION

Another potentially important but poorly understood issue in the biology of infections is the effect of hepatocyte proliferation on virus replication and on the evolution of liver disease. The hepatocyte, the major cell type of the liver and also the major target of virus infection, probably has a normal life span of at least 6 months (41, 79). This is also true of bile duct epithelial cells, the only other cell type in the liver for which there is clear evidence of viral infection (6, 48). Thus, hepadnavirus infection in a healthy liver occurs mostly within noncycling cells. However, even in a totally healthy individual, hepatocytes occasionally die and, in this situation, are replaced by proliferation of other mature hepatocytes. In a chronic HBV infection with very active liver disease, the half-life of hepatocytes may be reduced to as little as 2 weeks (96) or perhaps even less.

Cell culture experiments suggest that virus replication is suppressed in proliferating cells (e.g., reference 58), but the basis for suppression is unknown. The

◄───

Figure 2 Hepadnavirus replication. The pathway for viral DNA synthesis is shown (see text and reference 116 for details). Virus, shown at the top left (A), may contain either circular or linear, partially double-stranded DNA. Following infection, viral DNA is delivered to the nucleus, where the DNA is completed and converted to a CCC form (B). The CCC DNA acts as the template for transcription of viral mRNAs, including a large terminally redundant RNA known as the pregenome (see Fig. 1). The pregenome has the capacity to act as mRNA for both the viral nucleocapsid (core protein) and the viral RT. Translation of the RT is infrequent, but when it occurs, this protein binds a stem-loop region at the 5′ end of the RNA known as epsilon (C) (116, 149, 150). The complex is then assembled into viral nucleocapsids, where reverse transcription actually takes place. Initiation of minus-strand synthesis begins at a bulge in the stem-loop structure, from which four bases are copied, using a tyrosine-OH located in the N-terminal region of the RT as a primer (155, 166). The complex then translocates and base pairs to a homologous four-base sequence in the 3′-terminal redundancy of the pregenome, at a 10-to-12-base sequence known as DR1 (DR1 was so named because there is a copy of this sequence outside the terminal redundancy, about 50 to 200 bases upstream in different hepadnaviruses, denoted DR2). Reverse transcription then extends to the 5′ end of the minus strand as the pregenome is degraded by the viral RNase H. The capped 5′ end of the pregenome, 17 to 18 bases in length and extending through DR1, is retained and serves as the primer of second (plus)-strand synthesis (70, 71, 115, 158). Normally, priming occurs after a translocation event in which the primer is dissociated from DR1 and annealed to DR2 (D). The mechanism of translocation is unknown. Occasionally, stable translocation fails to occur and priming occurs at DR1 to produce linear viral genomes (E) (125). Following extension to the 5′ end of the minus strand, plus-strand synthesis jumps to the 3′ end to create a circular DNA molecule. Circularization is apparently facilitated by the short (8 to 9 bases) terminal redundancy on the minus strand (74). Plus-strand synthesis is never completed during virus assembly, and most virion DNA has a partially double-stranded, relaxed circular conformation, though some is linear. The RT, which served as the primer of reverse transcription, and the RNA, which primed plus-strand synthesis, remain attached to the DNA strands in mature virus.

fate of replication intermediates during cell cycling is also unclear, as is that of CCC DNA, which must either be delivered to daughter nuclei during mitosis or be reformed from replication intermediates in the cytoplasm. Thus, hepatocyte proliferation in response to random injury, and especially to immune-mediated injury, could have a major impact on the evolution of viral as well as cell populations within the liver. For instance, the immune response to viral antigens may lead to the selective proliferation of hepatocytes in which virus replication is reduced, or of hepatocytes which are infected with virus that contain "escape" mutations in immunodominant epitopes. Another consequence of such selective pressure may, for example, be the expansion of dedifferentiated, transformed hepatocytes (e.g., reference 160). Also, in a situation in which there is a prolonged period of hepatocyte destruction, the constant need for new hepatocytes may lead to production of hepatocytes via proliferation and differentiation of facultative stem cells, in addition to that provided by division of mature cells. Such facultative stem cells are believed to exist at low levels in the healthy liver and to have the ability to produce progenitor cells which can differentiate into either hepatocytes or bile duct epithelium (21, 28, 29, 135). Constant stimulation of progenitor cell proliferation may be important in the disease process leading to HCC, tumors possibly arising from incompletely differentiated progenitor cells. Current knowledge of the evolution of liver cancer from virus-infected cells is discussed in a later section. The evolution of viral populations in response to selective pressure (e.g., by the immune system) is discussed in more detail in chapter 9. The evolution of liver cell populations in response to injury is discussed in a number of reviews (e.g., reference 123).

CHRONIC INFECTION BY HBV

Liver Cancer Risk in Human Populations

In view of the relatively low cell turnover in the normal liver, and the potential of the virus to make large amounts of intracellular viral DNA without damaging the cell, it is hardly surprising that a large number of infections become chronic. Moreover, in view of the ability of the host to respond to viral antigens, it is equally expected that such infections can damage the liver. What is still surprising is that this damage, which is generally not always clinically apparent, can lead to such a high incidence of cancer. Nonetheless, considering the large number of chronically infected patients worldwide (ca. 300 million), it has been possible to obtain very strong epidemiological evidence that chronic infection with HBV is causally associated with HCC in humans. The International Agency for Research on Cancer (57) evaluated the evidence and concluded that "chronic infection with hepatitis B virus is carcinogenic to humans" and classified HBV as a group 1 carcinogen.

The epidemiological evidence includes the following. (i) In those areas of the world where more than 2% of the general population is chronically infected with HBV, there is also an increased incidence of and mortality from HCC (130). (ii) Since 1970, more than 65 case-control studies, using the presence or absence of HBsAg in serum as the indicator of chronic infection, have been conducted in high-, intermediate-, and low-incidence areas of the world (57). All have shown elevated odds ratios ranging from 5 to 55 and averaging about 10. That is, persons diagnosed with HCC are about 10 times more likely to have HBsAg in their serum than individuals from the same location without HCC. Neither matching nor adjusting for age and sex, nor increasing the sensitivity of the assay for surface antigen, has diminished the strength of this association. (iii) Fewer prospective studies (less than 20), in which HBsAg-positive individuals were followed over long periods of time, have been reported (57). Follow-up times in these studies ranged from 3 to 11 years. All such studies have reported much higher incidences of HCC in HBV carriers than in controls. Relative risks (the ratio of these incidences) have ranged from 5 to 148. That is, during the follow-up period, persons chronically infected with HBV were much more likely to develop HCC than individuals from the same population who were HBsAg negative (i.e., who lacked circulating HBsAg, the most commonly used marker for the chronic carrier state [see below]).

A priori, it was assumed that the risk of liver cancer would be the same in all carriers. However, it has become clear that this is not the case. Instead, the annual incidences of HCC in HBV carriers vary widely, from population to population. In Taiwan, Beasley and Hwang (3) recorded 474 cases of HCC per 100,000 persons per year among adult men 30 to 59 years of age at study entry. Similarly high rates have been observed among the Inuit population of Alaska (89), railway workers in Japan (111), and male populations in mainland China (76, 142). In contrast, rates among chronically infected blood donors in Japan have been considerably lower, ranging from 20 to $100/10^5$/year (35, 100, 137). Even lower annual incidence rates, in the range of 10 to $20/10^5$, were observed among chronic carrier blood donors in the United States (26, 105). Sub-Saharan Africa is one of the high-incidence HCC regions of the world, but no prospective studies of HBV carriers from that area have been reported. (We are currently conducting such a study among the members of the Senegalese army.) Insufficient data are available at this time to give a precise estimate of HCC incidence among these carriers, but it is clear that it is much lower than in China, Taiwan, or Alaska (Evans, unpublished data), i.e., less than $100/10^5$/year.

Various reasons for these population differences in HCC incidence might be considered, including different host genetics, differences in exposure to one or more environmental carcinogens, and duration of virus replication. With respect to the latter, it has been known for a long time that many carriers eventually stop making virus (that is, titers drop below levels detectable by Southern blot hybridization and sometimes even PCR assays), but continue to produce virus surface antigen (HBsAg), probably via expression from integrated viral DNA. Moreover, it has been clear that virus replication shows a strong association with chronic liver injury and that active liver disease abates when virus production ceases, even though HBsAg production continues (66, 94). Thus, one possible contributor to the variation in the risk of liver cancer among HBV carriers in different locations (and even within a population) is variation in the duration of viral replication and, therefore, liver injury. It has been recognized for many years that at the time of HCC diagnosis most individuals are negative for serum HBeAg, which is only produced during virus replication, and have low or undetectable levels of circulating or intrahepatic virus (8, 61, 72, 110). However, the relationship between duration of virus replication and HCC risk has not so far been reported for the different risk groups mentioned above. Prospective clinic-based studies of HCC risk among HBV carriers have been carried out in some populations (17, 20, 120), but because the subjects tend to enter clinic-based studies due to prior knowledge of their HBV status and/or symptoms of liver disease, asymptomatic carriers (who make up the majority of HBV carriers) are underrepresented. Therefore, it is impossible to determine from such studies whether it is liver disease or viral replication (or both) that is the etiologic factor for HCC. Other prospective studies of asymptomatic carriers (23, 148) have been limited by small sample sizes.

Recently, we compared the prevalence of viremia and liver damage among HBsAg-positive male members of a population in Haimen City, in the People's Republic of China, and HBsAg-positive members of a population in Senegal, West Africa. The two populations have an equivalent prevalence of HBsAg positivity. However, there is a very high incidence of HCC among HBsAg carriers in Haimen and a much lower incidence in Senegal. In particular, we have observed an age-adjusted HCC incidence of $774/10^5$/year among 9,392 HBsAg-positive men in Haimen. This is one of the highest rates reported in the world. In Senegal, we are following 15,000 members of the Senegalese military of whom 23% are HBsAg positive. As noted above, the HCC incidence in this population is less than $100/10^5$/year. To determine whether the difference in incidence of HCC was linked to virological and serological parameters in HBV carriers in these two populations, we carried out a detailed analysis using a randomly selected subsample of 285 adult male HBV carriers from the Haimen City cohort and 289 adult male HBV carriers from the Senegalese Army. HBsAg levels were determined in a quantitative enzyme-linked immunosorbent assay (ELISA). Viremia was assessed by Southern blot hybridization for virion DNA. Liver injury was evaluated by ELISA for glutathione S-transferase concentrations in the serum. This assay was employed because, unlike conventional assays for serum transaminases, results could be obtained with serum samples that had been frozen.

Comparison of the two populations revealed a similar prevalence of HBsAg at all ages, as expected. A dramatic difference was observed, however, when viremia among HBV carriers was assessed by Southern blot hybridization (Fig. 3). In the Senegalese group 15% were viremic in their 20s (i.e., had virus titers of $>5 \times 10^5$/ml of serum), and this declined to 3% in their 30s and 40s and 0% in their 50s. In the Chinese group a significantly higher prevalence of viremia and a less consistent reduction with age were seen. About 29% were positive in their 20s, 30% in their 30s, 24% in their 40s, and 21% in their 50s. Viral titers among those who were HBV DNA positive, however, were similar in the two populations. In the Chinese group the median titer was ca. 4×10^7 viral particles/ml of serum, and in the Senegal group it was ca. 8×10^7. Viral titers did, however, decline with age among HBV DNA positives in both populations. The presence of HBV DNA was significantly associated with the detection of glutathione S-transferase in the serum (odds ratio = 6.7; 95% confidence interval, 2.8–15.9), indicating that liver damage was 6.7 times more common among HBV DNA-positive subjects than among subjects in whom circulating virus was not detected.

Thus, these data are consistent with the idea that duration of high-titer virus production (and of chronic liver injury) is an important factor in HCC. To begin

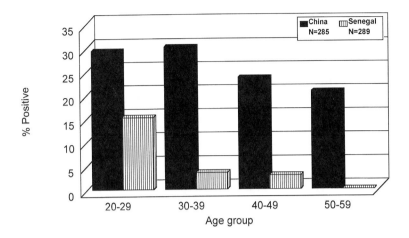

Figure 3 Prevalence of serum HBV DNA in HBV carriers from two different populations by age group. The presence of HBV DNA (virions) in serum was determined by standard filter DNA hybridization assays. The percent of carriers scoring positive in each age group, from 20 to 60 years, is shown. Further details are provided in the text.

to understand this correlation, it is first necessary to know why virus replication persists at a high level so much longer in some individuals than in others. This is an important issue if persistence increases the cancer risk. One hypothesis is that persistence of virus production over many decades is solely due to an inadequate host immune response. Another hypothesis, not exclusive of the first, is that chronic HBV infections that are most likely to result in liver cancer are those in which mutations have appeared in the predominant virus strain of the infected individual to facilitate escape from the host immune response (i.e., by reducing expression of one or more immunodominant epitopes). In some patients, loss in the capacity to produce HBeAg occurs as a result of mutations in the viral genome. Patients who stop making HBeAg but continue to replicate HBV (i.e., have HBV DNA detectable in serum by hybridization methods) and have elevated alanine aminotransferases (evidence of liver damage) usually have a mutation introducing a stop codon in the precore region of the genome (Fig. 1), predominantly at nucleotide 1896. Such patients appear to have a poorer prognosis with a greater risk of developing a more aggressive chronic hepatitis (46), presumably because this virus is better able to persist. The origin of these mutants is unclear. Some mutants may be present at the initiation of infection, while others appear to arise later, becoming prevalent due to immune selection (1, 2, 132). Prospective, population-based epidemiological studies of chronic carriers with and without precore mutants have not been reported. If such mutations consistently lead to more sustained liver damage, they may increase the incidence of HCC within a population. Comparisons of the prevalence of precore mutants in high-risk and low-risk populations of chronic carriers could provide a partial test of this hypothesis.

In summary, population-based studies carried out to date have begun to elucidate some aspects of the relationship between chronic infections and liver cancer. First, comparison of two populations suggests that HCC incidence may be associated with the duration of the productive stage of an infection. Perinatal transmission is more common in HBV-endemic areas of Asia (4, 126) and carries a much higher risk for prolonged viremia than the exposure later in infancy characteristic of endemic areas of Africa (81, 106). The higher prevalence of viremia observed in Chinese carriers, along with the correlation of viremia with a measure of liver damage, may explain the higher risk of HCC in Chinese HBV carriers. Nevertheless, HBV carriers in Senegal are still at substantial risk for development of HCC despite low levels of viral replication throughout most of adult life. Second, the correlation between duration of virus production and age of infection (often perinatal in China vs. childhood in Senegal) is consistent with the idea that the shutdown of viremia is dependent on a more vigorous immune response. This could be at least partially tested by comparing the level of viral antigen-primed T cells in blood samples collected from these two populations. Finally, follow-up studies of small numbers of patients with chronic hepatitis B, successfully treated with interferon (IFN), suggest that liver damage is ameliorated (66, 94). Nonetheless, the studies of the Haimen and Senegalese cohorts described above raise the possibility that antiviral therapy to eliminate replicating virus from adults that were infected as children will reduce but not eliminate their risk of liver cancer.

Unfortunately, the role of host genetics and environmental factors in these outcomes is still very poorly understood. Thus, conclusions about causes for the differences in the relative risk of HCC in different populations must still be evaluated with caution. Nonetheless, the effort is essential. For example, one obvious lesson from all population-based studies is that the incidence of liver cancer is always less than 100%, even for long-lived HBV carriers. One goal in studying these populations is to determine if all individuals in a population are at equal risk, or if there are high- and low-risk carriers. The work already done suggests that duration of viremia is a risk factor which varies between individuals in a population, but it is still not known if this is the case. That is, while the duration of infection within a population has been associated with an increase in HCC incidence, there is still no proof that the excess cases of HCC occurred in individuals with sustained viremia; alternatively, prolonged viremia and HCC may merely have a common cause. The hope is that more knowledge of HCC risk in individual carriers will be gained as prospective and retrospective studies of various populations continue. Ultimately, it will be necessary to connect the correlative information obtained from epidemiological studies to a more detailed understanding of the biology of chronic infection.

Infection in the Liver

A variety of studies have suggested that every hepatocyte in the liver is susceptible to hepadnavirus infection. Therefore, considering the high-titer viremia ($>10^9$ virus per ml) that characterizes the peak of transient and, presumably, the early stages of chronic infections, initial infection of the entire hepatocyte population is probably a normal event in chronic carriers. What is surprising is that infections (transient) can rapidly clear, in a matter of a few weeks after such a peak (59, 60, 103), and that chronic carriers can eventually stop producing virus detectable by current assays, even though HBsAg production continues. (In this situation, the individuals are still considered to be chronically infected, though they may not actually carry any infectious or replicating virus.) It is possible that similar mechanisms may underlie cessation of virus production in these two very different situations. Alternatively, some of the underlying mechanisms may be quite different.

The clearance of transient infections may involve suppression of virus replication within individual hepatocytes, immune-mediated killing of at least some of the infected cells, and proliferation of the remainder. Studies with mice carrying an HBV transgene that supports virus replication within the liver suggest that cytokines produced by the immune response, particularly tumor necrosis factor-alpha, alpha-IFN (IFN-α), and IFN-γ, induce degradation of viral proteins (43, 44) and RNAs (141) within hepatocytes. Intermediates in viral DNA synthesis, which are found in viral nucleocapsids, are also degraded. Degradation of viral RNA has also been observed following induction of an antiviral immune response by intramuscular injection of a plasmid expressing HBsAg, in a transgenic lineage that produces HBsAg in the liver (80). It is not known, however, if cytokines can induce degradation of CCC DNA within nondividing hepatocytes.

Thus, an unresolved issue is how CCC DNA is eliminated during clearance of transient infections and during cessation of virus production in chronic infections. Studies of DHBV-infected primary duck hepatocytes by Civitico and Locarnini (16) suggested that CCC DNA has a short half-life, on the order of 3 to 5 days, at least in nondividing cells. Therefore, suppression by cytokines of new viral DNA synthesis in the cytoplasm for only a few weeks should lead to curing of the infected cells due to the spontaneous loss of the 5 to 50 copies of CCC DNA. However, there is no evidence for a rapid loss of CCC DNA from the liver of "healthy" carriers when viral DNA synthesis is inhibited by antiviral chemotherapy (33, 78). This dichotomy may reflect a true difference between hepatocyte cultures and hepatocytes in situ. Alternatively, the inhibition of viral DNA synthesis achieved in the in vivo studies may have been inadequate to completely shut down formation of new CCC DNA. Irrespective, there are two other pathways by which CCC DNA may be lost. One, destruction of infected hepatocytes by cytotoxic T cells, is of obvious importance (see chapter 9). The other possibility, at least in theory, is loss of CCC DNA during mitosis. Hepatocytes rarely divide. However, these cells are induced to proliferate, during transient infections, to replace hepatocytes destroyed by the immune response, usually via apoptosis (see, e.g., reference 59). Thus, it is possible that CCC DNA, even though present in chromatin structures (93), is lost as the cell undergoes division. Whatever the mechanism for CCC DNA loss from hepatocytes, it is clear that such a loss does occur. Hepatocytes populating the liver at the end of a transient infection can arise from hepatocytes that had once been infected (60). Current findings do not support the alternative that infected hepatocytes were all destroyed and replaced by proliferation and maturation of an uninfected progenitor cell population. Thus, there are natural mechanisms for curing hepatocytes, or their progeny, of a hepadnavirus infection.

Cessation of virus production during long-term HBV infections (see following section) may be achieved by mechanisms similar to those involved in transient infections, especially if cessation is coincident with a spontaneous episode of acute hepatitis (16). Another possible mechanism is that integrated viral DNA that accumulates during long-term infections functions to shut down virus replication. Studies of integrated DNA in tumors suggests that this integrated DNA is generally rearranged, incomplete, or otherwise mutated so that it is not competent to initiate new cycles of virus replication (52, 91, 99, 121, 159). This may be because the linear viral DNA (Fig. 2) is a preferred precursor for integration, and even if the complete linear DNA were integrated it would, in this conformation, be unable to direct synthesis of the pregenomic RNA, which is also the mRNA for core and polymerase proteins. However, integrated DNA often appears to retain the capacity to direct synthesis of mRNAs for the viral envelope proteins, which are known inhibitors of CCC DNA synthesis for DHBV and are presumed to have a similar role for the other hepadnaviruses (Fig. 2). By inhibiting new CCC DNA formation, HBsAg could lead to a shutdown of virus production as CCC DNA is diluted by hepatocyte proliferation or by spontaneous degradation, as suggested by Civitico and Locarnini (16). Thus, the accumulation of integrated viral DNA may serve to reduce the viral load by curing some hepatocytes of replicating virus, and this may also facilitate clearance during a spontaneous activation of the immune response.

How Does Chronic Infection Cause Liver Cancer?

Persistent, albeit low-level, liver damage appears to be a characteristic of chronic, productive infections. The damage appears to be entirely immune mediated (see chapter 9), and there is no evidence that virus replication is intrinsically cytopathic. For instance, chronic infection in ducks following vertical transmission is essentially apathogenic and does not cause a consistent pattern of liver disease or appear to cause HCC in the normal lifetime of the duck. The idea that chronic liver damage may be a major factor in the ontogeny of liver cancer has been suggested by many studies. One of the more dramatic discoveries for those working with HBV was the finding that overexpression of one of the viral envelope proteins in the liver of transgenic mice leads to liver cancer (15). This protein, the largest of the three viral envelope proteins, accumulates in the endoplasmic reticulum of hepatocytes when overexpressed relative to the other envelope proteins. This accumulation ultimately causes hepatocyte death, leading to an increased rate of hepatocyte proliferation to maintain liver mass (14; see also reference 37). This increased rate of hepatocyte proliferation is then presumably a major contributor to the outgrowth of HCC in these transgenic mice.

However, proliferation by itself is not oncogenic, and other changes must also occur. Phenotypic changes in hepatocytes are in fact observed in response to the physical damage caused by the transgene, with selective outgrowth of immature and perhaps transformed hepatocytes which are no longer susceptible to injury and which ultimately progress to HCC (15, 47). Overexpression of the large envelope protein, however, is probably not a major source of liver damage in natural chronic hepadnavirus infections. Two other sources of hepatocyte damage in chronic hepadnavirus infection are usually envisioned. One is the chronic immune response to viral antigens which is a hallmark of these infections. This leads to cell death and compensatory proliferation, which presumably will contribute to the selective outgrowth of altered hepatocytes. The other is the effect, generally inapparent, of chronic infection of the hepatocyte population. Though infection is not usually considered to be intrinsically cytopathic, it is likely that many changes which may contribute to cellular transformation are induced in hepatocytes by the viral gene products (see, e.g., references 31, 53, 63, 77, 155). Based on many studies, it is now widely believed that immune-mediated damage is a driving force for hepatocyte proliferation; on the other hand, a combination of inapparent cytopathic effects caused by infection of the hepatocyte and more direct damage to the host cell as a byproduct of the immune response are probably responsible for the changes in gene expression that lead to unregulated cell proliferation.

Inflammation, which is an integral part of the immune response to HBV-infected hepatocytes, may play a direct role not just in liver cell killing and cell proliferation, but also in damage to DNA. Macrophages, especially Kupffer cells and granulocytes, produce superoxide and other reactive oxygen species as a result of phagocytosis of HBV virions and surface antigen particles (147) and damaged hepatocytes (47, 122). Reactive oxygen molecules in high concentration can kill cells and in lower concentration damage DNA. The reaction of oxygen free radicals with DNA most frequently results in C→T substitutions (42), but G→C transversions also occur (87). Accumulation of 8-hydroxydeoxyguanosine (8-OH-dg), although not an entirely specific marker for oxidative DNA damage, has been used as a biomarker of such damage. 8-OH-dg levels are elevated in human livers with chronic hepatitis, but not in liver tissues with cirrhosis or liver cancer (122). On the other hand, in the transgenic mouse model of liver cancer in which there is overproduction and retention of the large envelope protein of HBV, 8-OH-dg accumulates continuously in hepatocyte DNA throughout the life of the animal, apparently as a consequence of the activation of resident macrophages in the liver (47). Thus, the available information is compatible with oxidative DNA damage being a critical mutagenic agent in HBV-induced HCC, but the data are insufficient at this time to draw a firm conclusion.

Ongoing immune-mediated destruction of hepatocytes and oxidative DNA damage are thought to decrease if virus replication within the liver ceases or declines in the carrier, as it often does (see below). However, HCC often appears in carriers after virus replication has ceased, suggesting that neither virus replication nor active hepatitis, which stimulates hepatocyte proliferation, is needed during the final stages of the tumorigenic process (51). Virtually nothing is known about the evolution from normal to fully transformed hepatocytes. However, recent investigations with animal models and human tumors have begun to reveal some features of transformed cells.

Studies of viral pathogenesis and liver cancer in animal models have focused on the ground squirrel and woodchuck. Woodchucks chronically infected with WHV develop liver cancer with essentially 100% incidence by 2 to 4 years of age (65, 104). The incidence of liver cancer in infected ground squirrels is somewhat lower, about 50% in GSHV carriers over 4 years of age (84, 85; P. Marion, personal communication). Some evidence suggests that part of the difference between cancer development in woodchucks and ground squirrels is due to differences in the viruses. GSHV can establish chronic infection in the woodchuck, but the appearance of HCC takes about 18 months longer than in WHV-infected woodchucks (114).

In addition to differences in cancer incidence in different species, there may also be differences in the mechanisms underlying oncogenic transformation. Specifically, activation of a *myc* oncogene in association with nearby integration of WHV DNA is observed in about 90% of woodchuck HCCs (27, 32, 117, 145, 154, 157). About 10% of the time insertion is at c-*myc* or N-*myc*1. Of the remainder, about half occur at N-*myc*2 and half occur about 180 kb downstream of N-*myc*2. In contrast, c-*myc* activation is the common feature of liver cancer in ground squirrels, though not usually in association with GSHV integration (49, 138, 139). Thus, the two closely related viruses do not induce HCC through a shared mechanism for induction of host gene expression.

Activation of some common oncogene by virus integration has not so far been described in HCCs in HBV carriers. In rare instances, however, this has been seen. For instance, HBV DNA integration near the genes for the retinoic acid receptor and cyclin A have been reported (24, 151). In contrast to the rare evidence for oncogene activation via insertional mutagenesis, there is abundant evidence for mutation of p53 in human HCC. In some studies, about half the tumors have been found to contain mutant p53 (codon 249) (e.g., references 9, 11, 55, 56, 113). The high incidence of p53 mutation in HCCs from some HBV carriers has been attributed to a higher level of exposure of these patients to environmental carcinogens, to which some individuals may be especially susceptible (88). Some investigators have presented evidence which suggests that p53 inactivation or functional redirection may, in fact, be a feature of all HBV-associated HCCs. In particular, evidence has been reported that the product of the viral X gene can inhibit p53-induced apoptosis, possibly by binding directly to the p53 protein (152, 153). Thus, X may overcome the suppressive effects of p53 on cell division, particularly in cells that have sustained DNA damage. In contrast to the situation in human HCC, mutation of p53 is not a common feature of HCCs in infected woodchucks and ground squirrels (109).

Interestingly, insulin-like growth factor II (IGF II) expression has been reported to occur in about 50% of woodchuck HCCs, but not in normal liver (34, 108, 160). Recent cell culture experiments suggest that *myc* overexpression in hepatocytes leads to apoptosis, which in turn can be inhibited by IGF II (144, 161). IGF II expression was also detected in more than half of the HCCs from HBV carriers that were characterized in a recent study (25); however, there have been no reports so far that elevated expression of *myc* is common in human HCC.

Finally, the possibility cannot be ruled out that novel viral gene products have a role in carcinogenesis, as suggested by the observation that X and a truncated envelope protein containing preS2 and S domains (Fig. 2) can assume the role of transcriptional transactivators (67, 77).

Thus, considering the apparent differences in cancer incidence and in patterns of oncogene activation in the different species, it is likely that oncogenesis may follow somewhat different molecular routes. On the other hand, it seems likely that a driving force for liver cancer in all three hosts is the chronic liver damage caused by the hepadnavirus infection.

ANTIVIRAL THERAPY

While there is a great need to understand in more detail how liver cancers develop in even one of the hosts for hepadnavirus infection, it is even more essential to know whether the cancer incidence would be reduced, or the progression to liver cancer slowed, if the ongoing liver damage were stopped early enough. Virus replication and the associated liver damage can be suppressed, at least in some patients, by IFN-α administration (102). However, while it is clear that such therapy can arrest the clinical manifestations of hepatitis, it is not known if there is a change in the risk of developing HCC. Ideally, this issue might be studied with the animal models. In the woodchuck, it is possible that the age at which HCC arises can be increased by antiviral therapy (114). The ground squirrel may also be suitable for such studies because of the somewhat lower incidence and slower progression to HCC (83–85). However, this model has been less widely used, and a breeding colony for large-scale production of experimental animals has not been established as it has for the woodchuck (18). Ultimately, the relationship between the duration of ongoing liver damage and the cancer risk may be determined, in the first instance, from clinical studies of HBV-infected patients and epidemiological studies of selected populations of chronic carriers of HBV. The efficacy of antiviral therapy in reducing cancer risk may take longer to evaluate.

The rationale for developing new, more efficacious antiviral therapies is actually twofold. First, in some chronic carriers the liver damage caused by the immune response to replicating virus is, in itself, life-threatening (119). Second, as already noted, it is hoped that an early termination of virus replication, even in "healthy" carriers, may still reduce the risk of developing cirrhosis and liver cancer, but current therapies are of low efficacy. Generally, approaches to antiviral therapies fall into two categories. The first category includes those that stimulate or augment natural defense mechanisms, with the goal of stimulating the same responses that lead to virus clearance during transient infections. The second type of therapy includes those that attempt to eliminate the virus by directly blocking some step in virus reproduction. IFN-α therapy, which probably falls into the first category, has been in use for many years, though the response rate is relatively low, perhaps 20% after correcting for spontaneous remissions (54, 102). Not unexpectedly, this works best in patients with the most active hepatitis (i.e., those with the most active immune response to their infection). These are often patients who have acquired their infections as adults. Unfortunately, the majority of patients worldwide have infections acquired in early childhood. These individuals apparently have a lower grade of hepatitis which can eventually lead to cirrhosis and HCC, but which is usually not augmented by IFN treatment enough to facilitate clearance of replicating virus (75). Thus, additional approaches are needed to treat these individuals.

Therapy with a nucleoside analog inhibitor of virus replication appears likely, from preliminary results, to have the same limitation as IFN. That is, the most rapid decline in viral burden in the liver, at least for treatment periods of less than 12 months, appears to be achieved in individuals with active liver disease (33, 96). In theory, longer-term, perhaps life-long treatment with such inhibitors may eventually control and then eliminate the virus of all carriers. The problem with this idea is, not unexpectedly, the evolution of drug-resistant virus (73, 136). With HBV carriers, virus that is resistant to lamivudine (3TC) starts to grow out after a year or so of therapy. This problem may eventually be overcome by combination therapy with new analogs or with nucleoside analogs now under evaluation (e.g., references 38, 64, 118). A key issue here may be whether drug resistance develops as a consequence of mutations that appear during therapy, or if drug-resistant virus preexists, but can only become prevalent as drug-susceptible virus is slowly cleared from hepatocytes.

New approaches to immunotherapy may also prove useful, either alone or in conjunction with nucleoside therapy. Several recent papers, discussed earlier, indicate that the immune system can induce degradation of essentially all of the viral proteins and RNAs present in the liver of HBV transgenic mice, through pathways mediated by IFN-α and IFN-γ and tumor necrosis factor-alpha (44, 45, 141). A key point of these studies was that these pathways may be noncytocidal and therefore should not require an antigen-specific stimulus. This was established in a study in which it was shown that the down-regulation of HBV could be achieved by infection with lymphocytic choriomeningitis virus (43). This virus infects the liver, but infection is confined primarily to macrophages, not hepatocytes. Thus, delivery of a similar infectious but clinically unimportant virus to the liver of HBV carriers could provide a similar immune stimulus. The transient suppression of HBV replication during acute hepatitis delta virus and hepatitis A virus superinfections may be examples of such effects (22, 92). As already noted, suppression of viral RNA and protein expression in the liver of transgenic mice have also been achieved using a DNA vaccine to stimulate T- and B-cell responses to HBsAg (80). Cytotoxic T-lymphocyte killing of hepatocytes appeared minimal, and the global effect on the liver was apparently due, as in the earlier studies with HBV transgenic mice, to cytokines produced by the hepatic infiltrate.

However, while these immunologic approaches could have major implications for treatment of HBV carriers, it is important to remember the dichotomy between the apparently noncytocidal pathways identified in the mice and the obvious correlation between liver damage and responsiveness to IFN or nucleoside therapy in humans. As discussed earlier, in the context of how transient infections are resolved, a central issue is how CCC DNA is eliminated from the cell nucleus. If CCC DNA can be eliminated, noncytocidally, from nondividing hepatocytes, then the "bystander approach" observed with lymphocytic choriomeningitis virus in transgenic mice may indeed be efficacious. Alternatively, if hepatocyte death and/or proliferation is necessary to eliminate this DNA, down-regulating viral gene expression may actually be counterproductive, since infected hepatocytes may then no longer be targets for CTL killing. The result would be a reduction in liver disease activity, but an environment which would also favor virus survival. As Chisari and colleagues have pointed out, it is possible that the posttranscriptional suppression of virus gene expression by cytokines such as tumor necrosis factor-alpha and IFN is an adaptation of the virus to avoid the immune response and becomes a basis for virus clearance only when the immune response is strong enough (13).

OUTLOOK

The replication scheme of HBV has been understood, in general terms, for almost 16 years. However, efforts to understand the biology of transient and chronic infection have been difficult and are just beginning to elucidate some of the most pressing issues. There is now some hope, at least on theoretical grounds, that truly effective antiviral therapies may be available in the next few years. There is also reason to hope that an effective antiviral therapy would reduce the risk of liver cancer. On the other hand, there is still much that is unclear. The study of the role of environmental and genetic factors in cancer risk is still in its infancy. It is still not possible to predict which carriers will develop liver cancer and when it will first appear. And even if this knowledge were available, it is still not determined how many individuals could then be helped. Finally, the treatment of liver cancers that do appear, as with many solid tumors, is problematic. There is no effective chemotherapy for these tumors and surgery is only effective, in general, if the tumors are removed when very small. In theory, all of these issues would disappear if we had universal vaccination. However, the general question of how to treat other chronic viral infections will remain, and the current efforts to understand HBV will almost certainly have applicability to other virus infections.

Acknowledgments
We thank C. Seeger, J. Summers (University of New Mexico), and J. Taylor for helpful comments. Work in the authors' laboratories is supported by Public Health Service grants AI-18641, CA-40737, CA-42542, CA-06927, GM-35535, and RR-05509 from the National Institutes of Health and by an appropriation from the Commonwealth of Pennsylvania.

REFERENCES

1. **Akarca, U. S., S. Greene, and A. S. F. Lok.** 1994. Detection of hepatitis B virus mutant in asymptomatic HBsAg-positive family members. *Hepatology* **19:**1366–1370.
2. **Akarca, U. S., and A. S. F. Lok.** 1995. Naturally occurring hepatitis B virus core gene mutations. *Hepatology* **22:**50–60.
3. **Beasley, R. P., and L.-Y. Hwang.** 1991. Overview on the epidemiology of hepatocellular carcinoma, p. 532–535. *In* F. B. Hollinger, S. M. Lemon, and H. S. Margolis (ed.), *Viral Hepatitis and Liver Disease.* The Williams & Wilkins Co., Baltimore.
4. **Beasley, R. P., C. Trepo, C. E. Stevens, and W. Szmuness.** 1977. The e antigen and vertical transmission of hepatitis B surface antigen. *Am. J. Epidemiol.* **105:**94–98.
5. **Berquist, K. R., J. M. Peterson, B. L. Murphy, J. W. Ebert, J. E. Maynard, and R. H. Purcell.** 1973. Hepatitis B antigens in serum and liver of chimpanzees acutely infected with hepatitis B virus. *J. Infect. Dis.* **127:**648–652.
6. **Blum, H. E., A. T. Haase, and G. N. Vyas.** 1984. Molecular pathogenesis of hepatitis B virus infection: simultaneous detection of viral DNA and antigens in paraffin-embedded liver sections. *Lancet* **2:**771–775.
7. **Blumberg, B. S., B. J. S. Gerstley, D. A. Hungerford, W. T. London, and A. I. Sutnik.** 1967. A serum antigen (Australia antigen) inn Down's syndrome, leukemia and hepatitis. *Ann. Intern. Med.* **66:**924–931.
8. **Brechot, C.** 1987. Hepatitis B virus (HBV) and hepatocelular carcinoma. HBV status and its implication. *J. Hepatol.* **4:**269.
9. **Bressac, B., M. Kew, J. Wands, and M. Ozturk.** 1991. Selective G to T mutations of p53 gene in hepatocellular carcinoma from southern Africa. *Nature* **350:**429–431.

10. **Bruss, V., and K. Vieluf.** 1995. Functions of the internal pre-S domain of the large surface protein in hepatitis B virus particle morphogenesis. *J. Virol.* **69:**6652–6657.

11. **Buetow, K. H., V. C. Sheffield, M. Zhu, T. Zhou, F. M. Shen, O. Hino, M. Smith, B. J. McMahon, A. P. Lanier, W. T. London, A. G. Redeker, and S. Govindarajan.** 1992. Low frequency of p53 mutations observed in a diverse collection of primary hepatocellular carcinomas. *Proc. Natl. Acad. Sci. USA* **89:**9622–9626.

12. **Chisari, F. V.** 1995. Hepatitis B virus immunopathogenesis. *Annu. Rev. Immunol.* **13:** 29–60.

13. **Chisari, F. V., P. Filippi, J. Buras, A. McLachlan, H. Popper, C. A. Pinkert, R. D. Palmiter, and R. L. Brinster.** 1987. Structural and pathological effects of synthesis of hepatitis B virus large envelope polypeptide in transgenic mice. *Proc. Natl. Acad. Sci. USA* **84:**6909–6913.

14. **Chisari, F. V., K. Klopchin, T. Moriyama, C. Pasquinelli, H. A. Dunsford, S. Sell, C. A. Pinkert, R. L. Brinster, and R. D. Palmiter.** 1989. Molecular pathogenesis of hepatocellular carcinoma in hepatitis B virus transgenic mice. *Cell* **59:**1145–1156.

15. **Chu, C. M., Y. F. Liaw, C. C. Pao, and M. J. Huang.** 1989. The etiology of acute hepatitis superimposed upon previously unrecognized asymptomatic HBsAg carriers. *Hepatology* **9:**452–456.

16. **Civitico, G. M., and S. A. Locarnini.** 1994. The half-life of duck hepatitis B virus supercoiled DNA in congenitally infected primary hepatocyte cultures. *Virology* **203:** 81–89.

17. **Colombo, M., R. De Franchis, E. Del Ninno, A. Sangiovanni, C. De Fazio, M. Tommasini, M. Donato, A. Piva, V. Di Carlo, and N. Dioguardi.** 1991. Hepatocellular carcinoma in Italian patients with cirrhosis. *N. Engl. J. Med.* **325:**675–680.

18. **Concannon, P., P. Roberts, B. Ball, D. Schlafer, X. Yang, B. Baldwin, and B. Tennant.** 1997. Estrus, fertility, early embryo development and autologous embryo transfer in laboratory woodchucks (Marmota monax). *Lab. Anim. Sci.* **47:**63–74.

19. **Condreay, L. D., C. E. Aldrich, L. Coates, W. S. Mason, and T. T. Wu.** 1990. Efficient duck hepatitis B virus production by an avian liver tumor cell line. *J. Virol.* **64:** 3249–3258.

20. **Cottone, M., M. Turri, M. Caltagirone, P. Parisi, A. Orlando, G. Fiorentino, R. Virdone, G. Fusco, R. Grasso, R. G. Simonetti, and L. Pagliaro.** 1994. Screening for hepatocellular carcinoma in patients with Child's A cirrhosis: an 8-year prospective study by ultrasound and alphafetoprotein. *J. Hepatol.* **21:**1029–1034.

21. **Dabeva, M. D., G. Alpini, E. Hurston, and D. A. Shafritz.** 1993. Models for hepatic progenitor cell activation. *Proc. Soc. Exp. Biol. Med.* **204:**242–252.

22. **Davis, G. L., J. H. Hoofnagle, and J. G. Waggoner.** 1984. Acute type A hepatitis during chronic hepatitis B virus infection: association of depressed hepatitis B virus replication with appearance of endogenous alpha interferon. *J. Med. Virol.* **14:**141–147.

23. **deFranchis, R., G. Meucci, M. Vecchi, M. Tatarella, M. Colombo, N. E. Del, M. G. Rumi, M. F. Donato, and G. Ronchi.** 1993. The natural history of asymptomatic hepatitis B surface antigen carriers. *Ann. Intern. Med.* **118:**191–194.

24. **Dejean, A., L. Bougueleret, K. H. Grzeschik, and P. Tiollais.** 1986. Hepatitis B virus DNA integration in a sequence homologous to v-erb-A and steroid receptor genes in a hepatocellular carcinoma. *Nature* **322:**70–72.

25. **D'Errico, A., W. F. Grigioni, M. Fiorentino, P. Baccarini, E. Lamas, M. S. De, G. Gozzetti, A. M. Mancini, and C. Brechot.** 1994. Expression of insulin-like growth factor II (IGF-II) in human hepatocellular carcinomas: an immunohistochemical study. *Pathol. Int.* **44:**131–137.

26. **Dodd, R. Y., and N. Nath.** 1987. Increased risk for lethal forms of liver disease among HBsAg-positive blood donors in the United States. *J. Virol. Methods* **17:**81–94.

27. **Etiemble, J., C. Degott, C. A. Renard, G. Fourel, B. Shamoon, T. L. Vitvitski, T. Y. Hsu, P. Tiollais, C. Babinet, and M. A. Buendia.** 1994. Liver-specific expression and high oncogenic efficiency of a c-myc transgene activated by woodchuck hepatitis virus insertion. *Oncogene* **9:**727–737.

28. **Fausto, N.** 1990. Hepatocyte differentiation and liver progenitor cells. *Curr. Opin. Cell Biol.* **2:**1036–1042.

29. **Fausto, N., J. M. Lemire, and N. Shiojiri.** 1993. Cell lineages in hepatic development and the identification of progenitor cells in normal and injured liver. *Proc. Soc. Exp. Biol. Med.* **214:**237–241.

30. **Feitelson, M. A., and L.-X. Duan.** 1997. Hepatitis B virus X antigen in the pathogenesis of chronic infections and the development of hepatocellular carcinoma. *Am. J. Pathol.* **150:**1141–1157.

31. **Feitelson, M. A., M. Zhu, L. X. Duan, and W. T. London.** 1993. Hepatitis B x antigen and p53 are associated in vitro and in liver tissues from patients with primary hepatocellular carcinoma. *Oncogene* **8:**1109–1117.

32. **Fourel, G., J. Couturier, Y. Wei, F. Apiou, P. Tiollais, and M. A. Buendia.** 1994. Evidence for long-range oncogene activation by hepadnavirus insertion. *EMBO J.* **13:**2526–2534.

33. **Fourel, I., J. M. Cullen, J. Saputelli, C. E. Aldrich, P. Schaffer, D. Averett, J. Pugh, and W. S. Mason.** 1994. Evidence that hepatocyte turnover is required for rapid clearance of DHBV during antiviral therapy of chronically infected ducks. *J. Virol.* **12:**8321–8330.

34. **Fu, X.-X., C. Y. Su, Y. Lee, R. Hintz, L. Biempica, R. Snyder, and C. Rogler.** 1988. Insulinlike growth factor II expression and oval cell proliferation associated with hepatocarcinogenesis in woodchuck hepatitis virus carriers. *J. Virol.* **62:**3422–3430.

35. **Fukao, A.** 1985. An epidemiological study on relationship between hepatitis B virus and hepatocellular carcinoma. *Jpn. J. Gastroenterol.* **82:**232–238.

36. **Ganem, D.** 1991. Assembly of hepadnaviral virions and subviral particles. *Curr. Top. Microbiol. Immunol.* **168:**61–83.

37. **Gerber, M. A., F. Schaffner, and F. Paronetto.** 1972. Immuno-electron microscopy of hepatitis B antigen in liver. 1. *Proc. Soc. Exp. Biol. Med.* **140:**1334–1339.

38. **Gish, R. G., J. Y. Lau, L. Brooks, J. W. Fang, S. L. Steady, J. C. Imperial, K. R. Garcia, C. O. Esquivel, and E. B. Keeffe.** 1996. Ganciclovir treatment of hepatitis B virus infection in liver transplant recipients. *Hepatology* **23:**1–7.

39. **Gong, S. S., A. D. Jensen, and C. E. Rogler.** 1996. Loss and acquisition of duck hepatitis B virus integrations in lineages of LMH-D2 chicken hepatoma cells. *J. Virol.* **70:**2000–2007.

40. **Gong, S. S., A. D. Jensen, H. Wang, and C. E. Rogler.** 1995. Duck hepatitis B virus integrations in LMH chicken hepatoma cells: identification and characterization of new episomally derived integrations. *J. Virol.* **69:**8102–8108.

41. **Grisham, J. W.** 1962. A morphologic study of deoxyribonucleic acid synthesis and cell proliferation in regenerating rat liver; autoradiography with thymidine-H3. *Cancer Res.* **22:**842–849.

42. **Grosovsky, A. J., J. G. DeBoer, P. J. deJog, E. A. Drobetsky, and B. W. Glickman.** 1988. Base substitutions, frame shifts, and small deletions constitute ionizing radiation induced point mutations in mammalian cells. *Proc. Natl. Acad. Sci. USA* **85:**185–188.

43. **Guidotti, L. G., P. Borrow, M. V. Hobbs, B. Matzke, I. Gresser, M. B. Oldstone, and F. V. Chisari.** 1996. Viral cross talk: intracellular inactivation of the hepatitis B virus during an unrelated viral infection of the liver. *Proc. Natl. Acad. Sci. USA* **93:**4589–4594.

44. **Guidotti, L. G., T. Ishikawa, M. V. Hobbs, B. Matzke, R. Schreiber, and F. V. Chisari.** 1996. Intracellular inactivation of the hepatitis B virus by cytotoxic T lymphocytes. *Immunity* **4**:25–36.

45. **Guidotti, L. G., B. Matzke, H. Schaller, and F. V. Chisari.** 1995. High-level hepatitis B virus replication in transgenic mice. *J. Virol.* **69**:6158–6169.

46. **Hadziyannis, S., E. Tabor, E. Kaklamani, A. Tzonou, S. Stuver, N. Tassopoulos, N. Mueller, and D. Trichopoulos.** 1995. A case-control study of hepatitis B and C virus infections in the etiology of hepatocellular carcinoma. *Int. J. Cancer* **60**:627–631.

47. **Hagen, T. M., S. Huang, J. Curnutte, P. Fowler, V. Martinez, C. M. Wehr, B. N. Ames, and F. C. Chisari.** 1994. Extensive oxidative damage in hepatocytes of transgenic mice with chronic active hepatitis destined to develop chronic hepatitis. *Proc. Natl. Acad. Sci. USA* **91**:12808–12812.

48. **Halpern, M. S., J. M. England, D. T. Deery, D. J. Petcu, W. S. Mason, and K. L. Molnar-Kimber.** 1983. Viral nucleic acid synthesis and antigen accumulation in pancreas and kidney of Pekin ducks infected with duck hepatitis B virus. *Proc. Natl. Acad. Sci. USA* **80**:4865–4869.

49. **Hansen, L. J., B. C. Tennant, C. Seeger, and D. Ganem.** 1993. Differential activation of *myc* gene family members in hepatic carcinogenesis by closely related hepatitis B viruses. *Mol. Cell. Biol.* **13**:659–667.

50. **Heermann, K. H., U. Goldmann, W. Schwartz, T. Seyffarth, H. Baumgarten, and W. H. Gerlich.** 1984. Large surface proteins of hepatitis B virus containing the pre-s sequence. *J. Virol.* **52**:396–402.

51. **Henkler, F., N. Waseem, M. H. Golding, M. R. Alison, and R. Koshy.** 1995. Mutant p53 but not hepatitis B virus X protein is present in hepatitis B virus-related human hepatocellular carcinoma. *Cancer Res.* **55**:6084–6091.

52. **Hino, O., T. B. Shows, and C. E. Rogler.** 1986. Hepatitis B virus integration site in hepatocellular carcinoma at chromosome 17;18 translocation. *Proc. Natl. Acad. Sci. USA* **83**:8338–8342.

53. **Hohne, M., S. Schaefer, M. Seifer, M. A. Feitelson, D. Paul, and W. H. Gerlich.** 1990. Malignant transformation of immortalized transgenic hepatocytes after transfection with hepatitis B virus DNA. *EMBO J.* **9**:1137–1145.

54. **Hoofnagle, J. H., and A. M. DiBisceglie.** 1997. The treatment of chronic viral hepatitis. *N. Engl. J. Med.* **336**:347–356.

55. **Hosono, S., M. J. Chou, C. S. Lee, and C. Shih.** 1993. Infrequent mutation of p53 gene in hepatitis B virus positive primary hepatocellular carcinomas. *Oncogene* **8**:491–496.

56. **Hsia, C. C., D. J. Kleiner, C. A. Axiotis, B. A. Di, A. M. Nomura, G. N. Stemmermann, and E. Tabor.** 1992. Mutations of p53 gene in hepatocellular carcinoma: roles of hepatitis B virus and aflatoxin contamination in the diet [see comments]. *J. Natl. Cancer Inst.* **84**:1638–1641.

57. **International Agency for Research on Cancer.** 1994. *IARC Monographs on the Evaluation of Carcinogenic Risk to Humans*, vol. 59, *Hepatitis Viruses*. International Agency for Research on Cancer, Lyon, France.

58. **Jansen, R. W., L. C. Johnson, and D. R. Averett.** 1993. High-capacity in vitro assessment of anti-hepatitis B virus compound selectivity by a virion-specific polymerase chain reaction assay. *Antimicrob. Agents Chemother.* **37**:441–447.

59. **Jilbert, A. R., T.-T. Wu, J. M. England, P. de la M. Hall, N. Z. Carp, A. P. O'Connell, and W. S. Mason.** 1992. Rapid resolution of duck hepatitis B virus infections occurs after massive hepatocellular involvement. *J. Virol.* **66**:1377–1388.

60. **Kajino, K., A. R. Jilbert, J. Saputelli, C. E. Aldrich, J. Cullen, and W. S. Mason.**

1994. Woodchuck hepatitis virus infections: very rapid recovery after a prolonged viremia and infection of virtually every hepatocyte. *J. Virol.* **68**:5792–5803.

61. **Kashala, L. O., B. Conne, K. Y, P. C. Frei, P. H. Lambert, M. R. Kalengayi, and M. Essex.** 1992. Hepatitis B virus, alpha-fetoprotein synthesis, and hepatocellular carcinoma in Zaire. *Liver* **12**:330–340.

62. **Kawaguchi, T., K. Nomura, Y. Hirayama, and T. Kitagawa.** 1987. Establishment and characterization of a chicken hepatocellular carcinoma cell line, LMH. *Cancer Res.* **47**:4460–4464.

63. **Kekule, A. S., U. Lauer, M. Meyer, W. H. Caselmann, P. H. Hofschneider, and R. Koshy.** 1990. The preS2/S region of integrated hepatitis B virus DNA encodes a transcriptional transactivator. *Nature* **343**:457–461.

64. **Korba, B. E., H. Xie, K. N. Wright, W. E. Hornbuckle, J. L. Gerin, B. C. Tennant, and K. Y. Hostetler.** 1996. Liver-targeted antiviral nucleosides: enhanced antiviral activity of phosphatidyl-dideoxyguanosine versus dideoxyguanosine in woodchuck hepatitis virus infection in vivo. *Hepatology* **23**:958–963.

65. **Korba, B. E., F. V. Wells, B. Baldwin, P. J. Cote, B. C. Tennant, H. Popper, and J. L. Gerin.** 1989. Hepatocellular carcinoma in woodchuck hepatitis virus-infected woodchucks: presence of viral DNA in tumor tissue from chronic carriers and animals serologically recovered from acute infections. *Hepatology* **9**:461–470.

66. **Korenman, J., B. Baker, J. Waggoner, J. E. Everhart, A. M. Di Bisceglie, and J. H. Hoofnagle.** 1991. Long-term remission of chronic hepatitis B after alpha-interferon therapy. *Ann. Intern. Med.* **114**:629–634.

67. **Lauer, U., L. Weiss, M. Lipp, P. H. Hofschneider, and A. S. Kekule.** 1994. The hepatitis B virus preS2/St transactivator utilizes AP-1 and other transcription factors for transactivation. *Hepatology* **19**:23–31.

68. **Lenhoff, R. J., and J. Summers.** 1994. Construction of avian hepadnavirus variants with enhanced replication and cytopathicity in primary hepatocytes. *J. Virol.* **68:**5706–5713.

69. **Lenhoff, R. J., and J. Summers.** 1994. Coordinate regulation of replication and virus assembly by the large envelope protein of an avian hepadnavirus. *J. Virol.* **68:**4565–4571.

70. **Lien, J. M., C. E. Aldrich, and W. S. Mason.** 1986. Evidence that a capped oligoribonucleotide is the primer for duck hepatitis B virus plus-strand DNA synthesis. *J. Virol.* **57**:229–236.

71. **Lien, J. M., D. J. Petcu, C. E. Aldrich, and W. S. Mason.** 1987. Initiation and termination of duck hepatitis B virus DNA synthesis during virus maturation. *J. Virol.* **61:**3832–3840.

72. **Lin, T.-M., C.-J. Chen, S.-N. Lu, A.-S. Chang, Y.-C. Chang, S.-T. Hsu, J.-Y. Liu, Y.-F. Liaw, and W.-Y. Chang.** 1991. Hepatitis B virus e antigen and primary hepatocellular carcinoma. *Anticancer Res.* **11**:2063–2066.

73. **Ling, R., D. Mutimer, M. Ahmed, E. H. Boxall, E. Elias, G. M. Dusheiko, and T. J. Harrison.** 1996. Selection of mutations in the hepatitis B virus polymerase during therapy of transplant recipients with lamivudine. *Hepatology* **24**:711–713.

74. **Loeb, D. D., K. J. Gulya, and R. Tian.** 1997. Sequence identity of the terminal redundancies on the minus-strand DNA template is necessary but not sufficient for the template switch during hepadnavirus plus-strand DNA synthesis. *J. Virol.* **71:**152–160.

75. **Lok, A. S.** 1991. Alpha-interferon therapy for chronic hepatitis B virus infection in children and Oriental patients. *J. Gastroenterol. Hepatol.* **1**:15–17.

76. **London, W., A. Evans, K. McGlynn, K. Buetow, P. An, E. Gao, E. Ross, G.-C. Chen,**

and F.-M. Shen. 1995. Viral, host and environmental risk factors for hepatocellular carcinoma: a prospective study in Haimen City, China. *Intervirology* **38**:155–161.

77. **Luber, B., N. Arnold, M. Sturzl, M. Hohne, P. Schirmacher, U. Lauer, J. Wienberg, P. H. Hofschneider, and A. S. Kekule.** 1996. Hepatoma-derived integrated HBV DNA causes multi-stage transformation in vitro. *Oncogene* **12**:1597–1608.

78. **Luscombe, C., J. Pedersen, E. Uren, and S. Locarnini.** 1996. Long-term ganciclovir chemotherapy for congenital duck hepatitis B virus infection in vivo: effect on intra-hepatic-viral DNA, RNA, and protein expression. *Hepatology* **24**:766–773.

79. **MacDonald, R. A.** 1960. "Lifespan" of liver cells. *Arch. Intern. Med.* **107**:335–343.

80. **Mancini, M., H. Davis, P. Tiollais, and M. L. Michel.** 1996. DNA-based immunization against the envelope proteins of the hepatitis B virus. *J. Biotechnol.* **44**:47–57.

81. **Marinier, E., V. Barrois, B. Larouze, W. T. London, A. Cofer, L. Diakhate, and B. S. Blumberg.** 1985. Lack of perinatal transmission of hepatitis B virus infection in Senegal, West Africa. *J. Pediatr.* **106**:843–849.

82. **Marion, P. L., J. M. Cullen, R. R. Azcarrraga, D. M. Van, and W. S. Robinson.** 1987. Experimental transmission of duck hepatitis B virus to Pekin ducks and to domestic geese. *Hepatology* **1987**:724–731.

83. **Marion, P. L., L. S. Oshiro, D. C. Regnery, G. H. Scullard, and W. S. Robinson.** 1980. A virus in Beechey ground squirrels which is related to hepatitis B virus of man. *Proc. Natl. Acad. Sci. USA* **77**:2941–2945.

84. **Marion, P. L., H. Popper, R. R. Azcarraga, C. Steevens, M. J. V. Davelaar, G. Garcia, and W. S. Robinson.** 1987. Ground squirrel hepatitis virus and hepatocellular carcinoma, p. 337–348. *In* W. Robinson, K. Koike, and H. Will (ed.), *Hepadnaviruses*. A. R. Liss, Inc., New York.

85. **Marion, P. L., M. J. Van Davelaar, S. S. Knight, F. H. Salazar, G. Garcia, H. Popper, and W. S. Robinson.** 1986. Hepatocellular carcinoma in ground squirrels persistently infected with ground squirrel hepatitis virus. *Proc. Natl. Acad. Sci. USA* **83**:4543–4546.

86. **Mason, W. S., G. Seal, and J. Summers.** 1980. Virus of Pekin ducks with structural and biological relatedness to human hepatitis B virus. *J. Virol.* **36**:829–836.

87. **McBride, T. J., B. D. Preston, and L. A. Loeb.** 1991. Mutagenic spectrum resulting from DNA damage by oxygen radicals. *Biochemistry* **30**:207–213.

88. **McGlynn, K. A., E. A. Rosvold, E. D. Lustbader, Y. Hu, M. Clapper, T. Zhou, C. P. Wild, X.-L. Xia, A. Baffoe-Bonnie, D. Ofori-Adjei, G.-C. Chen, W. T. London, F.-M. Shen, and K. H. Buetow.** 1995. Susceptibility to hepatocellular carcinoma is associated with genetic variation in the enzymatic detoxification of aflatoxin B1. *Proc. Natl. Acad. Sci. USA* **92**:2384–2387.

89. **McMahon, B., S. Alberts, R. Wainwright, L. Bulkow, and A. Lanier.** 1990. Hepatitis B-related sequelae: prospective study in 1400 hepatitis B surface antigen-positive Alaska Native carriers. *Arch. Intern. Med.* **150**:1051–1054.

90. **Milich, D. R., J. E. Jones, J. L. Hughes, J. Price, A. K. Raney, and A. McLachlan.** 1990. Is a function of the secreted hepatitis B e antigen to induce immunologic tolerance in utero? *Proc. Natl. Acad. Sci. USA* **87**:6599–6603.

91. **Nagaya, T., T. Nakamura, T. Tokino, T. Tsurimoto, M. Imai, T. Mayumi, K. Kamino, K. Yamamura, and K. Matsubara.** 1987. The mode of hepatitis B virus DNA integration in chromosomes of human hepatocellular carcinoma. *Genes Dev.* **1**:773–782.

92. **Negro, F., B. E. Korba, B. Forzani, B. M. Baroudy, T. L. Brown, J. L. Gerin, and A. Ponzetto.** 1989. Hepatitis delta virus (HDV) and woodchuck hepatitis virus (WHV) nucleic acids in tissues of HDV-infected chronic WHV carrier woodchucks. *J. Virol.* **63**:1612–1618.

93. **Newbold, J. E., H. Xin, M. Tencza, G. Sherman, J. Dean, S. Bowden, and S. Locar-**

nini. 1995. The covalently closed duplex form of the hepadnavirus genome exists in situ as a heterogeneous population of viral minichromosomes. *J. Virol.* **69:**3350–3357.

94. **Niederau, C., T. Heintges, S. Lange, G. Goldmann, C. M. Niederau, L. Mohr, D. Haussinger.** 1996. Long-term follow-up of HBeAg-positive patients treated with interferon alfa for chronic hepatitis B. *N. Engl. J. Med.* **334:**1422–1427.

95. **Norder, H., J. W. Ebert, H. A. Fields, I. K. Mushahwar, and L. O. Magnius.** 1996. Complete sequencing of a gibbon hepatitis B virus genome reveals a unique genotype distantly related to the chimpanzee hepatitis B virus. *Virology* **218:**214–223.

96. **Nowak, M. A., S. Bonhoeffer, A. M. Hill, R. Boehme, H. C. Thomas, and H. McDade.** 1996. Viral dynamics in hepatitis B virus infection. *Proc. Natl. Acad. Sci. USA* **93:**4398–4402.

97. **Obert, S., B. B. Zachmann, E. Deindl, W. Tucker, R. Bartenschlager, and H. Schaller.** 1996. A splice hepadnavirus RNA that is essential for virus replication. *EMBO J.* **15:**2565–2574.

98. **O'Connell, A. P., M. K. Urban, and W. T. London.** 1983. Naturally occurring infection of Pekin duck embryos by duck hepatitis B virus. *Proc. Natl. Acad. Sci. USA* **80:**1703–1706.

99. **Ogston, C. W., G. J. Jonak, C. E. Rogler, S. M. Astrin, and J. Summers.** 1982. Cloning and structural analysis of integrated woodchuck hepatitis virus sequences from hepatocellular carcinomas of woodchucks. *Cell* **29:**385–394.

100. **Oshima, A., H. Tsukuma, T. Hiyama, I. Fujimoto, H. Yamano, and M. Tanaka.** 1984. Follow-up study of HBs Ag-positive blood donors with special reference to effect of drinking and smoking on development of liver cancer. *Int. J. Cancer* **34:**775–779.

101. **Ou, J. H., O. Laub, and W. J. Rutter.** 1986. Hepatitis B virus gene function: the precore region targets the core antigen to cellular membranes and causes the secretion of the e antigen. *Proc. Natl. Acad. Sci. USA* **83:**1578–1582.

102. **Perrillo, R. P., E. R. Schiff, G. L. Davis, H. C. Bodenheimer, K. Lindsay, J. Payne, J. L. Dienstag, C. O'Brein, C. Tamburro, I. M. Jacobson, R. Sampliner, D. Feit, J. Lefkowitch, M. Kuhns, C. Meschievitz, B. Sanghvi, J. Albrecht, and A. Gibas.** 1990. A randomized, controlled trial of interferon alfa-2b alone and after prednisone withdrawal for the treatment of chronic hepatitis B. *N. Engl. J. Med.* **323:**295–301.

103. **Ponzetto, A., P. J. Cote, E. C. Ford, R. H. Purcell, and J. L. Gerin.** 1984. Core antigen and antibody in woodchucks after infection with woodchuck hepatitis virus. *J. Virol.* **52:**70–76.

104. **Popper, H., L. Roth, R. H. Purcell, B. C. Tennant, and J. L. Gerin.** 1987. Hepatocarcinogenicity of the woodchuck hepatitis virus. *Proc. Natl. Acad. Sci. USA* **84:**866–870.

105. **Prince, A. M., and P. Alcabes.** 1982. The risk of development of hepatocellular carcinoma in hepatitis B virus carriers in New York. A preliminary estimate using death-records matching. *Hepatology* **2:**15S–20S.

106. **Prince, M. A., T. White, N. Pollock, J. Riddle, B. Brotman, and L. Richardson.** 1981. Epidemiology of hepatitis B infection in Liberian infants. *Infect. Immun.* **32:**675.

107. **Pugh, J., A. Zweidler, and J. Summers.** 1989. Characterization of the major duck hepatitis B virus core particle protein. *J. Virol.* **63:**1371–1376.

108. **Quignon, F., C. A. Renard, P. Tiollais, M. A. Buendia, and C. Transy.** 1996. A functional N-myc2 retroposon in ground squirrels: implications for hepadnavirus-associated carcinogenesis. *Oncogene* **12:**2011–2017.

109. **Rivkina, M. B., J. M. Cullen, W. S. Robinson, and P. L. Marion.** 1994. State of the p53 gene in hepatocellular carcinomas of ground squirrels and woodchucks with past and ongoing infection with hepadnaviruses. *Cancer Res.* **54:**5430–5437.

110. **Ryder, R. W., H. C. Whittle, A. B. Sanneh, A. B. Ajdukiewicz, S. Tulloch, and B.**

Yvonnet. 1992. Persistent hepatitis B virus infection and hepatoma in The Gambia, west Africa. A case-control study of 140 adults and their 603 family contacts. *Am. J. Epidemiol.* **136:**1122–1131.

111. **Sakuma, K., N. Saitoh, M. Kasai, H. Jitsukawa, L. Yoshino, M. Yamaguchi, K. Nobutomo, M. Yamumi, F. Tsuda, T. Komazawa, T. Nakamura, Y. Yoshida, and K. Okuka.** 1988. Relative risks of death due to liver disease among Japanese male adults having various statuses for hepatitis B s and c antigen/antibody in serum: a prospective study. *Hepatology* **8:**1642–1646.

112. **Schaller, H., and M. Fischer.** 1991. Transcriptional control of hepadnavirus gene expression. *Curr. Top. Microbiol. Immunol.* **168:**21–39.

113. **Scorsone, K. A., Y. Z. Zhou, J. S. Butel, and B. L. Slagle.** 1992. p53 mutations cluster at codon 249 in hepatitis B virus-positive hepatocellular carcinomas from China. *Cancer Res.* **52:**1635–1638.

114. **Seeger, C., B. Baldwin, W. E. Hornbuckle, A. E. Yeager, B. C. Tennant, P. Cote, L. Ferrell, D. Ganem, and H. E. Varmus.** 1991. Woodchuck hepatitis virus is a more efficient oncogenic agent than ground squirrel hepatitis virus in a common host. *J. Virol.* **65:**1673–1679.

115. **Seeger, C., D. Ganem, and H. E. Varmus.** 1986. Biochemical and genetic evidence for the hepatitis B virus replication strategy. *Science* **232:**477–484.

116. **Seeger, C., and W. S. Mason.** 1996. Reverse transcription and amplification of the hepatitis B virus genome, p. 815–831. *In* M. DePamphilis (ed.), *DNA Replication in Eukaryotic Cells.* Cold Spring Harbor Laboratory Press, Cold Spring Harbor, N.Y.

117. **Seeger, C., and W. S. Mason.** Chronic hepadnavirus infections of the woodchuck and duck. *In* R. Ahmed and I. Chen (ed.), *Persistent Viral Infections,* John Wiley & Sons, New York, in press.

118. **Shaw, T., S. S. Mok, and S. A. Locarnini.** 1996. Inhibition of hepatitis B virus DNA polymerase by enantiomers of penciclovir triphosphate and metabolic basis for selective inhibition of HBV replication by penciclovir. *Hepatology* **24:**996–1002.

119. **Sheen, I. S., Y. F. Liaw, D. I. Tai, et al.** 1985. Hepatic decompensation associated with hepatitis B e antigen clearance in chronic type B hepatitis. *Gastroenterology* **89:**732–735.

120. **Sherman, M., K. M. Peltekian, and C. Lee.** 1995. Screening for hepatocellular carcinoma in chronic carriers of hepatitis B virus: incidence and prevalence of hepatocellular carcinoma in a North American urban population. *Hepatology* **22:**432–438.

121. **Shih, C., K. Burke, M. J. Chou, J. B. Zeldis, C. S. Yang, C. S. Lee, K. J. Isselbacher, J. R. Wands, and H. M. Goodman.** 1987. Tight clustering of human hepatitis B virus integration sites in hepatomas near a triple-stranded region. *J. Virol.* **61:**3491–3498.

122. **Shimoda, R., M. Nagashima, M. Sakamoto, N. Yamaguchi, S. Hirohashi, J. Yokota, and H. Kasai.** 1994. Increased formation of oxidative DNA damage, 8-hydrodeoxyguanosine, in human livers with chronic hepatitis. *Cancer Res.* **54:**3171–3172.

123. **Sirica, A. E. (ed.).** 1992. *The Role of Cell Types in Hepatocarcinogenesis.* CRC Press, Boca Raton, Fla.

124. **Sprengel, R., E. F. Kaleta, and H. Will.** 1988. Isolation and characterization of a hepatitis B virus endemic in herons. *J. Virol.* **62:**932–937.

125. **Staprans, S., D. D. Loeb, and D. Ganem.** 1991. Mutations affecting hepadnavirus plus-strand DNA synthesis dissociate primer cleavage from translocation and reveal the origin of linear viral DNA. *J. Virol.* **65:**1255–1262.

126. **Stevens, C. E., R. A. Neurath, R. P. Beasley, and W. Szmuness.** 1979. HBeAg and anti-HBe detection by radioimmunoassay: correlation with vertical transmission of hepatitis B virus in Taiwan. *J. Med. Virol.* **3:**237–241.

127. **Summers, J., P. M. Smith, and A. L. Horwich.** 1990. Hepadnavirus envelope proteins regulate covalently closed circular DNA amplification. *J. Virol.* **64:**2819–2824.

128. **Summers, J., P. M. Smith, M. J. Huang, and M. S. Yu.** 1991. Morphogenetic and regulatory effects of mutations in the envelope proteins of an avian hepadnavirus. *J. Virol.* **65:**1310–1317.

129. **Summers, J., J. Smolec, and R. Snyder.** 1978. A virus similar to human hepatitis B virus associated with hepatitis and hepatoma in woodchucks. *Proc. Natl. Acad. Sci. USA* **75:**4533–4537.

130. **Szmuness, W., C. E. Stevens, H. Ikram, M. I. Much, E. J. Harley, and B. Hollinger.** 1978. Prevalence of hepatitis B virus infection and hepatocellular carcinoma in Chinese-Americans. *J. Infect. Dis.* **137:**822–829.

131. **Tagawa, M., W. S. Robinson, and P. L. Marion.** 1987. Duck hepatitis B virus replicates in the yolk sac of developing embryos. *J. Virol.* **61:**2273–2279.

132. **Takayanagi, M., S. Kkakumu, T. Ishikawa, Y. Higashi, K. Yoshioka, and T. Wakita.** 1993. Comparison of envelope and precore/core variants of hepatitis B virus (HBV) during chronic HBV infection. *Virology* **196:**138–145.

133. **Tennant, B. C., N. Mrosovsky, K. McLean, P. J. Cote, B. E. Korba, R. E. Engle, J. L. Gerin, J. Wright, G. R. Michener, E. Uhl, and J. M. King.** 1991. Hepatocellular carcinoma in Richardson's ground squirrels (Spermophilus richardsonii): evidence for association with hepatitis B-like virus infection. *Hepatology* **13:**1215–1221.

134. **Testut, P., C. A. Renard, O. Terradillos, T. L. Vitvitski, F. Tekaia, C. Degott, J. Blake, B. Boyer, and M. A. Buendia.** 1996. A new hepadnavirus endemic in arctic ground squirrels in Alaska. *J. Virol.* **70:**4210–4219.

135. **Thorgeirsson, S. S., R. P. Evarts, H. C. Bisgaard, K. Fujio, and Z. Hu.** 1993. Hepatic stem cell compartment: activation and lineage commitment. *Proc. Soc. Exp. Biol. Med.* **204:**253–260.

136. **Tipples, G. A., M. M. Ma, K. P. Fischer, V. G. Bain, N. M. Kneteman, and D. L. J. Tyrrell.** 1996. Mutation in HBV DNA-dependent DNA polymerase confers resistance to lamivudine in vivo. *Hepatology* **24:**714–717.

137. **Tokudome, S., M. Ikeda, K. Matsushita, Y. Maeda, and M. Yoshinara.** 1987. Hepatocellular carcinoma among female Japanese hepatitis B virus carriers. *Hepato-Gastroenterology* **34:**246–248.

138. **Transy, C., G. Fourel, W. S. Robinson, P. Tiollais, P. L. Marion, and M. A. Buendia.** 1992. Frequent amplification of c-myc in ground squirrel liver tumors associated with past or ongoing infection with a hepadnavirus. *Proc. Natl. Acad. Sci. USA* **89:** 3874–3878.

139. **Transy, C., C. A. Renard, and M. A. Buendia.** 1994. Analysis of integrated ground squirrel hepatitis virus and flanking host DNA in two hepatocellular carcinomas. *J. Virol.* **68:**5291–5295.

140. **Trueba, D., M. Phelan, J. Nelson, F. Beck, B. S. Pecha, R. J. Brown, H. E. Varmus, and D. Ganem.** 1985. Transmission of ground squirrel hepatitis virus to homologous and heterologous hosts. *Hepatology* **5:**435–439.

141. **Tsui, L. V., L. G. Guidotti, T. Ishikawa, and F. V. Chisari.** 1995. Posttranscriptional clearance of hepatitis B virus RNA by cytotoxic T lymphocyte-activated hepatocytes. *Proc. Natl. Acad. Sci. USA* **92:**12398–12402.

142. **Tu, J. T., R. N. Gao, D. H. Zhang, and B. C. Gu.** 1985. Hepatitis B virus and primary liver cancer on Chongming Island, People's Republic of China. *Natl. Cancer Inst. Monogr.* **69:**213–215.

143. **Tuttleman, J., C. Pourcel, and J. Summers.** 1986. Formation of the pool of covalently closed circular viral DNA in hepadnavirus infected cells. *Cell* **47:**451–460.

144. **Ueda, K., and D. Ganem.** 1996. Apoptosis is induced by N-*myc* expression in hepato-

cytes, a frequent event in hepadnavirus oncogenesis, and is blocked by insulin-like growth factor II. *J. Virol.* **70:**1375–1383.

145. **Ueda, K., Y. Wei, and D. Ganem.** 1996. Activation of N-myc2 gene expression by cis-acting elements of oncogenic hepadnaviral genomes: key role of enhancer II. *Virology* **217:**413–417.

146. **Urban, M. K., A. P. O'Connell, and W. T. London.** 1985. Sequence of events in natural infection of Pekin duck embryos with duck hepatitis B virus. *J. Virol.* **55:** 16–22.

147. **Vierucci, A., M. deMarino, E. Graziani, M. E. Rossi, W. T. London, and B. S. Blumberg.** 1983. A mechanism for liver cell injury in viral hepatitis: effects of hepatitis B virus on neutrophil function in vitro and in children with chronic hepatitis. *Pediatr. Res.* **17:**814–820.

148. **Villeneuve, J. P., M. Desrochers, R. C. Infante, B. Willems, G. Raymond, M. Bourcier, J. Cote, and G. Richer.** 1994. A long-term follow-up study of asymptomatic hepatitis B surface antigen-positive carriers in Montreal. *Gastroenterology* **106:** 1000–1005.

149. **Wang, G. H., and C. Seeger.** 1993. Novel mechanism for reverse transcription in hepatitis B viruses. *J. Virol.* **67:**6507–6512.

150. **Wang, G. H., F. Zoulim, E. H. Leber, J. Kitson, and C. Seeger.** 1994. Role of RNA in enzymatic activity of the reverse transcriptase of hepatitis B viruses. *J. Virol.* **68:** 8437–8442.

151. **Wang, J., X. Chenivesse, B. Henglein, and C. Brechot.** 1990. Hepatitis B virus integration in a cyclin A gene in a hepatocellular carcinoma. *Nature* **343:**555–557.

152. **Wang, X. W., K. Forrester, H. Yeh, M. A. Feitelson, J. R. Gu, and C. C. Harris.** 1994. Hepatitis B virus X protein inhibits p53 sequence-specific DNA binding, transcriptional activity, and association with transcription factor ERCC3. *Proc. Natl. Acad. Sci. USA* **91:**2230–2234.

153. **Wang, X. W., M. K. Gibson, W. Vermeulen, H. Yeh, K. Forrester, H. W. Sturzbecher, J. H. Hoeijmakers, and C. C. Harris.** 1995. Abrogation of p53-induced apoptosis by the hepatitis B virus X gene. *Cancer Res.* **55:**6012–6016.

154. **Wei, Y., J. Etiemble, G. Fourel, T. L. Vitvitski, and M. A. Buendia.** 1995. Hepadna virus integration generates virus-cell cotranscripts carrying 3' truncated X genes in human and woodchuck liver tumors. *J. Med. Virol.* **45:**82–90.

155. **Wei, Y., G. Fourel, A. Ponzetto, M. Silvestro, P. Tiollais, and M.-A. Buendia.** 1992. Hepadnavirus integration: mechanisms of activation of the N-*myc*2 retrotransposon in woodchuck liver tumors. *J. Virol.* **66:**5265–5276.

156. **Wei, Y., and D. Ganem.** 1996. Relationship between viral DNA synthesis and virion envelopment in hepatitis B viruses. *J. Virol.* **70:**6455–6458.

157. **Wei, Y., A. Ponzetto, P. Tiollais, and M. A. Buendia.** 1992. Multiple rearrangements and activated expression of c-myc induced by woodchuck hepatitis virus integration in a primary liver tumour. *Res. Virol.* **143:**89–96.

158. **Will, H., W. Reiser, T. Weimer, E. Pfaff, M. Buscher, R. Sprengel, R. Cattaneo, and H. Schaller.** 1987. Replication strategy of human hepatitis B virus. *J. Virol.* **61:**904–911.

159. **Yaginuma, K., H. Kobayashi, M. Kobayashi, T. Morishima, K. Matsuyama, and K. Koike.** 1987. Multiple integration site of hepatitis B virus DNA in hepatocellular carcinoma and chronic active hepatitis tissues from children. *J. Virol.* **61:**1808–1813.

160. **Yang, D., E. Alt, and C. E. Rogler.** 1993. Coordinate expression of N-myc 2 and insulin-like growth factor II in precancerous altered hepatic foci in woodchuck hepatitis virus carriers. *Cancer Res.* **53:**2020–2027.

161. **Yang, D., R. Faris, D. Hixson, S. Affigne, and C. E. Rogler.** 1996. Insulin-like growth

factor II blocks apoptosis of N-*myc2*-expressing woodchuck liver epithelial cells. *J. Virol.* **70**:6260–6268.

162. **Yang, W., W. S. Mason, and J. Summers.** 1996. Covalently closed circular viral DNA formed from two types of linear DNA in woodchuck hepatitis virus-infected liver. *J. Virol.* **70**:4567–4575.

163. **Yang, W., and J. Summers.** 1995. Illegitimate replication of linear hepadnavirus DNA through nonhomologous recombination. *J. Virol.* **69**:4029–4036.

164. **Zhihong, J., Z. Guolong, X. Shisong, K. Pingyuan, M. Lili, C. Hongtao, Q. Jianying, B. Quiju, and M. Kai.** 1988. An experimental transmission of woodchuck hepatitis virus to young chinese marmots. *Hepatology* **8**:371–373.

165. **Zoulim, F., J. Saputelli, and C. Seeger.** 1994. Woodchuck hepatitis virus X protein is required for viral infection in vivo. *J. Virol.* **68**:2026–2030.

166. **Zoulim, F., and C. Seeger.** 1994. Reverse transcription in hepatitis B viruses is primed by a tyrosine residue of the polymerase. *J. Virol.* **68**:6–13.

Human Tumor Viruses
Edited by Dennis J. McCance
© 1998 American Society for Microbiology

9 | Hepatitis B Virus Infection and Immunity
Jennifer A. Waters, Graham R. Foster,
Mark R. Thursz, and Howard C. Thomas

Hepatitis B virus (HBV) infection is a major public health problem, there being approximately 300 million people infected worldwide, and is a major cause of chronic liver disease and hepatocellular carcinoma. The majority of adults infected have a transient, acute hepatitis from which they completely recover and clear the virus. Rarely, the disease becomes fulminant and the patient may die. From 5 to 10% of adults become persistently infected and develop chronic liver disease.

When the virus is transmitted vertically from mother to child, the majority of the children become chronically infected (70), greatly increasing their risk of developing cirrhosis and hepatocellular carcinoma.

VIROLOGY

HBV is a small double-stranded DNA virus and is a member of the *Hepadnaviridae*, a related group of hepatotropic, enveloped viruses which includes woodchuck hepatitis virus, ground squirrel hepatitis virus, duck hepatitis virus, and heron hepatitis virus. This group shares many genetic and biological features.

The HBV genome is approximately 3.2 kb in size and contains four overlapping open reading frames (ORFs). These code for the envelope (hepatitis B surface antigen [HBsAg]), nucleocapsid antigens (hepatitis B core and e antigens [HBcAg and HBeAg, respectively]), DNA polymerase, and X protein (Fig. 1).

The envelope ORF of the virus encodes the large, middle, and major surface proteins, which all contain the same carboxyl terminus of 226 amino acids. The middle protein has an additional 55 amino acids designated the preS2 region, and the large protein has a further 120 amino acids designated the preS1 region. All of these proteins also occur in a glycosylated form.

The core gene encodes two overlapping polypeptides which lead to the expression of HBcAg and HBeAg. The longer of these polypeptides (precore/core) contains a signal peptide which directs it to the endoplasmic reticulum, where the amino- and carboxy-terminal ends are cleaved and HBeAg is secreted into the circulation (58). The smaller core polypeptide has a nucleophilic sequence at its carboxy-terminal end and associates with the RNA pregenome of the virus during the initial stage of the formation of nucleocapsids. The B-cell epitopes recognized on HBcAg and HBeAg are separate. Anti-HBc recognizes a single epitope around

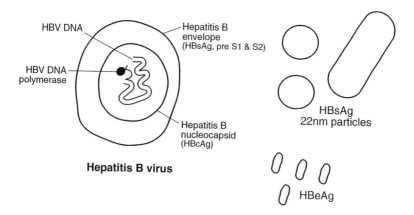

Figure 1 HBV. The virion is a 45-nm enveloped virus. Excess envelope protein is found in the serum as 22-nm lipoprotein particles. HBeAg is secreted by the infected cell as a soluble antigen.

amino acid 80 and is conformationally dependent. There are two B-cell epitopes in HBeAg; HBe1 is a linear determinant and maps to around amino acid 80, and HBe2, a conformational determinant, lies around amino acid 138 but requires the participation of the sequence between 10 and 140 (67).

The X ORF encodes a protein which has the ability to modify the transcription of both viral and host cell genes. The largest ORF codes for a bifunctional protein which has DNA polymerase and reverse transcriptase functions.

DISEASE SPECTRUM OF HBV INFECTION

Acute Hepatitis B Infection

Infection with HBV in adults usually results in an acute, self-limiting, inflammatory disease. High titers of immunoglobulin M (IgM) anti-HBc, a thymus-independent response, appear early in the course of acute HBV infection, together with HBsAg and HBeAg and a rise of liver transaminases, indicative of liver damage (Fig. 2). Anti-preS1 antibodies may also occur early in the course of infection, usually being detectable at clinical onset of the disease together with HBsAg and HBeAg. This antibody then becomes undetectable, and a late response follows which typically precedes the appearance of antibodies to the major envelope protein. Control of viral replication usually coincides with the detection of an anti-HBe response and loss of detectable HBeAg. This is followed by loss of HBsAg and the appearance of an antibody to the major surface protein, which corresponds to viral elimination.

Persistent Hepatitis B Infection

In some patients HBV is not successfully eliminated by the immune response and the patients become chronically infected.

The majority of babies born to HBeAg-positive mothers become infected, and over 90% of these develop a chronic carrier state (70). Although some cases have been documented of HBV infection during fetal life, it is thought that these infants probably receive a large inoculum of virus from maternal blood before or during birth and via close contact with secretions soon after birth (6, 70). HBcAg but not HBsAg has been detected in the cord blood of these neonates, suggesting that HBeAg crosses the placenta (33). The reason for the infants failing to clear the virus is unknown, but may relate to the induction of immune tolerance to HBeAg, since these children respond well to vaccination with HBsAg. Children born to HBeAg-negative/HBsAg-positive carrier mothers who are asymptomatic rarely become chronically infected (6). Evidence from a transgenic mouse model expressing HBeAg suggest that HBeAg may be tolerogenic and that this tolerance can be broken by immunization with an HBeAg peptide (46). A mechanism by which this tolerance may occur has been suggested by the demonstration of a reduced level of expression of major histocompatibility complex (MHC) class II antigen on the splenic dendritic cells in a transgenic mouse model expressing the whole HBV genome (1).

In marked contrast to the situation of infection at birth, young African children infected postnatally, more than half of them by 7 years of age, usually clear the virus; only 15% become persistently infected (19). Similarly, 5 to 10% of adults who become HBV infected fail to eliminate the virus and become persistently infected.

Four phases of persistent chronic infection can be recognized (Fig. 3). During the HBeAg-positive phase there is minimal liver damage. Anti-HBc is present and both anti-HBe and anti-HBs can sometimes be detected in low levels, complexed to antigen, up to 10 years before seroconversion (43, 44).

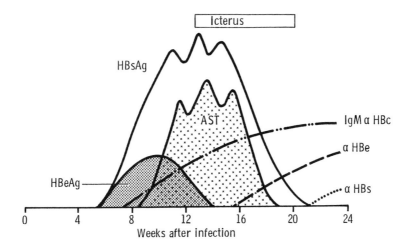

Figure 2 The natural history of acute HBV infection. HBsAg and HBeAg can be detected early in the serum, prior to the onset of clinical hepatitis. Antibody to preS1 can sometimes be detected during this early phase. AST (aspartate aminotransferase) is an indication of lysis of infected hepatocytes.

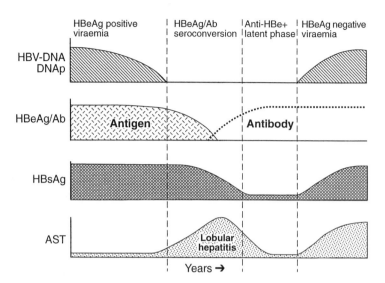

Figure 3 Natural history of persistent HBV infection after neonatal or adult exposure. During conversion from HBe antigenemia to the presence of anti-HBe, viremia ceases. The hepatitis associated with this seroconversion is the result of immune lysis of hepatocytes supporting HBV replication. This immune pressure may select an HBe-negative strain of HBV which causes viremia in the absence of HBeAg. AST, aspartate aminotransferase (an indication of lysis of hepatocytes).

As the infection continues, the inflammatory liver disease becomes more active. In the second phase, 5 to 15% of persistently infected patients undergo spontaneous HBeAg/antibody seroconversion each year, as do approximately 40% of those treated with interferon alpha (IFN-α) (9). This seroconversion event is usually accompanied by hepatocyte death, as indicated by raised liver enzymes.

The HBeAg/antibody seroconversion hepatitis results in clearance of hepatocytes supporting HBV replication and the appearance of uncomplexed anti-HBe antibodies. The sera of these patients do not contain virus particles detectable by dot-blot hybridization. However, in some cases, low levels of viremia can be detected by PCR. HBsAg detected in the serum of these patients may be encoded by integrated HBV DNA sequences. This phase represents immune control of the virus.

HBeAg-negative, anti-HBe-positive patients with persistent HBV replication represent a fourth phase of chronic infection with HBV which may be associated with a severe and progressive liver disease (22, 23). It was first recognized in the Mediterranean area, where it is more prevalent than in other European countries (23). Both in the Mediterranean and in Japan, where there is also a higher prevalence of HBeAg-negative chronic disease, neonatal and childhood infections are common, suggesting that the increased incidence may be due to the length of time of infection (22). However, genetic factors may also be important.

These patients are infected with a mutated HBV which is unable to encode

HBeAg (10). The most common mutation is a G-to-A change at nucleotide 1896, creating a novel translational stop codon in the precore region of the genome. This mutation is found as a very minor species in some patients with an HBeAg-positive infection. The mutant is selected at the time of seroconversion to anti-HBe in some patients. These patients usually become HBV DNA negative by hybridization initially, and chronic infection develops only after a long interval, presumably because the mutant virus escapes from immune control. The mutant virus can be shown to have a statistically higher rate of amino acid substitution in regions of the virus encoding known B-cell epitopes, which may also contain helper T-cell epitopes (14), possibly representing the emergence of variants that can replicate in the presence of these components of the immune response. The mutant virus is thought not to be more cytopathic than the wild-type virus, and the pathology is probably immune mediated.

IMMUNOPATHOLOGY

Acute Disease

HBV is not thought to be directly cytopathic; the liver damage in both the acute and the chronic disease is probably caused by the host immune response (12). This is supported by the histological picture of the infected liver in both the acute and chronic disease: a mononuclear inflammatory infiltrate is seen at the sites of viral replication and liver cell injury, and this consists mainly of T cells, both CD4$^+$ and CD8$^+$ cells (48). The number of CD4$^+$ and CD8$^+$ lymphocytes in liver biopsy tissue from infected patients is much greater than in normal liver biopsies (26).

By analogy to other viral infections (5), human leukocyte antigen (HLA) class I restricted virus-specific cytotoxic T lymphocytes (CTLs) are of prime importance in the elimination of virus-infected cells and so are implicated in the pathology of the disease. An acute HBV infection is accompanied by a vigorous major histocompatibility complex (MHC) class I restricted CTL response specific for the nucleocapsid (15, 59), envelope (53), and polymerase antigens (63). These cells are readily detectable in the peripheral blood following stimulation, in vitro, with peptides in bulk culture. Using this method a CTL response is generally detectable during the symptomatic phase of an acute infection; such response in some patients is long lasting, being still detectable after liver transaminases have returned to normal levels (63).

Stimulation of peripheral blood lymphocytes from an individual patient with peptides from the three viral antigens tested has demonstrated that the cytolytic response is often multispecific (63). In addition peripheral blood lymphocytes from one patient will often react to more than one MHC class I restricted peptide from a given antigen (53, 63). However, not all of the patients with the same HLA class I allele tested were able to respond to the same peptide epitopes (53, 63). Naturally occurring variants of HBV which act as T-cell receptor antagonists capable of inhibiting the CTL response to the wild-type epitope have been described (7). This appears to have led to a persistent infection in the presence of an effector T-cell response which in this instance was narrowly focused. Since there are usually more than one epitope of the virus recognized by the T-cell population, this mechanism must of necessity be rare.

The pathological effects of CTLs have been demonstrated in transgenic mouse models expressing either the major HBsAg or large HBsAg (3, 18, 49). Passive transfusion of HBsAg-specific CTLs showed that, in the early stages, the CTLs were directly responsible for apoptotic cell death (3). As the disease progressed further, necrotic damage was mediated by CTL-activated inflammatory cells. This damage was mediated by IFN-γ and tumor necrosis factor alpha (TNF-α) (18, 20).

Chronic Disease

Persistent infection in adults is associated with poor T-cell proliferative and cytolytic responses to HBV antigens. However, in persistent infection MHC class I-mediated cytolysis and its attendant inflammatory response probably have a role in the liver pathology of the disease. During seroconversion from HBeAg to antibody positivity, a nucleocapsid antigen-specific MHC class I restricted CTL response can be demonstrated by using the bulk stimulation techniques used for the acute disease, although this response is considerably lower than that found in the acute disease (85). Virus-specific CTLs can also be detected both in the liver, by cloning (40), and in the peripheral blood, using a split-well technique, during persistent infection and post-IFN-induced seroconversion (64, 80).

An increased T-cell proliferative response to the nucleocapsid antigen is also seen during periods of exacerbation of disease and during HBeAg/antibody seroconversion (16, 35, 36, 78). Both this and the increase in the CTL response suggest that CD4 and CD8 lymphocytes contribute to lysis of HBV-infected hepatocytes during seroconversion.

It is likely that there are many different reasons why the chronic carrier state develops after adult infection, and the liver pathology in chronic disease is probably the result of a dynamic equilibrium between the replication of the virus and the host's immune response. The replication of HBV has been shown to be inhibited by expression of the precore protein in vitro (39, 68) and in a transgenic mouse model (21). Hybrid nucleocapsids which contain HBcAg and the precore/core gene product with the signal sequence cleaved are unable to incorporate pregenomic RNA. The appearance of the precore mutant at seroconversion in some patients may lead to higher replication of the virus and an enhanced immune response, leading to elimination of HBV-infected hepatocytes (21). This is controversial as it has been suggested that increases in precore mutants occur following rather than preceding episodes of immune elimination (22). Also, patients receiving liver transplants for precore mutant-induced cirrhosis may revert to wild-type virus under the influence of immunosuppression. The situation is complicated by the occurrence of other mutations in the core region which may affect its antigenicity and result in its persistence (14).

ROLE OF IFN IN THE IMMUNE CLEARANCE OF HBV

The type I interferons are a family of closely related glycoproteins that are released by virally infected cells (60). The interferon family consists of at least 13 different IFN-α subtypes, one IFN-β subtype, and one IFN-ω subtype. All of the IFN-α subtypes and IFN-ω are released by leukocytes, and IFN-β is produced by fibroblasts (27, 32). The different subtypes are currently believed to have similar func-

tions, and the type I IFNs have antiviral, immunomodulatory, and antiproliferative effects.

Type I IFNs (usually IFN-α) have been used in the therapy of chronic HBV
infection for many years (reviewed in references 17 and 41). Therapy results in
viral clearance and seroconversion from HBeAg to anti-HBe in approximately
40% of cases. Viral clearance proceeds in two distinct phases. Initially there is an
antiviral response, characterized by a decrease in HBV replication and a fall in
serum HBV DNA. This is followed, in patients who respond to therapy, by immunological lysis of infected cells associated with a marked increase in the hepatitis
with a rise in serum transaminases and the development of antibodies against
HBeAg. Thus therapy of chronic HBV with type I IFNs is characterized by an
early fall in viral DNA followed by immune clearance of infected hepatocytes. It
is this latter immune clearance that is associated with long-term viral eradication—the antiviral effects of IFN alone are insufficient to cure the infection.

IFN-Induced Lysis of Infected Hepatocytes

Eradication of virally infected hepatocytes is essential for IFN-induced clearance
of chronic HBV. The mechanisms by which IFN induces lysis of infected cells are
still not completely understood, and a number of different effector pathways are
likely to be involved.

Enhancement of HLA expression

Recognition of virally infected cells by immunocytes (chiefly CTLs) requires the
presentation of viral antigens on the cell surface in association with HLA class I
antigens, and this presentation can be augmented by IFN treatment (8). Most
human solid tissues (including the liver) express very low levels of HLA antigens,
but expression is markedly increased by treatment with type I IFNs (24). Presentation of viral antigens by HLA class I antigens requires appropriate processing of
viral proteins, which involves cleavage of the proteins to peptides by the proteasome (38) and transport of the viral peptides into the endoplasmic reticulum by the
TAP transporter proteins TAP1 and TAP2 (membrane-spanning ATP-dependent
pumps) (77). In the endoplasmic reticulum the viral peptides are bound to HLA
class I antigens and β2-microglobulin before being expressed on the cell surface.
Studies in tissue culture systems have shown that type I IFNs increase the expression of all of these proteins; transcription of the proteasome, TAP, and β2-micro-
globulin genes is increased by type I IFNs (J. Trowsdale, personal communication),
and hence type I IFNs enhance the processing and presentation of viral proteins
to the cells of the immune system.

In chimpanzee models of acute HBV infection, the importance of viral antigen
presentation has been demonstrated by studies showing that viral clearance takes
place only when the hepatocytes express HLA class I antigens (61). Studies on
patient-derived material have confirmed that IFN treatment enhances HLA class
I antigen display on hepatocytes, and thus a major role for type I IFNs in viral
clearance is to augment CTL destruction of infected cells by enhancing the presentation of viral antigens.

Apoptosis of virally infected cells

Apoptosis is an active process involving autolysis of cells that are damaged or
redundant. The process (reviewed in reference 69) is important in development

and plays a key role in regulation of the immune system. In addition to its involvement in tissue homeostasis apoptosis is involved in the removal of damaged cells; i.e., cells that have been damaged by a wide variety of agents, such as UV light, heat, and chemicals, are destroyed by apoptosis. In cells infected with the influenza virus, apoptosis is one mechanism that appears to be involved in cell death (31), and there is some evidence to suggest that IFNs may be involved in this process (74). It is not yet known whether IFN-induced apoptosis plays a role in elimination of HBV-infected hepatocytes.

Activation of immunocytes by IFN

In addition to enhancing T-cell lysis of infected cells by augmenting HLA antigen display, type I IFNs also activate immunocytes directly. Natural killer (NK) cells lyse virus-infected cells by a mechanism that is still poorly understood, and some type I IFNs augment the activity of NK cells in tissue culture systems (57). It is likely that this contributes to the antiviral effects of the type I IFNs in chronic HBV, since there is an increase in the number of NK cells in the liver during IFN therapy (88).

Type I IFNs may also influence the antiviral immune response in less direct ways. HLA class II antigen expression on the surface of human macrophages may be increased by IFN-αI (65), and CTLs may be activated by type I IFNs (81). The precise role of these effects in viral clearance remains unknown, but it seems likely that these immunomodulatory effects assist in the development of immune-mediated clearance of HBV-infected cells.

IMMUNOGENETICS OF HBV INFECTION

One of the crucial issues in HBV infection is identification of the factors which determine the outcome of infection. Although it is clear that in some circumstances, for example the presence of the precore stop codon, the viral genomic variability plays an important role, we believe that host factors are also important in disease outcome. From early segregation analysis and from twin studies we know that there is a considerable contribution from host genes in determining whether an HBV infection becomes persistent (2). This misses the question of which genes contribute to the disease outcome.

The process of identifying the genes responsible for increased or decreased susceptibility to HBV infection uses methods which have been highly successful in other infectious diseases, such as malaria (30). Candidate polymorphisms have been identified from the known pathogenesis of the disease, and allele associations have been established through case-controlled studies. Although in the future it may be possible to generate new candidate polymorphisms for testing through genome scanning on affected sibling pairs, this technique has not yet been applied to hepatitis B infection.

It has been shown that the CD4$^+$ T-helper cell response directed against the nucleocapsid antigens of HBV is significantly greater in patients with acute self-limiting infection than in those with persistent infection. CD4$^+$ T-helper cells are stimulated by interaction of their T-cell receptor with an HLA class II molecule carrying antigenic peptides. It has therefore been postulated that persistent infection may be associated with specific MHC class II alleles which encode the HLA class II molecules. A number of groups have looked for HBV disease associations

Table 1 MHC class II and HBsAg persistence

Study (reference)	Country	N	Specificity	Odds ratio	Probability (P)
Jeannet, 1974 (34a)	China	96	None		
Patterson, 1977 (58a)[a]	USA	55	None		
Van Hattum, 1987 (79)	Netherlands	79	DRw6	0.4	0.02
			DQw1	3.5	0.001
Almarri and Batchelor, 1994 (2)	Qatar	34	DR2	0.1	0.013
			DR7	3.73	0.05

[a] Renal failure patients.

with MHC class II, and the results are summarized in Table 1. In our own study we used molecular MHC genotyping in a two-stage design in a Gambian population. We found a negative association with HLA-DRB1*1302, indicating that this allele probably offers resistance to persistent infection (76). The association of HLA-DRw6 with acute infection, which can be deduced from van Hattum's study (79), may also be due to HLA-DRB1*1302, as DRw6 is the serological supertype for this allele. The associations with DR2 and DR7 identified by Almarri and Batchelor (2) could not have been reproduced in our study as the frequency of these alleles is low in West Africa.

Antigen presentation via the HLA class II molecule is mainly performed by monocytes and macrophages. These cells may be activated by 1,25-dihydroxyvitamin D through a receptor present on the cell surface (28, 66). Recently, mutations in the vitamin D receptor gene have been identified. One of these, a T-to-C mutation at position 352, appears to be functionally important as it varies the rate of gene transcription (50). This mutation has been associated with variation in the outcome of leprosy and tuberculosis and also with osteoporosis in some populations. Activation of monocytes/macrophages using granulocyte-macrophage colony-stimulating factor appears to enhance HBV clearance, supporting the hypothesis that HBV elimination may be associated with polymorphisms in the vitamin D receptor gene. In a study in Gambian subjects it was shown that the T352C mutation was underrepresented in persistently infected HBsAg-positive subjects (odds ratio, 0.43 [95% confidence interval, 0.22–0.85]; $P = 0.01$).

CD8[+] CTLs appear to be active and numerous in acute, self-limiting HBV infection, whereas they are virtually undetectable in chronic hepatitis B infection. CTLs are stimulated by interaction of their T-cell receptors with HLA class I molecules bearing antigenic peptide epitopes. It has therefore been postulated that polymorphisms of MHC class I may be associated with the outcome of HBV infection. Although a number of studies have addressed this, no reproducible association has emerged.

TNF-α is an antiviral cytokine with a number of different modes of action, which include activation of macrophages and promotion of an intracellular antiviral state. In the HBV transgenic mouse, "replication" of HBV is inhibited by CTLs through expression of TNF-α, and injection of TNF-α directly into these mice temporarily terminates viral replication. The TNF-α gene promoter has a mutation at position -308 relative to the transcriptional start codon. At this position the rarer *TNF2* allele has been shown in functional studies to be associated with increased TNF-α production (86). This allele is associated with cerebral malaria, and it has been postulated that *TNF2* may be associated with clearance of HBV infection. However, in our own study, even after correcting for potential confounding factors such as the association of HBsAg carriage with cerebral malaria, we found that *TNF2* was associated with persistent infection (odds ratio, 1.48 [95% confidence limits, 1.01–2.24], $P < 0.05$). The explanation for this association is unclear but may relate to the role of the TNF-α receptor in apoptosis of activated T cells.

In the innate immune system a number of serum proteins bind nonspecifically to non-human sugar chains and opsonize pathogens for phagocytosis. Mannose binding lectin (MBL) is an example of this type of protein, and mutations in this gene have been associated with opsonization deficits and recurrent pyogenic infections (72). Three mutations occur in the MBL (71). Two of these interfere with structural elements in such a way that MBL levels are significantly reduced even in the heterozygous state. The functional significance of the third mutation at codon 52 is unclear, but it is this mutation which, in Europeans, is associated with persistent HBV infection (75). This association may be due to failure of opsonization, which is supported by the finding of a low serum opsonic index in patients with persistent HBV infection (51).

In summary, disease association studies have revealed a number of alleles which influence susceptibility to persistent infection. However, other outcomes of HBV infection have still to be studied. Furthermore, by analogy with malaria, where 14 susceptibility loci have been identified, a lot of work remains to be performed.

VACCINES

The small HBsAg, derived from infected serum or produced by recombinant DNA technology, has been shown to induce protective immunity against HBV infection in high-risk groups (73). There has been recent evidence that this molecule may contain the virus receptor binding site, although this awaits further confirmation (29). HBsAg vaccines are now in widespread use throughout the world. When these vaccines are used to prevent neonatal transmission from HBeAg-positive infected mothers, the best protection is achieved when given with hyperimmune globulin.

Although the majority of patients vaccinated with HBsAg are protected from further infection, 2.5 to 5% of immunocompetent adults do not respond, probably for genetic reasons. This percentage is greater in some high-risk groups of patients, such as hemodialysis patients, in whom the failure rate is approximately 40%; this probably represents a secondary immune deficiency due to the renal disease.

There has been considerable debate on whether the middle preS2-bearing and large preS1-bearing polypeptides should be added to existing vaccines. Experiments in mice suggested that both the preS1 and preS2 regions can recruit T-cell help independently for the production of anti-HBs (47), so their inclusion may lead to a lower nonresponse rate. Although the use of HBsAg alone has been shown to be protective, it has been suggested that preS1 is the region of the virus which binds to the hepatocyte during infection (54). Antibodies to this region might be virus neutralizing, and therefore inclusion of the preS1 region in a vaccine may be desirable.

GENETIC VARIATION OF HBsAg

HBsAg is a partially glycosylated transmembrane protein which is extensively cross-linked by disulfide bridges. The antibodies to HBsAg in convalescent-phase serum (>80%) bind predominantly to the common "a" determinant epitopes found in the hydrophilic region of the polypeptide between amino acids 124 and 147. The carboxy-terminal region of HBsAg has 14 cysteine residues, and the antigenicity of this protein is highly dependent on its conformational structure (13, 42). Using two cyclical peptide analogs of amino acids 124–137 and 139–147, and a panel of monoclonal antibodies binding to these regions, it has been possible to show in chimpanzees that antibody to the amino acids 124–137 region of HBsAg prevents infection (34, 82). Antibodies to this region, as well as to other epitopes on the surface antigen gene-encoded polypeptide, are present in the serum of patients convalescent from HBV infection and in normal subjects immunized with plasma-derived and recombinant DNA-produced vaccine (84).

The antigenic structure of the common "a" determinant is dependent on the conformation of the molecule. A gene mutation in this region which resulted in an amino acid change was first described in a child born to an HBeAg-positive mother (11). The child developed chronic liver disease despite passive and active immunization at birth and the development of high-titer anti-HBs. A point mutation at nucleotide position 587 resulted in an amino acid substitution from glycine to arginine at amino acid 145 of HBsAg. This change was shown to alter the common "a" determinant so that it was not recognized by the known protective anti-HBs response elicited by the current vaccines (83).

This particular mutation has been found in a number of vaccine recipients around the world and in liver transplant patients being treated with anti-HBs to prevent reinfection of their graft (45). The prevalence of mutant HBV has been studied in the United States in infants who had become HBV infected despite postexposure prophylaxis. Of the 80 infants in the study, 15 had amino acid changes, the most frequent of those being a glycine-to-arginine change at amino acid 145 (52). In a Singapore study of 345 vaccinated infants born to HBcAg-positive mothers, 12 of 41 infants who developed HBV infection had the glycine-to-arginine variant, 4 had other variant viruses, and 25 had the wild-type virus (56). This study demonstrates that the 145 variant is the most common variant to emerge so far.

This mutant virus has been shown to be transmissible in chimpanzees, confirming that this substitution does not affect transmissibility (55). Immunization of mice with the variant HBsAg elicited a high-titer antibody to the variant which bound poorly to the native protein. Similarly, immunization with the wild-type HBsAg elicited low-titer antibody to the variant while raising antibody with high titer to the native HBsAg (83). Hyperimmunization with yeast-derived wild-type vaccines protected chimpanzees against infection with the variant when the animals were infected shortly after the immunization (55).

Other variants of HBV which have amino acid changes in the common "a" determinant have now been described in patients and vaccinees with chronic liver disease. A mutation resulting in an amino acid change from lysine to glutamic acid at amino acid 141 has been described in an immunized child in The Gambia (37). In Singapore, another vaccinated child was infected with a variant which had a change at amino acid 144 (25). Insertion of extra amino acids has also been described in patients in Japan (87), China, and South Africa.

The effect on antigenicity of some changes in the amino acid sequence has been studied experimentally (4). Pro-142 is essential for full antigenicity, as are the cysteine molecules found in the "a" determinant (Cys-107, Cys-124, Cys-137, Cys-138, Cys-139, and Cys-149) (42). However, the effect of the mutations, other than the change from glycine to arginine at 145, found in infected carriers on the immunogenicity and antigenicity of the HBsAg remains to be formally evaluated.

SUMMARY

There have been major advances in our understanding of the life cycle of HBV, in particular the mode of replication. However, several important areas of the natural history, such as how genetic factors of both the patient and the virus affect the outcome of infection, remain to be elucidated. Important developments in control of replication by antiviral agents and/or modulation of the immune response are needed, along with improvements in the vaccine and the level of delivery.

REFERENCES

1. **Akbar, S. M. F., M. Onji, K. Inaba, K. Yamamura, and Y. Ohta.** 1993. Low responsiveness of hepatitis B virus-transgenic mice in antibody response to T-cell-dependent antigen: defect in antigen presenting activity of dendritic cell. *Immunology* **78:**468–475.
2. **Almarri, A., and J. R. Batchelor.** 1994. HLA and infection. *Lancet* **344:**1194–1195.
3. **Ando, K., L. G. Guidotti, S. Wirth, T. Ishikawa, G. Missale, T. Moriyama, R. D. Schreiber, H.-J. Schlict, S. Huang, and F. V. Chisari.** 1994. Class I restricted cytotoxic T lymphocytes are directly cytopathic for their target cells in vivo. *J. Immunol.* **152:**3245–3255.
4. **Ashton-Rickardt, P. G., and K. Murray.** 1989. Mutants of the hepatitis B virus surface antigen that define some antigenically essential residues in the immunodominant a region. *J. Med. Virol.* **29:**196–203.
5. **Askonas, B. A.** 1994. Immunopathology and virus infections, p. 1–9. *In* H. C. Thomas and J. A. Waters (ed.), *Immunology of Liver Disease*, vol. 21. *Immunology and Medicine.* Kluwer Academic Publishers, Dordrecht, The Netherlands.
6. **Beasley, R. P., and L. Hwang.** 1983. Postnatal infectivity of hepatitis B surface antigen-carrier mothers. *J. Infect. Dis.* **147:**185–190.
7. **Bertoletti, A., A. Sette, F. V. Chisari, A. Penna, M. Levrero, M. De Carli, F. Fiaccadori, and C. Ferrari.** 1994. Natural variants of cytotoxic epitopes are T-cell receptor antagonists for antiviral cytotoxic T cells. *Nature* **369:**407–410.
8. **Blackman, M., and A. Morris.** 1985. The effect of interferon treatment of targets on susceptibility to cytotoxic T cell killing: augmentation of allogeneic killing and virus specific killing relative to virus antigen expression. *Immunology* **56:**451–457.
9. **Brook, M. G., G. Chan, I. Yap, P. Karayiannis, A. M. Leuer, H. Jacyna, J. Main, and**

H. C. Thomas. 1989. Randomised controlled trial of lymphoblastoid interferon alpha in Europid men with chronic hepatitis B infection. *Br. Med. J.* **299:**652–656.

10. **Carman, W. F., M. R. Jacyna, S. Hadziyannis, P. Karayiannis, M. J. McGarvey, A. Makris, and H. C. Thomas.** 1989. Mutation preventing formation of hepatitis B e antigen in patients with chronic hepatitis B infection. *Lancet* **ii:**588–591.

11. **Carman, W. F., A. R. Zanetti, P. Karayiannis, J. Waters, G. Manzillo, E. Tanzi, A. J. Zuckerman, and H. C. Thomas.** 1990. Vaccine-induced escape mutant of hepatitis B virus. *Lancet* **326:**325–329.

12. **Dudley, F. J., R. A. Fox, and S. Sherlock.** 1972. Cellular immunity and hepatitis associated (Australia antigen) liver disease. *Lancet* **i:**723–725.

13. **Dreesman, G. R., F. B. Hollinger, R. M. McCombs, and J. L. Melnick.** 1973. Alteration of hepatitis B antigen (HBAg) determinants by reduction and alkylation. *J. Gen. Virol.* **19:**129–134.

14. **Ehata, T., M. Omata, O. Yokosuka, K. Hosoda, and M. Ohto.** 1992. Variations in codons 84–101 in the core nucleotide sequence correlate with hepatocellular injury in chronic hepatitis B virus infection. *J. Clin. Invest.* **89:**332–338.

15. **Ferrari, C., A. Bertoletti, A. Penna, A. Cavalli, A. Valli, G. Missale, M. Pilli, P. Fowler, T. Gluberti, F. V. Chisari, and F. Fiaccadori.** 1991. Identification of immunodominant T cell epitopes of the hepatitis B virus nucleocapsid antigen. *J. Clin. Invest.* **88:**214–222.

16. **Ferrari, C., A. Penna, A. Bertoletti, A. Valli, A. D. Antoni, T. Guiberti, A. Cavalli, M.-A. Petit, and F. Fiaccadori.** 1990. Cellular immune response to hepatitis B virus-encoded antigens in acute and chronic hepatitis B virus infection. *J. Immunol.* **145:**3442–3449.

17. **Foster, G., M. Jacyna, and H. Thomas.** 1994. Interferon treatment of hepatitis, p. 145–150. *In* G. Griffon (ed.), *Cytokines in Infection.* Baillieres, London.

18. **Gilles, P. N., G. Fey, and F. V. Chisari.** 1992. Tumor necrosis factor-alpha negatively regulates hepatitis B virus gene expression in transgenic mice. *J. Virol.* **66:**3955–3960.

19. **Greenfield, C., P. Karayiannis, B. M. Wankya, M. V. Shah, P. Tukei, S. Galpin, T. P. Jowett, and H. C. Thomas.** 1984. Aetiology of acute sporadic hepatitis in adults in Kenya. *J. Med. Virol.* **14:**357–362.

20. **Guidotti, L. G., S. Guilhot, and F. V. Chisari.** 1994. Interleukin 2 and interferon alpha/beta down-regulate hepatitis B virus gene expression in vivo by tumor necrosis factor-dependent and independent pathways. *J. Virol.* **68:**1265–1270.

21. **Guidotti, L. G., B. Matzke, C. Pasquinelli, J. M. Shoenberger, C. Rogler, and F. V. Chisari.** 1996. The hepatitis B virus (HBV) precore protein inhibits HBV replication in transgenic mice. *J. Virol.* **70:**7056–7061.

22. **Hadziyannis, S. J.** 1995. Hepatitis B e antigen negative chronic hepatitis B: from clinical recognition to pathogenesis and treatment. *Viral Hepatitis Rev.* **1:**7–36.

23. **Hadziyannis, S. J., H. M. Lieberman, G. G. Karvountzis, and D. Shafritz.** 1983. Analysis of liver disease, nuclear HBsAg, viral replication, and hepatitis B virus DNA in liver and serum of HBeAg vs. anti-HBe positive carriers of hepatitis B virus. *Hepatology* **3:**656–662.

24. **Harris, H., and T. Gill.** 1986. Expression of class I transplantation antigens. *Transplantation* **42:**109–117.

25. **Harrison, T. J., C.-J. Oon, and A. J. Zuckerman.** 1994. A novel antibody escape variant (Ala 144) of hepatitis B virus in an identical twin before selection in the mother, p. 248–251. *In* K. Nishioka, H. Suzuki, S. Mishiro, and T. Oda (ed.), *Viral Hepatitis and Liver Disease.* Springer, Tokyo.

26. **Hata, K., D. H. Van Theil, R. B. Heberman, and T. Whiteside.** 1991. Natural killer activity of human liver-derived lymphocytes in various hepatic diseases. *Hepatology* **14:**495–503.

27. **Hauptman, R., and P. Sweetly.** 1985. A novel class of human Type I interferons. *Nucleic Acids Res.* **13:**4739–4749.

28. **Hernandez-Frontera, E., and D. N. McMurray.** 1993. Dietary vitamin D affects cell-mediated hypersensitivity but not resistance to experimental pulmonary tuberculosis in guinea pigs. *Infect. Immun.* **61:**2116–2121.

29. **Hertogs, K., W. P. Leenders, E. Depla, W. C. C. De Bruin, L. Meheus, J. Raymackers, H. Moshage, and S. M. Yap.** 1993. Endonexin II, present on human liver plasma membranes, is a specific binding protein of small hepatitis B virus (HBV) envelope protein. *Virology* **197:**549–557.

30. **Hill, A. V. S., C. E. Allsopp, D. Kwiatkowski, N. Anstey, P. Twumasi, P. Rowe, S. Bennett, D. Brewster, A. McMichael, and B. Greenwood.** 1991. Common West African HLA antigens are associated with protection from severe malaria. *Nature* **352:**595–600.

31. **Hinshaw, V. S., C. W. Olsen, N. Dybdahl-Sissoko, and D. Evans.** 1994. Apoptosis: a mechanism of cell killing by influenza A and B viruses. *J. Virol.* **69:**3667–3673.

32. **Hiscott, J., K. Cantell, and C. Weissman.** 1984. Differential expression of human interferon genes. *Nucleic Acids Res.* **12:**3727–3746.

33. **Hsu, H.-Y., M.-H., Chang, K.-H. Hsieh, C.-Y. Lee, H.-H. Lee, L.-H. Hwang, P.-J. Chen, and D.-S. Chen.** 1992. Cellular immune response to HBcAg in mother-infant transmission of hepatitis B virus. *Hepatology* **15:**770–776.

34. **Iwarson, S., E. Tabor, H. C. Thomas, A. Goodall, J. Waters, P. Snoy, J. W.-K. Shih, and R. C. Gerety.** 1985. Neutralization of hepatitis B virus infectivity by a murine monoclonal antibody: an experimental study in the chimpanzee. *J. Med. Virol.* **16:**89–96.

34a. **Jeannet, M., and J. J. Farquet.** 1974. HLA antigens in asymptomatic chronic HBsAg carriers. *Lancet* **ii:**1383–1384.

35. **Jung, M. C., H. M. Diepolder, U. Spengler, E. A. Wierenga, R. Zachoval, R. M. Hoffmann, D. Eichenlaub, G. Frosner, H. Will, and G. R. Pape.** 1995. Activation of a heterogeneous hepatitis B (HB) core and e antigen-specific CD4+ T-cell population during seroconversion to anti-HBe and anti-HBs in hepatitis B virus infection. *J. Virol.* **69:**3358–3368.

36. **Jung, M. C., U. Spengler, W. Schraut, R. Hoffman, R. Zachoval, J. Eisenburg, D. Eichenlaub, G. Riethmuller, G. Paumgartner, H. W. L. Zeigler-Heitbrock, H. Will, and G. R. Pape.** 1991. Hepatitis B virus antigen-specific T-cell activation in patients with acute and chronic hepatitis B. *J. Hepatol.* **13:**310–317.

37. **Karthigesu, V., L. M. C. Allison, M. Fortuin, M. Mendy, H. C. Whittle, and C. R. Howard.** 1994. A novel hepatitis B virus variant in the sera of immunized children. *J. Gen. Virol.* **75:**443–448.

38. **Kelly, A., S. Powis, R. Glynne, E. Radley, S. Beck, and J. Trowsdale.** 1991. Second proteosome related gene in the human MHC class II region. *Nature* **353:**667–668.

39. **Lamberts, C., M. Nassal, I. Velhagen, H. Zentgraf, and C. H. Schroder.** 1993. Precore-mediated inhibition of hepatitis B virus progeny DNA synthesis. *J. Virol.* **67:**3756–3762.

40. **Lohr, H. F., G. Gerken, H.-J. Schliet, K. H. Meyer zum Buschenfelde, and B. Fleischer.** 1993. Low frequency of cytotoxic liver-infiltrating T lymphocytes specific for endogenous processed surface and core proteins in chronic hepatitis B. *J. Infect. Dis.* **168:**1133–1139.

41. **Lok, A.** 1993. Treatment of hepatitis B. *J. Viral Hepatitis.* **1:**91–102.

42. **Mangold, C. M., F. Unckell, M. Werr, and R. Streck.** 1995. Secretion and antigenicity of hepatitis B virus small envelope proteins lacking cysteines in the major antigenic region. *Virology* **211:**535–543.

43. **Maruyama, T., S. Iino, K. Koike, K. Yasuda, and D. R. Milich.** 1993. Serology of acute exacerbation in chronic hepatitis B virus infection. *Gastroenterology* **105:**1141–1151.

44. **Maruyama, T., A. McLachlan, S. Iino, K. Koike, K. Kurokawa, and D. R. Milich.** 1993. The serology of chronic hepatitis B infection revisited. *J. Clin. Invest.* **91:**2586–2595.

45. **McMahon, G., P. H. Ehlich, Z. A. Moustafa, L. A. McCarthy, D. Dottovio, M. D. Tolpin, P. I. Nadler, and L. Ostberg.** 1992. Genetic alterations in the gene encoding the major HBsAg: DNA and immunological analysis of recurrent HBsAg derive from monoclonal antibody-treated liver transplant patients. *Hepatology* **15:**757–766.

46. **Milich, D. R., J. Jones, J. Hughes, and T. Maruyama.** 1993. Role of T-cell tolerance in the persistence of hepatitis B virus infection. *J. Immunother.* **14:**226–233.

47. **Milich, D. R., A. McLachlan, F. V. Chisari, S. B. H. Kent, and G. B. Thornton.** 1986. Immune response to the pre-S(1) region of the hepatitis B surface antigen (HBsAg): a pre-S(1)-specific T cell response can bypass nonresponsiveness to the pre-S(2) and S regions of HBsAg. *J. Immunol.* **137:**315–322.

48. **Montano, L., F. Aranguibel, M. Boffil, A. H. Goodall, G. Janossy, and H. C. Thomas.** 1983. An analysis of the composition of the inflammatory infiltrate in autoimmune and hepatitis B virus induced chronic liver disease. *Hepatology* **3:**292–296.

49. **Moriyama, T., S. Guilhot, K. Klopchin, B. Moss, C. A. Pinkert, R. D. Palmiter, R. L. Brinster, O. Kanagawa, and F. V. Chisari.** 1990. Immunobiology and pathogenesis of hepatocellular injury in hepatitis B virus transgenic mice. *Science* **248:**361–364.

50. **Morrison, N. A., J. C. Qi, A. Tokita, P. J. Kelly, L. Crofts, T. V. Nguyen, P. N. Sambrook, and J. A. Eisman.** 1994. Prediction of bone density from vitamin D receptor allelles. *Nature* **367:**284–287.

51. **Munoz, L. A.** 1987. Serum opsonic activity and polymorphonuclear cell function in patients with chronic liver disease. Ph.D. thesis. University of London, London.

52. **Nainan, O. V., C. E. Stevens, and H. S. Margolis.** 1996. Hepatitis B virus (HBV) antibody resistant mutants: frequency and significance, abstr. 98. *In* M. Rizzetto, R. H. Purcell, J. L. Gerin, and G. Verme (ed.), IX Triennial International Symposium on Viral Hepatitis and Liver Disease, Rome, April, 1996. Edizioni Minerva Medica, Turin.

53. **Nayersina, R., P. Fowler, S. Guilhot, G. Missale, A. Cerny, H.-J. Schlict, A. Vitiello, R. Chesnut, J. L. Person, A. G. Redeker, and F. V. Chisari.** 1993. HLA-A2 restricted cytotoxic T lymphocyte response to multiple hepatitis B surface antigen epitopes during hepatitis B virus infection. *J. Immunol.* **150:**4659–4671.

54. **Neurath, A. R., S. B. H. Kent, N. Strick, and K. Parker.** 1986. Identification and chemical synthesis of a host cell receptor binding site on hepatitis B virus. *Cell* **46:**429–436.

55. **Ogata, N., R. H. Miller, K. G. Ishak, A. R. Zanetti, and R. H. Purcell.** 1994. Genetic and biological characterisation of two hepatitis B virus variants: a precore mutant implicated in fulminant hepatitis and a surface mutant resistant to immunoprophylaxis, p. 238–242. *In* K. Nishioka, H. Suzuki, S. Mishiro, and T. Oda (ed.), *Viral Hepatitis and Liver Disease.* Springer, Tokyo.

56. **Oon, C.-J., G.-K. Lim, Z. Ye, K.-T. Goh, K.-L. Tan, S.-L. Yo, E. Hopes, T. Harrison, and A. J. Zuckerman.** 1995. Molecular epidemiology of hepatitis B virus vaccine variants in Singapore. *Vaccine* **13:**699–702.

57. **Ortaldo, J. R., B. Hebermann, C. Harvey, P. Osheroff, Y.-C. E. Pan, B. Kelder, and S. Petska.** 1984. A species of human interferon that lacks the ability to boost natural killer cell activity. *Proc. Natl. Acad. Sci. USA* **81:**4926–4929.

58. **Ou, J.-H., O. Lamb, and W. J. Rutter.** 1986. Hepatitis B virus gene function: the precore region targets the core antigen to cellular membranes and causes the secretion of the e antigen. *Proc. Natl. Acad. Sci. USA* **83:**1578–1582.

58a. **Patterson, M. J., M. R. Hourani, and G. H. Mayor.** 1977. HLA antigens and hepatitis B. *N. Engl. J. Med.* **297:**1124.

59. **Penna, A., F. V. Chisari, A. Bertoletti, G. Missale, P. Fowler, T. Giuberti, F. Fiacca-dori, and C. Ferrari.** 1991. Cytotoxic T lymphocytes recognise an HLA-A2 restricted epitope within the hepatitis B virus nucleocapsid antigen. *J. Exp. Med.* **174:**1565–1570.

60. **Pestka, S., J. Langer, K. Zoon, and C. Samuel.** 1987. Interferons and their actions. *Annu. Rev. Biochem.* **56:**727–777.

61. **Pignatelli, M., J. Waters, D. Brown, A. Lever, S. Iwarson, Z. Schaff, R. Gerety, and H. C. Thomas.** 1986. HLA class I antigens on the hepatocyte membrane during recovery from acute hepatitis B virus infection and during interferon therapy in chronic hepatitis B virus infection. *Hepatology* **6:**349–353.

62. **Rehermann, B., C. Ferrari, C. Pasquinelli, and F. V. Chisari.** 1996. The hepatitis B virus persists for decades after patients' recovery from acute viral hepatitis despite active maintenance of a cytotoxic T-lymphocyte response. *Nat. Med.* **2:**1104–1108.

63. **Rehermann, B., P. Fowler, J. Sidney, J. Person, A. Redeker, M. Brown, B. Moss, A. Sette, and F. V. Chisari.** 1995. The cytotoxic T lymphocyte response to multiple hepatitis B virus polymerase epitopes during and after acute viral hepatitis. *J. Exp. Med.* **181:**1047–1058.

64. **Rehermann, B., D. Lau, J. H. Hoofnagle, and F. V. Chisari.** 1996. Cytotoxic T lymphocyte responsiveness after resolution of chronic hepatitis B virus infection. *J. Clin. Invest.* **97:**1655–1665.

65. **Rhodes, J., J. Ivanyi, and P. Cozens.** 1986. Antigen presentation by human monocytes: effects of modifying major histocompatibility complex class II antigen expression and interleukin I production using recombinant interferons and corticosteroids. *Eur. J. Immunol.* **16:**370–375.

66. **Rook, G. A., J. Taverne, C. Leveton, and J. Steele.** 1987. The role of gamma-interferon, vitamin D3 metabolites and tumour necrosis factor in the pathogenesis of tuberculosis. *Immunology* **62:**229–234.

67. **Salfeld, J., E. Pfaff, M. Noah, and H. Schaller.** 1989. Antigenic determinants and functional domains in core antigen and e antigen from hepatitis B virus. *J. Virol.* **63:**798–808.

68. **Scaglioni, P. P., M. Melegari, and J. R. Wands.** 1997. Posttranscriptional regulation of hepatitis B virus replication by the precore protein. *J. Virol.* **71:**345–353.

69. **Stellar, H.** 1995. Mechanisms and genes of cellular suicide. *Science* **267:**1445–1449.

70. **Stevens, C. E., R. P. Beasley, and J. W. C. Tsu.** 1975. Vertical transmission of hepatitis B antigen in Taiwan. *N. Engl. J. Med.* **292:**771–780.

71. **Sumiya, M., M. Super, P. Tabona, R. J. Levinski, T. Arai, M. W. Turner, and J. A. Summerfield.** 1992. Molecular basis of opsonic defect in immunodeficient children. *Lancet* **337:**1569–1570.

72. **Summerfield, J. A., S. Ryder, M. Sumiya, M. R. Thusz, A. Gorchein, M. A. Monteil, and M. W. Turner.** 1995. Mannose binding protein gene mutations associated with unusual and severe infections in adults. *Lancet* **345:**886–889.

73. **Szmuness, W., C. Stevens, E. Harley, M. S. E. Zang, W. R. Oleszko, D. C. William, R. Sadovsky, J. Morrison, and A. Kellner.** 1980. Hepatitis B vaccine. Demonstration of efficacy in a controlled clinical trial in a high-risk population. *N. Engl. J. Med.* **303:**833–841.

74. **Takizawa, T., R. Fukuda, T. Miyawaki, K. Ohashi, and Y. Nakanishi.** 1995. Activation of the apoptotic FAS antigen encoding gene upon influenza virus infection involving spontaneously produced interferon beta. *Virology* **209:**288–296.

75. **Thomas, H. C., G. R. Foster, M. Sumiya, D. McIntosh, M. W. Turner, and J. A. Summerfield.** A mannose binding protein gene mutation is associated with hepatitis B viral infections. *Lancet* **348:**1417–1419.

76. **Thursz, M., D. Kwiatkowski, C. E. M. Allsopp, B. M. Greenwood, H. C. Thomas, and A. V. S. Hill.** 1995. Association between an MHC class II allele and clearance of hepatitis B virus in the Gambia. *N. Engl. J. Med.* **332:**1065–1069.

77. **Trowsdale, J., I. Hanson, I. Mockbridge, S. Beck, A. Townsend, and A. Kelly.** 1990. Sequences encoded in the class II region of the MHC relate to the ABC family of transporters. *Nature* **348:**741–744.

78. **Tsai, S. L., P. J. Chen, M. Y. Lai, P. M. Yang, J. L. Sung, J. H. Huang, L. H. Hwang, T. H. Chang, and D. S. Chen.** 1992. Acute exacerbations of chronic type B hepatitis are accompanied by increased T cell responses to hepatitis B core and e antigens. *J. Clin. Invest.* **89:**87–96.

79. **Van Hattum, J., G. M. Schreuder, and S. W. Schalm.** 1987. HLA antigens in patients with various courses after hepatitis B virus infection. *Hepatology* **7:**11–14.

80. **Van Hecke, E., J. Paradijs, C. Molitor, C. Bastin, P. Pala, M. Slaoui, and G. Leroux-Roels.** 1994. Hepatitis B virus-specific cytotoxic T lymphocyte responses in patients with acute and chronic hepatitis B virus infection. *J. Hepatol.* **20:**514–523.

81. **von Hoegan, P.** 1995. Synergistic role of type I interferons in the induction of protective cytotoxic T lymphocytes. *Immunol. Lett.* **47:**157–162.

82. **Waters, J. A., S. E. Brown, M. W. Steward, C. R. Howard, and H. C. Thomas.** 1992. Analysis of the antigenic epitopes of hepatitis B surface antigen involved in the induction of a protective antibody response. *Virus Res.* **221:**1–12.

83. **Waters, J. A., M. Kennedy, P. Voet, P. Hauser, J. Petre, W. F. Carman, and H. C. Thomas.** 1992. Loss of the common 'a' determinant of hepatitis B surface antigen by a vaccine-induced escape mutant. *J. Clin. Invest.* **90:**2543–2547.

84. **Waters, J. A., S. M. O'Rourke, S. C. Richardson, G. Papaevangelou, and H. C. Thomas.** 1987. Qualitative analysis of the humoral immune response to the 'a' determinant of HBs antigen after inoculation with plasma derived or recombinant vaccine. *J. Med. Virol.* **21:**155–160.

85. **Waters, J. A., S. O'Rourke, H.-J. Schlict, and H. C. Thomas.** 1995. Cytotoxic T cell responses in patients with chronic hepatitis B virus infection undergoing HBe antigen/antibody seroconversion. *Clin. Exp. Immunol.* **102:**314–319.

86. **Wilson, A. G., N. deVries, F. Pociot, F. S. di Giovine, L. B. A. van der Putte, and G. W. Duff.** 1993. An allelic polymorphism within the tumour necrosis alpha promoter region is strongly associated with HLA A1, B8,DR3 alleles. *J. Exp. Med.* **177:**557–559.

87. **Yamamoto, K., M. Horikita, F. Tsuda, K. Itoh, Y. Akhane, S. Yotsumoto, H. Okamoto, and Y. Miyakawa.** 1994. Naturally occurring escape mutants of hepatitis B virus with various mutations in the S gene in carriers seropositive for antibody to hepatitis B surface antigen. *J. Virol.* **68:**2671–2676.

88. **Yoo, Y., B. Gavaler, K. Chen, T. L. Whiteside, and D. H. van Thiel.** 1990. The effect of recombinant interferon alfa on lymphocyte subpopulations and HLA DR expression on liver tissue of HBV positive individuals. *Clin. Exp. Immunol.* **82:**338–343.

Human Tumor Viruses
Edited by Dennis J. McCance
© 1998 American Society for Microbiology

10 | Hepatitis C Virus
David K. H. Wong and Bruce D. Walker

Hepatitis associated with transfusion of blood products continued to be a major problem despite the use of screening tests for hepatitis A and B in the early 1970s (3, 65, 126, 182). This disease, termed non-A, non-B (NANB) hepatitis, was usually characterized by a mild chronic clinical course that could potentially lead to cirrhosis, hepatocellular carcinoma (HCC), and liver failure (18, 71, 118, 147, 189). The facts that NANB could be transmitted to chimpanzees through an infectious inoculum (4, 64, 90, 215, 230) and that the inoculum remained infectious after passage through an 80-nm filter (22, 23, 80), but was sensitive to chloroform treatment (66), suggested that the causative agent of NANB was a virus with a lipid envelope. However, methods successfully employed to identify hepatitis A virus and hepatitis B virus were not successful in isolating this new virus (210). Using molecular biological techniques, a viral genome was cloned from the serum of a chimpanzee infected with NANB hepatitis (41). The virus was designated the hepatitis C virus (HCV) and has been shown to be the major causative agent of NANB hepatitis (1, 60, 131, 204, 261). Since then, HCV has emerged as a major cause of chronic liver disease with a prevalence of 1.8% in the United States (5).

VIROLOGY

Studies of HCV have been hampered by the lack of an easy animal or tissue culture model. The only reproducible animal model for HCV infection is the chimpanzee (254), but these animals are expensive and access to the animals is not easily available. Several groups have been working on transgenic mouse models of HCV, and the structural proteins of the virus have been successfully expressed in the liver of these mice (112, 160). Other groups have reported success in propagating HCV in various tissue culture systems (46, 96, 101, 107, 158, 163, 214, 216, 271). However, there is still no cell culture system that can reliably support unlimited, high-titer replication of the virus. Furthermore, demonstration of HCV replication requires the use of strand-specific reverse transcriptase-PCR to detect antigenomic RNA, the presumed replicative intermediate. This procedure is still technically difficult and can easily lead to false-positive results (135, 137, 217, 262). Without these experimental systems, fundamental questions, such as whether HCV is a cytopathic virus, have yet to be answered conclusively.

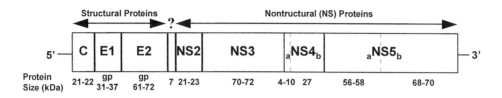

Figure 1   Structure and function of the various HCV genomic regions and proteins. See Table 1.

Despite the above limitations, a large amount has been learned about HCV. Much of the knowledge has been inferred from comparative analyses of its genomic sequence and hydrophobicity profile with those of other known viruses (42, 153) and the use of established heterologous eukaryotic expression systems, such as vaccinia virus, Sindbis virus, or baculovirus expression systems, to study viral assembly. The caveat from these studies is that potential problems may exist, as studies of processing pathways and kinetics in eukaryotic transient expression assays may not accurately reflect HCV-infected cells. For example, *trans*-acting factors, which are important for temporal regulation of RNA synthesis for other viruses, are concentration dependent and may be much lower in HCV-infected cells as compared to the levels achieved by the different expression systems (136). However, the similar results obtained from such diverse systems suggest that the data authentically represent HCV life cycle events.

HCV has been classified as the third genus of the *Flaviviridae* family, which, together with pestiviruses and flaviviruses, are single-stranded RNA viruses with a positive polarity and share a similar genomic structure and organization (153, 193). The HCV genome contains approximately 9,500 nucleotides organized as noncoding regions at both the 5′ and 3′ ends, flanking a single open reading frame that codes for a precursor polyprotein of just over 3,000 amino acids (aa) (Fig. 1 and Table 1) (2, 33, 37, 39, 42, 92, 98, 106, 164, 170–173, 233, 257, 266). In vitro experiments suggest that the polyprotein is processed co- and posttranslationally by both the host cellular proteases and virus-encoded proteases into a series of putative structural proteins (core and envelope) and nonstructural proteins (proteases, helicases, and RNA polymerases) (74, 207). Each of these regions will be discussed below.

Noncoding Regions

The noncoding regions are the most conserved regions of the HCV genome and likely play critical roles in regulation of the viral life cycle. As such, they are potentially useful sites for use in diagnostic tests and targeted therapeutic interventions (174, 252, 261). The 5′-untranslated region (5′UTR) likely plays a role in the initiation of translation of the HCV polypeptide. Translation of RNA molecules is usually initiated by ribosome scanning of the 5′ end of capped RNA. However, the 5′UTR of HCV is probably not capped. Its 341 nucleotides are predicted, based on computer modeling, to have a secondary structure consisting of four stem-loop structures. This secondary structure is similar to that seen in picornaviruses, where translation of uncapped RNA is initiated through an internal ribosome entry site (IRES) (100, 177, 250, 256). The 5'UTR of HCV has also been shown to have an IRES (25, 91). This region of HCV is very well conserved among the different HCV genotypes (28), and the few reported variants have generally been found within the single-stranded regions or in areas that do not affect the stability of the basic secondary structure (220, 223, 224, 250).

The 3'UTRs in other RNA viruses have been shown to be important for RNA synthesis and genome packaging during viral replication (249). Initial reports suggested that the HCV genome terminated in a poly(U) or poly(A) tract. More recent studies have shown that the 3'UTR extends further and is now thought to contain the following elements: 5'—stop codon—30 nucleotides (variable by genotype)—homopolymeric poly(U) tract—polypyrimidine stretch consisting of mainly U interspersed with C residues—highly conserved 98-base sequence—3' (123, 235, 236, 267). The latter conserved region has been predicted to form three stable stem-loop structures (21). As with the 5'UTR, variants reported in this conserved region generally do not affect the stability of the predicted secondary structure (123, 224, 267). The important role that the 3'UTR plays in HCV replication is highlighted by the recent success in constructing infectious cDNA clones only when the extended 3'UTR sequence was included (122, 268).

Structural Proteins

The structural proteins are located in the first 746 aa at the N terminus of the HCV precursor polyprotein, which is subsequently cleaved into the core protein and two putative envelope proteins, E1 and E2. Processing of the structural pro-

Table 1 HCV genomic regions and proteins[a]

Genome region/protein	Location of mature protein[b] (aa)	Function(s)
5'UTR	Not translated	Initiation of translation, replication[c]
C	1–191	Structural, encapsidation of viral RNA[c]
E1	192–383	Structural, receptor binding, cell entry[c]
E2	384–746	Structural, receptor binding, cell entry[c]
p7	747–809	Unknown
NS2	810–1026	Part of NS2-3 protease
NS3	1027–1657	NS2-3 protease, serine protease, helicase, NTPase
NS4A	1658–1711	Cofactor for NS3 serine protease
NS4B	1712–1972	Replicase component[c]
NS5A	1973–2420	Replicase component[c]
NS5B	2421–3011	RNA-dependent RNA polymerase
3'UTR	Not translated	Replication[c], packaging of viral genome[c]

[a] See Fig. 1. NTPase, nucleotide triphosphatase.
[b] Based on HCV-H strain amino acid sequence (140).
[c] Putative function based on comparisons with other viruses.

teins is facilitated by hydrophobic regions at the C terminus of each protein, which serve as signal sequences for directing the core protein to the cytosolic side of the endoplasmic reticulum (ER) membrane and translocating E1 and E2 inside the lumen of the ER, where they are accessible to cleavage by the host signal peptidases (11, 82, 134, 143, 199).

The core protein is derived from the N-terminal 191 aa and plays a role in viral encapsidation. Its intracellular distribution has been confirmed by immuno-fluorescence studies, which demonstrate a punctate pattern consistent with an ER distribution as well as a perinuclear distribution (78, 144, 159, 207, 229). The composition of the core protein includes a highly basic region at the N terminus that is capable of binding nonspecifically to the negatively charged HCV RNA, resulting in the formation of the nucleocapsid (199). This nucleocapsid nature of the core protein has been confirmed by immunoelectron microscopic studies of detergent-stripped virions (232). The core protein also has a DNA binding motif, and in vitro studies have shown translocation of the core protein to the nucleus (30, 134, 144, 229). Although this would have implications for the potential role that the core protein may play in the pathogenesis of HCC, translocation of the core protein into the nucleus has not been observed in infected hepatocytes. Other preliminary studies have also suggested that the core protein may affect host cell regulation and thus enhance the potential for malignant transformation, although this remains to be shown conclusively (185–188).

The E1 glycoprotein is a putative envelope protein, based on comparisons with the other members of the *Flaviviridae* family. The function of E2 was controversial, as this genomic region corresponds to an envelope glycoprotein in pestiviruses and a nonstructural protein (NS1) in flaviviruses (42). The functional status of E2 is further clouded by in vitro findings of the E2 precursors, E2-p7 and E2-NS2 (140, 207, 208). Although confirmation of the function of these proteins by direct analysis of purified HCV virions remains to be performed, HCV is more closely related to the pestiviruses and E2 is believed to be more likely an envelope glycoprotein. The function of p7 is also unknown, but it may have a role in viral assembly (157).

E1 and E2 are directed to the ER by C-terminal hydrophobic signal sequences, where they are cleaved and undergo glycosylation at six potential N-glycosylation sites in E1 and 11 potential sites in E2 (82, 201, 225). Truncation experiments show that the C-terminal portion of E2 is important for anchoring E2 to the membrane as well as associating E2 with E1 (157, 225). After cleavage to mature proteins, E1 and E2 are anchored to the ER membrane as noncovalently linked heterodimers (74, 82, 167, 184, 199).

Nonstructural proteins

The nonstructural (NS) proteins are located downstream of the structural proteins and constitute approximately 75% of the viral polyprotein by size. The functions of some these proteins were first suggested by genetic alignments with flaviviruses and pestiviruses, and further data have been obtained by in vitro expression systems. They have been shown to form a complex associated with the ER of stably transfected mammalian cells, suggesting a role in viral replication and assembly

(84). However, their NS status and function await confirmation by direct character-ization within virions and infected cells. Cleavage of the NS proteins is mediated by virus-encoded proteases, but the efficiency of these processes is at least partially dependent on microsomal membranes, suggesting that host cofactors may be re-quired (72, 176).

The NS2 protein is cleaved from p7 by the host signal peptidase and from NS3 by a zinc-dependent metalloprotease (73, 83, 141, 157, 176). The precise mech-anism for this viral protease is still unclear but is known to reside within aa 827–1207, spanning most of NS2 and the N terminus of NS3 (58, 72).

The NS3 protein is predicted to have a diverse number of functions based on conserved sequence motifs, including a serine protease, a helicase, and a nu-cleotide triphosphatase, as well as contributing to the NS2-3 metalloprotease. A trypsin-like serine protease was initially predicted based on the presence of three highly conserved residues representing the catalytic triad of the serine protease family within the N-terminal one-third of the NS3 protein: His-1083, Asp-1107, and Ser-1165 (54, 141, 153). Substitution at any of these three sites results in loss of the protease activity (12, 47, 77, 83, 141). This serine protease is distinct from the NS2-3 protease, as it is responsible for all NS protein cleavages downstream of NS2-3 (12, 57, 73, 83, 88, 141, 244) and its inactivation does not affect NS2-3 cleavage (72, 83). It is likely an essential protein, as its flavivirus homolog is indispensable for viral growth (34). The efficiency of this enzyme is greatly en-hanced by its interaction with the NS4A protein (13, 57, 58, 77, 121, 142, 202, 240). The crystal structure of NS3 has been resolved, and the initial analysis suggested the presence of a trypsin-like fold (145). However, the biochemical evidence was more consistent with a chymotrypsin-like protease (77), and another crystal struc-ture analysis that included the NS4A protein was also more consistent with a chymotrypsin-like protease (116).

The HCV helicase has been mapped to a region within aa 1209–1608 in the NS3 protein (47, 114, 115, 228), and it has been shown to unwind RNA-RNA substrates in a 3′-to-5′ direction (76, 94). This is supported by crystal structure analysis showing the presence of RNA binding and nucleotide triphosphatase domains within the NS3 protein (270). Helicases function in the life cycle of RNA viruses by playing a role in double-stranded RNA strand separation during viral replication and transcription. These enzymes usually have intrinsic nucleotide triphosphatase activity, allowing them to utilize nucleotide triphosphate as an energy source.

The NS4 protein is cleaved into two smaller proteins, NS4A and NS4B. The NS4A protein is 54 aa long and contains a hydrophilic C terminus and a hydropho-bic central core. As described above, NS4A functions as a cofactor for the NS3 serine protease. This occurs through the association of the small central hydropho-bic core of NS4A (57, 142, 245) with the N terminus of the NS3 protein (58, 121, 202). X-ray crystallography shows that NS4A intercalates within a β sheet of the NS3 enzyme core, altering its structure (116). The function of the NS4B protein is still unknown.

The NS5 protein is also cleaved into two smaller proteins, NS5A and NS5B. The NS5A is a phosphoprotein (7, 190, 241) whose function is still unknown. Comparative studies suggest that NS5A is important in HCV replication. Recent epidemiological studies from Japan suggested the presence of an area that might affect sensitivity to interferon-alpha therapy (36, 55, 56, 132, 200). However, this finding has not been confirmed by studies from outside Japan (113, 226, 275).

The NS5B protein is predicted to contain an RNA-dependent RNA polymerase, based on conserved sequence homologies (124, 153, 179). Indeed, RNA polymerase activity has been reported in several in vitro systems (15, 273). The enzyme is thought to mediate RNA synthesis through a copy-back mechanism where a hairpin-like RNA duplex is generated, allowing use of the 3'-terminal nucleotide of the template RNA as the primer (15).

Virion Morphology

Despite the large body of knowledge about the function of the various HCV proteins, visualization of the HCV virion by immunoelectron microscopy has only recently been reported (103, 211). Prior to this, a number of physical characteristics had already been inferred. The infectious particle was smaller than 80 nm, based on filtration studies (22, 23, 80), and it was hypothesized to contain a lipid envelope since it could be inactivated by chloroform treatment (66). Sucrose gradient analyses showed that infectious particles could be found in either a low-density (1.06 to 1.13 g/ml) or a high-density (1.17 to 1.25 g/ml) fraction in plasma (85, 103, 105, 242). The higher-density fraction has subsequently been shown to consist solely of HCV RNA and nucleocapsid (217), whereas the lower-density fraction has been shown to be HCV associated with low-density lipoproteins (242) and immunoglobulins (43, 85, 242).

Genomic Variation

RNA viruses, in general, have a high mutation rate because random errors introduced by the RNA-dependent RNA polymerase during replication remain uncorrected due to the lack of a proofreading function in that enzyme. However, the accumulation of mutations within the HCV genome is not random, as some errors may lead to lethal mutations and hence not detected. Other changes at the nucleotide level may be silent at the amino acid level due to the redundant nature of the genetic code, but might disrupt the secondary structure of the RNA.

Comparisons of HCV isolates obtained from different individuals have confirmed that significant differences exist at both the nucleotide level and the amino acid level. To categorize these differences, a phylogenetic tree with four branches (I through IV) was initially proposed by Okamoto et al. (172). This provisional system was based on the order in which HCV sequences were described and did not properly reflect the relationship between the various isolates. A more recent classification system has been proposed based on the sequence alignments of a large number of isolates (35, 218, 219). Using this system, isolates were categorized, based on sequence homology, as belonging to different types (<72%) or subtypes (75% to 86%) or to the same subtype (>86%) (218). Although sequencing of the entire HCV genome is the gold standard, it was found that comparison of the 5'UTR, core, NS3, or NS5 regions yielded similar classifications (35, 222). To date, at least six major types have been characterized, but more probably exist (29, 221).

Sequence alignment analyses of isolates from the same individuals have revealed that although the overall sequence homology is high (>98%), an area of tremendous variability exists in the N terminus of the E2 envelope glycoprotein, spanning residues 384 to 410. This area has been designated the hypervariable region 1 (HVR1) (110, 157, 258, 260). A second hypervariable region (HVR2), approximately 7 aa long, has also been observed among 1b strains (81, 197). This has given rise to the concept that HCV exists as a swarm of closely related yet heterogeneous sequences termed quasispecies, rather than a population of uniform viral sequences (62, 148, 171, 237). This region seems to have a remarkably high tolerance for different secondary structures (239), and variations in the HVR have been noted to accumulate rapidly over time (108, 109, 165). The accumulation of mutations at the HVR1 may be driven by pressure from the humoral immune system. Several groups have shown that mutations develop after an antibody response to the HVR1 is detected (111, 206, 212, 213, 253, 260, 276). Conversely, mutations in the HVR1 do not accumulate when an antibody response fails to develop (130, 169, 253). Further evidence is provided by the finding that mutations do not accumulate in the HVR of HCV that has been maintained in long-term culture in vitro, where immune pressure is absent (163). However, it is still uncertain if this represents the selection of an HCV variant that is particularly well adapted for tissue culture, or if this truly demonstrates the lack of mutation in the absence of a humoral immune response.

The geographical distribution of the various HCV genotypes is being studied by various groups. Types 1, 2, and 3 have a worldwide distribution, with types 1a and 1b predominating in North America; type 4 predominates in Africa, type 5 in South Africa, and type 6 in Asia (24, 29, 53, 146, 152, 219, 227, 231). The clinical significance of the various genotypes is controversial, with some groups reporting that type 1b strains are associated with a higher prevalence of cirrhosis and HCC (178, 181, 183) as well as a poorer response to interferon therapy (38, 53, 69, 86, 93, 149, 219, 274). However, other studies suggest that other potentially confounding factors, such as viral load and duration of infection, may be more dominant factors (17, 120, 265). The importance of sequence variability on the design of diagnostic tests, such as primer selection for PCR, is more evident (162).

IMMUNE RESPONSE TO HCV

The role of the host immune response in the pathogenesis of HCV infection is still not clear. Histologic evaluation of the liver from persons with chronic HCV infection has shown that diffuse and focal aggregates of lymphoid inflammatory cells characterize this disease (19). Immunohistochemical studies have shown that HCV antigens are widespread in the liver within the cytoplasm of hepatocytes and also within the infiltrating lymphoid aggregates. These lymphoid aggregates are collections of $CD4^+$ and $CD8^+$ T cells as well as $CD20^+$ B cells (20). A diffuse increased expression of HLA class I and ICAM-1 in the membranes of the hepatocytes has also been noted (9, 10).

Although these findings suggest an activated host immune response, the functional role of this immune response cannot be addressed by histologic studies. Functional studies are difficult to perform due to many technical limitations. Ideally, serial liver biopsy specimens should be examined, especially in subjects during the earliest stages of acute infection with HCV and in those who clear the infection spontaneously, presumably due to a successful immune response. However, these subjects are difficult to identify since a large proportion of individuals are only mildly symptomatic or asymptomatic. Furthermore, the risk associated with the liver biopsy procedure makes it unethical to obtain multiple liver biopsy specimens, especially in those subjects who clear HCV infection. Some investigators have studied the immune response to HCV that is present within the peripheral blood. However, the magnitude of the immune response in peripheral blood is believed to be much lower than that found within the site of infection in the liver. Studies are further hampered by the size and genetic heterogeneity of HCV. Currently, most studies are designed to detect immune responses to short segments of a reference HCV isolate that may be quite dissimilar in sequence to the autologous isolate. With these caveats in mind, a picture of the host immune response to HCV is emerging.

The principal cells of the humoral immune response are B cells and plasma cells, which secrete antibodies. Antibodies are useful in that they can potentially bind and neutralize the virus outside the cell before viral entry. Early observations in chimpanzees and humans suggested that HCV infection does not confer immunity to subsequent HCV exposure and infection (59, 133). More recent studies suggest that neutralizing antibodies to HCV do exist, but that the critical neutralization domain is within the HVR1 of the E2 glycoprotein (62, 194, 213, 277). The lack of immunity after HCV infection might be explained by the finding that neutralizing antibodies that develop during the course of infection are isolate specific and lead to the selection of isolates that are not recognized by the antibody (61). At present, it is uncertain whether other neutralization domains exist.

CD8$^+$ T cells are one arm of the cellular immune response. These T cells are potentially important since they have the ability to recognize virus-infected cells (8, 151, 255) and can respond either directly, by lysis of the infected cell, or indirectly, by secreting cytokines that inhibit viral replication (40, 75). HCV-specific CD8$^+$ cytotoxic T lymphocytes (CTL) have been isolated from liver biopsy specimens (127–129, 166, 263) as well as from peripheral blood (14, 32, 104, 117, 191, 192) of subjects with chronic HCV infection. These studies suggest that an individual can generate a CTL response that is targeted at multiple epitopes simultaneously. However, the epitopes are heterogeneous and there is no evidence of a single immunodominant epitope targeted by many individuals (263). The intrahepatic CTL have been shown to secrete the cytokines interferon-gamma, tumor necrosis factor-alpha, granulocyte-macrophage colony-stimulating factor, interleukin-8 (IL-8), and IL-10 in an antigen- and HLA class I-specific manner (128). Despite this CTL response, HCV infection persists. Mutations at the targeted epitopes may contribute to viral persistence, but this has yet to be shown conclusively.

CD4$^+$ T cells form another arm of the cellular immune response. Their in vitro activity is measured by the degree of proliferation in response to a specific antigen. Activation of CD4$^+$ T cells can lead to amplification of both the B-cell and CTL immune responses. HCV-specific CD4$^+$ T cells have been isolated from peripheral blood of chronic HCV carriers as well as in those whose infection has resolved. These cells have been reported to proliferate after in vitro stimulation with a variety of HCV proteins, particularly the core (89, 99, 138), NS3 (67, 89, 155), and NS4 (154) proteins. Preliminary studies suggest that in those with resolved infection, the CD4$^+$ T cells secrete cytokines with a Th1 profile (IL-2 and interferon-gamma), which may augment a CTL response, whereas in those with nonresolving infection, the CD4$^+$ T cells secrete cytokines with a Th2 profile (IL-4 and IL-10), which may augment a B-cell response over a CTL response (248). Some of these studies suggest that CD4$^+$ response to the NS3 protein is associated with resolution of HCV infection (50, 51), but other studies have shown NS3 responses even in chronic carriers (269).

It is clear that HCV infection is able to persist in the face of the above immune responses. Although some studies have shown that a detectable CTL response is associated with a modest reduction in viral load (87, 166), other studies have not confirmed this finding (191, 192, 263). This chronic inflammatory response may have an adverse effect, as it might contribute to hepatic damage (102, 139). Since HCV can tolerate a large degree of genetic variation, mutations at sites targeted by the immune response may be one mechanism by which HCV evades the immune response. This has been shown for the humoral immune response to the HVR1 of the E2 glycoprotein and for CTL responses in chimpanzees (259). However, there is no conclusive evidence for immune escape from a CTL response or a CD4$^+$ T-cell response in humans. Determining what constitutes a sufficient immune response and the viral factors that contribute to immune evasion will be important considerations for the development of immune-based therapies.

NATURAL HISTORY

To fully evaluate the natural history of an infection, one would need to accurately establish the time of onset in a well-defined cohort and follow each infected individual to the end of the disease course without the influence of treatment which might modify the natural history. These results would then have to be compared with the natural history of an appropriately matched control group. Such studies are particularly difficult to conduct in HCV infection since a large proportion of acute infections are only mildly symptomatic (156) and the majority of infections are diagnosed in the chronic phase. Furthermore, the vast majority of individuals develop chronic, slowly progressive lifelong infection, making it logistically difficult to follow the course of infection. Many studies that have described the sequelae of chronic HCV infection have been conducted on selected patients, who may have come to clinical attention initially because they had a more severe form of HCV disease (63, 246).

Prospective data from cohorts with a known onset of infection have come from posttransfusion NANB hepatitis studies (49, 95, 125, 205, 247). These studies were initiated before HCV testing was available, so the definition of posttransfusion hepatitis was defined as an acute elevation of serum alanine or aspartate aminotransferase levels. Although the majority of these subjects were subsequently shown to be infected with HCV (1, 60, 131, 204, 261), these studies did not include those individuals who were infected but did not have abnormal aminotransferase levels. Overall, these studies show that chronic HCV infection could lead to the development of cirrhosis in 10% to 20%, HCC in less than 1.5%, and an overall liver-related mortality of under 6% after 14 to 18 years' follow-up.

Retrospective data are also available from two cohorts who were exposed to HCV through immunoglobulin products (52, 180). In a cohort of 232 Irish women who were assessed 17 years after exposure to HCV type 1b (180), serum aminotransferase levels were normal in 37.6% and cirrhosis was found in 2.4%. These studies confirm that HCV infection results in a slowly progressive disease with minimal sequelae in the first 2 decades.

Epidemiologic studies have suggested a link between HCV and HCC, with the time from infection to development of HCC estimated to be on the order of 30 years (119). HCV appears to be the dominant factor associated with HCC in areas with intermediate rates of HCC (Japan, Italy, Spain), whereas hepatitis B virus is the dominant factor in areas with higher rates of HCC (Senegal, Hong Kong, Taiwan) (26, 27, 45, 79, 168, 195, 234, 251, 272). The precise mechanism by which HCV causes HCC is unknown. Analysis of liver tissue has demonstrated the presence of HCV RNA sequences in both the malignant tissue and the surrounding normal tissue in subjects with HCC (70, 196, 238). There is no evidence that HCV produces a DNA replicative intermediate that might integrate with the host genome. Several preliminary reports suggest that some of the HCV proteins may alter the host cell regulation and hence increase the malignant potential. The core protein, in particular, has been reported to have activity in regulating cellular proto-oncogenes (186), repressing p53 transcription (188), inhibiting apoptosis in Chinese hamster ovary cell lines (187), and transforming primary rat embryo fibroblasts (185). The NS3 protein has also been shown to localize to the nucleus when it is not associated with the NS4A protein (161), as well as to have the ability to transform NIH 3T3 cells (198). Finally, the NS5A protein has a nuclear localization signal that is able to translocate heterologous proteins to the nucleus (97). However, the significance of these studies in the pathogenesis of HCV-related HCC remains to be clarified. HCV may also play an indirect role in the pathogenesis of HCC through the associated chronic inflammation and cirrhosis (48, 175, 209). A study of hepatic tissue from subjects with chronic HCV infection has also suggested that microangiogenesis is increased compared to non-HCV controls (150).

THERAPY

Therapy for chronic viral infections is still a relatively new field. Great strides have been made in understanding the virology and immunology of the human immunodeficiency virus (HIV), and therapeutic regimens have recently achieved suppression of viral replication to undetectable levels. However, eradication of HIV is still an elusive goal (44, 68, 264). Therapeutic trials for NANB hepatitis using interferon-alpha were initiated in the 1980s before diagnostic tests for HCV were available. Using serum levels of aminotransaminases (alanine aminotransferase and aspartate aminotransferase) as markers of hepatic damage, interferon-alpha was observed to result in normalization of aminotransaminases in 50% of patients during treatment, but half of these responders relapsed at the end of 6 months of therapy (243). More recent studies using the sustained loss of detectable HCV RNA as an endpoint have shown that a durable response, lasting several years after withdrawal of therapy, can occur but is seen in less than 10% of treated patients (31). This long-term response rate is probably enhanced by increasing the duration of treatment from 6 months to 12 months (6). Other studies suggest that the addition of the nucleoside analog ribavirin also shows promise in increasing the response rate to interferon-alpha therapy (16, 203). Even though the proportion of sustained responders is modest at present, it represents a breakthrough in terms of our ability to eradicate a chronic viral infection.

SUMMARY

A great deal has been learned about HCV despite many obstacles. HCV is known to be a single-stranded RNA virus with a positive polarity. The basic structural organization of the HCV genome consists of a series of structural proteins (core and envelopes) and nonstructural proteins (proteases, helicases, RNA polymerases). These translated proteins are flanked by noncoding regions which are important in initiation of protein translation (5′UTR) and viral replication (3′UTR). The HCV genome is able to tolerate a tremendous amount of genetic heterogeneity, especially in the envelope proteins where hypervariable regions have been reported. Much work remains to be done in confirming the roles the various HCV proteins play in the HCV life cycle and their role in the pathogenesis of disease.

The humoral and cellular immune response to HCV has been well documented. Neutralizing antibodies have been reported, but they seem to be isolate specific as the critical neutralization domain is in the HVR1 of the E2 glycoprotein, a region of extreme variability. $CD8^+$ and $CD4^+$ T-cell responses are also present in persons with chronic HCV infection.

The development of infectious cDNA clones is a great stride forward as it will permit the use of a standard infectious inoculum without quasispecies. However, many questions and challenges remain. In vitro systems that allow high-titer HCV replication and an easy way of measuring HCV replication still need to be developed. Also, what factors contribute to viral persistence in the face of the host immune response, what constitutes a sufficient immune response, what aspects of the immune response contribute to protection as opposed to pathology, and what implications this has for vaccine development need to be addressed.

REFERENCES

1. **Aach, R. D., C. E. Stevens, F. B. Hollinger, J. W. Mosley, D. A. Peterson, P. E. Taylor, R. G. Johnson, L. H. Barbosa, and G. J. Nemo.** 1991. Hepatitis C virus infection in post-transfusion hepatitis. An analysis with first- and second-generation assays. *N. Engl. J. Med.* **325:**1325–1329.
2. **Adams, N. J., R. W. Chamberlain, L. A. Taylor, F. Davidson, C. K. Lin, R. M. Elliott, and P. Simmonds.** 1997. Complete coding sequence of hepatitis C virus genotype 6a. *Biochem. Biophys. Res. Commun.* **234:**393–396.
3. **Alter, H. J., P. V. Holland, A. G. Morrow, R. H. Purcell, S. M. Feinstone, and Y. Moritsugu.** 1975. Clinical and serological analysis of transfusion-associated hepatitis. *Lancet* **2:**838–841.
4. **Alter, H. J., R. H. Purcell, P. V. Holland, and H. Popper.** 1978. Transmissible agent in non-A, non-B hepatitis. *Lancet* **1:**459–463.
5. **Alter, M. J.** 1997. Epidemiology of hepatitis C. *Hepatology* **26:**62S–65S.
6. **Anonymous.** 1997. National Institutes of Health Consensus Development Conference Panel statement: management of hepatitis C. *Hepatology* **26:**2S–10S.

7. **Asabe, S. I., Y. Tanji, S. Satoh, T. Kaneko, K. Kimura, and K. Shimotohno.** 1997. The N-terminal region of hepatitis C virus-encoded NS5A is important for NS4A-dependent phosphorylation. *J. Virol.* **71**:790–796.

8. **Bachmann, M. F., T. M. Kundig, H. Hengartner, and R. M. Zinkernagel.** 1997. Protection against immunopathological consequences of a viral infection by activated but not resting cytotoxic T cells: T cell memory without "memory T cells"? *Proc. Natl. Acad. Sci. USA* **94**:640–645.

9. **Ballardini, G., P. Groff, F. Giostra, R. Francesconi, D. Zauli, B. Gianpaolo, M. Lenzi, F. Cassani, and F. Bianchi.** 1994. Hepatocellular expression of HLA-A, B, C molecules predicts primary response to interferon in patients with chronic hepatitis C. *Am. J. Clin. Pathol.* **102**:746–751.

10. **Ballardini, G., P. Groff, P. Pontisso, F. Giostra, R. Francesconi, M. Lenzi, D. Zauli, A. Alberti, and F. B. Bianchi.** 1995. Hepatitis C virus (HCV) genotype, tissue HCV antigens, hepatocellular expression of HLA-A,B,C, and intercellular adhesion-1 molecules. Clues to pathogenesis of hepatocellular damage and response to interferon treatment in patients with chronic hepatitis C. *J. Clin. Invest.* **95**:2067–2075.

11. **Barba, G., F. Harper, T. Harada, M. Kohara, S. Goulinet, Y. Matsuura, G. Eder, Z. Schaff, M. J. Chapman, T. Miyamura, and C. Brechot.** 1997. Hepatitis C virus core protein shows a cytoplasmic localization and associates to cellular lipid storage droplets. *Proc. Natl. Acad. Sci. USA* **94**:1200–1205.

12. **Bartenschlager, R., L. Ahlborn-Laake, J. Mous, and H. Jacobsen.** 1993. Nonstructural protein 3 of the hepatitis C virus encodes a serine-type proteinase required for cleavage at the NS3/4 and NS4/5 junctions. *J. Virol.* **67**:3835–3844.

13. **Bartenschlager, R., V. Lohmann, T. Wilkinson, and J. O. Koch.** 1995. Complex formation between the NS3 serine-type proteinase of the hepatitis C virus and NS4A and its importance for polyprotein maturation. *J. Virol.* **69**:7519–7528.

14. **Battegay, M., J. Fikes, A. M. Di Bisceglie, P. A. Wentworth, A. Sette, E. Celis, W.-M. Ching, A. Grakoui, C. M. Rice, K. Kurokohchi, J. A. Berzofsky, J. H. Hoofnagle, S. M. Feinstone, and T. Akatsuka.** 1995. Patients with chronic hepatitis C have circulating cytotoxic T cells which recognize hepatitis C virus-encoded peptides binding to HLA-A2.1 molecules. *J. Virol.* **69**:2462–2470.

15. **Behrens, S.-E., L. Tomei, and R. De Francesco.** 1996. Identification and properties of the RNA-dependent RNA polymerase of hepatitis C virus. *EMBO J.* **15**:12–22.

16. **Bellobuono, A., L. Mondazzi, S. Tempini, E. Silini, F. Vicari, and G. Ideo.** 1997. Ribavirin and interferon-alpha combination therapy vs interferon-alpha alone in the retreatment of chronic hepatitis C: a randomized clinical trial. *J. Viral Hepatitis* **4**:185–191.

17. **Benvegnu, L., P. Pontisso, D. Cavalletto, F. Noventa, L. Chemello, and A. Alberti.** 1997. Lack of correlation between hepatitis C virus genotypes and clinical course of hepatitis C virus-related cirrhosis. *Hepatology* **25**:211–215.

18. **Berman, M., H. J. Alter, K. G. Ishak, R. H. Purcell, and E. A. Jones.** 1979. The chronic sequelae of non-A, non-B hepatitis. *Ann. Intern. Med.* **91**:1–6.

19. **Bianchi, L., V. J. Desmet, H. Popper, P. J. Scheuer, L. M. Aledort, and P. D. Berk.** 1987. Histologic patterns of liver disease in hemophiliacs, with special reference to morphologic characteristics of non-A, non-B hepatitis. *Semin. Liver Dis.* **7**:203–209.

20. **Blight, K., R. R. Lesniewski, J. T. LaBrooy, and E. J. Gowans.** 1994. Detection and distribution of hepatitis C-specific antigens in naturally infected liver. *Hepatology* **20**:553–557.

21. **Blight, K. J., and C. M. Rice.** 1997. Secondary structure determination of the conserved 98-base sequence at the 3′ terminus of hepatitis C virus genome RNA. *J. Virol.* **71**:7345–7352.

22. **Bradley, D. W., J. E. Maynard, H. Popper, E. H. Cook, J. W. Ebert, K. A. McCaustland, C. A. Schable, and H. A. Fields.** 1983. Posttransfusion non-A, non-B hepatitis: physicochemical properties of two distinct agents. *J. Infect. Dis.* **148:**254–265.

23. **Bradley, D. W., K. A. McCaustland, E. H. Cook, C. A. Schable, J. W. Ebert, and J. E. Maynard.** 1985. Posttransfusion non-A, non-B hepatitis in chimpanzees. Physicochemical evidence that the tubule-forming agent is a small, enveloped virus. *Gastroenterology* **88:**773–779.

24. **Brechot, C., and D. Kremsdorf.** 1993. Genetic variation of the hepatitis C virus (HCV) genome: random events or a clinically relevant issue? *J. Hepatol.* **17:**265–268.

25. **Brown, E. A., H. Zhang, L. H. Ping, and S. M. Lemon.** 1992. Secondary structure of the 5′ nontranslated regions of hepatitis C virus and pestivirus genomic RNAs. *Nucleic Acids Res.* **20:**5041–5045.

26. **Bruix, J., J. M. Barrera, X. Calvet, G. Ercilla, J. Costa, J. M. Sanchez-Tapias, M. Ventura, M. Vall, M. Bruguera, and C. Bru.** 1989. Prevalence of antibodies to hepatitis C virus in Spanish patients with hepatocellular carcinoma and hepatic cirrhosis [see comments] *Lancet* **ii:**1004–1006.

27. **Bukh, J., R. H. Miller, M. C. Kew, and R. H. Purcell.** 1993. Hepatitis C virus RNA in southern African blacks with hepatocellular carcinoma. *Proc. Natl. Acad. Sci. USA* **90:**1848–1851.

28. **Bukh, J., R. H. Purcell, and R. H. Miller.** 1992. Sequence analysis of the 5′ noncoding region of hepatitis C virus. *Proc. Natl. Acad. Sci. USA* **89:**4942–4946.

29. **Bukh, J., R. H. Purcell, and R. H. Miller.** 1993. At least 12 genotypes of hepatitis C virus predicted by sequence analysis of the putative E1 gene of isolates collected worldwide. *Proc. Natl. Acad. Sci. USA* **90:**8234–8238.

30. **Bukh, J., R. H. Purcell, and R. H. Miller.** 1994. Sequence analysis of the core gene of 14 hepatitis C virus genotypes. *Proc. Natl. Acad. Sci. USA* **91:**8239–8243.

31. **Carithers, R. L., Jr., and S. S. Emerson.** 1997. Therapy of hepatitis C: meta-analysis of interferon alfa-2b trials. *Hepatology* **26:**83S–88S.

32. **Cerny, A., J. G. McHutchison, C. Pasquinelli, M. E. Brown, M. A. Brothers, B. Grabscheid, P. Fowler, M. Houghton, and F. V. Chisari.** 1995. Cytotoxic T lymphocyte response to hepatitis C virus-derived peptides containing the HLA A2.1 binding motif. *J. Clin. Invest.* **95:**521–530.

33. **Chamberlain, R. W., N. Adams, A. A. Saeed, P. Simmonds, and R. M. Elliott.** 1997. Complete nucleotide sequence of a type 4 hepatitis C virus variant, the predominant genotype in the Middle East. *J. Gen. Virol.* **78:**1341–1347.

34. **Chambers, T. J., R. C. Weir, A. Grakoui, D. W. McCourt, J. F. Bazan, R. J. Fletterick, and C. M. Rice.** 1990. Evidence that the N-terminal domain of nonstructural protein NS3 from yellow fever virus is a serine protease responsible for site-specific cleavages in the viral polyprotein. *Proc. Natl. Acad. Sci. USA* **87:**8898–8902.

35. **Chan, S.-W., F. McOmish, E. C. Holmes, B. Dow, J. F. Peutherer, E. Follett, P. L. Yap, and P. Simmonds.** 1992. Analysis of a new hepatitis C virus type and its phylogenetic relationship to existing variants. *J. Gen. Virol.* **73:**1131–1141.

36. **Chayama, K., A. Tsubota, M. Kobayashi, K. Okamoto, M. Hashimoto, Y. Miyano, H. Koike, I. Koida, Y. Arase, S. Saitoh, Y. Suzuki, N. Murashima, K. Ikeda, and H. Kumada.** 1997. Pretreatment virus load and multiple amino acid substitutions in the interferon sensitivity-determining region predict the outcome of interferon treatment in patients with chronic genotype 1b hepatitis C virus infection [see comments] *Hepatology* **25:**745–749.

37. **Chayama, K., A. Tsubota, I. Koida, Y. Arase, S. Saitoh, K. Ikeda, and H. Kumada.** 1994. Nucleotide sequence of hepatitis C virus (type 3b) isolated from a Japanese patient with chronic hepatitis C. *J. Gen. Virol.* **75:**3623–3628.

38. Chemello, L., L. Cavalletto, F. Noventa, P. Bonetti, C. Casarin, E. Bernardinello, P. Pontisso, C. Donada, P. Casarin, and F. Belussi. 1995. Predictors of sustained response, relapse and no response in patients with chronic hepatitis C treated with interferon-alpha. *J. Viral Hepatitis* **2**:91–96.

39. Chen, P. J., M. H. Lin, K. F. Tai, P. C. Liu, C. J. Lin, and D. S. Chen. 1992. The Taiwanese hepatitis C virus genome: sequence determination and mapping the 5′ termini viral genomic and antigenomic RNA. *Virology* **188**:102–113.

40. Chisari, F. V. 1997. Cytotoxic T cells and viral hepatitis. *J. Clin. Invest.* **99**:1472–1477.

41. Choo, Q.-L., G. Kuo, A. J. Weiner, L. R. Overby, D. W. Bradley, and M. Houghton. 1989. Isolation of a cDNA clone derived from a blood-borne non-A, non-B viral hepatitis B genome. *Science* **244**:359–362.

42. Choo, Q.-L., K. H. Richman, J. H. Han, K. Berger, C. Lee, C. Dong, C. Gallegos, D. Coit, A. Medina-Selby, P. J. Barr, A. J. Weiner, D. W. Bradley, G. Kuo, and M. Houghton. 1991. Genetic organization and diversity of the hepatitis C virus. *Proc. Natl. Acad. Sci. USA* **88**:2451–2455.

43. Choo, S. H., H. S. So, J. M. Cho, and W. S. Ryu. 1995. Association of hepatitis C virus particles with immunoglobulin: a mechanism for persistent infection. *J. Gen. Virol.* **76**:2337–2341.

44. Chun, T.-W., L. Stuyver, S. B. Mizell, L. A. Ehler, J. M. Mican, M. Baseler, A. L. Lloyd, M. A. Nowak, and A. S. Fauci. 1997. Presence of an inducible HIV-1 latent reservoir during highly active antiretroviral therapy. *Proc. Natl. Acad. Sci. USA* **94**:13193–13197.

45. Colombo, M., R. de Franchis, E. Del Ninno, A. Sangiovanni, C. De Fazio, M. Tommasini, M. F. Donato, A. Piva, V. Di Carlo, and N. Dioguardi. 1991. Hepatocellular carcinoma in Italian patients with cirrhosis [see comments]. *N. Engl. J. Med.* **325**:675–680.

46. Cribier, B., C. Schmitt, A. Bingen, A. Kirn, and F. Keller. 1995. In vitro infection of peripheral blood mononuclear cells by hepatitis C virus. *J. Gen. Virol.* **76**:2485–2491.

47. D'Souza, E. D. A., K. Grace, D. V. Sangar, D. J. Rowlands, and B. E. Clarke. 1995. *In vitro* cleavage of hepatitis C virus polyprotein substrates by purified recombinant NS3 protease. *J. Gen. Virol.* **76**:1729–1736.

48. De Mitri, M. S., K. Poussin, P. Baccarini, P. Pontisso, A. D'Errico, N. Simon, W. Grigioni, A. Alberti, M. Beaugrand, and E. Pisi. 1995. HCV-associated liver cancer without cirrhosis [see comments]. *Lancet* **345**:413–415.

49. Di Bisceglie, A. M., Z. D. Goodman, K. G. Ishak, J. H. Hoofnagle, J. J. Melpolder, and H. J. Alter. 1991. Long-term clinical and histopathological follow-up of chronic posttransfusion hepatitis. *Hepatology* **14**:969–974.

50. Diepolder, H. M., J.-T. Gerlach, R. Zachoval, R. M. Hoffmann, M.-C. Jung, E. A. Wierenga, S. Scholz, T. Santantonio, M. Houghton, S. Southwood, A. Sette, and G. R. Pape. 1997. Immunodominant CD4+ T-cell epitope within nonstructural protein 3 in acute hepatitis C virus infection. *J. Virol.* **71**:6011–6019.

51. Diepolder, H. M., R. Zachoval, R. M. Hoffmann, E. A. Wierenga, T. Santantonio, M. C. Jung, D. Eichenlaub, and G. R. Pape. 1995. Possible mechanism involving T-lymphocyte response to non-structural protein 3 in viral clearance in acute hepatitis C virus infection. *Lancet* **346**:1006–1007.

52. Dittmann, S., M. Roggendorf, J. Durkop, M. Wiese, B. Lorbeer, and F. Deinhardt. 1991. Long-term persistence of hepatitis C virus antibodies in a single source outbreak. *J. Hepatol.* **13**:323–327.

53. Dusheiko, G., H. Schmilovitz-Weiss, D. Brown, F. McOmish, P. L. Yap, S. Sherlock, N. McIntyre, and P. Simmonds. 1994. Hepatitis C virus genotypes: an investigation of type-specific differences in geographic origin and disease. *Hepatology* **19**:13–18.

54. **Eckart, M. R., M. Selby, F. Masiarz, C. Lee, K. Berger, K. Crawford, C. Kuo, G. Kuo, M. Houghton, and Q. L. Choo.** 1993. The hepatitis C virus encodes a serine protease involved in processing of the putative nonstructural proteins from the viral polyprotein precursor. *Biochem. Biophys. Res. Commun.* **192:**399–406.

55. **Enomoto, N., I. Sakuma, Y. Asahina, M. Kurosaki, T. Murakami, C. Yamamoto, N. Izumi, F. Marumo, and C. Sato.** 1995. Comparison of full-length sequences of interferon-sensitive and resistant hepatitis C virus 1b. Sensitivity to interferon is conferred by amino acid substitutions in the NS5A region. *J. Clin. Invest.* **96:**224–230.

56. **Enomoto, N., I. Sakuma, Y. Asahina, M. Kurosaki, T. Murakami, C. Yamamoto, Y. Ogura, N. Izumi, F. Marumo, and C. Sato.** 1996. Mutations in the nonstructural protein 5A gene and response to interferon in patients with chronic hepatitis C virus 1b infection. *N. Engl. J. Med.* **334:**77–81.

57. **Failla, C., L. Tomei, and R. De Francesco.** 1994. Both NS3 and NS4A are required for proteolytic processing of hepatitis C virus nonstructural proteins. *J. Virol.* **68:** 3753–3760.

58. **Failla, C., L. Tomei, and R. De Francesco.** 1995. An amino-terminal domain of the hepatitis C virus NS3 protease is essential for interaction with NS4A. *J. Virol.* **69:** 1769–1777.

59. **Farci, P., H. J. Alter, S. Govindarajan, D. C. Wong, R. Engle, R. R. Lesniewski, I. K. Mushahwar, S. M. Desai, R. H. Miller, and N. Ogata.** 1992. Lack of protective immunity against reinfection with hepatitis C virus. *Science* **258:**135–140.

60. **Farci, P., H. J. Alter, D. Wong, R. H. Miller, J. W. Shih, B. Jett, and R. H. Purcell.** 1991. A long-term study of hepatitis C virus replication in non-A, non-B hepatitis [see comments]. *N. Engl. J. Med.* **325:**98–104.

61. **Farci, P., H. J. Alter, D. C. Wong, R. H. Miller, S. Govindarajan, R. Engle, M. Shapiro, and R. H. Purcell.** 1994. Prevention of hepatitis C virus infection in chimpanzees after antibody-mediated in vitro neutralization. *Proc. Natl. Acad. Sci. USA* **91:** 7792–7796.

62. **Farci, P., A. Shimoda, D. Wong, T. Cabezon, D. De Gioannis, A. Strazera, Y. Shimizu, M. Shapiro, H. J. Alter, and R. H. Purcell.** 1996. Prevention of hepatitis C virus infection in chimpanzees by hyperimmune serum against the hypervariable region 1 of the envelope 2 protein. *Proc. Natl. Acad. Sci. USA* **93:**15394–15399.

63. **Fattovich, G., G. Giustina, F. Degos, F. Tremolada, G. Diodati, P. Almasio, F. Nevens, A. Solinas, D. Mura, J. T. Brouwer, H. Thomas, C. Njapoum, C. Casarin, P. Bonetti, P. Fuschi, J. Basho, A. Tocco, A. Bhalla, R. Galassini, F. Noventa, S. W. Schalm, and G. Realdi.** 1997. Morbidity and mortality in compensated cirrhosis type C: a retrospective follow-up study of 384 patients [see comments]. *Gastroenterology* **112:**463–472.

64. **Feinstone, S. M., H. J. Alter, H. P. Dienes, Y. Shimizu, H. Popper, D. Blackmore, D. Sly, W. T. London, and R. H. Purcell.** 1981. Non-A, non-B hepatitis in chimpanzees and marmosets. *J. Infect. Dis.* **144:**588–598.

65. **Feinstone, S. M., A. Z. Kapikian, R. H. Purcell, H. J. Alter, and P. V. Holland.** 1975. Transfusion-associated hepatitis not due to viral hepatitis type A or B. *N. Engl. J. Med.* **292:**767–770.

66. **Feinstone, S. M., K. B. Mihalik, T. Kamimura, H. J. Alter, W. T. London, and R. H. Purcell.** 1983. Inactivation of hepatitis B virus and non-A, non-B hepatitis by chloroform. *Infect. Immun.* **41:**816–821.

67. **Ferrari, C., A. Valli, L. Galati, A. Penna, P. Scaccaglia, T. Giuberti, C. Schianchi, G. Missale, M. G. Marin, and F. Fiaccadori.** 1994. T-cell response to structural and nonstructural hepatitis C virus antigens in persistent and self-limited hepatitis C virus infections. *Hepatology* **19:**286–295.

68. Finzi, D., M. Hermankova, T. Pierson, L. M. Carruth, C. Buck, R. E. Chaisson, T. C. Quinn, K. Chadwick, J. Margolick, R. Brookmeyer, J. Gallant, M. Markowitz, D. D. Ho, D. D. Richman, and R. F. Siliciano. 1997. Identification of a reservoir for HIV-1 in patients of highly active antiretroviral therapy. *Science* **278:**1295–1300.

69. Garson, J. A., S. Brillanti, K. Whitby, M. Foli, R. Deaville, C. Masci, M. Miglioli, and L. Barbara. 1995. Analysis of clinical and virological factors associated with response to alpha interferon therapy in chronic hepatitis C. *J. Med. Virol.* **45:**348–353.

70. Gerber, M. A., Y. S. Shieh, K. S. Shim, S. N. Thung, A. J. Demetris, M. Schwartz, G. Akyol, and S. Dash. 1992. Detection of replicative hepatitis C virus sequences in hepatocellular carcinoma. *Am. J. Pathol.* **141:**1271–1277.

71. Gilliam, J. H., 3d, K. R. Geisinger, and J. E. Richter. 1984. Primary hepatocellular carcinoma after chronic non-A, non-B post-transfusion hepatitis. *Ann. Intern. Med.* **101:**794–795.

72. Grakoui, A., D. W. McCourt, C. Wychowski, S. M. Feinstone, and C. M. Rice. 1993. A second hepatitis C virus-encoded proteinase. *Proc. Natl. Acad. Sci. USA* **90:** 10583–10587.

73. Grakoui, A., D. W. McCourt, C. Wychowski, S. M. Feinstone, and C. M. Rice. 1993. Characterization of the hepatitis C virus-encoded serine proteinase: determination of proteinase-dependent polyprotein cleavage sites. *J. Virol.* **67:**2832–2843.

74. Grakoui, A., C. Wychowski, C. Lin, S. M. Feinstone, and C. M. Rice. 1993. Expression and identification of hepatitis C virus polyprotein cleavage products. *J. Virol.* **67:**1385–1395.

75. Guidotti, L. G., T. Ishikawa, M. V. Hobbs, B. Matzke, R. Schreiber, and F. V. Chisari. 1996. Intracellular inactivation of the hepatitis B virus by cytotoxic T lymphocytes. *Immunity* **4:**25–36.

76. Gwack, Y., D. W. Kim, J. H. Han, and J. Choe. 1996. Characterization of RNA binding activity and RNA helicase activity of the hepatitis C virus NS3 protein. *Biochem. Biophys. Res. Commun.* **225:**654–659.

77. Hahm, B., D. S. Han, S. H. Back, O.-K. Song, M.-J. Cho, C.-J. Kim, K. Shimotohno, and S. K. Jang. 1995. NS3-4A of hepatitis C virus is a chymotrypsin-like protease. *J. Virol.* **69:**2534–2539.

78. Harada, S., Y. Watanabe, K. Takeuchi, T. Suzuki, T. Katayama, Y. Takebe, I. Saito, and T. Miyamura. 1991. Expression of processed core protein of hepatitis C virus in mammalian cells. *J. Virol.* **65:**3015–3021.

79. Haydon, G. H., L. M. Jarvis, P. Simmonds, D. J. Harrison, O. J. Garden, and P. C. Hayes. 1997. Association between chronic hepatitis C infection and hepatocellular carcinoma in a Scottish population. *Gut* **40:**128–132.

80. He, L. F., D. Alling, T. Popkin, M. Shapiro, H. J. Alter, and R. H. Purcell. 1987. Determining the size of non-A, non-B hepatitis virus by filtration. *J. Infect. Dis.* **156:** 636–640.

81. Hijikata, M., N. Kato, Y. Ootsuyama, M. Nakagawa, S. Ohkoshi, and K. Shimotohno. 1991. Hypervariable regions in the putative glycoprotein of hepatitis C virus. *Biochem. Biophys. Res. Commun.* **175:**220–228.

82. Hijikata, M., N. Kato, Y. Ootsuyama, M. Nakagawa, and K. Shimotohno. 1991. Gene mapping of the putative structural region of the hepatitis C virus genome by *in vitro* processing analysis. *Proc. Natl. Acad. Sci. USA* **88:**5547–5551.

83. Hijikata, M., H. Mizushima, T. Akagi, S. Mori, N. Kakiuchi, N. Kato, T. Tanaka, K. Kimura, and K. Shimotohno. 1993. Two distinct proteinase activities required for the processing of a putative nonstructural precursor protein of hepatitis C virus. *J. Virol.* **67:**4665–4675.

84. Hijikata, M., H. Mizushima, Y. Tanji, Y. Komoda, Y. Hirowatari, T. Akagi, N.

Kato, K. Kimura, and K. Shimotohno. 1993. Proteolytic processing and membrane association of putative nonstructural proteins of hepatitis C virus. *Proc. Natl. Acad. Sci. USA* **90**:10773–10777.

85. **Hijikata, M., Y. K. Shimizu, H. Kato, A. Iwamoto, J. W. Shih, H. J. Alter, R. H. Purcell, and H. Yoshikura.** 1993. Equilibrium centrifugation studies of hepatitis C virus: evidence for circulating immune complexes. *J. Virol.* **67**:1953–1958.

86. **Hino, K., S. Sainokami, K. Shimoda, S. Iino, Y. Wang, H. Okamoto, Y. Miyakawa, and M. Mayumi.** 1994. Genotypes and titers of hepatitis C virus for predicting response to interferon in patients with chronic hepatitis C. *J. Med. Virol.* **42**:299–305.

87. **Hiroishi, K., H. Kita, M. Kojima, H. Okamoto, T. Moriyama, T. Kaneko, T. Ishikawa, S. Ohnishi, T. Aikawa, N. Tanaka, Y. Yazaki, K. Mitamura, and M. Imawari.** 1997. Cytotoxic T lymphocyte response and viral load in hepatitis C virus infection. *Hepatology* **25**:705–712.

88. **Hirowatari, Y., M. Hijikata, Y. Tanji, H. Nyunoya, H. Mizushima, K. Kimura, T. Tanaka, N. Kato, and K. Shimotohno.** 1993. Two proteinase activities in HCV polypeptide expressed in insect cells using baculovirus vector. *Arch. Virol.* **133**:349–356.

89. **Hoffmann, R. M., H. M. Diepolder, R. Zachoval, F. Zwiebel, M.-C. Jung, S. Scholz, H. Nitschko, G. Riethmuller, and G. R. Pape.** 1995. Mapping of immunodominant CD4 + T lymphocyte epitope of hepatitis C virus antigens and their relevance during the course of chronic infection. *Hepatology* **21**:632–638.

90. **Hollinger, F. B., G. L. Gitnick, R. D. Aach, W. Szmuness, J. W. Mosley, C. E. Stevens, R. L. Peters, J. M. Weiner, J. B. Werch, and J. J. Lander.** 1978. Non-A, non-B hepatitis transmission in chimpanzees: a project of the transfusion-transmitted viruses study group. *Intervirology* **10**:60–68.

91. **Honda, M., E. A. Brown, and S. M. Lemon.** 1996. Stability of a stem-loop involving the initiator AUG controls the efficiency of internal initiation of translation on hepatitis C virus RNA. *RNA* **2**:955–968.

92. **Honda, M., S. Kaneko, U. Masashi, K. Kobayashi, and S. Murakami.** 1992. Sequence comparisons for a hepatitis C virus genome RNA isolated from a patient with liver cirrhosis. *Gene* **120**:317–318.

93. **Honda, M., S. Kaneko, A. Sakai, M. Unoura, S. Murakami, and K. Kobayashi.** 1994. Degree of diversity of hepatitis C virus quasispecies and progression of liver disease. *Hepatology* **20**:1144–1151.

94. **Hong, Z., E. Ferrari, J. Wright-Minogue, R. Chase, C. Risano, G. Seelig, C. G. Lee, and A. D. Kwong.** 1996. Enzymatic characterization of hepatitis C virus NS3/4A complexes expressed in mammalian cells by using the herpes simplex virus amplicon system. *J. Virol.* **70**:4261–4268.

95. **Hopf, U., B. Moller, D. Kuther, R. Stemerowicz, H. Lobeck, A. Ludtke-Handjery, E. Walter, H. E. Blum, M. Roggendorf, and F. Deinhardt.** 1990. Long-term follow-up of posttransfusion and sporadic chronic hepatitis non-A, non-B and frequency of circulating antibodies to hepatitis C virus (HCV). *J. Hepatol.* **10**:69–76.

96. **Iacovacci, S., A. Manzin, S. Barca, M. Sargiacomo, A. Serafino, M. B. Valli, G. Macioce, H. J. Hassan, A. Ponzetto, M. Clementi, C. Peschle, and G. Carloni.** 1997. Molecular characterization and dynamics of hepatitis C virus replication in human fetal hepatocytes infected *in vitro*. *Hepatology* **26**:1328–1337.

97. **Ide, Y., L. Zhang, M. Chen, G. Inchauspe, C. Bahl, Y. Sasaguri, and R. Padmanabhan.** 1996. Characterization of the nuclear localization signal and subcellular distribution of hepatitis C virus nonstructural protein NS5A. *Gene* **182**:203–211.

98. **Inchauspe, G., S. Zebedee, D.-H. Lee, M. Sugitani, M. Nasoff, and A. M. Prince.** 1991. Genomic structure of the human prototype strain H of hepatitis C virus: com-

parison with American and Japanese isolates. *Proc. Natl. Acad. Sci. USA* **88:** 10292–10296.

99.   **Iwata, K., T. Wakita, A. Okumura, K. Yoshioka, M. Takayanagi, J. R. Wands, and S. Kakumu.** 1995. Interferon gamma production by peripheral blood lymphocytes to hepatitis C virus core protein in chronic hepatitis C infection. *Hepatology* **22:** 1057–1064.

100. **Jackson, R. J., M. T. Howell, and A. Kaminski.** 1990. The novel mechanism of initiation of picornavirus RNA translation. *Trends Biochem. Sci.* **15:**477–483.

101. **Jacob, J. R., K. H. Burk, J. W. Eichberg, G. R. Dreesman, and R. E. Lanford.** 1990. Expression of infectious viral particles by primary chimpanzee hepatocytes isolated during the acute phase of non-A, non-B hepatitis. *J. Infect. Dis.* **161:**1121–1127.

102. **Jin, Y., L. Fuller, M. Carreno, K. Zucker, D. Roth, V. Esquenazi, T. Karatzas, S. J. Swanson, A. G. Tzakis, and J. Miller.** 1997. The immune reactivity role of HCV-induced liver infiltrating lymphocytes in hepatocellular damage. *J. Clin. Immunol.* **17:**140–153.

103. **Kaito, M., S. Watanabe, K. Tsukiyama-Kohara, K. Yamaguchi, Y. Kobayashi, M. Konishi, M. Yokoi, S. Ishida, S. Suzuki, and M. Kohara.** 1994. Hepatitis C virus particle detected by immunoelectron microscopic study. *J. Gen. Virol.* **75:**1755–1760.

104. **Kaneko, T., I. Nakamura, H. Kita, K. Hiroishi, T. Moriyama, and M. Imawari.** 1996. Three new cytotoxic T cell epitopes identified within the hepatitis C virus nucleoprotein. *J. Gen. Virol.* **77:**1305–1309.

105. **Kanto, T., N. Hayashi, T. Takehara, H. Hagiwara, E. Mita, M. Oshita, K. Katayama, A. Kasahara, H. Fusamoto, and T. Kamada.** 1995. Serial density analysis of hepatitis C virus particle populations in chronic hepatitis C patients treated with interferon-alpha. *J. Med. Virol.* **46:**230–237.

106. **Kato, N., M. Hijikata, Y. Ootsuyama, M. Nakagawa, S. Ohkoshi, T. Sugimura, and K. Shimotohno.** 1990. Molecular cloning of the human hepatitis C virus genome from Japanese patents with non-A, non-B hepatitis. *Proc. Natl. Acad. Sci. USA* **87:** 9524–9528.

107. **Kato, N., T. Nakazawa, T. Mizutani, and K. Shimotohno.** 1995. Susceptibility of human T-lymphotropic virus type I infected cell line MT-2 to hepatitis C virus infection. *Biochem. Biophys. Res. Commun.* **206:**863–869.

108. **Kato, N., Y. Ootsuyama, S. Ohkoshi, T. Nakazawa, H. Sekiya, M. Hijikata, and K. Shimotohno.** 1992. Characterization of hypervariable regions in the putative envelope protein of hepatitis C virus. *Biochem. Biophys. Res. Commun.* **189:**119–127.

109. **Kato, N., Y. Ootsuyama, H. Sekiya, S. Ohkoshi, T. Nakazawa, M. Hijikata, and K. Shimotohno.** 1994. Genetic drift in hypervariable region 1 of the viral genome in persistent hepatitis C virus infection. *J. Virol.* **68:**4776–4784.

110. **Kato, N., Y. Ootsuyama, T. Tanaka, M. Nakagawa, T. Nakazawa, K. Muraiso, S. Ohkoshi, M. Hijikata, and K. Shimotohno.** 1992. Marked sequence diversity in the putative envelope proteins of hepatitis C viruses. *Virus Res.* **22:**107–123.

111. **Kato, N., H. Sekiya, Y. Ootsuyama, T. Nakazawa, M. Hijikata, S. Ohkoshi, and K. Shimotohno.** 1993. Humoral immune response to hypervariable region 1 of the putative envelope glycoprotein (gp70) of hepatitis C virus. *J. Virol.* **67:**3923–3930.

112. **Kawamura, T., A. Furusaka, M. J. Koziel, R. T. Chung, T. C. Wang, E. V. Schmidt, and T. J. Liang.** 1997. Transgenic expression of hepatitis C virus structural proteins in the mouse. *Hepatology* **25:**1014–1021.

113. **Khorsi, H., S. Castelain, A. Wyseur, J. Izopet, V. Canva, A. Rombout, D. Capron, J. P. Capron, F. Lunel, L. Stuyver, and G. Duverlie.** 1997. Mutations of hepatitis C virus 1b NS5A 2209–2248 amino acid sequence do not predict the response to recombinant interferon-alfa therapy in French patients. *J. Hepatol.* **27:**72–77.

114. **Kim, D. W., Y. Gwack, J. H. Han, and J. Choe.** 1995. C-terminal domain of the hepatitis C virus NS3 protein contains an RNA helicase activity. *Biochem. Biophys. Res. Commun.* **215:**160–166.

115. **Kim, D. W., Y. Gwack, J. H. Han, and J. Choe.** 1997. Towards defining a minimal functional domain for NTPase and RNA helicase activities of the hepatitis C virus NS3 protein. *Virus Res.* **49:**17–25.

116. **Kim, J. L., K. A. Morgenstern, C. Lin, T. Fox, M. D. Dwyer, J. A. Landro, S. P. Chambers, W. Markland, C. A. Lepre, E. T. O'Malley, S. L. Harbeson, C. M. Rice, M. A. Murcko, P. R. Caron, and J. A. Thomson.** 1996. Crystal structure of the hepatitis C virus NS3 protease domain complexed with a synthetic NS4A cofactor peptide. *Cell* **87:**343–355. (Erratum, *Cell* **89:**159, 1997.)

117. **Kita, H., K. Hiroishi, T. Moriyama, H. Okamoto, T. Kaneko, S. Ohnishi, Y. Yazaki, and M. Imawari.** 1995. A minimal and optimal cytotoxic T cell epitope within hepatitis C virus nucleoprotein. *J. Gen. Virol.* **76:**3189–3193.

118. **Kiyosawa, K., Y. Akahane, A. Nagata, Y. Koike, and S. Furuta.** 1982. The significance of blood transfusion in non-A, non-B chronic liver disease in Japan. *Vox Sang.* **43:** 45–52.

119. **Kiyosawa, K., T. Sodeyama, E. Tanaka, Y. Gibo, K. Yoshizawa, Y. Nakano, S. Furuta, Y. Akahane, K. Nishioka, and R. H. Purcell.** 1990. Interrelationship of blood transfusion, non-A, non-B hepatitis and hepatocellular carcinoma: analysis by detection of antibody to hepatitis C virus. *Hepatology* **12:**671–675.

120. **Kobayashi, M., H. Kumada, K. Chayama, Y. Arase, S. Saitou, A. Tsubota, I. Koida, K. Ikeda, M. Hashimoto, and S. Iwasaki.** 1994. Prevalence of HCV genotype among patients with chronic liver diseases in the Tokyo metropolitan area. *J. Gastroenterol.* **29:**583–587.

121. **Koch, J. O., V. Lohmann, U. Herian, and R. Bartenschlager.** 1996. In vitro studies on the activation of the hepatitis C virus NS3 proteinase by the NS4A cofactor. *Virology* **221:**54–66.

122. **Kolykhalov, A. A., E. V. Agapov, K. J. Blight, K. Mihalik, S. M. Feinstone, and C. M. Rice.** 1997. Transmission of hepatitis C by intrahepatic inoculation with transcribed RNA. *Science* **277:**570–574.

123. **Kolykhalov, A. A., S. M. Feinstone, and C. M. Rice.** 1996. Identification of a highly conserved sequence element at the 3' terminus of hepatitis C virus genome RNA. *J. Virol.* **70:**3363–3371.

124. **Koonin, E. V.** 1991. The phylogeny of RNA-dependent RNA polymerases of positive-strand RNA viruses. *J. Gen. Virol.* **72:**2197–2206.

125. **Koretz, R. L., H. Abbey, E. Coleman, and G. Gitnick.** 1993. Non-A, non-B post-transfusion hepatitis. Looking back in the second decade [see comments]. *Ann. Intern. Med.* **119:**110–115.

126. **Koretz, R. L., S. C. Suffin, and G. L. Gitnick.** 1976. Post-transfusion chronic liver disease. *Gastroenterology* **71:**797–803.

127. **Koziel, M. J., D. Dudley, N. Afdhal, Q.-L. Choo, M. Houghton, R. Ralston, and B. D. Walker.** 1993. Hepatitis C virus (HCV)-specific cytotoxic T lymphocytes recognize epitopes in the core and envelope proteins of HCV. *J. Virol.* **67:**7522–7532.

128. **Koziel, M. J., D. Dudley, N. Afdhal, A. Grakoui, C. M. Rice, Q.-L. Choo, M. Houghton, and B. D. Walker.** 1995. HLA class I-restricted cytotoxic T lymphocytes specific for hepatitis C virus. Identification of multiple epitopes and characterization of patterns of cytokine release. *J. Clin. Invest.* **96:**2311–2321.

129. **Koziel, M. J., D. Dudley, J. T. Wong, J. Dienstag, M. Houghton, R. Ralston, and B. D. Walker.** 1992. Intrahepatic cytotoxic T lymphocytes specific for hepatitis C virus in persons with chronic hepatitis. *J. Immunol.* **149:**3339–3344.

130. **Kumar, U., J. Monjardino, and H. C. Thomas.** 1994. Hypervariable region of hepatitis C virus envelope glycoprotein (E2/NS1) in an agammaglobulinemic patient [see comments]. *Gastroenterology* **106:**1072–1075.

131. **Kuo, G., Q.-L. Choo, H. J. Alter, G. L. Gitnick, A. G. Redeker, R. H. Purcell, T. Miyamura, J. L. Dienstag, M. J. Alter, C. E. Stevens, G. E. Tegtmeier, F. Bonino, M. Colombo, W.-S. Lee, C. Kuo, K. Berger, J. R. Shuster, L. R. Overby, D. W. Bradley, and M. Houghton.** 1989. An assay for circulating antibodies to a major etiologic virus of human non-A, non-B hepatitis. *Science* **244:**362–364.

132. **Kurosaki, M., N. Enomoto, T. Murakami, I. Sakuma, Y. Asahina, C. Yamamoto, T. Ikeda, S. Tozuka, N. Izumi, F. Marumo, and C. Sato.** 1997. Analysis of genotypes and amino acid residues 2209 to 2248 of the NS5A region of hepatitis C virus in relation to the response to interferon-beta therapy [see comments]. *Hepatology* **25:** 750–753.

133. **Lai, M. E., A. P. Mazzoleni, F. Argiolu, S. De Virgilis, A. Balestrieri, R. H. Purcell, A. Cao, and P. Farci.** 1994. Hepatitis C virus in multiple episodes of acute hepatitis in polytransfused thalassaemic children. *Lancet* **343:**388–390.

134. **Lanford, R. E., L. Notvall, D. Chavez, R. White, G. Frenzel, C. Simonsen, and J. Kim.** 1993. Analysis of hepatitis C virus capsid, E1, and E2/NS1 proteins expressed in insect cells. *Virology* **197:**225–235.

135. **Lanford, R. E., C. Sureau, J. R. Jacob, R. White, and T. R. Fuerst.** 1994. Demonstration of in vitro infection of chimpanzee hepatocytes with hepatitis C virus using strand-specific RT/PCR. *Virology* **202:**606–614.

136. **Lemm, J. A., T. Rumenapf, E. G. Strauss, J. H. Strauss, and C. M. Rice.** 1994. Polypeptide requirements for assembly of functional Sindbis virus replication complexes: a model for the temporal regulation of minus and plus-strand RNA synthesis. *EMBO J.* **13:**2925–2934.

137. **Lerat, H., F. Berby, M. A. Trabaud, O. Vidalin, M. Major, C. Trepo, and G. Inchauspe.** 1996. Specific detection of hepatitis C virus minus strand RNA in hematopoietic cells. *J. Clin. Invest.* **97:**845–851.

138. **Leroux-Roels, G., C. A. Esquivel, R. DeLeys, L. Stuyver, A. Elewaut, J. Philippe, I. Desombere, J. Paradijs, and G. Maertens.** 1996. Lymphoproliferative responses to hepatitis C virus core, E1, E2, and NS3 in patients with chronic hepatitis C infection treated with interferon alfa. *Hepatology* **23:**8–16.

139. **Liaw, Y.-F., C.-S. Lee, S. L. Tsai, B.-W. Liaw, T.-C. Chen, I.-S. Sheen, and C.-M. Chu.** 1995. T-cell-mediated autologous hepatotoxicity in patients with chronic hepatitis C virus infection. *Hepatology* **22:**1368–1373.

140. **Lin, C., B. D. Lindenbach, B. M. Pragai, D. W. McCourt, and C. M. Rice.** 1994. Processing in the hepatitis C virus E2-NS2 region: identification of p7 and two distinct E2-specific products with different C termini. *J. Virol.* **68:**5063–5073.

141. **Lin, C., B. M. Pragai, A. Grakoui, J. Xu, and C. M. Rice.** 1994. Hepatitis C virus NS3 serine proteinase: *trans*-cleavage requirements and processing kinetics. *J. Virol.* **68:**8147–8157.

142. **Lin, C., J. A. Thomson, and C. M. Rice.** 1995. A central region in the hepatitis C virus NS4A protein allows formation of an active NS3-NS4A serine proteinase complex in vivo and in vitro. *J. Virol.* **69:**4373–4380.

143. **Lo, S.-Y., M. J. Selby, and J.-H. Ou.** 1996. Interaction between hepatitis C virus core protein and E1 envelope protein. *J. Virol.* **70:**5177–5182.

144. **Lo, S. Y., F. Masiarz, S. B. Hwang, M. M. Lai, and J. H. Ou.** 1995. Differential subcellular localization of hepatitis C virus core gene products. *Virology* **213:**455–461.

145. **Love, R. A., H. E. Parge, J. A. Wickersham, Z. Hostomsky, N. Habuka, E. W. Moomaw, T. Adachi, and Z. Hostomska.** 1996. The crystal structure of hepatitis C NS3

proteinase reveals a trypsin-like fold and a structural zinc binding site. *Cell* **87:** 331–342.

146. **Mahaney, K., V. Tedeschi, G. Maertens, A. M. Di Bisceglie, J. Vergalla, J. H. Hoofnagle, and R. Sallie.** 1994. Genotypic analysis of hepatitis C virus in American patients [see comments]. *Hepatology* **20:**1405–1411.

147. **Mannucci, P. M., M. Colombo, and M. Rizzetto.** 1982. Nonprogressive course of non-A, non-B chronic hepatitis in multitransfused hemophiliacs. *Blood* **60:**655–658.

148. **Martell, M., J. I. Esteban, J. Quer, J. Genesca, A. J. Weiner, R. Esteban, J. Guardia, and J. Gomez.** 1992. Hepatitis C virus (HCV) circulates as a population of different but closely related genomes: quasispecies nature of HCV genome distribution. *J. Virol.* **66:**3225–3229.

149. **Martinot-Peignoux, M., P. Marcellin, M. Pouteau, C. Castelnau, N. Boyer, M. Poliquin, C. Degott, I. Descombes, V. Le Breton, and V. Milotova.** 1995. Pretreatment serum hepatitis C virus RNA levels and hepatitis C virus genotype are the main and independent prognostic factors of sustained response to interferon alfa therapy in chronic hepatitis C. *Hepatology* **22:**1050–1056.

150. **Mazzanti, R., L. Messerini, L. Monsacchi, G. Buzzelli, A. L. Zignego, M. Foschi, M. Monti, G. Laffi, L. Morbidelli, O. Fantappie, F. Bartoloni Saint Omer, and M. Ziche.** 1997. Chronic viral hepatitis induced by hepatitis C but not hepatitis B virus infection correlates with increased liver angiogenesis. *Hepatology* **25:**229–234.

151. **McMichael, A. J., F. M. Gotch, G. R. Noble, and P. A. Beare.** 1983. Cytotoxic T-cell immunity to influenza. *N. Engl. J. Med.* **309:**13–17.

152. **McOmish, F., P. L. Yap, B. C. Dow, E. A. Follett, C. Seed, A. J. Keller, T. J. Cobain, T. Krusius, E. Kolho, and R. Naukkarinen.** 1994. Geographical distribution of hepatitis C virus genotypes in blood donors: an international collaborative survey. *J. Clin. Microbiol.* **32:**884–892.

153. **Miller, R. H., and R. H. Purcell.** 1990. Hepatitis C virus shares amino acid sequence similarity with pestivirus and flavivirus as well as members of two plant virus supergroups. *Proc. Natl. Acad. Sci. USA* **87:**2057–2061.

154. **Minutello, M. A., P. Pileri, D. Unutmaz, S. Censini, G. Kuo, M. Houghton, F. Bonino, and S. Abrignani.** 1993. Compartmentalization of T lymphocytes to the site of disease: intrahepatic CD4+ T cells specific for the protein NS4 of hepatitis C virus in patients with chronic hepatitis C. *J. Exp. Med.* **178:**17–25.

155. **Missale, G., R. Bertoni, V. Lamonaca, A. Valli, M. Massari, C. Mori, M. G. Rumi, M. Houghton, F. Fiaccadori, and C. Ferrari.** 1996. Different clinical behaviors of acute hepatitis C virus infection are associated with different vigor of the anti-viral cell-mediated immune response. *J. Clin. Invest.* **98:**706–714.

156. **Mitsui, T., K. Iwano, K. Masuko, C. Yamazaki, H. Okamoto, F. Tsuda, T. Tanaka, and S. Mishiro.** 1992. Hepatitis C virus infection in medical personnel after needlestick accident [see comments]. *Hepatology* **16:**1109–1114.

157. **Mizushima, H., M. Hijikata, S. Asabe, M. Hirota, K. Kimura, and K. Shimotohno.** 1994. Two hepatitis C virus glycoprotein E2 products with different C termini. *J. Virol.* **68:**6215–6222.

158. **Mizutani, T., N. Kato, M. Ikeda, K. Sugiyama, and K. Shimotohno.** 1996. Long-term human T-cell culture system supporting hepatitis C virus replication. *Biochem. Biophys. Res. Commun.* **227:**822–826.

159. **Moradpour, D., C. Englert, T. Wakita, and J. R. Wands.** 1996. Characterization of cell lines allowing tightly regulated expression of hepatitis C virus core protein. *Virology* **222:**51–63.

160. **Moriya, K., H. Yotsuyanagi, Y. Shintani, H. Fujie, K. Ishibashi, Y. Matsuura, T.**

Miyamura, and K. Koike. 1997. Hepatitis C virus core protein induces hepatic steatosis in transgenic mice. *J. Gen. Virol.* **78:**1527–1531.

161. **Muramatsu, S., S. Ishido, T. Fujita, M. Itoh, and H. Hotta.** 1997. Nuclear localization of the NS3 protein of hepatitis C virus and factors affecting the localization. *J. Virol.* **71:**4954–4961.

162. **Nagayama, R., F. Tsuda, H. Okamoto, Y. Wang, T. Mitsui, T. Tanaka, Y. Miyakawa, and M. Mayumi.** 1993. Genotype dependence of hepatitis C virus antibodies detectable by the first-generation enzyme-linked immunosorbent assay with C100-3 protein. *J. Clin. Invest.* **92:**1529–1533.

163. **Nakajima, N., M. Hijikata, H. Yoshikura, and Y. K. Shimizu.** 1996. Characterization of long-term cultures of hepatitis C virus. *J. Virol.* **70:**3325–3329.

164. **Nakao, H., H. Okamoto, H. Tokita, T. Inoue, H. Iizuka, G. Pozzato, and S. Mishiro.** 1996. Full-length genomic sequence of a hepatitis C virus genotype 2c isolate (BEBE1) and the 2c-specific PCR primers. *Arch. Virol.* **141:**701–704.

165. **Nakazawa, T., N. Kato, Y. Ootsuyama, H. Sekiya, T. Fujioka, A. Shibuya, and K. Shimotohno.** 1994. Genetic alteration of the hepatitis C virus hypervariable region obtained from an asymptomatic carrier. *Int. J. Cancer* **56:**204–207.

166. **Nelson, D. R., C. G. Marousis, G. L. Davis, C. M. Rice, J. T. Wong, M. Houghton, and J. Y. N. Lau.** 1997. The role of hepatitis C virus-specific cytotoxic T lymphocytes in chronic hepatitis C. *J. Immunol.* **158:**1473–1481.

167. **Nishihara, T., C. Nozaki, H. Nakatake, K. Hoshiko, M. Esumi, N. Hayashi, K. Hino, F. Hamada, K. Mizuno, and T. Shikata.** 1993. Secretion and purification of hepatitis C virus NS1 glycoprotein produced by recombinant baculovirus-infected insect cells. *Gene* **129:**207–214.

168. **Nishioka, K., J. Watanabe, S. Furuta, E. Tanaka, S. Iino, H. Suzuki, T. Tsuji, M. Yano, G. Kuo, and Q. L. Choo.** 1991. A high prevalence of antibody to the hepatitis C virus in patients with hepatocellular carcinoma in Japan. *Cancer* **67:**429–433.

169. **Odeberg, J., Z. Yun, A. Sonnerborg, K. Bjoro, M. Uhlen, and J. Lundeberg.** 1997. Variation of hepatitis C virus hypervariable region 1 in immunocompromised patients. *J. Infect. Dis.* **175:**938–943.

170. **Okamoto, H., N. Kanai, and S. Mishiro.** 1992. Full-length nucleotide sequence of a Japanese hepatitis C virus isolate (HC-J1) with high homology to USA isolates. *Nucleic Acids Res.* **20:**6410.

171. **Okamoto, H., M. Kojima, S.-I. Okada, H. Yoshizawa, H. Iizuka, T. Tanaka, E. E. Muchmore, D. A. Paterson, Y. Ito, and S. Mishiro.** 1992. Genetic drift of hepatitis C virus during an 8.2-year infection in a chimpanzee: variability and stability. *Virology* **190:**894–899.

172. **Okamoto, H., K. Kurai, S. Okada, K. Yamamoto, H. Lizuka, T. Tanaka, S. Fukuda, F. Tsuda, and S. Mishiro.** 1992. Full-length sequence of a hepatitis C virus genome having poor homology to reported isolates: comparative study of four distinct genotypes. *Virology* **188:**331–341.

173. **Okamoto, H., S. Okada, Y. Sugiyama, K. Kurai, H. Iizuka, A. Machida, Y. Miyakawa, and M. Mayumi.** 1991. Nucleotide sequence of the genomic RNA of hepatitis C virus isolated from a human carrier: comparison with reported isolates for conserved and divergent regions. *J. Gen. Virol.* **72:**2697–2704.

174. **Okamoto, H., S. Okada, Y. Sugiyama, T. Tanaka, Y. Sugai, Y. Akahane, A. Machida, S. Mishiro, H. Yoshizawa, and Y. Miyakawa.** 1990. Detection of hepatitis C virus RNA by a two-stage polymerase chain reaction with two pairs of primers deduced from the 5′-noncoding region. *Jpn. J. Exp. Med.* **60:**215–222.

175. **Pateron, D., N. Ganne, J. C. Trinchet, M. H. Aurosseau, F. Mal, C. Meicler, E. Coderc, P. Reboullet, and M. Beaugrand.** 1994. Prospective study of screening for

hepatocellular carcinoma in Caucasian patients with cirrhosis [see comments]. *J. Hepatol.* **20**:65–71.

176. **Pieroni, L., E. Santolini, C. Fipaldini, L. Pacini, G. Migliaccio, and N. La Monica.** 1997. In vitro study of the NS2-3 protease of hepatitis C virus. *J. Virol.* **71**:6373–6380.

177. **Pilipenko, E. V., A. P. Gmyl, S. V. Maslova, Y. V. Svitkin, A. N. Sinyakov, and V. I. Agol.** 1992. Prokaryotic-like cis elements in the cap-independent internal initiation of translation on picornavirus RNA. *Cell* **68**:119–131.

178. **Pistello, M., F. Maggi, L. Vatteroni, N. Cecconi, F. Panicucci, G. P. Bresci, L. Gambardella, M. Taddei, A. Bionda, and M. Tuoni.** 1994. Prevalence of hepatitis C virus genotypes in Italy. *J. Clin. Microbiol.* **32**:232–234.

179. **Poch, O., I. Sauvaget, M. Delarue, and N. Tordo.** 1989. Identification of four conserved motifs among the RNA dependent polymerase encoding elements. *EMBO J.* **8**:3867–3874.

180. **Power, J. P., E. Lawlor, F. Davidson, E. C. Holmes, P. L. Yap, and P. Simmonds.** 1995. Molecular epidemiology of an outbreak of infection with hepatitis C virus in recipients of anti-D immunoglobulin [see comments]. *Lancet* **345**:1211–1213.

181. **Pozzato, G., M. Moretti, F. Franzin, L. S. Croce, C. Tiribelli, T. Masayu, S. Kaneko, M. Unoura, and K. Kobayashi.** 1991. Severity of liver disease with different hepatitis C viral clones. *Lancet* **338**:509. (Letter.)

182. **Prince, A. M., B. Brotman, G. F. Grady, W. J. Kuhns, C. Hazzi, R. W. Levine, and S. J. Millian.** 1974. Long-incubation post-transfusion hepatitis without serological evidence of exposure to hepatitis-B virus. *Lancet* **2**:241–246.

183. **Qu, D., J. S. Li, L. Vitvitski, S. Mechai, F. Berby, S. P. Tong, F. Bailly, Q. S. Wang, J. L. Martin, and C. Trepo.** 1994. Hepatitis C virus genotypes in France: comparison of clinical features of patients infected with HCV type I and type II. *J. Hepatol.* **21**:70–75.

184. **Ralston, R., K. Thudium, K. Berger, C. Kuo, B. Gervase, J. Hall, M. Selby, G. Kuo, M. Houghton, and Q.-L. Choo.** 1993. Characterization of hepatitis C virus envelope glycoprotein complexes expressed by recombinant vaccinia viruses. *J. Virol.* **67**:6753–6761.

185. **Ray, R. B., L. M. Lagging, K. Meyer, and R. Ray.** 1996. Hepatitis C virus core protein cooperates with *ras* and transforms primary rat embryo fibroblasts to tumorigenic phenotype. *J. Virol.* **70**:4438–4443.

186. **Ray, R. B., L. M. Lagging, K. Meyer, R. Steele, and R. Ray.** 1995. Transcriptional regulation of cellular and viral promoters by the hepatitis C virus core protein. *Virus Res.* **37**:209–220.

187. **Ray, R. B., K. Meyer, and R. Ray.** 1996. Suppression of apoptotic cell death by hepatitis C virus core protein. *Virology* **226**:176–182.

188. **Ray, R. B., R. Steele, K. Meyer, and R. Ray.** 1997. Transcriptional repression of p53 promoter by hepatitis C virus core protein. *J. Biol. Chem.* **272**:10983–10986.

189. **Realdi, G., A. Alberti, M. Rugge, A. M. Rigoli, F. Tremolada, L. Schivazappa, and A. Ruol.** 1982. Long-term follow-up of acute and chronic non-A, non-B post-transfusion hepatitis: evidence of progression to liver cirrhosis. *Gut* **23**:270–275.

190. **Reed, K. E., J. Xu, and C. M. Rice.** 1997. Phosphorylation of the hepatitis C virus NS5A protein in vitro and in vivo: properties of the NS5A-associated kinase. *J. Virol.* **71**:7187–7197.

191. **Rehermann, B., K.-M. Chang, J. G. McHutchison, R. Kokka, M. Houghton, and F. V. Chisari.** 1996. Quantitative analysis of the peripheral blood cytotoxic T lymphocyte response in patients with chronic hepatitis C virus infection. *J. Clin. Invest.* **98**:1432–1440.

192. **Rehermann, B., K.-M. Chang, J. G. McHutchison, R. Kokka, M. Houghton, C. M.**

Rice, and F. V. Chisari. 1996. Differential cytotoxic T-lymphocyte responsiveness to the hepatitis B and C viruses in chronically infected patients. *J. Virol.* **70**:7092–7102.

193. **Rice, C. M.** 1996. Flaviviridae: the viruses and their replication, p. 931–960. *In* B. N. Fields, D. M. Knipe, and P. M. Howley (ed.), *Fields Virology*, 3rd ed. Raven Press, New York.

194. **Rosa, D., S. Campagnoli, C. Moretto, E. Guenzi, L. Cousens, M. Chin, C. Dong, A. J. Weiner, J. Y. Lau, Q.-L. Choo, D. Chien, P. Pileri, M. Houghton, and S. Abrignani.** 1996. A quantitative test to estimate neutralizing antibodies to the hepatitis C virus: cytofluorimetric assessment of envelope glycoprotein 2 binding to target cells. *Proc. Natl. Acad. Sci. USA* **93**:1759–1763.

195. **Saito, I., T. Miyamura, A. Ohbayashi, H. Harada, T. Katayama, S. Kikuchi, Y. Watanabe, S. Koi, M. Onji, and Y. Ohta.** 1990. Hepatitis C virus infection is associated with the development of hepatocellular carcinoma. *Proc. Natl. Acad. Sci. USA* **87**:6547–6549.

196. **Saito, K., D. Sullivan, Y. Haruna, N. D. Theise, S. N. Thung, and M. A. Gerber.** 1997. Detection of hepatitis C virus RNA sequences in hepatocellular carcinoma and its precursors by microdissection polymerase chain reaction. *Arch. Pathol. Lab. Med.* **121**:400–403.

197. **Saito, S., N. Kato, M. Hijikata, T. Gunji, M. Itabashi, M. Kondo, K. Tanaka, and K. Shimotohno.** 1996. Comparison of hypervariable regions (HVR1 and HVR2) in positive-and negative-stranded hepatitis C virus RNA in cancerous and non-cancerous liver tissue, peripheral blood mononuclear cells and serum from a patient with hepatocellular carcinoma. *Int. J. Cancer* **67**:199–203.

198. **Sakamuro, D., T. Furukawa, and T. Takegami.** 1995. Hepatitis C virus nonstructural protein NS3 transforms NIH 3T3 cells. *J. Virol.* **69**:3893–3896.

199. **Santolini, E., G. Migliaccio, and N. La Monica.** 1994. Biosynthesis and biochemical properties of the hepatitis C virus core protein. *J. Virol.* **68**:3631–3641.

200. **Sato, C., and N. Enomoto.** 1996. Specific hepatitis C virus NS5A sequences determine the outcome of interferon treatment. *Gastroenterology* **111**:1152–1154.

201. **Sato, K., H. Okamoto, S. Aihara, Y. Hoshi, T. Tanaka, and S. Mishiro.** 1993. Demonstration of sugar moiety on the surface of hepatitis C virions recovered from the circulation of infected humans. *Virology* **196**:354–357.

202. **Satoh, S., Y. Tanji, M. Hijikata, K. Kimura, and K. Shimotohno.** 1995. The N-terminal region of hepatitis C virus nonstructural protein 3 (NS3) is essential for stable complex formation with NS4A. *J. Virol.* **69**:4255–4260.

203. **Schalm, S. W., B. E. Hansen, L. Chemello, A. Bellobuono, J. T. Brouwer, O. Weiland, L. Cavalletto, R. Schvarcz, G. Ideo, and A. Alberti.** 1997. Ribavirin enhances the efficacy but not the adverse effects of interferon in chronic hepatitis C. Meta-analysis of individual patient data from European centers. *J. Hepatol.* **26**:961–966.

204. **Schmidt, W. N., P. Wu, J. Cederna, F. A. Mitros, D. R. LaBrecque, and J. T. Stapleton.** 1997. Surreptitious hepatitis C virus (HCV) infection detected in the majority of patients with cryptogenic chronic hepatitis and negative HCV antibody tests. *J. Infect. Dis.* **176**:27–33.

205. **Seeff, L. B., Z. Buskell-Bales, E. C. Wright, S. J. Durako, H. J. Alter, F. L. Iber, F. B. Hollinger, G. L. Gitnick, R. G. Knodell, R. P. Perrillo, C. E. Stevens, and C. G. Hollingsworth.** 1992. Long-term mortality after transfusion-associated non-A, non-B hepatitis. *N. Engl. J. Med.* **327**:1906–1911.

206. **Sekiya, H., N. Kato, Y. Ootsuyama, T. Nakazawa, K. Yamauchi, and K. Shimotohno.** 1994. Genetic alterations of the putative envelope proteins encoding region of the hepatitis C virus in the progression to relapsed phase from acute hepatitis: humoral immune response to hypervariable region 1. *Int. J. Cancer* **57**:664–670.

207. **Selby, M. J., Q.-L. Choo, K. Berger, G. Kuo, E. Glazer, M. Eckart, C. Lee, D. Chien, C. Kuo, and M. Houghton.** 1993. Expression, identification and subcellular localization of the proteins encoded by the hepatitis C viral genome. *J. Gen. Virol.* **74:** 1103–1113.

208. **Selby, M. J., E. Glazer, F. Masiarz, and M. Houghton.** 1994. Complex processing and protein: protein interactions in the E2: NS2 region of HCV. *Virology* **204:**114–122.

209. **Shieh, Y. S., C. Nguyen, M. V. Vocal, and H. W. Chu.** 1993. Tumor-suppressor p53 gene in hepatitis C and B virus-associated human hepatocellular carcinoma. *Int. J. Cancer* **54:**558–562.

210. **Shih, J. W., J. I. Mur, and H. J. Alter.** 1986. Non-A, non-B hepatitis: advances and unfulfilled expectations of the first decade. *Prog. Liver Dis.* **8:**433–452.

211. **Shimizu, Y. K., S. M. Feinstone, M. Kohara, R. H. Purcell, and H. Yoshikura.** 1996. Hepatitis C virus: detection of intracellular virus particles by electron microscopy [see comments]. *Hepatology* **23:**205–209.

212. **Shimizu, Y. K., M. Hijikata, A. Iwamoto, H. J. Alter, R. H. Purcell, and H. Yoshikura.** 1994. Neutralizing antibodies against hepatitis C virus and the emergence of neutralization escape mutant viruses. *J. Virol.* **68:**1494–1500.

213. **Shimizu, Y. K., H. Igarashi, T. Kiyohara, T. Cabezon, P. Farci, R. H. Purcell, and H. Yoshikura.** 1996. A hyperimmune serum against a synthetic peptide corresponding to the hypervariable region 1 of hepatitis C virus can prevent viral infection in cell cultures. *Virology* **223:**409–412.

214. **Shimizu, Y. K., A. Iwamoto, M. Hijikata, R. H. Purcell, Yoshikura, and H.** 1992. Evidence for in vitro replication of hepatitis C virus genome in a human T-cell line. *Proc. Natl. Acad. Sci. USA* **89:**5477–5481.

215. **Shimizu, Y. K., A. J. Weiner, J. Rosenblatt, D. C. Wong, M. Shapiro, T. Popkin, M. Houghton, H. J. Alter, and R. H. Purcell.** 1990. Early events in hepatitis C virus infection of chimpanzees. *Proc. Natl. Acad. Sci. USA* **87:**6441–6444.

216. **Shimizu, Y. K., and H. Yoshikura.** 1994. Multicycle infection of hepatitis C virus in cell culture and inhibition by alpha and beta interferons. *J. Virol.* **68:**8406–8408.

217. **Shindo, M., A. M. Di Bisceglie, T. Akatsuka, T. L. Fong, L. Scaglione, M. Donets, J. H. Hoofnagle, and S. M. Feinstone.** 1994. The physical state of the negative strand of hepatitis C virus RNA in serum of patients with chronic hepatitis C. *Proc Natl Acad Sci USA* **91:**8719–8723.

218. **Simmonds, P., A. Alberti, H. J. Alter, F. Bonino, D. W. Bradley, C. Brechot, J. T. Brouwer, S. W. Chan, K. Chayama, and D. S. Chen.** 1994. A proposed system for the nomenclature of hepatitis C viral genotypes. *Hepatology* **19:**1321–1324. (Letter.)

219. **Simmonds, P., E. C. Holmes, T.-A. Cha, S.-W. Chan, F. McOmish, B. Irvine, E. Beall, P. L. Yap, J. Kolberg, and M. S. Urdea.** 1993. Classification of hepatitis C virus into six major genotypes and a series of subtypes by phylogenetic analysis of the NS-5 region. *J. Gen. Virol.* **74:**2391–2399.

220. **Simmonds, P., F. McOmish, P. L. Yap, S. W. Chan, C. K. Lin, G. Dusheiko, A. A. Saeed, and E. C. Holmes.** 1993. Sequence variability in the 5′ non-coding region of hepatitis C virus: identification of a new virus type and restrictions on sequence diversity. *J. Gen. Virol.* **74:**661–668.

221. **Simmonds, P., J. Mellor, T. Sakuldamrongpanich, C. Nuchaprayoon, A. S. Tanprasert, E. C. Holmes, and D. B. Smith.** 1996. Evolutionary analysis of variants of hepatitis C virus found in South-East Asia: comparison with classifications based upon sequence similarity. *J. Gen. Virol.* **77:**3013–3024.

222. **Simmonds, P., D. B. Smith, F. McOmish, P. L. Yap, J. Kolberg, M. S. Urdea, and E. C. Holmes.** 1994. Identification of genotypes of hepatitis C virus by sequence comparisons in the core, E1 and NS-5 regions. *J. Gen. Virol.* **75:**1053–1061.

223. **Smith, D. B., J. Mellor, L. M. Jarvis, F. Davidson, J. Kolberg, M. Urdea, P. L. Yap, and P. Simmonds.** 1995. Variation of the hepatitis C virus 5′ non-coding region: implications for secondary structure, virus detection and typing. The International HCV Collaborative Study Group. *J. Gen. Virol.* **76:**1749–1761.

224. **Smith, D. B., and P. Simmonds.** 1997. Characteristics of nucleotide substitution in the hepatitis C virus genome: constraints on sequence change in coding regions at both ends of the genome. *J. Mol. Evol.* **45:**238–246.

225. **Spaete, R. R., D. Alexander, M. E. Rugroden, Q. L. Choo, K. Berger, K. Crawford, C. Kuo, S. Leng, C. Lee, and R. Ralston.** 1992. Characterization of the hepatitis C virus E2/NS1 gene product expressed in mammalian cells. *Virology* **188:**819–830.

226. **Squadrito, G., F. Leone, M. Sartori, B. Nalpas, P. Berthelot, G. Raimondo, S. Pol, and C. Brechot.** 1997. Mutations in the nonstructural 5A region of hepatitis C virus and response of chronic hepatitis C to interferon alfa. *Gastroenterology* **113:**567–572.

227. **Stuyver, L., W. van Arnhem, A. Wyseur, F. Hernandez, E. Delaporte, and G. Maertens.** 1994. Classification of hepatitis C viruses based on phylogenetic analysis of the envelope 1 and nonstructural 5B regions and identification of five additional subtypes. *Proc. Natl. Acad. Sci. USA* **91:**10134–10138.

228. **Suzich, J. A., J. K. Tamura, F. Palmer-Hill, P. Warrener, A. Grakoui, C. M. Rice, S. M. Feinstone, and M. S. Collett.** 1993. Hepatitis C virus NS3 protein polynucleotide-stimulated nucleoside triphosphatase and comparison with the related pestivirus and flavivirus enzymes. *J. Virol.* **67:**6152–6158.

229. **Suzuki, R., Y. Matsuura, T. Suzuki, A. Ando, J. Chiba, S. Harada, I. Saito, and T. Miyamura.** 1995. Nuclear localization of the truncated hepatitis C virus core protein with its hydrophobic C terminus deleted. *J. Gen. Virol.* **76:**53–61.

230. **Tabor, E., R. J. Gerety, J. A. Drucker, L. B. Seeff, J. H. Hoofnagle, D. R. Jackson, M. April, L. F. Barker, and G. Pineda-Tamondong.** 1978. Transmission of non-A, non-B hepatitis from man to chimpanzee. *Lancet* **i:**463–466.

231. **Takada, N., S. Takase, A. Takada, and T. Date.** 1993. Differences in the hepatitis C virus genotypes in different countries. *J. Hepatol.* **17:**277–283.

232. **Takahashi, K., S. Kishimoto, H. Yoshizawa, H. Okamoto, A. Yoshikawa, and S. Mishiro.** 1992. p26 protein and 33-nm particle associated with nucleocapsid of hepatitis C virus recovered from the circulation of infected hosts. *Virology* **191:**431–434.

233. **Takamizawa, A., C. Mori, I. Fuke, S. Manabe, S. Murakami, J. Fujita, E. Onishi, T. Andoh, I. Yoshida, and H. Okayama.** 1991. Structure and organization of the hepatitis C virus genome isolated from human carriers. *J. Virol.* **65:**1105–1113.

234. **Takano, S., O. Yokosuka, F. Imazeki, M. Tagawa, and M. Omata.** 1995. Incidence of hepatocellular carcinoma in chronic hepatitis B and C: a prospective study of 251 patients. *Hepatology* **21:**650–655.

235. **Tanaka, T., N. Kato, M.-J. Cho, K. Sugiyama, and K. Shimotohno.** 1996. Structure of the 3′ terminus of the hepatitis C virus genome. *J. Virol.* **70:**3307–3312.

236. **Tanaka, T., N. Kato, M. J. Cho, and K. Shimotohno.** 1995. A novel sequence found at the 3′ terminus of hepatitis C virus genome. *Biochem. Biophys. Res. Commun.* **215:**744–749.

237. **Tanaka, T., N. Kato, M. Nakagawa, Y. Ootsuyama, M. J. Cho, T. Nakazawa, M. Hijikata, Y. Ishimura, and K. Shimotohno.** 1992. Molecular cloning of hepatitis C virus genome from a single Japanese carrier: sequence variation within the same individual and among infected individuals. *Virus Res.* **23:**39–53.

238. **Tang, L., Y. Tanaka, N. Enomoto, F. Marumo, and C. Sato.** 1995. Detection of hepatitis C virus RNA in hepatocellular carcinoma by in situ hybridization. *Cancer* **76:**2211–2216.

239. **Taniguchi, S., H. Okamoto, M. Sakamoto, M. Kojima, F. Tsuda, T. Tanaka, E.**

Munekata, E. E. Muchmore, D. A. Peterson, and S. Mishiro. 1993. A structurally flexible and antigenically variable N-terminal domain of the hepatitis C virus E2/NS1 protein: implication for an escape from antibody. *Virology* **195**:297–301.

240. Tanji, Y., M. Hijikata, S. Satoh, K. Takasi, and K. Shimotohno. 1995. Hepatitis C virus-encoded nonstructural protein NS4A has versatile functions in viral protein processing. *J. Virol.* **69**:1575–1581.

241. Tanji, Y., T. Kaneko, S. Satoh, and K. Shimotohno. 1995. Phosphorylation of hepatitis C virus-encoded nonstructural protein NS5A. *J. Virol.* **69**:3980–3986.

242. Thomssen, R., S. Bonk, and A. Thiele. 1993. Density heterogeneities of hepatitis C virus in human sera due to the binding of beta-lipoproteins and immunoglobulins. *Med. Microbiol. Immunol.* **182**:329–334.

243. Tine, F., S. Magrin, A. Craxı, and L. Pagliaro. 1991. Interferon for non-A, non-B chronic hepatitis: a meta-analysis of randomised clinical trials. *J. Hepatol.* **13**:192–199.

244. Tomei, L., C. Failla, E. Santolini, R. De Francesco, and N. La Monica. 1993. NS3 is a serine protease required for processing of hepatitis C virus polyprotein. *J. Virol.* **67**:4017–4026.

245. Tomei, L., C. Failla, R. L. Vitale, E. Bianchi, and R. De Francesco. 1996. A central hydrophobic domain of the hepatitis C virus NS4A protein is necessary and sufficient for the activation of the NS3 protease. *J. Gen. Virol.* **77**:1065–1070.

246. Tong, M. J., N. S. El-Farra, A. R. Reikes, and R. L. Co. 1995. Clinical outcomes after transfusion-associated hepatitis C. *N. Engl. J. Med.* **332**:1463–1466.

247. Tremolada, F., C. Casarin, A. Alberti, C. Drago, A. Tagger, M. L. Ribero, and G. Realdi. 1992. Long-term follow-up of non-A, non-B (type C) post-transfusion hepatitis. *J. Hepatol.* **16**:273–281.

248. Tsai, S. L., Y. F. Liaw, M. H. Chen, C. Y. Huang, and G. C. Kuo. 1997. Detection of type 2-like T-helper cells in hepatitis C virus infection: implications for hepatitis C virus chronicity. *Hepatology* **25**:449–458.

249. Tsuchihara, K., T. Tanaka, M. Hijikata, S. Kuge, H. Toyoda, A. Nomoto, N. Yamamoto, and K. Shimotohno. 1997. Specific interaction of polypyrimidine tract-binding protein with the extreme 3'-terminal structure of the hepatitis C virus genome, the 3'X. *J. Virol.* **71**:6720–6726.

250. Tsukiyama-Kohara, K., N. Iizuka, M. Kohara, and A. Nomoto. 1992. Internal ribosome entry site within hepatitis C virus RNA. *J. Virol.* **66**:1476–1483.

251. Tsukuma, H., T. Hiyama, S. Tanaka, M. Nakao, T. Yabuuchi, T. Kitamura, K. Nakanishi, I. Fujimoto, A. Inoue, H. Yamazaki, and T. Kawashima. 1993. Risk factors for hepatocellular carcinoma among patients with chronic liver disease. *N. Engl. J. Med.* **328**:1797–1801.

252. Umlauft, F., D. T. Wong, P. J. Oefner, P. A. Underhill, R. C. Cheung, T. L. Wright, A. A. Kolykhalov, K. Gruenewald, and H. B. Greenberg. 1996. Hepatitis C virus detection by single-round PCR specific for the 3' noncoding region. *J. Clin. Microbiol.* **34**:2552–2558.

253. van Doorn, L. J., I. Capriles, G. Maertens, R. DeLeys, K. Murray, T. Kos, H. Schellekens, and W. Quint. 1995. Sequence evolution of the hypervariable region in the putative envelope region E2/NS1 of hepatitis C virus is correlated with specific humoral immune responses. *J. Virol.* **69**:773–778.

254. Walker, C. M. 1997. Comparative features of hepatitis C virus infection in humans and chimpanzees. *Springer Semin. Immunopathol.* **19**:85–98.

255. Walter, E. A., P. D. Greenberg, M. J. Gilbert, R. J. Finch, K. S. Watanabe, E. D. Thomas, and S. R. Riddell. 1995. Reconstitution of cellular immunity against cytomegalovirus in recipients of allogeneic bone marrow by transfer of T-cell clones from the donor. *N. Engl. J. Med.* **333**:1038–1044.

256. **Wang, C., P. Sarnow, and A. Siddiqui.** 1993. Translation of human hepatitis C virus RNA in cultured cells is mediated by an internal ribosome-binding mechanism. *J. Virol.* **67:**3338–3344.

257. **Wang, Y., H. Okamoto, F. Tsuda, R. Nagayama, Q. M. Tao, and S. Mishiro.** 1993. Prevalence, genotypes, and an isolate (HC-C2) of hepatitis C virus in Chinese patients with liver disease. *J. Med. Virol.* **40:**254–260.

258. **Weiner, A. J., M. J. Brauer, J. Rosenblatt, K. H. Richman, J. Tung, K. Crawford, F. Bonino, G. Saracco, Q. L. Choo, and M. Houghton.** 1991. Variable and hypervariable domains are found in the regions of HCV corresponding to the flavivirus envelope and NS1 proteins and the pestivirus envelope glycoproteins. *Virology* **180:**842–848.

259. **Weiner, A. J., A. L. Erickson, J. Kansopon, K. Crawford, E. Muchmore, A. L. Hughes, M. Houghton, and C. M. Walker.** 1995. Persistent hepatitis C virus infection in a chimpanzee is associated with emergence of a cytotoxic T lymphocyte escape variant. *Proc. Natl. Acad. Sci. USA* **92:**2755–2759.

260. **Weiner, A. J., H. M. Geysen, C. Christopherson, J. E. Hall, T. J. Mason, G. Saracco, F. Bonino, K. Crawford, C. D. Marion, and K. A. Crawford.** 1992. Evidence for immune selection of hepatitis C virus (HCV) putative envelope glycoprotein variants: potential role in chronic HCV infections. *Proc. Natl. Acad. Sci. USA* **89:**3468–3472.

261. **Weiner, A. J., G. Kuo, D. W. Bradley, F. Bonino, G. Saracco, C. Lee, J. Rosenblatt, Q.-L. Choo, and M. Houghton.** 1990. Detection of hepatitis C viral sequences in non-A, non-B hepatitis. *Lancet* **335:**1–3.

262. **Willems, M., H. Moshage, and S. H. Yap.** 1993. PCR and detection of negative HCV RNA strands. *Hepatology* **17:**526. (Letter; comment.)

263. **Wong, D. K. H., D. Dudley, N. Afdhal, J. Dienstag, C. M. Rice, L. Wang, M. Houghton, B. D. Walker, and M. J. Koziel.** 1998. Liver-derived CTL in hepatitis C virus infection: breadth and specificity of responses in a cohort of persons with chronic infection. *J. Immunol.* **160:**1479–1488.

264. **Wong, J. K., M. Hezareh, H. F. Gunthard, D. V. Havlir, C. C. Ignacio, C. A. Spina, and D. D. Richman.** 1997. Recovery of replication-competent HIV despite prolonged suppression of plasma viremia. *Science* **27/8:**1291–1295.

265. **Yamada, M., S. Kakumu, K. Yoshioka, Y. Higashi, K. Tanaka, T. Ishikawa, and M. Takayanagi.** 1994. Hepatitis C virus genotypes are not responsible for development of serious liver disease. *Dig. Dis. Sci.* **39:**234–239.

266. **Yamada, N., K. Tanihara, M. Mizokami, K. Ohba, A. Takada, M. Tsutsumi, and T. Date.** 1994. Full-length sequence of the genome of hepatitis C virus type 3a: comparative study with different genotypes. *J. Gen. Virol.* **75:**3279–3284.

267. **Yamada, N., K. Tanihara, A. Takada, T. Yorihuzi, M. Tsutsumi, H. Shimomura, T. Tsuji, and T. Date.** 1996. Genetic organization and diversity of the 3′ noncoding region of the hepatitis C virus genome. *Virology* **223:**255–261.

268. **Yanagi, M., R. H. Purcell, S. U. Emerson, and J. Bukh.** 1997. Transcripts from a single full-length cDNA clone of hepatitis C virus are infectious when directly transfected into the liver of a chimpanzee. *Proc. Natl. Acad. Sci. USA* **94:**8738–8743.

269. **Yang, P. M., L. H. Hwang, M. Y. Lai, W. L. Huang, Y. D. Chu, W. K. Chi, B. L. Chiang, J. H. Kao, P. J. Chen, and D. S. Chen.** 1995. Prominent proliferative response of peripheral blood mononuclear cells to a recombinant non-structural (NS3) protein of hepatitis C virus in patients with chronic hepatitis C. *Clin. Exp. Immunol.* **101:**272–277.

270. **Yao, N., T. Hesson, M. Cable, Z. Hong, A. D. Kwong, H. V. Le, and P. C. Weber.** 1997. Structure of the hepatitis C virus RNA helicase domain. *Nat. Struct. Biol.* **4:**463–467.

271. **Yoo, B. J., M. J. Selby, J. Choe, B. S. Suh, S. H. Choi, J. S. Joh, G. J. Nuovo, H. S.**

Lee, M. Houghton, and J. H. Han. 1995. Transfection of a differentiated human hepatoma cell line (Huh7) with in vitro-transcribed hepatitis C virus (HCV) RNA and establishment of a long-term culture persistently infected with HCV. *J. Virol.* **69:** 32–38.

272. **Yu, M. C., J. M. Yuan, R. K. Ross, and S. Govindarajan.** 1997. Presence of antibodies to the hepatitis B surface antigen is associated with an excess risk for hepatocellular carcinoma among non-Asians in Los Angeles County, California. *Hepatology* **25:** 226–228. (Erratum, *Hepatology* **25:**1298, 1997.)

273. **Yuan, Z. H., U. Kumar, H. C. Thomas, Y. M. Wen, and J. Monjardino.** 1997. Expression, purification, and partial characterization of HCV RNA polymerase. *Biochem. Biophys. Res. Commun.* **232:**231–235.

274. **Zein, N. N., J. Rakela, E. L. Krawitt, K. R. Reddy, T. Tominaga, and D. H. Persing.** 1996. Hepatitis C virus genotypes in the United States: epidemiology, pathogenicity, and response to interferon therapy. *Ann. Intern. Med.* **125:**634–639.

275. **Zeuzem, S., J. H. Lee, and W. K. Roth.** 1997. Mutations in the nonstructural 5A gene of European hepatitis C virus isolates and response to interferon alfa [see comments]. *Hepatology* **25:**740–744.

276. **Zibert, A., W. Kraas, H. Meisel, G. Jung, and M. Roggendorf.** 1997. Epitope mapping of antibodies directed against hypervariable region 1 in acute self-limiting and chronic infections due to hepatitis C virus. *J. Virol.* **71:**4123–4127.

277. **Zibert, A., E. Schreier, and M. Roggendorf.** 1995. Antibodies in human sera specific to hypervariable region 1 of hepatitis C virus can block viral attachment. *Virology* **208:**653–661.

Human Tumor Viruses
Edited by Dennis J. McCance
© 1998 American Society for Microbiology

11 | Human T-Cell Lymphotropic Viruses
Corliss L. Newman
and Joseph D. Rosenblatt

Human T-cell lymphotropic virus type I (HTLV-I) and type II (HTLV-II) comprise a unique family of closely related human retroviruses. These two viruses, along with simian T-cell lymphoma virus, make up the group of retroviruses known as the primate T-cell leukemia/lymphoma viruses. HTLV-I is associated with several human diseases, including adult T-cell leukemia/lymphoma (ATL) and HTLV-I-associated myelopathy (HAM) or tropical spastic paresis (TSP). HTLV-II has been isolated from intravenous drug abusers (IVDAs) and patients with several neoplastic and neurologic diseases, but is not yet convincingly associated with any specific disorder. Both HTLV-I and HTLV-II share unique regulatory genes which regulate transcription and viral mRNA processing. In this chapter, we review the epidemiology, disease associations, molecular biology, and pathogenic mechanisms of HTLV types I and II.

EPIDEMIOLOGY OF HTLV-I

HTLV-I was first isolated in 1980 from a T-cell line derived from a patient thought to have a "cutaneous" T-cell lymphoma (133). Antibodies against HTLV-I were subsequently found to be prevalent in Southern Japan and several equatorial regions. ATL, a unique form of T-cell lymphoma originally described by Takatsuki and coworkers, is endemic in these areas of HTLV-I antibody prevalence. Further studies showed antibodies against HTLV-I to be present in patients with ATL and described integration of HTLV-I into ATL cells. In 1985 a separate neurologic disease, TSP, was also found to be serologically associated with infection by HTLV-I (35). Similar neurologic syndromes in temperate zones (such as Japan) were subsequently associated with presence of HTLV-I antibodies. This led to the subsequent designation of this disease as HAM (HTLV-associated myelopathy) (127).

Association of HTLV-I with two distinct disease processes, a malignancy and a chronic neurologic disorder, spurred worldwide surveys for prevalence of HTLV-I and associated diseases. Endemic areas for HTLV-I infection include Japan, the Caribbean basin, parts of Africa, parts of the Middle East, Australia, and Melanesia (23, 39, 40, 71, 113, 162, 170, 180, 181, 188, 189). HTLV-I virus genetics have been studied in these endemic areas of infection. Genetics of HTLV-I appear to vary somewhat by geographic area but not by disease type (85, 86).

Based on patterns of genetic drift, it has been proposed that HTLV-I originated in the Indo-Malay region or Asia and then spread to Africa. It has been hypothesized that European adventurers and slave traders disseminated the infection from Africa (and perhaps from the Indo-Malay region) to the New World (the Americas) and Japan (36, 37, 148). HTLV-I appears to be widespread in Central and South America, but is not indigenous to North America or Europe. A seroepidemiological survey of HTLV-I infection in two cities in China did not indicate significant HTLV-I prevalence there (128). In the United States most cases of HTLV-I infection (and ATL and HAM/TSP) are found in the Southeast and among Caribbean immigrants (10, 13).

Transmission of HTLV-I seems to require prolonged, close contact between individuals. Transfer of infected T cells is likely required. The three major reported routes of HTLV-I transmission include sexual intercourse, blood product transmission, and mother-to-child transfer. Male-to-female transmission of HTLV-I has been potentially linked with the length of the sexual relationship as well as viral load (79). Although rare intrauterine transfer of virus has been documented (87), studies of mother-to-child transfer indicate that the major mechanism of infection is breast-feeding, and not intrauterine or perinatal infection with HTLV-I (61, 124). Longer duration of breast-feeding (>6 months) correlates with a higher infection rate. Babies breast-fed for a short term (<6 months) have been shown to have rates of infection similar to bottle-fed babies (171). Increased risk of transmission in breast-fed babies is associated with increased viral load and increased antibody titers in the mother (59, 60, 186). However, high antibody titers in the mothers of bottle-fed children are associated with a decreased risk of transmission (59). Based upon these studies, it is hypothesized that passive maternal antibodies are protective against infection. These antibodies decline at 6 to 12 months after birth. Thus, babies who are breast-fed beyond 6 months would theoretically be at higher risk, especially in mothers with higher viral loads (manifested by higher antibody titers) (171). Of note, in vitro and in vivo transmission of HTLV-I has been shown to be inhibited by passive immunization with antibody against HTLV-I (151, 171).

EPIDEMIOLOGY OF HTLV-II

A permanently immortalized T-cell line, known as the Mo T-cell line, was established in 1976 from the spleen of a patient thought to have an unusual "T-cell variant" of hairy-cell leukemia. Kalyanaraman et al. subsequently showed that this cell line contained a second highly related retrovirus distinct from any previous isolates of HTLV-I (77). Ultimately, the virus in the Mo T-cell line was found to be the first isolate of a new and different type of human retrovirus, now known as HTLV-II (16, 77). The virus, HTLV-II$_{MO}$, was first cloned by Chen and colleagues at UCLA (16). Subsequently, a second isolate, thought to represent a unique subtype, HTLV-II$_{NRA}$, was isolated from a patient with a concurrent presentation of a B-cell hairy-cell leukemia and a CD8$^+$ T-cell malignancy resembling large granular-cell leukemia (143, 144). Three subtypes of HTLV-II have now been reported (21, 51).

The prevalence of HTLV-II is not as well characterized as that of HTLV-I. Populations where HTLV-II appears endemic include certain Native American populations of the United States, Central America, and parts of South America (15, 27, 70, 95, 96, 142). HTLV-II seropositivity has also been reported in pygmies in Cameroon and Zaire (29, 45–47). Infection with HTLV-II is highly prevalent among IVDAs in parts of North America, Europe, and Southeast Asia (21, 32, 95, 96). In 1986, the prevalence of HTLV-II infection in IVDAs in Queens, New York, was 18% (137). Fukushima and colleagues have shown that 60% of IVDAs in South Vietnam were seropositive for HTLV-II (32). Routes of transmission are thought to be similar to those of HTLV-I (32, 79, 124), but have not been adequately studied.

DIAGNOSIS

The most commonly used screening test for HTLV-I/HTLV-II is an enzyme-linked immunosorbent assay (ELISA) which uses a viral lysate-based antigen derived from HTLV-I infection of human T cells (18). Since extensive homology exists between HTLV-I and HTLV-II, serologic assays that use whole virus antigen cannot distinguish between the two. After an initial positive ELISA is obtained, supplemental testing, generally using a Western blot (immunoblot) (although an indirect fluorescent-antibody assay or radioimmunoprecipitation assay can also be used), is performed to confirm presence of HTLV-I/HTLV-II antibodies. Most supplemental assays also cannot distinguish between HTLV-I and HTLV-II, so further definitive testing must be done. Methods successfully used to distinguish HTLV-I and HTLV-II include PCR (20, 95, 147) and enzyme immunoassays using virus-specific synthetic peptides derived from HTLV-I and HTLV-II (182).

GENETIC VARIABILITY

Unlike human immunodeficiency virus (HIV), which has significant genetic differences between isolates, the various HTLV-I strains show a high degree of genetic conservation. Isolates of HTLV-I from Japan, the West Indies, the Americas, and Africa share greater than or equal to 97% genetic homology (37, 85, 104). The most divergent variants of HTLV-I (isolated from populations in Melanesia) still show approximately 92% homology compared with a prototype Japanese HTLV-I (40).

HTLV-II shares approximately 65% homology with the HTLV-I genome (158). The long terminal repeat (LTR) sequences of HTLV-I and HTLV-II share the lowest homology, while certain regions of pX (see Fig. 3) share the highest homology (78%) (54, 158, 160). HTLV-II$_{MO}$ (HTLV-IIa) and HTLV-II$_{NRA}$ (HTLV-IIb) are not highly divergent, with approximately 95% genetic homology (51, 94). Recently, a third strain, HTLV-IIc, has been identified (21). Nucleotide sequences of the *env* and LTR regions of this new strain are more closely related to HTLV-IIa, while the Tax protein of this strain is similar to that of HTLV-IIb (21).

DISEASES ASSOCIATED WITH HTLV-I

ATL

Historically, ATL (adult T-cell leukemia/lymphoma) was the first disease associated with HTLV-I infection. The clinical syndrome of ATL was first described by Takatsuki, Uchiyama, and colleagues in 1977. This is a proliferative disorder of T-cells (leukemia and/or lymphoma) with frequent clinical features of lymphade-

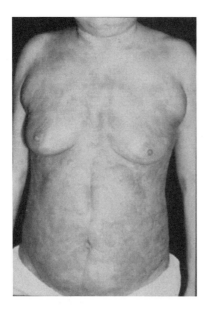

Figure 1 Skin involvement in a Japanese patient with ATL (courtesy of Masao Tomanaga).

nopathy, hypercalcemia, skin involvement (Fig. 1), and hepatosplenomegaly (172). The blood contains abnormal lymphocytes with characteristic convoluted nuclei (178), also called "flower cells" (Fig. 2). These cells are generally CD4-positive T cells and show increased expression of the alpha chain of the high-affinity interleukin-2 (IL-2) receptor (55, 183). ATL has a long incubation period, approximately 20 to 40 years from the time of infection with HTLV-I. The lifetime risk of ATL in a carrier of HTLV-I is estimated to be 2 to 4% (175), and most individuals with ATL appear to have acquired HTLV-I infection in childhood.

Due to variation in the clinical features of disease, ATL has been classified by Shimoyama and colleagues into four subtypes, as follows (161). (i) Smoldering ATL is characterized by 5% or more abnormal T cells in the peripheral blood with

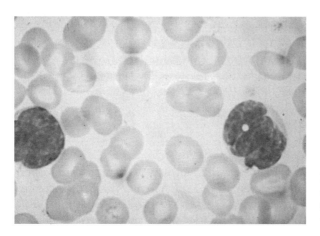

Figure 2 Typical lymphocytes with convoluted nuclei seen in ATL; also called "flower cells."

a normal total lymphocyte count. Serum lactate dehydrogenase levels may be elevated. There is no hypercalcemia, no lymphadenopathy, no ascites or pleural effusions, and no involvement of the liver, spleen, central nervous system (CNS), bone, or gastrointestinal tract. In cases of less than 5% abnormal T cells in the peripheral blood, at least one histologically proven skin or pulmonary lesion should be present. (ii) Chronic ATL is characterized by an absolute lymphocytosis (4×10^9/liter or more) with a T-cell lymphocytosis of more than 3.5×10^9/liter. Generally, 5% or more abnormal T cells are seen in the peripheral blood. Lactate dehydrogenase may be increased up to twice the normal limit. Patients may have lymphadenopathy, hepatomegaly, splenomegaly, and skin and/or pulmonary lesions. No hypercalcemia, no ascites or pleural effusion, and no involvement of the CNS, bone, or gastrointestinal tract should be present in a patient classified as having chronic ATL. (iii) Lymphoma-type ATL is characterized by lymphadenopathy in the absence of lymphocytosis. Lymph node involvement with ATL cells must be histologically proven. Generally, less than 1% abnormal T cells are present in the peripheral blood. Finally, (iv) acute-type ATL encompasses the remaining patients. They generally have a leukemic manifestation as well as lymphadenopathy and/or extranodal involvement. Hypercalcemia, lytic bone lesions, and visceral involvement are frequent.

The malignant T cells of ATL are mature (terminal deoxynucleotide transferase negative), activated $CD4^+$ cells with elevated expression of the alpha chain of the IL-2 receptor (55, 183, 184). Integration of the HTLV-I provirus into the cellular genome is monoclonal (82, 192), indicating that the malignant T cells are monoclonal and originate from a single HTLV-I-infected T cell. The site of integration, although constant for a given patient, will vary and does not appear to be important in pathogenesis of ATL.

Patients with ATL are immunocompromised. Opportunistic pathogens causing *Pneumocystis carinii* pneumonia, cytomegalovirus infection, cryptococcal meningitis, and disseminated fungal infections have been described (13). Severe infections with *Strongyloides stercoralis* have also been found in patients with ATL (130, 132). A Japanese study in a region endemic for *Strongyloides* compared patients with a negative history of strongyloidiasis ($n = 38$) versus patients with a history of past infection ($n = 52$) versus patients with current strongyloidiasis ($n = 91$) and found that the prevalences of HTLV-I infection were 18.4%, 61.5%, and 73.6%, respectively (150). Another Japanese study showed that 58.3% of patients with strongyloidiasis were HTLV-I positive, and of these, 66.6% had monoclonal integration of HTLV-I proviral DNA in their blood lymphocytes (121). These and other studies have led several researchers to propose that *Strongyloides* infection may be a cofactor in development of ATL (121, 131, 150).

Acute ATL has a poor prognosis, with a median survival of 6.2 months (161). Various treatment regimens have been tried, including single-agent or combination chemotherapy (13, 24, 101), interferons (IFNs) (24, 110, 173), and monoclonal antibodies against the IL-2 receptor (184). Taguchi and colleagues report a regimen of CHOP (cyclophosphamide, doxorubicin, vincristine, and prednisone) followed by etoposide, vindesine, ranimustine, and mitoxantrone with complete response and partial response of 35.8% and 38.3%, respectively (169). More recently, treatment of ATL using a regimen of the antiretroviral zidovudine along with IFN-α has shown complete or partial responses in patients, with some patients achieving long remissions (43, 56). Whether the antiretroviral effect of zidovudine is important in controlling ATL is not known.

Smoldering and chronic ATL have a longer course. In 1991, Shimoyama and colleagues projected 2- and 4-year survival rates for chronic ATL to be 52.4% and 26.9%, and projected survival rates for smoldering ATL to be 77.7% and 62.8%, respectively (161). Side effects of treatment for indolent disease may currently outweigh any benefit.

The differential diagnosis of ATL includes other T-cell malignancies: non-Hodgkin's lymphoma, mycosis fungoides (cutaneous T-cell lymphoma), Sezary syndrome, and T-cell chronic lymphocytic leukemia. Laboratory studies that help to confirm the diagnosis of ATL include positive HTLV-I serology, hypercalcemia, negative staining for terminal deoxynucleotidyl transferase, and flow cytometry of the ATL cells, which express both CD4 and CD25 (IL-2 receptor) antigens. Definitive diagnosis of the disease is demonstrated by the presence of mono-clonally or oligoclonally integrated HTLV-I in the malignant T-cell clones of patients seropositive for HTLV-I.

HAM/TSP

Tropical spastic paresis (TSP), renamed HTLV-I-associated myelopathy (HAM), is the second major disease clearly associated with HTLV-I infection. HAM is a chronic progressive demyelinating disease which affects the spinal cord and white matter of the CNS (1, 7, 125). It generally occurs in adulthood and is heralded by an insidious gait disturbance, often with symptoms of weakness and stiffness in the lower extremities. This may be accompanied by paresthesias, heaviness, and persistent low back pain. Spasticity is generally moderate to severe. Lower extremities are affected to a much greater degree than upper extremities (38, 49). Over time, bowel and bladder incontinence occur (38, 49, 156). Disease progression is variable. In one series, after a mean period of 14.4 years (range, 1 to 30 years), 34% of patients could walk with minor difficulty, 40% of patients could walk with difficulty using a cane or crutches, and 26% of patients were bedridden and unable to walk, even with assistance (139).

Magnetic resonance imaging studies in HAM patients can be normal, or show atrophy of the spinal cord and nonspecific lesions of the brain (38, 49, 99, 138, 176). HTLV-I antibodies are found in the serum and cerebrospinal fluid (CSF) of patients with HAM (127). Most patients have elevated immunoglobulin G and oligoclonal bands in the CSF (74, 100). Typical ATL-like lymphocytes (flower cells) may be found in the CSF and are also commonly detected in the peripheral blood (127). As opposed to ATL, HTLV-I-infected lymphocyte populations in HAM are oligoclonal or polyclonal (rather than monoclonal).

Interestingly, in patients diagnosed with HAM, the incubation period between initial infection and the onset of myelopathy is much shorter for patients who acquire the infection via blood transfusion than in patients infected via breast-feeding or by sexual transmission (48, 126, 152). This may, in part, relate to viral load at the time of transmission. The coexistence of both HAM and ATL has been rarely reported (81, 97, 190).

Differential diagnosis of HAM includes multiple sclerosis (especially primary progressive multiple sclerosis) (44), toxic neuropathies (such as cassava neurotoxicity or lathyrism in tropical countries), and/or malnutrition (140). Alternative considerations include infections such as HIV and syphilis and other spinal cord (such as transverse myelitis) or CNS disorders.

Pathologic findings in HAM have been published in various case reports (1, 7, 72, 91, 116, 125, 149). Gross appearance of the brain has been described as normal or atrophied. Various degrees of brain parenchymal degeneration have been described, with reactive astrocytosis and perivascular mononuclear cell infiltration. The spinal cord has been described as showing varying degrees of atrophy. The classic feature of spinal cord pathology in HAM is a chronic progressive inflammatory process preferentially involving the white and gray matter of the spinal cord. In patients with a short duration of disease (up to 3 years), lesions tend to be more inflammatory in character, and mononuclear cell infiltration into both gray and white matter, as well as adjacent parenchymal tissues, is observed. In patients with a long duration of disease (9 or more years), lesions tend to be primarily degenerative, with little or no monocytic or lymphocytic cell infiltration. In all cases, variable degrees of demyelinization are seen, predominantly in the lower thoracic and upper lumbar segments of the spinal cord.

Recently, Levin, Jacobson, and collegues analyzed a spinal cord biopsy specimen from a patient with HAM (99). They found infiltration of the leptomeninges and adjacent spinal cord parenchyma by mononuclear cells which were predominately CD8$^+$ T lymphocytes and monocytes. Functional studies of T-cell lines derived from the specimen indicated that these cells had HTLV-I-specific cytotoxic activity. The authors hypothesized that an immune response to HTLV-I infection may play a role in the development of HAM. Bhigjee and colleagues, along with Moore and colleagues, also found a predominance of CD8$^+$ T cells in spinal cord lesions and postulated an immune mechanism underlying the pathogenesis of HAM (7, 116).

Results of treatment for HAM have been disappointing. Generally, most agents produce only short-term results. Older treatment regimens include corticosteroids (125, 127), plasmapheresis (109), cyclophosphamide (108), and IFN-α therapy (73, 157). Danazol, an attenuated androgen found useful in certain prednisone-responsive diseases, has been tried, with reported improvement in both gait and bladder function (53). Other experimental approaches include anti-TAC antibodies (antibodies directed against the alpha chain of the IL-2 receptor) (174) or antiretroviral therapies (50, 155).

Other

Other suspected sequelae of HTLV-I infection include arthropathy (122), uveitis and other ocular manifestations (115), polymyositis (117), infectious dermatitis in children (93), and pulmonary disorders (107). A possible association between HTLV-I infection, myasthenia gravis, and transverse myelitis has been reported (65). Kompoliti and colleagues reported a patient with HAM, Sjogren's syndrome, and lymphocytic pneumonitis (84). HTLV-I is postulated to play a role in mycosis fungoides and its leukemic variant, Sezary syndrome. The presence of HTLV-I-like Tax, Rex, and Pol (see below) sequences, as well as virus-like particles, has been described in studies of peripheral blood mononuclear cells from patients with mycosis fungoides and Sezary syndrome (41, 129, 194).

DISEASES ASSOCIATED WITH HTLV-II

HTLV-II has not been definitively linked to any specific disorder. However, suspected sequelae of HTLV-II infection include CD8 suppressor cell lymphocytosis or leukemia (so called "large granular-cell leukemia") (102, 145), myositis or abnormal creatine kinase levels (145), and possible HTLV-II-associated neurodegenerative diseases such as spastic paraparesis with variable degrees of cerebellar or cerebral atrophy (52, 62, 75, 141). Maytal and colleagues reported a case of progressive nemaline-rod myopathy in a patient coinfected with HIV-I and HTLV-II. They proposed that HIV patients with myopathies should be tested for HTLV-II, and that HTLV-II coinfection may play a role in pathogenesis (111). Zucker-Franklin and colleagues reported a patient with mycosis fungoides who was found to have HTLV-II infection, with HTLV-II provirus integration detected in cultures of the patient's lymphocytes (195).

MOLECULAR BIOLOGY OF HTLV

The 9-kb genome of HTLV-I (153) encodes the viral structural proteins (Gag and Env), protease, Pol proteins (reverse transcriptase and integrase), and several regulatory proteins from the pX region (Fig. 3), the best characterized of which are Tax and Rex (summarized by Ferreira and colleagues [27]). HTLV-I and HTLV-II genomes are similarly organized (103, 153). The two LTRs located at the 5' and 3' ends of the viral genome contain the viral promoter and transcriptional regulatory elements, including RNA sequences involved in regulation of splicing and mRNA processing (Fig. 3). The LTR, gag, pol, and env genome regions are typical of retroviruses. The retroviral gag genome encodes viral core proteins, while env encodes envelope glycoproteins (179). Of note, the open reading frames (ORFs) of gag, pro, and pol are arranged at a −1 position relative to one another, and translation of gag, pro, and pol gene products has been shown to require two frameshift events (Fig. 3) (103).

The pX region, located at the 3' end of the genome, contains at least four ORFs (153). Proteins encoded by mRNA from the HTLV-I pX region include Tax (14, 26), Rex, p21rex (83), p12, p13, and p30 (88, 89). Corresponding proteins encoded by the pX region in HTLV-II include p11, p10, Tax, Rex, p28, and p22/p20 (17). An excellent recent review of the molecular biology of HTLV-I and HTLV-II has been written by Franchini (28).

Tax

Tax, a nuclear phosphoprotein, is a viral transactivator which functions to regulate HTLV transcription. It acts upon the Tax-responsive elements (TRE-1 and TRE-2) located in the U3 region of the viral LTR (14, 26, 164). Tax does not bind directly to these elements, but rather activates other transcriptional factors which bind to TRE-1 and TRE-2 (76, 105, 167). These transcriptional factors include the cAMP-

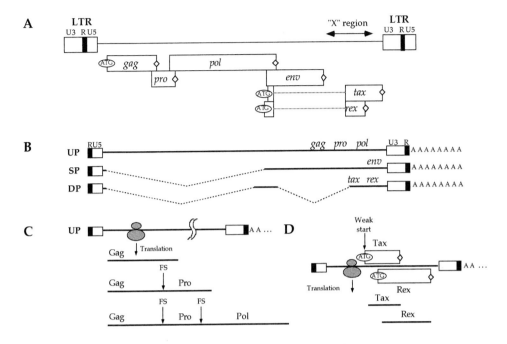

Figure 3 Molecular features of HTLV. (A) Genomic organization. LTRs and ORFs are depicted. Translational start (ATG) and stop (diamonds) codons are indicated. (B) HTLV produces multiple RNA species by splicing. ORFs encoded by the different RNA species are indicated. Solid lines represent exons and discontinuous lines represent introns. UP, unspliced; SP, singly spliced; DP, doubly spliced. (C) Translation of polycistronic messages by ribosomal frameshifting (FS). (D) Translation of polycistronic messages by readthrough of a weak translational ATG. (Used with permission from reference 27.)

responsive element (CRE)-binding protein (CREB) and CRE modulator (CREM), which are members of the CRE binding proteins and activating transcription factor (CREB/ATF) family. CREB and CREM have been shown to bind, in conjunction with Tax, to TRE-1 (167). Other transcriptional factors implicated in regulation of transcription include Sp1, TIF-1, Ets1, Myb, and THP, which have been shown to interact with TRE-2 (11, 12, 105, 106, 191).

The 21-bp enhancer element in the LTR, designated TRE-1, contains a CRE sequence which can be bound by several members of the CREB/ATF family of proteins. The CREB/ATF family includes leucine zipper proteins such as activating transcription factors (ATF), Tax-responsive-element binding proteins (TREB), CREB, CREM, and the 21-bp binding proteins (summarized by Yoshida [191]). How these proteins interact with Tax to increase transcription is not well understood. One study has demonstrated that Tax can activate both HTLV-I CRE and cellular CREs, but that this activation occurs through mechanisms differentially dependent on CREB phosphorylation (92). The ability of Tax to enhance CREB-mediated induction of a cellular CRE was dependent on CREB phosphorylation,

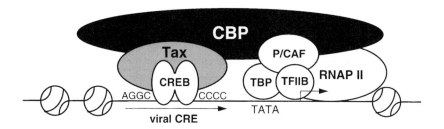

Figure 4 Model of Tax recruitment of the coactivator CBP to the HTLV-I promoter. For simplicity, only one of the three 21-bp repeats is shown. Tax in association with CREB and the viral CRE creates a high-affinity binding site that anchors CBP to the viral promoter. Once bound, CBP activates HTLV-I transcription through chromatin remodeling and recruitment of the general transcription machinery. TFIIB, transcription factor IIB; P/CAF, p300/CBP-associated factor; RNAP II, RNA polymerase II. (Used with permission from reference 42.)

while Tax appeared to induce HTLV-I CRE independent of phosphorylation. CREB binding protein (CBP), thought to be a coactivator of CREB-mediated transcription, has been shown to bind both phosphorylated and unphosphorylated HTLV-I CREB in the presence of Tax (92). Recent studies have shown that the KIX domain of CBP can bind with Tax, and that Tax can promote binding of the KIX domain to DNA-bound CREB complexes (42). A schema of Tax-mediated HTLV-I transcription is shown in Fig. 4.

The three 21-bp sequences of the HTLV LTR are essential for transactivation by Tax (31); the presence of only one of these TRE-1 sequences is not adequate for efficient Tax activation (159). One of the 21-bp units is separated by a spacer sequence, TRE-2. When TRE-2 is included in the promoter region, even one 21-bp sequence becomes sufficient to elicit Tax transactivation. Therefore, the presence of both the three TRE-1 sequences and TRE-2 in the HTLV LTR implies that HTLV gene expression can be activated by different signals (summarized by Yoshida [191]).

Tax has been termed a "promiscuous" transactivator since it also transactivates multiple cellular genes via different transcriptional pathways. Cellular genes that appear to be activated by Tax include those encoding IL-2, IL-2R-alpha, granulocyte-macrophage colony-stimulating factor (GM-CSF), IL-1, IL-3, IL-6, platelet-derived growth factor, tumor growth factor β1 (TGF-β1), tumor necrosis factor beta (TNF-β), vimentin, nerve growth factor, major histocompatibility complex class 1, and PTHrP (parathyroid hormone-related protein) (28). Tax also has been found to interact with serum responsive factor (p67sre) to activate the serum response element SRE, which induces transcription of the immediate-early genes c-*fos*, *egr-1*, and *egr-2* (30). Tax appears to interact with several cellular transcription factors, and these interactions then lead to enhanced transcription of various cellular genes by unknown mechanisms.

Tax binds directly to several of the NF-κB proteins (28, 120). The NF-κB protein group I (p50, p65, RelA, c-Rel, v-Rel, and RelB) and group 2 (NF-*k*B2) appear to be involved in transactivation of several cellular genes, including IL-2, TGF-β, GM-CSF, IL-2R-alpha chain, c-*myc*, vimentin, TNF-β, and HIV LTR enhancers (28, 191). Tax has been shown to bind the NF-κB inhibitor protein IκB-alpha and to dissociate IκB-alpha/NF-κB complexes. This may lead to enhanced transcription of NF-κ-B-regulated genes (168), and these gene products may, in turn, activate other cellular pathways. An example of this was demonstrated for Jak tyrosine kinase activation, where Tax activation of NF-κB induced IL-6, which, in turn, led to Jak2 phosphorylation and subsequent activation (187). Indeed, Jak kinases and STAT (signal transducer and activation of transcription) proteins (Jak-STAT pathway) have been shown to be constitutively activated in some HTLV-I-transformed T cells (114).

Rex

Rex is a nuclear phosphoprotein which regulates the balance of single-spliced and double-spliced versus unspliced mRNA. It appears to localize primarily to the nucleolus of HTLV-infected cells (123, 163). Rex enhances expression of unspliced genomic viral RNA encoding Gag/Pol proteins and single-spliced mRNA encoding the envelope gene. Rex suppresses expression of double-spliced mRNAs, such as mRNAs for Tax, Rex, and the other proteins of the pX region (57, 68, 69). Rex acts upon the HTLV mRNA via the Rex-responsive element (Rex RE), a *cis*-acting element located in the U3/R region of the 3′ viral LTR of HTLV-I (3, 154), and in the R/U5 region of the 5′ LTR of HTLV-II (8). Other *cis*-acting elements, *cis*-acting repressive sequences (CRS), within the HTLV viral RNA inhibit LTR-directed gene expression, and Rex RE allows Rex to overcome this inhibitory effect (8). Recently, the nuclear proteins hnRNP I and hnRNP A1 have been shown to bind to CRS and may be involved in Rex regulation of RNA processing (9). HTLV-II Rex has also been shown to directly inhibit early spliceosome assembly (2). Recent experiments indicate that, additionally, Rex may play a role in inhibiting transcription from the HTLV-II LTR and, possibly by this mechanism, lead to an establishment of latent infection (185).

Like Tax, Rex also seems to affect host cellular processes. It appears to enhance the expression of IL-2R-alpha chain (78) and may possibly aid Tax in promoting IL-2 gene expression as well (112). HTLV-I Rex has recently been shown to bind prothymosin-alpha, a nuclear protein thought to be associated with cellular proliferation (90). However, the significance of this binding is not known. The eukaryotic translation initiation factor 5A (eIF-5A) appears to function as a cellular cofactor in Rex activation. Indeed, Katahira and colleagues showed that *trans*-dominant inhibition of wild-type Rex could be reversed by eIF-5A overexpression (80).

Other Genes of the pX Region

Other proteins encoded by the pX region are not well characterized with regard to their function in HTLV infection or effects upon the host cell gene expression. HTLV-I p12 protein shares structural similarity with the bovine papillomavirus E5 protein. Both proteins localize to the cellular endomembranes and are thought to interact with the H⁺ vacuolar ATPase complex which is involved in proton transport into cellular organelles (summarized by Franchini [28]). The p12 protein has also been shown to bind to IL-2R-β and IL-2Rγ$_c$ chains and decrease their surface expression (119). It is hypothesized that p12 may play a role in altering cellular responses to IL-2 and other cytokines (119) and thereby may be important in T-cell transformation.

PATHOGENESIS

Although many questions remain unanswered, a variety of mechanisms are thought to play a role in pathogenesis of the diseases associated with HTLV-I infection. HTLV-I preferentially infects T cells, with CD4$^+$ cells being predominantly infected, although some infected CD8$^+$ cells are also found (134, 136). In contrast to HTLV-I, HTLV-II preferentially infects CD8$^+$ cells (64). How these cells are targeted for infection is unclear. Neither the cellular receptor nor potential "coreceptors" for HTLV-I or HTLV-II have been isolated.

In ATL, the leukemic cells demonstrate a clonal population of infected T cells (192). This implies that ATL arises from malignant transformation of a single cell. HTLV-I does not carry oncogenes and is not known to activate cellular oncogenes via viral insertion (179). Indeed, clonal cells from different ATL patients show different regions of proviral insertion. Thus it is hypothesized that a "second hit" may be required for leukemic transformation of HTLV-I-infected T cells (63). A variety of chromosomal abnormalities have been described in ATL, including abnormalities of chromosome 14 involving the T-cell receptor locus (4).

In contrast to ATL, the infected T cells present in HAM/TSP are polyclonal. Interestingly, however, HAM/TSP patients and their seropositive family members have been found to harbor one or more subpopulations of clonally proliferating cells infected with HTLV-I (33). HTLV-I is detected in peripheral blood mononuclear cells and in cell cultures of spinal fluid from HAM/TSP patients (6). There are two major hypotheses for pathogenesis in HAM/TSP (summarized by Hollsberg and Hafler [63]). The first hypothesis is that HTLV-I infects glial cells in the CNS and subsequently induces a cytotoxic immune response against these cells, leading to demyelination (116). The second hypothesis is that HTLV-I infection activates autoreactive T cells which then cause autoimmune destruction within the CNS (and perhaps in other areas) (63). What is known is that cytotoxic T cells infected with HTLV-I do appear to retain their specificity and cytolytic capacity. Interestingly, these cells also seem to have the ability to proliferate in the absence of antigen stimulation and to have a prolonged survival (25).

Muller and colleagues have shown that elevated levels of autoantibodies reacting with several nuclear and cytoplasmic antigens are present in ATL and HAM/TSP patients. They found lower prevalence and levels of such antibodies in asymptomatic HTLV-I carriers, but these antibodies were directed only against a restricted set of antigens (118). Other unusual autoantibodies have been described in a patient with myelopathy linked to HTLV-II infection (146). Thymic T cells, unlike mature T cells, are not directly susceptible to HTLV-I-induced activation but, instead, require a comitogen (19). However, the fact that early T cells can be activated may support an autoimmune mechanism of disease in HAM/TSP. Of note, CNS lesions in HAM/TSP have been compared (via magnetic resonance imaging studies) and found to be similar in appearance to lesions in multiple sclerosis and CNS lesions (presumed to be CNS vasculitis) in patients with collagen-vascular disorders (44). Multiple sclerosis and CNS vasculitis are thought to be autoimmune-mediated diseases.

Lehky and colleagues studied autopsy specimens from three patients with HAM/TSP (98). In situ hybridization using HTLV-I Tax RNA probe detected cells in the spinal cord and cerebellum which contained HTLV-I RNA. Perivascular infiltrates were present in these specimens, but the HTLV-I RNA was detected deeper within the CNS tissue and not in the infiltrates. Some of the infected cells were identified as astrocytes (98). This finding may support the first hypothesis (see above) regarding pathogenesis in HAM/TSP. Autopsy specimens from HAM/TSP patients with a short history of disease show spinal cord parenchymal inflammation with lymphocytes and monocyte/macrophages present (72). An HTLV-I-transformed cell line has been shown to release the monocyte chemoattractant MIP-Iα (5). Perhaps chemoattractants released by HTLV-I infected cells play a role in HAM/TSP and other related disorders.

Finally, it has been demonstrated that supernatants of HTLV-I-infected cells inhibit endothelial cell growth and enhance fibroblast growth. The supernatants were found to contain high levels of TNF-β, which was subsequently found to be responsible for this observed effect. Supernatant activity was not dependent upon concentrations of HTLV-I viral particles and was minimally affected by anti-HTLV-I antibodies (193). Increased production of TNF-β (or other cytokines) in HTLV-I-infected cells may therefore also play a role in pathogenesis in HAM/TSP.

Cellular mechanisms affected by Tax and Rex are thought to play a role in HTLV-I pathogenesis in ATL and HAM/TSP, although to what degree these mechanisms play any role is unclear. Tax transactivation of the PTHrP gene, for example, may account for the humoral hypercalcemia commonly seen in ATL (22). Perhaps Tax transactivation—by inducing production of cytokines (such as PTHrP, various interleukins, etc.) which in turn activate osteoclasts to cause bone resorption—is responsible for the lytic bone lesions of ATL. Tax induction of these cytokines may also play a role in the demyelination and tissue destruction seen in HAM/TSP. However, interestingly, Tax mRNA expression per infected cell is lower in ATL patients than in HAM/TSP patients or in asymptomatic carriers of HTLV-I (34). Thus, pathogenesis is probably not related to absolute levels of Tax present.

Tax induces expression of OX40 and gp34 (OX40 ligand) (58). OX40 and gp34, members of the TNF receptor (TNF-R) and TNF family, respectively, have been demonstrated to directly mediate adhesion of activated T cells to vascular endothelial cells (66). They are postulated to play a role in trafficking and homing of HTLV-I-infected cells (67, 177). Of note, this receptor/ligand system has also been demonstrated to play a role in the T-cell-dependent humoral response (166).

Like HIV, HTLV appears to escape immune detection and destruction. The mechanism for this is not completely understood. During infection with most viruses, an immune response develops against virus-encoded envelope proteins. Several immune mechanisms may be activated, including complement activation. It is known that HIV-1 virions acquire complement control proteins (proteins which regulate complement activity) CD55 and CD59 from host cells. These proteins associate with external membrane proteins of HIV-1 and may account for viral resistance to complement. A recent study indicates that HTLV-I and human cytomegalovirus virions also incorporate CD55 and CD59 proteins and suggests that HTLV virions may use this method to escape immune destruction (165).

HTLV-II infection has not been definitively associated with any disease process. Therefore, one cannot comment on pathogenesis of HTLV-II infection. However, the molecular biology of HTLV-II is strikingly similar to that of HTLV-I. The prevalence of HTLV-II is quite high in certain populations, as discussed previously. Therefore, why is HTLV-I associated with specific diseases while HTLV-II is not? There are differences in cellular responses to HTLV-I and HTLV-II infection. For instance, it has been shown that HTLV-I and HTLV-II infection exhibit different dysregulatory effects on memory and naive T-cell populations (135). While our knowledge of the molecular biology of HTLV-II has increased over the years, there are still many questions to be answered regarding HTLV-II infection and pathogenesis.

CONCLUSION

HTLV-I and II are unique primate T-cell leukemia/lymphoma viruses which infect humans. HTLV-I infection occurs worldwide and appears to correspond to migration patterns of early humans. HTLV-I infection occurs predominantly in Native American populations of the Americas and in IVDAs. HTLV-II infection is not associated with any specific human disorder. HTLV-I infection, however, has been linked to several disorders, especially ATL and HAM/TSP. HTLV-I and HTLV-II have a similar genetic organization and regulation and share several unique transregulatory proteins. The pathogenesis of HTLV infection is unclear, though many mechanisms appear to play a role and numerous hypotheses exist.

REFERENCES

1. **Akizuki, S., M. Setoguchi, O. Nakazato, S. Yoshida, Y. Higuchi, S. Yamamoto, and T. Okajima.** 1988. An autopsy case of human T-lymphotropic virus type I-associated myelopathy. *Hum. Pathol.* **19:**988–990.

2. **Bakker, A., X. Li, C. T. Ruland, D. W. Stephens, A. C. Black, and J. D. Rosenblatt.** 1996. Human T-cell leukemia virus type 2 Rex inhibits pre-mRNA splicing in vitro at an early stage of spliceosome formation. *J. Virol.* **70:**5511–5518.

3. **Ballaun, C., G. K. Farrington, M. Dobrovnik, J. Rusche, J. Hauber, and E. Bohnlein.** 1991. Functional analysis of human T-cell leukemia virus type I *rex*-response element: direct RNA binding of Rex protein correlates with in vivo activity. *J. Virol.* **65:**4408–4413.

4. **Berger, R.** 1991. Chromosomal abnormalities in T-cell malignant lymphoma. *Bull. Cancer* (Paris) **78:**283–290.

5. **Bertini, R., W. Luini, S. Sozzani, B. Bottazzi, P. Ruggiero, D. Boraschi, D. Saggioro, L. Chieco-Bianchi, P. Proost, J. van Damme, and A. Mantovani.** 1995. Identification of MIP-1 alpha/LD78 as a monocyte chemoattractant released by the HTLV-I-transformed cell line MT4. *AIDS Res. Hum. Retroviruses* **11:**155–160.

6. **Bhagavati, S., G. Ehrlich, R. W. Kula, S. Kwok, J. Sninsky, V. Udani, and B. J. Poiesz.** 1988. Detection of human T-cell lymphoma/leukemia virus type I DNA and antigen in spinal fluid and blood of patients with chronic progressive myelopathy. *N. Engl. J. Med.* **318:**1141–1147.

7. **Bhigjee, A. I., C. A. Wiley, W. Wachsman, T. Amenomori, D. Pirie, P. L. Bill, and**

I. **Windsor.** 1991. HTLV-I-associated myelopathy: clinicopathologic correlation with localization of provirus to spinal cord. *Neurology* **41**:1990–1992.

8. **Black, A. C., I. S. Chen, S. Arrigo, C. T. Ruland, T. Allogiamento, E. Chin, and J. D. Rosenblatt.** 1991. Regulation of HTLV-II gene expression by Rex involves positive and negative cis-acting elements in the 5′ long terminal repeat. *Virology* **181**:433–444.

9. **Black, A. C., J. Luo, C. Watanabe, S. Chun, A. Bakker, J. K. Fraser, J. P. Morgan, and J. D. Rosenblatt.** 1995. Polypyrimidine tract-binding protein and heterogeneous nuclear ribonucleoprotein A1 bind to human T-cell leukemia virus type 2 RNA regulatory elements. *J. Virol.* **69**:6852–6858.

10. **Blayney, D. W., E. S. Jaffe, W. A. Blattner, J. Cossman, M. Robert-Guroff, D. L. Longo, P. A. Bunn, Jr., and R. C. Gallo.** 1983. The human T-cell leukemia/lymphoma virus associated with American adult T-cell leukemia/lymphoma. *Blood* **62**:401–405.

11. **Bosselut, R., J. F. Duvall, A. Gegonne, M. Bailly, A. Hemar, J. Brady, and J. Ghysdael.** 1990. The product of the c-ets-1 proto-oncogene and the related Ets2 protein act as transcriptional activators of the long terminal repeat of human T-cell leukemia virus HTLV-I. *EMBO J.* **9**:3137–3144.

12. **Bosselut, R., F. Lim, P. C. Romond, J. Frampton, J. Brady, and J. Ghysdael.** 1992. Myb protein binds to multiple sites in the human T cell lymphotropic virus type 1 long terminal repeat and transactivates LTR-mediated expression. *Virology* **186:** 764–769.

13. **Bunn, P. A., Jr., G. P. Schechter, E. Jaffe, D. Blayney, R. C. Young, M. J. Matthews, W. Blattner, S. Broder, M. Robert-Guroff, and R. C. Gallo.** 1983. Clinical course of retrovirus-associated adult T-cell lymphoma in the United States. *N. Engl. J. Med.* **309**:257–264.

14. **Cann, A. J., J. D. Rosenblatt, W. Wachsman, N. P. Shah, and I. S. Chen.** 1985. Identification of the gene responsible for human T-cell leukaemia virus transcriptional regulation. *Nature* **318**:571–574.

15. **Cartier, L., F. Araya, J. L. Castillo, V. Zaninovic, M. Hayami, T. Miura, J. Imai, S. Sonoda, H. Shiraki, K. Miyamoto, and K. Tajima.** 1993. Southernmost carriers of HTLV-I/II in the world. *Jpn. J. Cancer Res.* **84**:1–3.

16. **Chen, I. S., J. McLaughlin, J. C. Gasson, S. C. Clark, and D. W. Golde.** 1983. Molecular characterization of genome of a novel human T-cell leukaemia virus. *Nature* **305**: 502–505.

17. **Ciminale, V., D. M. D'Agostino, L. Zotti, G. Franchini, B. K. Felber, and L. Chieco-Bianchi.** 1995. Expression and characterization of proteins produced by mRNAs spliced into the X region of the human T-cell leukemia/lymphotropic virus type II. *Virology* **209**:445–456.

18. **Constantine, N. T.** 1993. Serologic tests for the retroviruses: approaching a decade of evolution. *AIDS* **7**:1–13. (Editorial.)

19. **Dumontet, C., M. D. Dodon, L. Gazzolo, and D. Gerlier.** 1988. Human T-cell leukemia virus type I-induced proliferation of human thymocytes requires the presence of a comitogen. *Cell Immunol.* **112**:391–401.

20. **Ehrlich, G. D., J. B. Glaser, K. LaVigne, D. Quan, D. Mildvan, J. J. Sninsky, S. Kwok, L. Papsidero, and B. J. Poiesz.** 1989. Prevalence of human T-cell leukemia/lymphoma virus (HTLV) type II infection among high-risk individuals: type-specific identification of HTLVs by polymerase chain reaction. *Blood* **74:** 1658–1664.

21. **Eiraku, N., P. Novoa, M. da Costa Ferreira, C. Monken, R. Ishak, O. da Costa Ferreira, S. W. Zhu, R. Lorenco, M. Ishak, V. Azvedo, J. Guerreiro, M. P. de Oliveira, P. Loureiro, N. Hammerschlak, S. Ijichi, and W. M. Hall.** 1996. Identification and

characterization of a new and distinct molecular subtype of human T-cell lymphotropic virus type 2. *J. Virol.* **70:**1481–1492.

22. **Ejima, E., J. D. Rosenblatt, J. Ou, and D. Prager.** 1995. Parathyroid hormone-related protein gene expression and human T cell leukemia virus-1 infection. *Miner Electrolyte Metab.* **21:**143–147.

23. **Englebrecht, S., E. J. van Rensburg, and B. A. Robson.** 1996. Sequence variation and subtyping of human and simian T-cell lymphotropic virus type I strains from South Africa. *J. Acquired Immune Defic. Syndr. Hum. Retrovirol.* **12:**298–302.

24. **Ezaki, K., M. Hirano, R. Ohno, K. Yamada, K. Naito, Y. Hirota, S. Shirakawa, and K. Kimura.** 1991. A combination trial of human lymphoblastoid interferon and bestrabucil (KM2210) for adult T-cell leukemia-lymphoma. *Cancer* **68:**695–698.

25. **Faller, D. V., M. A. Crimmins, and S. J. Mentzer.** 1988. Human T-cell leukemia virus type I infection of CD4$^+$ or CD8$^+$ cytotoxic T-cell clones results in immortalization with retention of antigen specificity. *J. Virol.* **62:**2942–2950.

26. **Felber, B. K., H. Paskalis, C. Kleinman-Ewing, F. Wong-Staal, and G. N. Pavlakis.** 1985. The pX protein of HTLV-I is a transcriptional activator of its long terminal repeats. *Science* **229:**675–679.

27. **Ferreira, O. C., Jr., V. Planelles, and J. D. Rosenblatt.** 1997. Human T-cell leukemia viruses: epidemiology, biology, and pathogenesis. *Blood Rev.* **11:**91–104.

28. **Franchini, G.** 1995. Molecular mechanisms of human T-cell leukemia/lymphotropic virus type I infection. *Blood* **86:**3619–3639.

29. **Froment, A., E. Delaporte, M. C. Dazza, and B. Larouze.** 1993. HTLV-II among pygmies from Cameroon. *AIDS Res. Hum. Retroviruses* **9:**707. (Letter.)

30. **Fujii, M., H. Tsuchiya, T. Chuhjo, T. Akizawa, and M. Seiki.** 1992. Interaction of HTLV-I Tax1 with p67SRF causes the aberrant induction of cellular immediate early genes through CArG boxes. *Genes Dev.* **6:**2066–2076.

31. **Fujisawa, J., M. Seiki, M. Sato, and M. Yoshida.** 1986. A transcriptional enhancer sequence of HTLV-I is responsible for trans-activation mediated by p40^x HTLV-I. *EMBO J.* **5:**713–718.

32. **Fukushima, Y., H. Takahashi, W. W. Hall, T. Nakasone, S. Nakata, P. Song, D. Dinh Duc, B. Hien, X. Q. Nguyen, T. Ngoc Trinh, K. Nishioka, K. Kitamura, K. Komuro, A. Vahlne, and M. Honda.** 1995. Extraordinary high rate of HTLV type II seropositivity in intravenous drug abusers in South Vietnam. *AIDS Res. Hum. Retroviruses* **11:**637–645.

33. **Furukawa, Y., J. Fujisawa, M. Osame, M. Toita, S. Sonoda, R. Kubota, S. Ijichi, and M. Yoshida.** 1992. Frequent clonal proliferation of human T-cell leukemia virus type 1 (HTLV-I)-infected T cells in HTLV-I-associated myelopathy (HAM-TSP). *Blood* **80:**1012–1016.

34. **Furukawa, Y., M. Osame, R. Kubota, M. Tara, and M. Yoshida.** 1995. Human T-cell leukemia virus type-I (HTLV-I) Tax is expressed at the same level in infected cells of HTLV-I-associated myelopathy or tropical spastic paraparesis patients as in asymptomatic carriers but at a lower level in adult T-cell leukemia cells. *Blood* **85:** 1865–1870.

35. **Gessain, A., F. Barin, J. C. Vernant, O. Gout, L. Maurs, A. Calender, and G. de The.** 1985. Antibodies to human T-lymphotropic virus type-I in patients with tropical spastic paraparesis. *Lancet* **ii:**407–410.

36. **Gessain, A., E. Boeri, R. Yanagihara, R. C. Gallo, and G. Franchini.** 1993. Complete nucleotide sequence of a highly divergent human T-cell leukemia (lymphotropic) virus type I (HTLV-I) variant from Melanesia: genetic and phylogenetic relationship to HTLV-I strains from other geographical regions. *J. Virol.* **67:**1015–1023.

37. **Gessain, A., R. C. Gallo, and G. Franchini.** 1992. Low degree of human T-cell

leukemia/lymphoma virus type I genetic drift in vivo as a means of monitoring viral transmission and movement of ancient human populations. *J. Virol.* **66:**2288–2295.

38. **Gessain, A., and O. Gout.** 1992. Chronic myelopathy associated with human T-lymphotropic virus type I (HTLV-I). *Ann. Intern. Med.* **117:**933–946.

39. **Gessain, A., V. Herve, D. Jeannel, B. Garin, C. Mathiot, and G. de-The.** 1993. HTLV-I but not HTLV-2 found in pygmies from Central African Republic. *J. Acquired Immune Defic. Syndr.* **6:**1373–1374. (Letter; comment.)

40. **Gessain, A., R. Yanagihara, G. Franchini, R. M. Garruto, C. L. Jenkins, A. B. Ajdukiewicz, R. C. Gallo, and D. C. Gajdusek.** 1991. Highly divergent molecular variants of human T-lymphotropic virus type I from isolated populations in Papua New Guinea and the Solomon Islands. *Proc. Natl. Acad. Sci. USA* **88:**7694–7698.

41. **Ghosh, S. K., J. T. Abrams, H. Terunuma, E. C. Vonderheid, and E. DeFreitas.** 1994. Human T-cell leukemia virus type I tax/rex DNA and RNA in cutaneous T-cell lymphoma. *Blood* **84:**2663–2671.

42. **Giebler, H. A., J. E. Loring, K. van Orden, M. A. Colgin, J. E. Garrus, K. W. Escudero, A. Brauweiler, and J. K. Nyborg.** 1997. Anchoring of CREB binding protein to the human T-cell leukemia virus type I promoter: a molecular mechanism of Tax transactivation. *Mol. Cell. Biol.* **17:**5156–5164.

43. **Gill, P. S., W. Harrington, Jr., M. H. Kaplan, R. C. Ribeiro, J. M. Bennett, H. A. Liebman, M. Bernstein-Singer, B. M. Espina, L. Cabral, S. Allen, S. Kornblau, M. C. Pike, and A. M. Levine.** 1995. Treatment of adult T-cell leukemia-lymphoma with a combination of interferon alfa and zidovudine. *N. Engl. J. Med.* **332:**1744–1748.

44. **Godoy, A. J., J. Kira, K. Hasuo, and I. Goto.** 1995. Characterization of cerebral white matter lesions of HTLV-I-associated myelopathy/tropical spastic paraparesis in comparison with multiple sclerosis and collagen-vasculitis: a semiquantitative MRI study. *J. Neurol. Sci.* **133:**102–111.

45. **Goubau, P., J. Desmyter, J. Ghesquiere, and B. Kascreka.** 1992. HTLV-II among pygmies. *Nature* **359:**201. (Letter.)

46. **Goubau, P., H. F. Liu, G. G. De Lange, A. M. Vandamme, and J. Desmyter.** 1993. HTLV-II seroprevalence in pygmies across Africa since 1970. *AIDS Res. Hum. Retroviruses* **9:**709–713.

47. **Goubau, P., A. Vandamme, K. Beuselinck, and J. Desmyter.** 1996. Proviral HTLV-I and HTLV-II in the Efe pygmies of northeastern Zaire. *J. Acquired Immune Defic. Syndr. Hum. Retrovirol.* **12:**208–209. (Letter.)

48. **Gout, O., M. Baulac, A. Gessain, F. Semah, F. Saal, J. Peries, C. Cabrol, C. Foucault-Fretz, D. Laplane, F. Sigaux, and G. de The.** 1990. Rapid development of myelopathy after HTLV-I infection acquired by transfusion during cardiac transplantation. *N. Engl. J. Med.* **322:**383–388.

49. **Gout, O., A. Gessain, F. Bolgert, F. Saal, E. Tournier-Lasserve, J. Lasneret, C. Caudie, P. Brunet, G. De-The, F. Lhermitte, and O. Lyon-Caen.** 1989. Chronic myelopathies associated with human T-lymphotropic virus type I. A clinical, serologic, and immunovirologic study of ten patients in France. *Arch. Neurol.* **46:**255–260.

50. **Gout, O., A. Gessain, M. Iba-Zizen, S. Kouzan, F. Bolgert, G. de The, and O. Lyon-Caen.** 1991. The effect of zidovudine on chronic myelopathy associated with HTLV-I. *J. Neurol.* **238:**108–109. (Letter.)

51. **Hall, W. W., H. Takahashi, C. Liu, M. H. Kaplan, O. Scheewind, S. Ijichi, K. Nagashima, and R. C. Gallo.** 1992. Multiple isolates and characteristics of human T-cell leukemia virus type II. *J. Virol.* **66:**2456–2463.

52. **Harrington, W. J., Jr., W. Sheremata, B. Hjelle, D. K. Dube, P. Bradshaw, S. K. Foung, S. Snodgrass, G. Toedter, L. Cabral, and B. Poiesz.** 1993. Spastic ataxia

associated with human T-cell lymphotropic virus type II infection. *Ann. Neurol.* **33:** 411–414.

53. **Harrington, W. J., Jr., W. A. Sheremata, S. R. Snodgrass, S. Emerson, S. Phillips, and J. R. Berger.** 1991. Tropical spastic paraparesis/HTLV-I-associated myelopathy (TSP/HAM): treatment with an anabolic steroid danazol. *AIDS Res. Hum. Retroviruses* **7:**1031–1034.

54. **Haseltine, W. A., J. Sodroski, R. Patarca, D. Briggs, D. Perkins, and F. Wong-Staal.** 1984. Structure of 3′ terminal region of type II human T lymphotropic virus: evidence for new coding region. *Science* **225:**419–421.

55. **Hattori, T., T. Uchiyama, T. Toibana, K. Takatsuki, and H. Uchino.** 1981. Surface phenotype of Japanese adult T-cell leukemia cells characterized by monoclonal antibodies. *Blood* **58:**645–647.

56. **Hermine, O., D. Bouscary, A. Gessain, P. Turlure, V. Leblond, N. Franck, A. Buzyn-Veil, B. Rio, E. Macintyre, F. Dreyfus, and A. Bazarbachi.** 1995. Brief report: treatment of adult T-cell leukemia-lymphoma with zidovudine and interferon alfa. *N. Engl. J. Med.* **332:**1749–1751.

57. **Hidaka, M., J. Inoue, M. Yoshida, and M. Seiki.** 1988. Post-transcriptional regulator (rex) of HTLV-I initiates expression of viral structural proteins but suppresses expression of regulatory proteins. *EMBO J.* **7:**519–523.

58. **Higashimura, N., N. Takasawa, Y. Tanaka, M. Nakamura, and K. Sugamura.** 1996. Induction of OX40, a receptor of gp34, on T cells by trans-acting transcriptional activator, Tax, of human T-cell leukemia virus type I. *Jpn. J. Cancer Res.* **87:**227–231.

59. **Hino, S., S. Katamine, T. Miyamoto, H. Doi, Y. Tsuji, T. Yamabe, J. E. Kaplan, D. L. Rudolph, and R. B. Lal.** 1995. Association between maternal antibodies to the external envelope glycoprotein and vertical transmission of human T-lymphotropic virus type I. Maternal anti-env antibodies correlate with protection in non-breast-fed children. *J. Clin. Invest.* **95:**2920–2925.

60. **Hino, S., H. Sugiyama, H. Doi, T. Ishimaru, T. Yamabe, Y. Tsuji, and T. Miyamoto.** 1987. Breaking the cycle of HTLV-I transmission via carrier mothers' milk. *Lancet* **ii:** 158–159. (Letter.)

61. **Hino, S., K. Yamaguchi, S. Katamine, H. Sugiyama, T. Amagasaki, K. Kinoshita, Y. Yoshida, H. Doi, Y. Tsuji, and T. Miyamoto.** 1985. Mother-to-child transmission of human T-cell leukemia virus type-I. *Jpn. J. Cancer Res.* **76:**474–480.

62. **Hjelle, B., O. Appenzeller, R. Mills, S. Alexander, N. Torrez-Martinez, R. Jahnke, and G. Ross.** 1992. Chronic neurodegenerative disease associated with HTLV-II infection. *Lancet* **339:**645–646.

63. **Hollsberg, P., and D. A. Hafler.** 1993. Seminars in medicine of the Beth Israel Hospital, Boston. Pathogenesis of diseases induced by human lymphotropic virus type I infection. *N. Engl. J. Med.* **328:**1173–1182.

64. **Ijichi, S., M. B. Ramundo, H. Takahashi, and W. W. Hall.** 1992. In vivo cellular tropism of human T cell leukemia virus type II (HTLV-II). *J. Exp. Med.* **176:**293–296.

65. **Ijichi, T., Y. Adachi, A. Nishio, T. Kanaitsuka, T. Ohtomo, and M. Nakamura.** 1995. Myasthenia gravis, acute transverse myelitis, and HTLV-I. *J. Neurol. Sci.* **133:**194–196.

66. **Imura, A., T. Hori, K. Imada, T. Ishikawa, Y. Tanaka, M. Maeda, S. Imamura, and T. Uchiyama.** 1996. The human OX40/gp34 system directly mediates adhesion of activated T cells to vascular endothelial cells. *J. Exp. Med.* **183:**2185–2195.

67. **Imura, A., T. Hori, K. Imada, S. Kawamata, Y. Tanaka, S. Imamura, and T. Uchiyama.** 1997. OX40 expressed on fresh leukemic cells from adult T-cell leukemia patients mediates cell adhesion to vascular endothelial cells: implication for the possible involvement of OX40 in leukemic cell infiltration. *Blood* **89:**2951–2958.

68. **Inoue, J., M. Seiki, and M. Yoshida.** 1986. The second pX product p27 x-III of HTLV-I is required for age gene expression. *FEBS Lett.* **209:**187–190.

69. **Inoue, J., M. Yoshida, and M. Seiki.** 1987. Transcriptional (p40x) and post-transcriptional (p27x-III) regulators are required for the expression and replication of human T-cell leukemia virus type I genes. *Proc. Natl. Acad. Sci. USA* **84:**3653–3657.

70. **Ishak, R., W. J. Harrington, Jr., V. N. Azevedo, M. Eiraku, M. O. Ishak, J. F. Guerreiro, S. B. Santos, T. Kubo, C. Monken, S. Alexander, and W. W. Hall.** 1995. Identification of human T cell lymphotropic virus type IIa infection in the Kayapo, an indigenous population of Brazil. *AIDS Res. Hum. Retroviruses* **11:**813–821.

71. **Ishida, T., K. Yamamoto, K. Omoto, M. Iwanaga, T. Osato, and Y. Hinuma.** 1985. Prevalence of a human retrovirus in native Japanese: evidence for a possible ancient origin. *J. Infect.* **11:**153–157.

72. **Iwasaki, Y.** 1990. Pathology of chronic myelopathy associated with HTLV-I infection (HAM/TSP). *J. Neurol. Sci.* **96:**103–123.

73. **Izumo, S., I. Goto, Y. Itoyama, T. Okajima, S. Watanabe, Y. Kuroda, S. Araki, M. Mori, S. Nagataki, S. Matsukura, T. Akamine, M. Nakagawa, I. Yamamoto, and M. Osame.** 1996. Interferon-alpha is effective in HTLV-I-associated myelopathy: a multicenter, randomized, double-blind, controlled trial. *Neurology* **46:**1016–1021.

74. **Jacobson, S., A. Gupta, D. Mattson, E. Mingioli, and D. E. McFarlin.** 1990. Immunological studies in tropical spastic paraparesis. *Ann. Neurol.* **27:**149–156.

75. **Jacobson, S., T. Lehky, M. Nishimura, S. Robinson, D. E. McFarlin, and S. Dhib-Jalbut.** 1993. Isolation of HTLV-II from a patient with chronic, progressive neurological disease clinically indistinguishable from HTLV-I-associated myelopathy/tropical spastic paraparesis. *Ann. Neurol.* **33:**392–396.

76. **Jeang, K. T., I. Boros, J. Brady, M. Radonovich, and G. Khoury.** 1988. Characterization of cellular factors that interact with the human T-cell leukemia virus type I p40x-responsive 21-base-pair sequence. *J. Virol.* **62:**4499–4509.

77. **Kalyanaraman, V. S., M. G. Sarngadharan, M. Robert-Guroff, I. Miyoshi, D. Golde, and R. C. Gallo.** 1982. A new subtype of human T-cell leukemia virus (HTLV-II) associated with a T-cell variant of hairy cell leukemia. *Science* **218:**571–573.

78. **Kanamori, H., N. Suzuki, H. Siomi, T. Nosaka, A. Sato, H. Sabe, M. Hatanaka, and T. Honjo.** 1990. HTLV-I p27rex stabilizes human interleukin-2 receptor alpha chain mRNA. *EMBO J.* **9:**4161–4166.

79. **Kaplan, J. E., R. F. Khabbaz, E. L. Murphy, S. Hermansen, C. Roberts, R. Lal, W. Heneine, D. Wright, L. Matijas, R. Thomson, D. Rudolph, W. M. Switzer, S. Kleinman, M. Busch, and G. B. Schreiber.** 1996. Male-to-female transmission of human T-cell lymphotropic virus types I and II: association with viral load. The Retrovirus Epidemiology Donor Study Group. *J. Acquired Immune Defic. Syndr. Hum. Retrovirol.* **12:**193–201.

80. **Katahira, J., T. Ishizaki, H. Sakai, A. Adachi, K. Yamamoto, and H. Shida.** 1995. Effects of translation initiation factor eIF-5A on the functioning of human T-cell leukemia virus type I Rex and human immunodeficiency virus Rev inhibited *trans* dominantly by a Rex mutant deficient in RNA binding. *J. Virol.* **69:**3125–3133.

81. **Kawai, H., Y. Nishida, M. Takagi, K. Nakamura, K. Masuda, S. Saito, and A. Shirakami.** 1989. HTLV-I-associated myelopathy with adult T-cell leukemia. *Neurology* **39:**1129–1131.

82. **Kinoshita, K., T. Amagasaki, S. Ikeda, J. Suzuyama, K. Toriya, K. Nishino, M. Tagawa, M. Ichimaru, K. Kamihira, Y. Yamada, S. Momita, M. Kusano, T. Morikawa, S. Fujita, Y. Ueda, N. Ito, and M. Yoshida.** 1985. Preleukemic state of adult T cell leukemia: abnormal T lymphocytosis induced by human adult T cell leukemia-lymphoma virus. *Blood* **66:**120–127.

83. **Kiyokawa, T., M. Seiki, S. Iwashita, K. Imagawa, F. Shimizu, and M. Yoshida.** 1985. p27x-III and p21x-III, proteins encoded by the pX sequence of human T-cell leukemia virus type I. *Proc. Natl. Acad. Sci. USA.* **82:**8359–8363.

84. **Kompoliti, A., B. Gage, L. Sharma, and J. C. Daniels.** 1996. Human T-cell lymphotropic virus type 1-associated myelopathy, Sjogren syndrome, and lymphocytic pneumonitis. *Arch. Neurol.* **53:**940–942.

85. **Komurian, F., F. Pelloquin, and G. de The.** 1991. In vivo genomic variability of human T-cell leukemia virus type I depends more upon geography than upon pathologies. *J. Virol.* **65:**3770–3778.

86. **Komurian-Pradel, F., F. Pelloquin, S. Sonoda, M. Osame, and G. de The.** 1992. Geographical subtypes demonstrated by RFLP following PCR in the LTR region of HTLV-I. *AIDS Res. Hum. Retroviruses* **8:**429–434.

87. **Komuro, A., M. Hayami, H. Fujii, S. Miyahara, and M. Hirayama.** 1983. Vertical transmission of adult T-cell leukaemia virus. *Lancet* **ii:**240. (Letter.)

88. **Koralnik, I. J., J. Fullen, and G. Franchini.** 1993. The p12I, p13II, and p30II proteins encoded by human T-cell leukemia/lymphotropic virus type I open reading frames I and II are localized in three different cellular compartments. *J. Virol.* **67:**2360–2366.

89. **Koralnik, I. J., A. Gessain, M. E. Klotman, A. Lo Monico, Z. N. Berneman, and G. Franchini.** 1992. Protein isoforms encoded by the pX region of human T-cell leukemia/lymphotropic virus type I. *Proc. Natl. Acad. Sci. USA* **89:**8813–8817.

90. **Kubota, S., Y. Adachi, T. D. Copeland, and S. Oroszlan.** 1995. Binding of human prothymosin alpha to the leucine-motif/activation domains of HTLV-I Rex and HIV-I Rev. *Eur. J. Biochem.* **233:**48–54.

91. **Kuroda, Y., and H. Sugihara.** 1991. Autopsy report of HTLV-I-associated myelopathy presenting with ALS-like manifestations. *J. Neurol. Sci.* **106:**199–205.

92. **Kwok, R. P., M. E. Laurance, J. R. Lundblad, P. S. Goldman, H. Shih, L. M. Connor, S. J. Marriott, and R. H. Goodman.** 1996. Control of cAMP-regulated enhancers by the viral transactivator Tax through CREB and the co-activator CBP. *Nature* **380:** 642–646.

93. **La Grenade, L., R. A. Schwartz, and C. K. Janniger.** 1996. Childhood dermatitis in the tropics: with special emphasis on infective dermatitis, a marker for infection with human T-cell leukemia virus-I. *Cutis* **58:**115–118.

94. **Lee, H., K. B. Idler, P. Swanson, J. J. Aparicio, K. K. Chin, J. P. Lax, M. Nguyen, T. Mann, G. Leckie, A. Zanetti, G. Marinucci, I. S. Y. Chen, and J. D. Rosenblatt.** 1993. Complete nucleotide sequence of HTLV-II isolate NRA: comparison of envelope sequence variation of HTLV-II isolates from U.S. blood donors and U.S. and Italian i.v. drug users. *Virology* **196:**57–69.

95. **Lee, H., P. Swanson, V. S. Shorty, J. A. Zack, J. D. Rosenblatt, and I. S. Chen.** 1989. High rate of HTLV-II infection in seropositive i.v. drug abusers in New Orleans. *Science* **244:**471–475.

96. **Lee, H. H., P. Swanson, J. D. Rosenblatt, I. S. Chen, W. C. Sherwood, D. E. Smith, G. E. Tegtmeier, L. P. Fernando, C. T. Fang, M. Osame, and S. H. Kleinman.** 1991. Relative prevalence and risk factors of HTLV-I and HTLV-II infection in US blood donors. *Lancet* **337:**1435–1439.

97. **Lee, J. W., E. P. Fox, P. Rodgers-Johnson, C. J. Gibbs, Jr., E. DeFreitas, A. Manns, W. Blattner, J. Cotelingam, P. Piccardo, C. Mora, J. Safar, P. Liberski, E. Sausville, J. Trepel, and B. S. Kramer.** 1989. T-cell lymphoma, tropical spastic paraparesis, and malignant fibrous histiocytoma in a patient with human T-cell lymphotropic virus, type I. *Ann. Intern. Med.* **110:**239–241.

98. **Lehky, T. J., C. H. Fox, S. Koenig, M. C. Levin, N. Flerlage, S. Izumo, E. Sato, C. S. Raine, M. Osame, and S. Jacobson.** 1995. Detection of human T-lymphotropic

virus type I (HTLV-I) tax RNA in the central nervous system of HTLV-I-associated myelopathy/tropical spastic paraparesis patients by in situ hybridization. *Ann. Neurol.* **37**:167–175.

99. **Levin, M. C., T. J. Lehky, A. N. Flerlage, D. Katz, D. W. Kingma, E. S. Jaffe, J. D. Heiss, N. Patronas, H. F. McFarland, and S. Jacobson.** 1997. Immunologic analysis of a spinal cord-biopsy specimen from a patient with human T-cell lymphotropic virus type I-associated neurologic disease. *N. Engl. J. Med.* **336**:839–845.

100. **Link, H., M. Cruz, A. Gessain, O. Gout, G. de The, and S. Kam-Hansen.** 1989. Chronic progressive myelopathy associated with HTLV-I: oligoclonal IgG and anti-HTLV-I IgG antibodies in cerebrospinal fluid and serum. *Neurology* **39**:1566–1572.

101. **Lofters, W., M. Campbell, W. N. Gibbs, and B. D. Cheson.** 1987. 2'-Deoxycoformycin therapy in adult T-cell leukemia/lymphoma. *Cancer* **60**:2605–2608.

102. **Loughran, T. P., Jr., T. Coyle, M. P. Sherman, G. Starkebaum, G. D. Ehrlich, F. W. Ruscetti, and B. J. Poiesz.** 1992. Detection of human T-cell leukemia/lymphoma virus, type II, in a patient with large granular lymphocyte leukemia. *Blood* **80**:1116–1119.

103. **Mador, N., A. Panet, and A. Honigman.** 1989. Translation of *gag, pro,* and *pol* gene products of human T-cell leukemia virus type 2. *J. Virol.* **63**:2400–2404.

104. **Malik, K. T., J. Even, and A. Karpas.** 1988. Molecular cloning and complete nucleotide sequence of an adult T cell leukaemia virus/human T cell leukaemia virus type I (ATLV/HTLV-I) isolate of Caribbean origin: relationship to other members of the ATLV/HTLV-I subgroup. *J. Gen. Virol.* **69**:1695–1710.

105. **Marriott, S. J., I. Boros, J. F. Duvall, and J. N. Brady.** 1989. Indirect binding of human T-cell leukemia virus type I *tax1* to a responsive element in the viral long terminal repeat. *Mol. Cell. Biol.* **9**:4152–4160.

106. **Marriott, S. J., P. F. Lindholm, K. M. Brown, S. D. Gitlin, J. F. Duvall, M. F. Radonovich, and J. N. Brady.** 1990. A 36-kilodalton cellular transcription factor mediates an indirect interaction of human T-cell leukemia/lymphoma virus type I TAX1 with a responsive element in the viral long terminal repeat. *Mol. Cell. Biol.* **10**:4192–4201.

107. **Maruyama, I., S. Mori, M. Kawabata, and M. Osame.** 1992. [Bronchopneumonopathy in HTLV-I associated myelopathy (HAM) and non-HAM HTLV-I carriers]. *Nippon Kyobu Shikkan Gakkai Zasshi* **30**:775–779.

108. **Matsuo, H., T. Nakamura, K. Shibayama, K. Nagasato, M. Tsujihata, and S. Nagataki.** 1989. Long-term follow-up of immunomodulation in treatment of HTLV-I-associated myelopathy. *Lancet* i:790. (Letter.)

109. **Matsuo, H., T. Nakamura, M. Tsujihata, I. Kinoshita, A. Satoh, I. Tomita, S. Shirabe, K. Shibayama, and S. Nagataki.** 1988. Plasmapheresis in treatment of human T-lymphotropic virus type-I associated myelopathy. *Lancet* ii:1109–1113.

110. **Matsushima, M., A. Yoneyama, T. Nakamura, M. Higashihara, Y. Yatomi, A. Tanabe, T. Ohashi, H. Oka, and K. Nakahara.** 1987. A first case of complete remission of beta-interferon sensitive adult T-cell leukemia. *Eur. J. Haematol.* **39**:282–287.

111. **Maytal, J., S. Horowitz, S. Lipper, B. Poiesz, C. Y. Wang, and F. P. Siegal.** 1993. Progressive nemaline rod myopathy in a woman coinfected with HIV-I and HTLV-2. *Mt. Sinai J. Med.* **60**:242–246.

112. **McGuire, K. L., V. E. Curtiss, E. L. Larson, and W. A. Haseltine.** 1993. Influence of human T-cell leukemia virus type I *tax* and *rex* on interleukin-2 gene expression. *J. Virol.* **67**:1590–1599.

113. **Meytes, D., B. Schochat, H. Lee, G. Nadel, Y. Sidi, M. Cerney, P. Swanson, M. Shaklai, Y. Kilim, M. Elgat, E. Chin, Y. Danon, and J. D. Rosenblatt.** 1990. Serological and molecular survey for HTLV-I infection in a high-risk Middle Eastern group. *Lancet* **336**:1533–1535.

114. **Migone, T. S., J. X. Lin, A. Cereseto, J. C. Mulloy, J. J. O'Shea, G. Franchini, and**

W. J. Leonard. 1995. Constitutively activated Jak-STAT pathway in T cells transformed with HTLV-I. *Science* **269**:79–81.

115. **Mochizuki, M., K. Tajima, T. Watanabe, and K. Yamaguchi.** 1994. Human T lymphotropic virus type 1 uveitis. *Br. J. Ophthalmol.* **78**:149–154.

116. **Moore, G. R., U. Traugott, L. C. Scheinberg, and C. S. Raine.** 1989. Tropical spastic paraparesis: a model of virus-induced, cytotoxic T-cell-mediated demyelination? *Ann. Neurol.* **26**:523–530.

117. **Morgan, O. S., P. Rodgers-Johnson, C. Mora, and G. Char.** 1989. HTLV-I and polymyositis in Jamaica. *Lancet* ii:1184–1187.

118. **Muller, S., G. Boire, M. Ossondo, V. Ricchiuti, D. Smadja, J. C. Vernant, and S. Ozden.** 1995. IgG autoantibody response in HTLV-I-infected patients. *Clin. Immunol. Immunopathol.* **77**:282–290.

119. **Mulloy, J. C., R. W. Crownley, J. Fullen, W. J. Leonard, and G. Franchini.** 1996. The human T-cell leukemia/lymphotropic virus type I p12^I protein binds the interleukin-2 receptor β and γ$_c$ chains and affects their expression on the cell surface. *J. Virol.* **70**: 3599–3605.

120. **Murakami, T., H. Hirai, T. Suzuki, J. Fujisawa, and M. Yoshida.** 1995. HTLV-I Tax enhances NF-kappa B2 expression and binds to the products p52 and p100, but does not suppress the inhibitory function of p100. *Virology* **206**:1066–1074.

121. **Nakada, K., K. Yamaguchi, S. Furugen, T. Nakasone, K. Nakasone, Y. Oshiro, M. Kohakura, Y. Hinuma, M. Seiki, M. Yoshida, E. Matutes, D. Catovsky, T. Ishii, and K. Takatsuki.** 1987. Monoclonal integration of HTLV-I proviral DNA in patients with strongyloidiasis. *Int. J. Cancer.* **40**:145–148.

122. **Nishioka, K., T. Nakajima, T. Hasunuma, and K. Sato.** 1993. Rheumatic manifestation of human leukemia virus infection. *Rheum. Dis. Clin. North Am.* **19**:489–503.

123. **Nosaka, T., H. Siomi, Y. Adachi, M. Ishibashi, S. Kubota, M. Maki, and M. Hatanaka.** 1989. Nucleolar targeting signal of human T-cell leukemia virus type I rex-encoded protein is essential for cytoplasmic accumulation of unspliced viral mRNA. *Proc. Natl. Acad. Sci. USA* **86**:9798–9802.

124. **Nyambi, P. N., Y. Ville, J. Louwagie, I. Bedjabaga, E. Glowaczower, M. Peeters, D. Kerouedan, M. Dazza, B. Larouze, G. van der Groen, and E. Delaporte.** 1996. Mother-to-child transmission of human T-cell lymphotropic virus types I and II (HTLV-I/II) in Gabon: a prospective follow-up of 4 years. *J. Acquired Immune Defic. Syndr. Hum. Retrovirol.* **12**:187–192.

125. **Ohama, E., Y. Horikawa, T. Shimizu, T. Morita, K. Nemoto, H. Tanaka, and F. Ikuta.** 1990. Demyelination and remyelination in spinal cord lesions of human lymphotropic virus type I-associated myelopathy. *Acta Neuropathol.* (Berlin) **81**:78–83.

126. **Osame, M., S. Izumo, A. Igata, M. Matsumoto, T. Matsumoto, S. Sonoda, M. Tara, and Y. Shibata.** 1986. Blood transfusion and HTLV-I associated myelopathy. *Lancet* ii:104–105. (Letter.)

127. **Osame, M., M. Matsumoto, K. Usuku, S. Izumo, N. Ijichi, H. Amitani, M. Tara, and A. Igata.** 1987. Chronic progressive myelopathy associated with elevated antibodies to human T-lymphotropic virus type I and adult T-cell leukemialike cells. *Ann. Neurol.* **21**:117–122.

128. **Pan, X. Z., Z. D. Qiu, N. Chein, J. W. Gold, W. D. Hardy, Jr., E. Zuckerman, Q. N. Wang, and D. Armstrong.** 1991. A seroepidemiological survey of HTLV-I infection in Shanghai and Chongqing cities in China. *AIDS* **5**:782–783. (Letter.)

129. **Pancake, B. A., D. Zucker-Franklin, and E. E. Coutavas.** 1995. The cutaneous T cell lymphoma, mycosis fungoides, is a human T cell lymphotropic virus-associated disease. A study of 50 patients. *J. Clin. Invest.* **95**:547–554.

130. **Phelps, K. R., S. S. Ginsberg, A. W. Cunningham, E. Tschachler, and H. Dosik.** 1991.

Case report: adult T-cell leukemia/lymphoma associated with recurrent strongyloides hyperinfection. *Am. J. Med. Sci.* **302**:224–228.

131. **Plumelle, Y., C. Gonin, A. Edouard, B. J. Bucher, L. Thomas, A. Brebion, and G. Panelatti.** 1997. Effect of Strongyloides stereoralis infection and eosinophilia on age at onset and prognosis of adult T-cell leukemia. *Am. J. Clin. Pathol.* **107**:81–87.

132. **Plumelle, Y., N. Pascaline, D. Nguyen, G. Panelatti, A. Jouannelle, H. Jouault, and M. Imbert.** 1993. Adult T-cell leukemia-lymphoma: a clinico-pathologic study of twenty-six patients from Martinique. *Hematol. Pathol.* **7**:251–262.

133. **Poiesz, B. J., F. W. Ruscetti, A. F. Gazdar, P. A. Bunn, J. D. Minna, and R. C. Gallo.** 1980. Detection and isolation of type C retrovirus particles from fresh and cultured lymphocytes of a patient with cutaneous T-cell lymphoma. *Proc. Natl. Acad. Sci. USA* **77**:7415–7419.

134. **Popovic, M., G. Lange-Wantzin, P. S. Sarin, D. Mann, and R. C. Gallo.** 1983. Transformation of human umbilical cord blood T cells by human T-cell leukemia/lymphoma virus. *Proc. Natl. Acad. Sci. USA* **80**:5402–5406.

135. **Prince, H. E., J. York, S. M. Owen, and R. B. Lal.** 1995. Spontaneous proliferation of memory (CD45RO+) and naive (CD45RO−) subsets of CD4 cells and CD8 cells in human T lymphotropic virus (HTLV) infection: distinctive patterns for HTLV-I versus HTLV-II. *Clin. Exp. Immunol.* **102**:256–261.

136. **Richardson, J. H., A. J. Edwards, J. K. Cruickshank, P. Rudge, and A. G. Dalgleish.** 1990. In vivo cellular tropism of human T-cell leukemia virus type 1. *J. Virol.* **64**:5682–5687.

137. **Robert-Guroff, M., S. H. Weiss, J. A. Giron, A. M. Jennings, H. M. Ginzburg, I. B. Margolis, W. A. Blattner, and R. C. Gallo.** 1986. Prevalence of antibodies to HTLV-I, -II, and -III in intravenous drug abusers from an AIDS endemic region. *JAMA* **255**:3133–3137.

138. **Rodgers-Johnson, P., O. S. Morgan, C. Mora, P. Sarin, M. Ceroni, P. Piccardo, R. M. Garruto, C. J. Gibbs, Jr., and D. C. Gajdusek.** 1988. The role of HTLV-I in tropical spastic paraparesis in Jamaica. *Ann. Neurol.* **23**(Suppl.):S121–S126.

139. **Roman, G. C., and L. N. Roman.** 1988. Tropical spastic paraparesis. A clinical study of 50 patients from Tumaco (Colombia) and review of the worldwide features of the syndrome. *J. Neurol. Sci.* **87**:121–138.

140. **Roman, G. C., P. S. Spencer, and B. S. Schoenberg.** 1985. Tropical myeloneuropathies: the hidden endemias. *Neurology* **35**:1158–1170.

141. **Rosenblatt, J. D.** 1993. Human T-lymphotropic virus types I and II. *West. J. Med.* **158**:379–384.

142. **Rosenblatt, J. D., and A. C. Black.** 1994. Human T-cell leukemia viruses: HTLV-2, p. 686–695. *In* R. G. Webster (ed.), *Encyclopedia of Virology*. Saunders Scientific Publications, W.B. Saunders Company, London.

143. **Rosenblatt, J. D., J. V. Giorgi, D. W. Golde, J. B. Ezra, A. Wu, C. D. Winberg, J. Glaspy, W. Wachsman, and I. S. Chen.** 1988. Integrated human T-cell leukemia virus II genome in CD8+ T cells from a patient with "atypical" hairy cell leukemia: evidence for distinct T and B cell lymphoproliferative disorders. *Blood* **71**:363–369.

144. **Rosenblatt, J. D., D. W. Golde, W. Wachsman, J. V. Giorgi, A. Jacobs, G. M. Schmidt, S. Quan, J. C. Gasson, and I. S. Chen.** 1986. A second isolate of HTLV-II associated with atypical hairy-cell leukemia. *N. Engl. J. Med.* **315**:372–377.

145. **Rosenblatt, J. D., S. Plaeger-Marshall, J. V. Giorgi, P. Swanson, I. S. Chen, E. Chin, H. J. Wang, M. Canavaggio, M. A. Hausner, and A. C. Black.** 1990. A clinical, hematologic, and immunologic analysis of 21 HTLV-II-infected intravenous drug users. *Blood* **76**:409–417. (Erratum, *Blood* **76**:1901, 1991.)

146. **Rosenblatt, J. D., P. Tomkins, M. Rosenthal, A. Kacena, G. Chan, R. Valderama,**

W. Harrington, Jr., E. Saxton, A. Diagne, J. Q. Zhao, R. T. Mitsuyasu, and R. H. Weisbart. 1992. Progressive spastic myelopathy in a patient co-infected with HIV-I and HTLV-II: autoantibodies to the human homologue of rig in blood and cerebrospinal fluid. *AIDS* **6**:1151–1158.

147. Rosenblatt, J. D., J. A. Zack, I. S. Chen, and H. Lee. 1990. Recent advances in detection of human T-cell leukemia viruses type I and type II infection. *Nat. Immun. Cell. Growth Regul.* **9**:143–149.

148. Saksena, N. K., M. P. Sherman, R. Yanagihara, D. K. Dube, and B. J. Poiesz. 1992. LTR sequence and phylogenetic analyses of a newly discovered variant of HTLV-I isolated from the Hagahai of Papua New Guinea. *Virology* **189**:1–9.

149. Sasaki, S., T. Komori, S. Maruyama, M. Takeishi, and Y. Iwasaki. 1990. An autopsy case of human T lymphotropic virus type I-associated myelopathy (HAM) with a duration of 28 years. *Acta Neuropathol.* (Berlin) **81**:219–222.

150. Sato, Y., and Y. Shiroma. 1989. Concurrent infections with Strongyloides and T-cell leukemia virus and their possible effect on immune responses of host. *Clin. Immunol. Immunopathol.* **52**:214–224.

151. Sawada, T., Y. Iwahara, K. Ishii, H. Taguchi, H. Hoshino, and I. Miyoshi. 1991. Immunoglobulin prophylaxis against milkborne transmission of human T cell leukemia virus type I in rabbits. *J. Infect. Dis.* **164**:1193–1196.

152. Saxton, E. H., H. Lee, P. Swanson, I. S. Chen, C. Ruland, E. Chin, D. Aboulafia, R. Delamarter, and J. D. Rosenblatt. 1989. Detection of human T-cell leukemia/lymphoma virus type I in a transfusion recipient with chronic myelopathy. *Neurology* **39**:841–844.

153. Seiki, M., S. Hattori, Y. Hirayama, and M. Yoshida. 1983. Human adult T-cell leukemia virus: complete nucleotide sequence of the provirus genome integrated in leukemia cell DNA. *Proc. Natl. Acad. Sci. USA* **80**:3618–3622.

154. Seiki, M., J. Inoue, M. Hidaka, and M. Yoshida. 1988. Two cis-acting elements responsible for posttranscriptional transregulation of gene expression of human T-cell leukemia virus type I. *Proc. Natl. Acad. Sci. USA* **85**:7124–7128.

155. Sheremata, W. A., D. Benedict, D. C. Squilacote, A. Sazant, and E. DeFreitas. 1993. High-dose zidovudine induction in HTLV-I-associated myelopathy: safety and possible efficacy. *Neurology* **43**:2125–2129.

156. Shibasaki, H., C. Endo, Y. Kuroda, R. Kakigi, K. Oda, and S. Komine. 1988. Clinical picture of HTLV-I associated myelopathy. *J. Neurol. Sci.* **87**:15–24.

157. Shibayama, K., T. Nakamura, K. Nagasato, S. Shirabe, M. Tsujihata, and S. Nagataki. 1991. Interferon-alpha treatment in HTLV-I-associated myelopathy. Studies of clinical and immunological aspects. *J. Neurol. Sci.* **106**:186–192.

158. Shimotohno, K., Y. Takahashi, N. Shimizu, T. Gojobori, D. W. Golde, I. S. Chen, M. Miwa, and T. Sugimura. 1985. Complete nucleotide sequence of an infectious clone of human T-cell leukemia virus type II: an open reading frame for the protease gene. *Proc. Natl. Acad. Sci. USA* **82**:3101–3105.

159. Shimotohno, K., M. Takano, T. Teruuchi, and M. Miwa. 1986. Requirement of multiple copies of a 21-nucleotide sequence in the U3 regions of human T-cell leukemia virus type I and type II long terminal repeats for trans-acting activation of transcription. *Proc. Natl. Acad. Sci. USA* **83**:8112–8116.

160. Shimotohno, K., W. Wachsman, Y. Takahashi, D. W. Golde, M. Miwa, T. Sugimura, and I. S. Chen. 1984. Nucleotide sequence of the 3′ region of an infectious human T-cell leukemia virus type II genome. *Proc. Natl. Acad. Sci. USA* **81**:6657–6661.

161. Shimoyama, M. 1991. Diagnostic criteria and classification of clinical subtypes of adult T-cell leukaemia-lymphoma. A report from the Lymphoma Study Group (1984–87). *Br. J. Haematol.* **79**:428–437.

162. **Sidi, Y., D. Meytes, B. Shohat, E. Fenig, Y. Weisbort, H. Lee, J. Pinkhas, and J. D. Rosenblatt.** 1990. Adult T-cell lymphoma in Israeli patients of Iranian origin. *Cancer* **65:**590–593.

163. **Siomi, H., H. Shida, S. H. Nam, T. Nosaka, M. Maki, and M. Hatanaka.** 1988. Sequence requirements for nucleolar localization of human T cell leukemia virus type I pX protein, which regulates viral RNA processing. *Cell* **55:**197–209.

164. **Sodroski, J. G., C. A. Rosen, and W. A. Haseltine.** 1984. Trans-acting transcriptional activation of the long terminal repeat of human T lymphotropic viruses in infected cells. *Science* **225:**381–385.

165. **Spear, G. T., N. S. Lurain, C. J. Parker, M. Ghassemi, G. H. Payne, and M. Saiffuddin.** 1995. Host cell-derived complement control proteins CD55 and CD59 are incorporated into the virions of two unrelated enveloped viruses. Human T cell leukemia/lymphoma virus type I (HTLV-I) and human cytomegalovirus (HCMV). *J. Immunol.* **155:**4376–4381.

166. **Stuber, E., and W. Strober.** 1996. The T cell-B cell interaction via OX40-OX40L is necessary for the T cell-dependent humoral immune response. *J. Exp. Med.* **183:**979–989.

167. **Suzuki, T., J. I. Fujisawa, M. Toita, and M. Yoshida.** 1993. The trans-activator tax of human T-cell leukemia virus type I (HTLV-I) interacts with cAMP-responsive element (CRE) binding and CRE modulator proteins that bind to the 21-base-pair enhancer of HTLV-1. *Proc. Natl. Acad. Sci. USA* **90:**610–614.

168. **Suzuki, T., H. Hirai, T. Murakami, and M. Yoshida.** 1995. Tax protein of HTLV-I destabilizes the complexes of NF-kappa B and I kappa B-alpha and induces nuclear translocation of NF-kappa B for transcriptional activation. *Oncogene* **10:**1199–1207.

169. **Taguchi, H., K. I. Kinoshita, K. Takatsuki, M. Tomonaga, K. Araki, N. Arima, S. Ikeda, K. Uozumi, H. Kohno, F. Kawano, H. Kikuchi, H. Takahashi, K. Tamura, S. Chiyoda, H. Tsuda, H. Nishimura, T. Hosokawa, H. Matsuzaki, S. Momita, O. Yamada, and I. Miyoshi.** 1996. An intensive chemotherapy of adult T-cell leukemia/lymphoma: CHOP followed by etoposide, vindesine, ranimustine, and mitoxantrone with granulocyte colony-stimulating factor support. *J. Acquired Immune Defic. Syndr. Hum. Retrovirol.* **12:**182–186.

170. **Tajima, K., and S. Tominaga.** 1985. Epidemiology of adult T-cell leukemia/lymphoma in Japan. *Curr. Top. Microbiol. Immunol.* **115:**53–66.

171. **Takahashi, K., T. Takezaki, T. Oki, K. Kawakami, S. Yashiki, T. Fujiyoshi, K. Usuku, N. Mueller, M. Osame, K. Miyata, Y. Nagata, and S. Sonoda.** 1991. Inhibitory effect of maternal antibody on mother-to-child transmission of human T-lymphotropic virus type I. The Mother-to-Child Transmission Study Group. *Int. J. Cancer* **49:**673–677.

172. **Takatsuki, K., K. Yamaguchi, F. Kawano, H. Nishimura, M. Seiki, and M. Yoshida.** 1985. Clinical aspects of adult T-cell leukemia/lymphoma. *Curr. Top. Microbiol. Immunol.* **115:**89–97.

173. **Tamura, K., S. Makino, Y. Araki, T. Imamura, and M. Seita.** 1987. Recombinant interferon beta and gamma in the treatment of adult T-cell leukemia. *Cancer* **59:**1059–1062.

174. **Tendler, C. L., S. J. Greenberg, W. A. Blattner, A. Manns, E. Murphy, T. Fleisher, B. Hanchard, O. Morgan, J. D. Burton, D. L. Nelson, and T. A. Waldmann.** 1990. Transactivation of interleukin 2 and its receptor induces immune activation in human T-cell lymphotropic virus type I-associated myelopathy: pathogenic implications and a rationale for immunotherapy. *Proc. Natl. Acad. Sci. USA* **87:**5218–5222.

175. **Tokudome, S., O. Tokunaga, Y. Shimamoto, Y. Miyamoto, I. Sumida, M. Kikuchi, M. Takeshita, T. Ikeda, K. Fujiwara, M. Yoshihara, T. Yanagawa, and M. Nishi-**

zumi. 1989. Incidence of adult T-cell leukemia/lymphoma among human T-lymphotropic virus type I carriers in Saga, Japan. *Cancer Res.* **49:**226–228.

176. **Tournier-Lasserve, E., O. Gout, A. Gessain, M. T. Iba-Zizen, O. Lyon-Caen, F. Lhermitte, and G. de-The.** 1987. HTLV-I, brain abnormalities on magnetic resonance imaging, and relation with multiple sclerosis. *Lancet* **ii:**49–50. (Letter.)

177. **Uchiyama, T.** 1996. ATL and HTLV-I: in vivo cell growth of ATL cells. *J. Clin. Immunol.* **16:**305–314.

178. **Uchiyama, T., J. Yodoi, K. Sagawa, K. Takatsuki, and H. Uchino.** 1977. Adult T-cell leukemia: clinical and hematologic features of 16 cases. *Blood* **50:**481–492.

179. **Varmus, H.** 1988. Retroviruses. *Science* **240:**1427–1435.

180. **Vidal, A. U., A. Gessain, M. Yoshida, R. Mahieux, K. Nishioka, F. Tekaia, L. Rosen, and G. De The.** 1994. Molecular epidemiology of HTLV type I in Japan: evidence for two distinct ancestral lineages with a particular geographical distribution. *AIDS Res. Hum. Retroviruses* **10:**1557–1566.

181. **Vidal, A. U., A. Gessain, M. Yoshida, F. Tekaia, B. Garin, B. Guillemain, T. Schulz, R. Farid, and G. De The.** 1994. Phylogenetic classification of human T cell leukaemia/lymphoma virus type I genotypes in five major molecular and geographical subtypes. *J. Gen. Virol.* **75:**3655–3666.

182. **Viscidi, R. P., P. M. Hill, S. J. Li, E. H. Cerny, D. Vlahov, H. Farzadegan, N. Halsey, G. D. Kelen, and T. C. Quinn.** 1991. Diagnosis and differentiation of HTLV-I and HTLV-II infection by enzyme immunoassays using synthetic peptides. *J. Acquired Immune Defic. Syndr.* **4:**1190–1198.

183. **Waldmann, T. A., W. C. Greene, P. S. Sarin, C. Saxinger, D. W. Blayney, W. A. Blattner, C. K. Goldman, K. Bongiovanni, S. Sharrow, J. M. Depper, W. Leonard, T. Uchiyama, and R. C. Gallo.** 1984. Functional and phenotypic comparison of human T cell leukemia/lymphoma virus positive adult T cell leukemia with human T cell leukemia/lymphoma virus negative Sezary leukemia, and their distinction using anti-Tac. Monoclonal antibody identifying the human receptor for T cell growth factor. *J. Clin. Invest.* **73:**1711–1718.

184. **Waldmann, T. A., J. D. White, C. K. Goldman, L. Top, A. Grant, R. Bamford, E. Roessler, I. D. Horak, S. Zaknoen, C. Kasten-Sportes, R. England, E. Horak, B. Mishra, M. Kipre, P. Hale, T. A. Fleisher, R. P. Junghans, E. S. Jaffe, and D. L. Nelson.** 1993. The interleukin-2 receptor: a target for monoclonal antibody treatment of human T-cell lymphotrophic virus I-induced adult T-cell leukemia. *Blood* **82:**1701–1712.

185. **Watanabe, C. T., J. D. Rosenblatt, A. Bakker, J. P. Morgan, J. Luo, S. Chun, and A. C. Black.** 1996. Negative regulation of gene expression from the HTLV type II long terminal repeat by Rex: functional and structural dissociation from positive posttranscriptional regulation. *AIDS Res. Hum. Retroviruses.* **12:**535–546.

186. **Wiktor, S. Z., E. J. Pate, E. L. Murphy, T. J. Palker, E. Champegnie, A. Ramlal, B. Cranston, B. Hanchard, and W. A. Blattner.** 1993. Mother-to-child transmission of human T-cell lymphotropic virus type I (HTLV-I) in Jamaica: association with antibodies to envelope glycoprotein (gp46) epitopes. *J. Acquired Immune Defic. Syndr.* **6:**1162–1167.

187. **Xu, X., S. H. Kang, O. Heidenreich, M. Okerholm, J. J. O'Shea, and M. I. Nerenberg.** 1995. Constitutive activation of different Jak tyrosine kinases in human T cell leukemia virus type 1 (HTLV-1) tax protein or virus-transformed cells. *J. Clin. Invest.* **96:**1548–1555.

188. **Yanagihara, R., V. R. Nerukar, and A. B. Ajdukiewicz.** 1991. Comparison between strains of human T lymphotropic virus type I isolated from inhabitants of the Solomon Islands and Papua New Guinea. *J. Infect. Dis.* **164:**443–449.

189. **Yanagihara, R., V. R. Nerurkar, R. M. Garruto, M. A. Miller, M. E. Leon-Monzon, C. L. Jenkins, R. C. Sanders, P. P. Liberski, M. P. Alpers, and D. C. Gajdusek.** 1991. Characterization of a variant of human T-lymphotropic virus type I isolated from a healthy member of a remote, recently contacted group in Papua New Guinea. *Proc. Natl. Acad. Sci. USA* **88:**1446–1450.

190. **Yasui, C., T. Fukaya, H. Koizumi, H. Kobayashi, and A. Ohkawara.** 1991. HTLV-I-associated myelopathy in a patient with adult T-cell leukemia. *J. Am. Acad. Dermatol.* **24:**633–637.

191. **Yoshida, M.** 1995. HTLV-I oncoprotein Tax deregulates transcription of cellular genes through multiple mechanisms. *J. Cancer Res. Clin. Oncol.* **121:**521–528.

192. **Yoshida, M., M. Seiki, K. Yamaguchi, and K. Takatsuki.** 1984. Monoclonal integration of human T-cell leukemia provirus in all primary tumors of adult T-cell leukemia suggests causative role of human T-cell leukemia virus in the disease. *Proc. Natl. Acad. Sci. USA* **81:**2534–2537.

193. **Yu, F., Y. Itoyama, J. Kira, K. Fujihara, T. Kobayashi, T. Kitamoto, A. Suzumura, N. Yamamoto, Y. Nakajima, and I. Goto.** 1994. TNF-beta produced by human T lymphotropic virus type I-infected cells influences the proliferation of human endothelial cells and fibroblasts. *J. Immunol.* **152:**5930–5938.

194. **Zucker-Franklin, D., E. E. Coutavas, M. G. Rush, and D. C. Zouzias.** 1991. Detection of human T-lymphotropic virus-like particles in cultures of peripheral blood lymphocytes from patients with mycosis fungoides. *Proc. Natl. Acad. Sci. USA* **88:**7630–7634.

195. **Zucker-Franklin, D., W. C. Hooper, and B. L. Evatt.** 1992. Human lymphotropic retroviruses associated with mycosis fungoides: evidence that human T-cell lymphotropic virus type II (HTLV-II) as well as HTLV-I may play a role in the disease. *Blood* **80:**1537–1545.

INDEX†

---

† Page numbers followed by *f* or *t* indicate a
figure or table, respectively.